Sexually Transmitted Diseases

Vaccines, Prevention and Control

To our supportive wives Elizabeth and Donna,
and children Lindsey, Martin, Kendall and Logan

Sexually Transmitted Diseases

Vaccines, Prevention and Control

Edited by

Lawrence R. Stanberry and David I. Bernstein

Division of Infectious Diseases
Children's Hospital Medical Center
Cincinnati
Ohio, USA

ACADEMIC PRESS

A Harcourt Science and Technology Company

San Diego   San Francisco   New York   Boston
London   Sydney   Tokyo

Academic Press
A Harcourt Science and Technology Company
24–28 Oval Road, London, NW1 7DX, UK
http://www.hbuk.co.uk/ap/

Academic Press
525 B Street, Suite 1900, San Diego, California 92101-4495, USA
http://www.apnet.com

ISBN 0-12-663330-4

A catalogue for this book is available from the British Library

Typeset by Bibliocraft Ltd, Dundee
Printed in Great Britain by The University Press, Cambridge

00 01 02 03 04 05 CU 9 8 7 6 5 4 3 2 1

CONTENTS

CONTRIBUTORS

Vagan A. Akovbian Central Research Institute of Skin and Venereal Diseases, Korolenko 3/4, Room 410, 107076 Moscow, Russia

David I. Bernstein Division of Infectious Diseases, University of Cincinnati College of Medicine, Children's Hospital Medical Center, 3333 Burnet Avenue, Cincinnati, OH 45229, USA

Karl R. Beutner Solano Dermatology Associates, 127 Hospital Drive, Suite 204, Vallejo, CA 94589, USA

Frank M. Biro Division of Adolescent Medicine, University of Cincinnati College of Medicine, Children's Hospital Medical Center, 3333 Burnet Avenue, Cincinnati, OH 45229-3039, USA

Robert C. Brunham University of British Columbia, Centre for Disease Control, 2077, 655 West 12th Avenue, Vancouver, British Columbia, Canada V5Z 4R4

Sheila S. Cohen Central Psychiatric Clinic and Department of Pediatrics, University of Cincinnati College of Medicine, Children's Hospital Medical Center, 3333 Burnet Avenue, Cincinnati, OH 45229-3039, USA

Morris D. Cooper Department of Medical Microbiology and Immunology, Southern Illinois University School of Medicine, PO Box 19230, Springfield, IL 62794-1220, USA

Anthony L. Cunningham Centre for Virus Research, Westmead Millenium Institute, Westmead Hospital, University of Sydney, Westmead, NSW 2145, Australia

Carolyn D. Deal Division of Bacterial Products, Center for Biologics Evaluation and Research, Food and Drug Administration, Bld 29, Rm 130, 29 Lincoln Drive, Bethesda, MD 20892, USA

Dominic E. Dwyer Centre for Virus Research, Westmead Millenium Institute and Centre for Infectious Diseases and Microbiology, Westmead Hospital, University of Sydney, Westmead, NSW 2145, Australia

Claudia Estcourt Academic Unit of Sexual Health Medicine, Sydney Hospital, GPO Box 1614, Sydney, NSW 2001, Australia

Mikhail Gomberg Viral Urogenital Infections Laboratory, Central Research Institute of Skin and Venereal Diseases, Korolenko 3/4, Room 410, 107076 Moscow, Russia

Penelope J. Hitchcock Sexually Transmitted Diseases Branch, National Institute of Allergies and Infectious Diseases, 6700-B Rockledge Drive, MSC 7630, Room 3112, Bethesda, MD 20892-7630, USA

Joseph Kelaghan Contraception and Reproductive Health Branch, Center for Population Research, National Institute of Child Health and Human Development, 6100 Executive Blvd, Bethesda, MD 20892, USA

Anna A. Koubanova Central Research Institute of Skin and Venereal Diseases, Korolenko 3/4, Room 410, 107076 Moscow, Russia

Bonnie J. Mathieson Office of AIDS Research, National Institute of Allergy and Infectious Diseases, 31/4C, Bethesda, MD 20892, USA

Grant McClarty Dept of Medical Microbiology, University of Manitoba, Basic Medical Sciences Building, 5th Floor, 730 Williams Ave, Winnipeg, Manitoba, Canada R3E 0W3

Gregg N. Milligan Division of Infectious Diseases, University of Cincinnati College of Medicine, Children's Hospital Medical Center, 3333 Burnet Avenue, Cincinnati, OH 45229, USA

Adrian Mindel Academic Unit of Sexual Health Medicine, Sydney Hospital, GPO Box 1614, Sydney, NSW 2001, Australia

Susan L. Rosenthal Division of Adolescent Medicine, University of Cincinnati College of Medicine, Children's Hospital Medical Center, 3333 Burnet Avenue, Cincinnati, OH 45229, USA

Jessica L. Severson University of Texas Medical Branch, Department of Microbiology, 301 University Boulevard, Galveston, TX 77555, USA

John D. Shanley Division of Infectious Diseases, Department of Medicine, University of Connecticut Health Center, Room L2060, Farmington, CT 06030, USA

Lawrence R. Stanberry Division of Infectious Diseases, University of Cincinnati College of Medicine, Children's Hospital Medical Center, 3333 Burnet Avenue, Cincinnati, OH 45229, USA

Stephen K. Tyring University of Texas Medical Branch, Departments of Dermatology and Microbiology, 301 University Boulevard, Galverston, TX 77555, USA

Pierre Vandepapelière SmithKline Beecham Biologicals, Immunotherapeutics, Rue de l'Institut 89, B-1330, Rixensart, Belgium

Jonathan M. Zenilman Division of Infectious Diseases, Johns Hopkins University School of Medicine, Ross Building, Room 1165, 720 Rutland Avenue, Baltimore, MD 21250, USA

Arie J. Zuckerman WHO Collaborating Centre for Reference Research on Viral Diseases, Royal Free and University College Medical School, Rowland Hill Street, London, NW3 2PF, UK

Jane N. Zuckerman Academic Unit of Travel Medicine and Vaccines, and Clinical Trials Centre, Royal Free and University College Medical School, Rowland Hill Street, London, NW3 2PF, UK

FOREWORD

In recent decades we have learned much about sexually transmitted infections. We now have a better understanding of the associated diseases, the nature of the pathogens, how to diagnose and how to treat. We have seen new agents and new diseases emerge. Sexually transmitted diseases (STDs) once thought to be rare are now listed among the most common of human infections. Other sexually transmitted infections, once thought to be relatively innocuous, are now known to cause potentially fatal disease. Most of all, we have learned that the simplistic approach of diagnosis and treatment of symptomatic individuals will not control these infections.

It has become obvious that the only efficient method of controlling STDs will involve a major commitment to preventive measures. This was an important theme from the Institue of Medicine report on sexually transmitted diseases, 'The Hidden Epidemic' (Thomas R. Eng, William T. Butler, editors, National Academic Press, 1997). That report recommended 'an effective national system for STD prevention be established in the United States'. It also emphasized that we are not yet willing to commit funds to support research on prevention or to implement preventive efforts. The report pointed out that the annual cost of selected STDs and their complications in the United States in 1994 was approximately $10b, but in 1995 only $230m was committed to prevention.

Clearly we must identify more effective means of preventing STDs. This book therefore fills a need by presenting a timely set of state of the art reviews on this subject. It begins with chapters defining the scope of the problem and some overviews of genital tract physiology and muscosal immunity as they may be related to STD prevention. A second series of reviews cover non-specific measures of preventing STDs through use of vaginal microbicides, barrier methods, or through behavioral intervention aimed at reducing STD risks.

The main portion of the book presents the current status of research aimed at developing vaccines for important STD pathogens. These chapters are written by acknowledged experts in the field and review past and current vaccine strategies, summarize the current state of vaccine development, and discuss progress towards successful vaccine development. This is an important book which is appearing at a most propitious moment. The prevention of STDs will be a key part of giving humankind a healthier start into this new millennium. Hopefully, some of the blueprints for those programs will be included in these chapters.

Julius Schachter, PhD
President, American Sexually Transmitted Disease Association
and
Professor of Laboratory Medicine, University of California at San Francisco

I

EPIDEMIOLOGY, PHYSIOLOGY AND IMMUNOLOGY

1

GLOBAL EPIDEMIOLOGY OF SEXUALLY TRANSMITTED DISEASES

ANTHONY L. CUNNINGHAM[*],
ADRIAN MINDEL[†]
AND
DOMINIC E. DWYER[*]

[*]Centre for Virus Research, Westmead Millenium Institute,
Westmead Hospital, University of Sydney, Australia

[†]Academic Unit, Sydney Sexual Health Centre,
University of Sydney, Australia

INTRODUCTION

As would be expected from the diverse cultures and sexual mores throughout the world, the epidemiology of sexually transmitted disease is highly variable in distribution and changing in different ways in different regions. The factors that influence these differences in prevalence and incidence are the nature of the sexually transmitted disease itself, whether it is curable or incurable by anti-microbials, or is preventable or non-preventable by vaccines. The availability of highly developed health-care networks in western industrialized countries, compared with those of developing countries, influences the epidemiology in a number of ways, through ease of transmission, availability of diagnostic facilities and drugs, transmissibility of behaviour modification messages, and levels of education allowing receptiveness to these messages. The availability of new diagnostic tests has allowed the definition of large reservoirs of asympto-matic infection which have led to marked changes in our understanding of the epidemiology of these diseases. However, in the western industrialized coun-tries there are also marked differences according to race, socioeconomic status, sexual preference and, recently, the influence of drugs and prostitution, which create similar (or worse) conditions for the spread of these diseases than in many developing countries. The data available to measure the epidemiology of sexually transmitted diseases are limited even in many western industrialized countries, and are often only available through infrequent sampling in devel-oping countries. Global comparisons of epidemiology are important in that they allow cross-comparison of the factors influencing spread and of optimal strategies for control, allowing adaptation of the latter to the unique cultural characteristics and health-care systems of each country.

THE HUMAN IMMUNODEFICIENCY VIRUSES

Infection with the human immunodeficiency virus type 1 (HIV) and the development of the acquired immunodeficiency syndrome (AIDS) is one of the major epidemics of the latter part of the 20th century. Nearly all countries in the world have been affected, although there are significant differences in the epidemiology between and within countries. The first clinical description of AIDS were made in the United States in 1981, with the recognition of unusual clusters of Kaposi's sarcoma and *Pneumocystis carinii* pneumonia in young homosexual men (CDC, 1981; Masur *et al.*, 1980; Gottlieb *et al.* 1981; Joag *et al.*, 1996; Pavlakis 1997). Over the next few years, AIDS was also reported in other populations, including injecting drug users and hemophiliacs, with expansion of the clinical spectrum of associated diseases (Poon *et al.*, 1983; Holmberg *et al.*, 1989; Mastro *et al.*, 1994). It was then found in blood transfusion recipients, adults from Central Africa, infants born to mothers with AIDS or who were injecting drug users, and the sexual partners of risk group members (Curran *et al.*, 1984; Jaffe *et al.*, 1984; Dickover *et al.*, 1996; Schreiber *et al.*, 1996; Luzuriaga and Sullivan, 1997; Mertens and Piot, 1997; Piot *et al.*, 1984; Scott *et al.*, 1984). Although AIDS was first recognized in North America and Europe, the major foci of HIV infection are subSaharan Africa and Asia.

VIROLOGY OF HIV

HIV (previously called HTLV-3, LAV and ARV) was first isolated in 1983 (Barre-Sinoussi *et al.*, 1983; Gallo *et al.*, 1984) and confirmed by physical, electron microscopy and sequence analysis to be a retrovirus. The lentivirus genus of the viral family Retroviridae includes HIV-1 and 2, the simian immunodeficiency viruses, and other viruses affecting various animal species. Lentiviruses are characterized by long incubation periods prior to the onset of disease, clinically progressive disease leading to cachexia and death, a diversity of organ systems affected, and the failure of the infected host to recover from infection. Based on genomic structure and sequence homology, the lentiviruses that infect primates are classified into five groups at approximately equal phylogenetic distances from each other (Joag *et al.*, 1996):

1. HIV-1 of humans, SIVCPZ of chimpanzees
2. HIV-2 of humans, SIVMAC from macaques, SIVSMM from sooty mangabey monkeys
3. SIVAGM from African green monkeys
4. SIVMMD from mandrills
5. SIVSYK from Syke's monkeys.

TRANSMISSION OF HIV

HIV is transmitted in three major ways: through sexual intercourse, in blood, and from mother to child. It is mainly a sexually transmitted disease, with sexual transmission accounting for over 75% of all HIV infections worldwide. Initially, in developed countries homosexual or bisexual male to male (sometimes combined with injecting drug use) transmission was the major risk factor identified, leading to a male predominance. Heterosexual transmission and injecting drug use have now increased in many of these countries. In many other parts of the world heterosexual transmission predominated from the outset of the epidemic, meaning that the male to female ratio was approximately equal. Although the probability of HIV infection during a single sexual exposure from an infected partner and in the absence other cofactors is low (< 0.003), it is significantly increased if other risk factors are present, including the presence of sexually transmitted diseases (Holmberg et al., 1989; Mastro et al., 1994). People newly infected with HIV or with untreated AIDS usually have a high viral load, and are generally more infectious than people stably infected for a long time.

Mother to child transmission of HIV includes transmission during pregnancy and delivery, and via breastfeeding. Approximately 90% of HIV infected children acquire the infection vertically from their mothers. The seropositivity rate of pregnant women varies throughout the world, with figures ranging from 6% to 30% of pregnant women in subSaharan Africa being seropositive, compared to 0.17% of childbearing women in the USA (Luzuriaga and Sullivan, 1997). It is estimated that half of new infections worldwide occur in women of childbearing age, which will lead to a marked increase in vertical HIV infection. In developing nations this has begun to reverse gains in infant and childhood morbidity and mortality previously realized through nutrition and vaccine programs. The transmission rate of HIV-1 to the fetus is between 13 and 48% of infected mothers (significantly more than with HIV-2), with the rate affected by the level of maternal viremia and the use of antiretroviral therapy in the mother and child (Dickover et al., 1996).

Early in the HIV epidemic, recipients of blood or blood products were a significant source of HIV infections in developed countries. Approximately one-third were hemophiliacs, with the rest transfusion-associated cases. Most of these infections occurred before 1985, when HIV antibody screening of all blood products became available in the United States and other countries. HIV transmission by blood products still occurs in countries where such routine screening is not available. In the USA, the estimated risk of giving blood during an infectious window period (seroconversion) where blood has passed screening tests for HIV antibodies is estimated as 1 in 493 000 (95% CI, 202 000–2 778 000) (Schreiber et al., 1996).

HIV transmission via contaminated blood from the use of shared injecting drug equipment is a major problem in both developed and developing countries. In Europe, approximately 44% of recent AIDS cases are due to injecting drug use, especially in urban areas of Italy, Spain, Switzerland and some Eastern European countries (Mertens and Piot, 1997). Large numbers of HIV-1 infected injecting drug users are now seen in the developing countries of Asia. High HIV infection rates in injecting drug users are also seen in some urban populations of South American countries.

Nosocomial transmission of HIV has been described both within and outside the health-care setting. For example, large outbreaks of HIV have occurred in children in Russia and Romania from the use of shared syringes, improperly sterilized needles and transfusions of unscreened blood (Hersh *et al.*, 1991; Mertens and Piot, 1997). It is likely that contaminated instruments and needles contributes to HIV transmission in many developing countries. Exposure to HIV-infected blood or other body fluids also poses an occupational risk for health-care and laboratory workers (Gerberding, 1995). Transmission in the health-care setting is usually caused by needlestick injuries, although other exposures to HIV-infected blood or products also pose a risk. Occasional cases of female to female donor insemination, tattooing, allograft and organ transplantation, non-parental bloodborne, saliva and biting, casual contact, health-care worker to patient, etc. transmissions have been described (Chu and Curran, 1997).

Apart from the recognized modes of transmission, other factors contribute at local, regional and national levels to the spread of HIV. These include the prevalence of other disease that facilitates HIV infection, e.g. ulcerative (herpes simplex, chancroid) and non-ulcerative (gonorrhea, chlamydia) genital diseases, increasing urbanization and shifts of rural populations to urban areas, forced mobilization for economic reasons or to avoid warfare and famine, changes in social and sexual mores, and lack or breakdown of medical and education services (Decosas and Adrien, 1997; Mertens and Piot, 1997).

THE HIV EPIDEMICS

As of 15 November 1998, a cumulative total of 1 987 217 cases of AIDS in adults and children have been reported to the World Health Organisation (WHO) since the onset of the AIDS pandemic (WHO, 1998). The reported cases of AIDS in each region includes:

> Africa 706 318
> The Americas 915 755 (USA 691 697)
> Asia 108 738
> Europe 211 352
> Oceania 9084.

Allowing for incomplete and delayed reporting, and underdiagnosis, WHO has estimated that by the end of 1998 approximately 13.9 million AIDS cases in adults and children (2.0 million infected vertically) had occurred worldwide since the beginning of the pandemic. Globally there are three men infected for every two women, although by the year 2000 the number of new infections among women may be closer to that among men. It is estimated that there are currently 33.4 million people living with HIV/AIDS, of which 5.8 million were newly infected during 1998, including 590 000 children. This amounts to 16 000 new infections per day. The estimated number of adults and children with HIV/AIDS as of the end of 1998, by region, are expressed in Figure 1.1, and these figures suggest that HIV is still spreading rapidly. However, there is evidence that a stabilization and perhaps a decline in the prevalence of HIV infection is taking place in some industrialized countries, including North America, Western Europe and Australasia, as well as in some high-prevalence

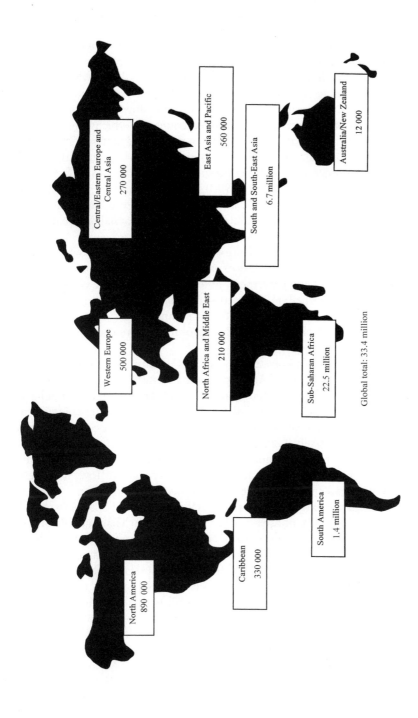

FIGURE 1.1 World map showing, by region, the estimated number of adults and children (global total 33.4 million) living with HIV/AIDS as at end 1998 (adapted from the **UNAIDS/WHO Working Group on Global HIV/AIDS and STD Surveillance**, *Weekly Epidemiological Record* 1999, 73: 373–380).

Western Europe
500 000

Central/Eastern Europe and Central Asia
270 000

East Asia and Pacific
560 000

South and South-East Asia
6.7 million

Australia/New Zealand
12 000

North Africa and Middle East
210 000

Sub-Saharan Africa
22.5 million

Global total: 33.4 million

North America
890 000

Caribbean
330 000

South America
1.4 million

areas of East and Central Africa. For the year 2000, WHO currently projects a cumulative total of close to 40 million HIV infections in men, women and children, of which 90% are in developing countries, and a total of over 10 million adult AIDS cases (WHO, 1997).

Although HIV has spread rapidly around the globe, its spread has different characteristics in different regions. In Western Europe, North America and Australasia, HIV infection has predominantly affected men who have sex with men, and injecting drug users, together with their sexual partners, with the proportion of heterosexually acquired infection slowly increasing. However, differences exist between some developed countries, which is reflected in high HIV infection rates in homosexual men in the United States (52% in the years 1993–1995), Australasia (89%) and England (70%), compared to Italy and Spain, where injecting drug users provide the majority of notifications (68% and 68%, respectively) (WHO, 1997). Various modes of transmission have been described in the former USSR and other Eastern European countries, including infecting drug use, sexual spread and nosocomial outbreaks in infants and children resulting from unsafe medical practices.

In South America most infection occurred initially in homosexual or bisexual men, but increasing heterosexual transmission, principally among bisexual men and their female partners, and among female sex workers and their clients, is now described. HIV infection among injecting drug users is a rapidly growing problem, particularly in Argentina and Brazil. In most of the Caribbean, heterosexual transmission is the major form of transmission (Mertens and Piot, 1997).

SubSaharan Africa constitutes 75% of the estimated global total of AIDS cases, with East and Central Africa providing 50–60% of cases despite accounting for only 15% of the total population. Very high rates of infection have been seen in countries such as Tanzania, Uganda and Zaire. The epidemic is continuing to evolve in Western and Southern Africa, and in selected countries HIV prevalence is relatively low, eg Cameroon, Benin and Gabon. Heterosexual intercourse is a prominent mode of transmission in Africa. In North Africa and the Middle East prevalence rates of HIV infection are low, although data concerning prevalence and modes of transmission are lacking.

Although the extensive spread of HIV in South and Southeast Asia began only in the mid-1980s, progression has been rapid. In Thailand HIV has spread in waves among injecting drug users, commercial sex workers, their clients and regular partners of clients (Weniger et al., 1991). Rapidly rising seroconversion rates towards the end of the 1980s, followed by a relative decline and stabilization around 1991–1992, have occurred in injecting drug users in Bangkok (Kitayaporn et al., 1994). Rapid HIV spread is occurring in India, although obtaining an accurate picture is difficult. In some regions (eg Manipur) injecting drug use is a major contributor; in other states (eg Maharashtra, Goa) high rates are seen in commercial sex workers and their clients. Rapid spread of HIV has occurred in countries close to Thailand, including Myanmar, Cambodia, Vietnam and Malaysia. In Yunnan Province in China there is large outbreak of HIV infection in injecting drug users, and spread to adjoining provinces is likely. Over 200 000 cases of HIV infection have been reported in China (Xiwen Z, personal communication)

A large proportion of the reported AIDS cases in Japan were in hemophiliacs transfused with HIV-infected blood products in the early to mid-1980s, although sexual transmission is now well described. In Papua New Guinea, where prevalence rates are increasing quickly, more than 70% of cases are acquired heterosexually, whereas in neighboring Australia most cases are in homosexual or bisexual men, with a male to female ratio of 17:1. In contrast, rates of AIDS in New Zealand are declining, and approximately 90% of cases occur in homosexual or bisexual men.

HIV-2

HIV-2 was first described from Senegal in 1985, and is predominantly seen in Western African nations (eg Senegal, Guinea-Bissau, Gambia, Cape Verde, the Côte d'Ivoire etc) and in countries with colonial links to Western Africa, eg Portugal and France (Barin et al., 1985; Clavel et al., 1987). Other countries with links to Portugal, eg Brazil, Angola, Southwest India and Mozambique, also report HIV-2 infections (Marlink, 1996). Other countries in Europe, USA and Australasia report occasional HIV-2 cases in immigrants (and their partners) from Western Africa (Smallman-Raynor and Cliff, 1991; Downie et al., 1992; O'Brien et al., 1992). The modes of transmission of HIV-2 are similar to those of HIV-1, although the clinical progression and rates of vertical transmission are less. Dual infections with HIV-1 and HIV-2 occur, and pose problems for some diagnostic tests.

THE MOLECULAR EPIDEMIOLOGY OF HIV

Following the publication of the first full-length HIV sequences from Europe, Africa and the USA, and SIV sequences from non-human primates, it was realized that there was significant polymorphism at the nucleotide and amino acid level (Ratner et al., 1985; Wain-Hobson et al., 1985). A number of mechanisms contribute to this genetic variability, including lack of proof-reading or the generation of errors by RNA polymerases, resulting in high error rates, and enhanced by the high viral turnover that occurs at all stages of disease. Other sources of genetic diversity include the presence of insertions and deletions, G–A hypermutations, and recombination. Over the last 10–12 years the variation in HIV-1 has allowed it (and HIV-2) to be broadly classified into different subtypes (Myers et al., 1995). Recombination between different subtypes occurs, and a number of subtypes are in fact recombinant viruses. The HIV-1 subtypes (also called clades, genotypes or strains) are classified into two groups, the major or main (M) group, containing subtypes A–J, and the outlier (O) group (Myers et al., 1995). The maximum intrasubtype nucleotide variation is typically less than 15%, but intersubtype variation at the nucleotide level is between 20% and 30%. In unrooted phylogenetic trees all subtypes are approximately equidistant, suggesting that they may have moved out from a hypothetical central common source, although no published sequence is an obvious candidate HIV-1 protovirus.

The global dispersion of different subtypes is not uniform (Figure 1.2). The overwhelming majority of isolates from the USA, Haiti, Western European and

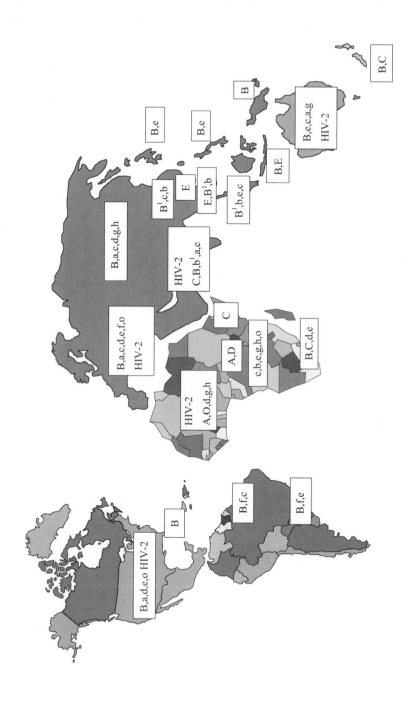

FIGURE 1.2 HIV-1 subtypes throughout the world are divided into the main (M) group (subtypes A-H) and the outlier or O group. The subtype in capitals reflects the likely or known subtype. HIV-2 subtypes are not displayed.

Australasia are subtype B (Slobod *et al.*, 1994; Myers *et al.*, 1995; Dwyer *et al.*, 1996, 1997). Subtype B predominates in Brazil and other South American countries, with some subtypes F and C also present (Louwagie *et al.*, 1994; Campodonico *et al.*, 1996). Although Western Europe has subtype B almost exclusively, in Eastern Europe subtypes A, D, C, F, G and B also occur (Chaix *et al.*, 1993; Bobkov *et al.*, 1994; Dumitrescu *et al.*, 1994; Clewley *et al.*, 1996). In subSaharan Africa subtypes A and D predominate, although in some Central African countries (eg Gabon, Central African Republic, Cameroon) at least six other subtypes are circulating (Albert *et al.*, 1992; Murphy *et al.*, 1993; Delaporte *et al.*, 1996; Heyndrickx *et al.*, 1996; Lasky *et al.*, 1997). In Southern Africa, the Horn of Africa and Western India subtype C is most common, with evidence that the Indian subtype C viruses came originally from Southern Africa (Grez *et al.*, 1994; Van Harmelen *et al.*, 1997). Subtype E is now dominant in Thailand and has spread to surrounding countries, whereas subtype B is more frequently found in other countries of East Asia and Australasia (Kalish *et al.*, 1994, 1995; Weniger *et al.*, 1994). A variant of subtype B (with a different envelope V3 loop tip tetrapeptide motif), first described in injecting drug users in Bangkok, has now spread to other injecting drug-using populations in Southeast Asia (Weniger *et al.*, 1994; Kalish *et al.*, 1995). In contrast to the variation seen in countries with more mature HIV epidemics, less genetic variability is seen in countries where the epidemic is more recent. Examples include subtype E strains in Thailand, C in India and F in Brazil; it is likely that each of these represents an outgrowth on a massive scale of a single epidemiologic clone. With most subtypes more heterogeneous clusters of genetically related viruses exist in Africa, suggesting that these strains were dispersed to these countries from Africa; transmission of a single epidemiologic clone would explain the bottle-neck reduction in diversity in the new epidemics.

It has become apparent that the genetic variation exhibited by HIV is significantly contributed to by recombination between different subtypes. This was first recognized with HIV-1 MAL, a Zairian HIV-1 sequence, where phylogenetic analyses of regions of the full length of HIV-1 MAL genome showed it to be a recombinant of subtypes A and D (Robertson *et al.*, 1995). Similar recombination has been observed with a number of other subtypes, including the Thai subtype E (an A/E recombinant), a recombinant B/F from Brazil, and the Ugandan/Russian subtype G, an A/G recombinant virus (Myers *et al.*, 1995, Gao *et al.*, 1996). Genetic subtypes of HIV-2 exist, phylogenetically related to various SIV, but widespread population studies of HIV-2 subtypes are yet to be performed (Gao *et al.*, 1994).

HUMAN T-LYMPHOTROPIC VIRUS TYPE I (HTLV-I)

The spread of HTLV-1 into human populations, presumably from primates (given the close phylogenetic relationship between HTLV and STLV), appears to have occurred much earlier than for HIV-1 and HIV-2. The virus differs from HIV, being highly cell associated with relatively low concentrations in blood, body fluids and genital secretions. It also exhibits much less genetic variability, being distributed as a major worldwide strain, the cosmopolitan type and Asian and African strains (Saksena and Bolton, 1998). Like HIV, transmission may be vertical (most commonly via breast milk), parenteral (including blood transfusion and injecting drug use) and sexual, male to female being at least four times commoner than the converse, but occurs mainly via infected lymphocytes (Kajiyama *et al.*, 1986; Tsuji *et al.*, 1990; Stuver *et al.*, 1993; Sullivan *et al.*, 1991). The seroprevalence of HTLV-1 varies markedly throughout the world, with high endemicity in southern Japan, Central and West Africa, the Caribbean basin, Southern India, Melanesia and some parts of South America. In Japan overall HTLV-1 coverage is 0.1%, but with pockets of up to 50% (Hinuma *et al.*, 1982). However, the importance of sexual transmission has been difficult to distinguish in each region. Nevertheless, in Japan clear evidence for sexual transmission has been demonstrated, and in New Guinea indirect evidence shows a parity-related increase in seroprevalence in migrant but not indigenous women (Brabin *et al.*, 1989). HTLV-1 transmission in injecting drug users has been demonstrated in London, but not in the attendees at STD clinics (Loveday and Mercey, 1993). However, in Peru high prevalences of antibody to HTLV-1 were found among female sex workers, homosexuals and patients with sexually transmitted diseases, with a strong correlation with the presence of genital ulcers and intercourse during menses (Zurita *et al.*, 1997).

Only 1% of infected patients develop T-cell leukemia or spastic paraparesis, with a latency period of up to 40 years (Cann and Chen, 1996).

GENITAL HERPES

Herpes simplex virus type 2 (HSV-2) usually affects the genital or adjacent areas, with transmission being sexual. In contrast, herpes simplex virus type 1 (HSV-1) usually causes oropharyngeal infection, with transmission being by direct contact. However both viruses *can* cause oropharyngeal or genital infections. HSV-1 is becoming an increasingly common cause of primary genital infection in most Western countries (up to 50%), but is less commonly a cause of recurrent infection (Corey, 1990). HSV-2 is the major cause of initial genital herpes (50–95% of cases) in many countries, and the predominant cause of recurrent genital herpes (90–98%). It varies markedly in its distribution in different countries and in different populations within those countries.

After initial infection of the skin or mucosa the virus is transmitted by sensory nerves to the dorsal root ganglia, or to trigeminal ganglia for HSV-2 and HSV-1, respectively, where it becomes latent. Reactivation of both viruses in the oral and genital areas is frequent, and more often asymptomatic than symptomatic (Wald *et al.*, 1997). Only approximately 20% of patients infected with HSV-2 recognize genital lesions. Another 60% can be taught to recognize typical or atypical lesions, leaving 20% with true asymptomatic shedding (Koutsky *et al.*, 1992). Patients shed virus frequently, eg in the first 2 years after initial infection with HSV-2 women in Seattle, USA, shed at a mean of 6% of all days and men at 4% of all days (Wald *et al.*, 1997). Shedding occurs from the vulva, cervix, urethral and perianal skin in women, and the penile and urethral skin in men. Furthermore, it is now known that most transmission of HSV-2 occurs via asymptomatic shedding, although the greatest risk of transmission at any one time is during primary infection. The annual rate of transmission between partners serodiscordant for genital herpes appears to vary from 4 to 30%, depending upon whether it is male–female or female–male, and whether the recipient has had previous HSV infection or not (Mertz et al., 1992). Both HSV-1 and HSV-2 infections are associated with significant complications, such as neonatal infection and death, meningitis or encephalitis, recurrent rashes and eye disease.

Infection with either HSV-1 or HSV-2 is usually associated with an antibody response, which reflects the presence of latent virus within the body and the likelihood of subsequent asymptomatic or symptomatic shedding from oral or genital mucosae. However, until a decade ago it was difficult serologically to differentiate infection by HSV-2 and HSV-1, because of the antigenic similarity between most of the proteins shared by these two closely related viruses. One of the most divergent proteins between the two viruses is glycoprotein G (gG), which consists of approximately 238 amino acids in HSV-1 and 699 amino acids in HSV-2. Furthermore, patients infected with HSV-2 have type-specific antibodies to the immunodominant epitopes of gG2 which are not present in gG1 (Lee et al., 1985) and which can be measured by ELISA or Western blot (Ho et al., 1992, 1993). ELISAs based on glycoprotein G2 or Western blots incorporating this protein (and other type-specific proteins or protein epitopes) are now sufficiently reliable to provide estimates of the incidence and prevalence of HSV-2 infection, and are obviously more accurate than comparisons of clinically overt genital herpes or virus isolation (Ashley et al., 1988).

HSV-2 seroprevalence around the world has been compared in antenatal populations and in STD clinic populations, although very few population-based studies have been performed. However, in the United States the National Health and Nutrition Examination Surveys (NHANES) II and III of 1976–80 and 1988–1994, respectively, allowed the accurate estimation of age-specific seroprevalence nationwide (Johnson et al., 1989; Fleming et al., 1997). In randomly sampled American populations aged 12 or older there was a 30% increase in HSV-2 seroprevalence from 16% to 22% (male 10% and female 26%) in the last decade. This is surprising in view of the emphasis on safe sex campaigns during the era of spread of HIV.

In various populations, including general surveys, college students, blood donors at antenatal clinics and STD clinics, a consistent cluster of factors that influence the acquisition of HSV-2 has been recognized. These include age, ethnic origin, socioeconomic group or educational level, gender, geographic variables and sexual exposure, including age of first exposure, total number of sexual partners, and obviously a partner with herpes (Brienig et al., 1990; Gibson et al., 1990; Cowan et al., 1996; Wald et al., 1997; Johnson et al., 1989; Eberhardt-Phillips et al., 1998; Mindel et al unpublished). Prior HSV-1 exposure may delay and decrease the prevalence of HSV2 infection.

The lifetime number of sexual partners is one of the most important risk factors for HSV-2 infection, and therefore HSV-2 antibodies have often been used as a marker for sexually transmitted diseases (Nahmias et al., 1980; Basset et al., 1994). However, HSV-2 antibody may not always be a good marker for frequency of exposure in age cohorts where there has been little circulation of HSV-2 (eg in the 21-year-old cohort in Christchurch, NZ) (Eberhardt-Phillips et al., 1998).

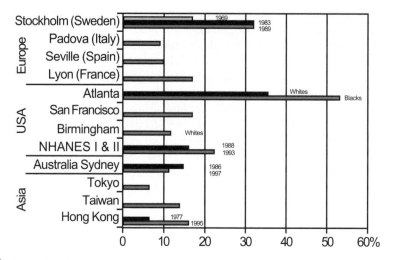

FIGURE 1.3a

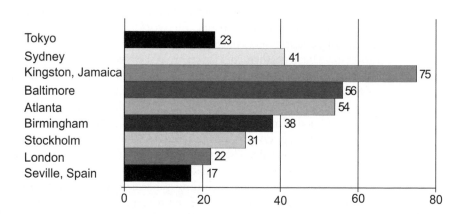

FIGURE 1.3b

FIGURE 1.3 Worldwide seroepidemiology for HSV2, based on gG2 ELISA or Western blot in antenatal populations (which may slightly overestimate the proportion of infected men), except in the **NHANES II and III and Hong Kong** studies. The **NHANES** studies were random population-based surveys and the Hong Kong studies were obtained from consecutive sera from attendees at antenatal clinics in Hong Kong and Southern China. The Sydney antenatal studies were supported by similar figures in blood bank populations. Similar populations were sampled in the serial studies shown (eg for Stockholm). Data from Nahmias *et al.*, 1980; Forsgren *et al.*, 1994; Cunningham *et al.*, 1993; Lo *et al.*, 1998.

A comparison of seroprevalence studies in pregnant women performed at approximately the same time from the southern USA, Italy, Spain, Sweden, France, Iceland, Japan, Taiwan and Australia showed that the highest seroprevalences were in the USA (especially in black American antenatal patients in Atlanta), followed by Sweden, France, Australia, Italy and Taiwan, and was lowest in Japan (Nahmias et al., 1990; Figure 1.3). Much lower HSV-2 seroprevalences have been observed in cohorts of 16-year-old Swedes and 21-year-old New Zealanders (3–4%) than in the USA (12% in a 12–22-year-old cohort in Cincinnati (Andersson-Ellstrom A. et al., 1995; Eberhardt-Phillips et al., 1998; Rosenthal et al., 1997). Nevertheless, in Sweden seroprevalences increased rapidly over 15 years of surveillance of a cohort of 14-year-olds (0.4% to 22%) (Christenson et al., 1992).

In blood donors (who are usually asked to declare whether they have had past sexually transmitted diseases) prevalence ranged from a mean of 8% in London to 13% in Sydney (Field et al., unpublished observations). The seroprevalence in STD clinic attendees from Japan, Spain, the UK, Sweden and New Zealand ranged from 21% to 26%, and in the USA from 40% to 57% (Nahmias et al., 1990; Perkins et al., 1996). In STD clinic attendees in Sydney, Australia, HSV-2 seroprevalence was 40% in 1984–85 (Cunningham et al., 1993), and in 1990–91 the seroprevalence in a group of patients at the Sydney Sexual Health Clinic was much higher, at 64% (Bassett et al., 1994). Female prostitutes have high levels of HSV-2 seropositivity worldwide (79% in Tokyo and the USA, 72% in Sydney, Australia, 96% in Dakar, Senegal). The HSV-2 seroprevalence among homosexual men is also proportional to the number of lifetime partners, is increased in HIV-positive men, and varies geographically within Western countries from 83% in San Francisco to 24% in Japan and 22% in Spain (Nahmias et al., 1990). In a Sydney STD clinic 57% of all homosexual men were positive, but this increased to 78% of HIV antibody-positive men (Cunningham, Field and Ho, unpublished observations).

Surveys of various populations around the world at different times have shown marked changes in seroprevalence. As mentioned above, the NHANES II and III studies provide the best data for the USA, showing a 30% increase between 1976 and 1980 and 1988–1994, mostly in young whites in their teens and 20s. Random consecutive serosurveys in pregnant women also showed increases in Reykjavik, Iceland (4% to 23%, 1979–1985); Lyon, France (11% to 17%, 1978–1985); Stockholm, Sweden (17% to 32%, 1969–1983); and Hong Kong (6% to 15%, 1977–1995) (Forsgren et al., 1994; Lo et al., 1998; Nahmias et al., 1990). However, no increase was observed between surveys of similar populations in 1983 and 1989 in Sweden (Forsgren et al., 1994) and in 1986 and 1996 in Sydney (14.5% to 12.1%) (Cunningham et al., 1993, Mindel et al., unpublished) perhaps reflecting a greater awareness of sexual transmission of HIV during that decade. Nevertheless, several studies have shown that condoms are only partially protective against HSV-2 transmission.

The incidence of neonatal herpes varies markedly worldwide, from 20–50/100 000 live births in the USA (Seattle/Birmingham) to 11/100 000 in Australia, 6.5/100 000 in Sweden, and approximately 1.65/100 000 in the UK. HSV-2 seroconversion and genital shedding in late pregnancy is a major risk factor, but HSV-1 infection is as common as HSV-2 in Australia and the UK (Whitely, 1993;

Forsgren *et al.*, 1990; Brown *et al.*, 1997) Tookey and Garland, 1992. The HSV-2 seroconversion rate during pregnancy may also vary widely. In Seattle, USA, the seroconversion rate during pregnancy was 1.9% pa, in San Francisco 0.6% pa and in Sydney 0.34% in 2040 women, correlating with seroprevalences of 28%, 16% and 12%, respectively (Boucher *et al.*, 1990; Brown *et al.*, 1997; Mindel *et al* unpublished). HSV-2 antibody seropositivity is also a risk factor for HIV acquisition, in both the USA and East Africa (Handsfield *et al.*, 1991; Holmberg *et al.*, 1988), probably through increased transmissibility of HIV in the presence of genital ulceration and infiltrating T cells (Cunningham *et al.*, 1985).

CYTOMEGALOVIRUS

After initial infection with cytomegalovirus, which is usually asymptomatic and mostly occurs in childhood, the virus remains latent in leukocytes of the marrow and blood (probably granulocyte–macrophage precursors and mono-cytes) for life, reactivating relatively infrequently after immunosuppression or in advancing AIDS (Kondo and Mocarski, 1996). Chronic active infection and reinfection may occur in certain circumstances. Therefore, the epidemiology of cytomegalovirus is best measured by IgG antibody, reflecting the prevalence and incidence of infection, and by viral shedding in saliva, genital secretions, urine and blood, as an indicator of active infection and therefore likely to result in transmission. The virus infects and reactivates in salivary glands, renal tissue and cervical mucosa, and can be transmitted via saliva, urine, semen or cervical secretions (Britt and Alford 1996).

Infection with cytomegalovirus is considered endemic and ubiquitous throughout the world, with the seroprevalence among young adults ranging from 40 to 100% (Figure 1.4). The higher rates occur in developing countries and in those of lower socioeconomic status in the developed industrialized countries (Ho, 1990). The virus is acquired at various stages during life, in utero via transplacental infection, at term via cervical shedding, during breastfeeding, during infancy, through saliva and urine in daycare centres, by sexual transmission (both heterosexual and male homosexual), and from infant or young child to susceptible parent in adulthood, completing the cycle through infection of susceptible pregnant women and infection of the fetus in utero (Ho, 1990).

Sexual transmission of cytomegalovirus among heterosexual women attending STD clinics, in prostitutes and homosexual men, has been shown by high rates of cervical shedding in women and higher seroprevalence in homosexual men in STD clinic attendees and prostitutes. There is a marked variability in the rate of cervical shedding according to country and technique used. For example, using PCR in Taiwan, prevalences of CMV DNA reached 35–39% for prostitutes and STD clinic attendees, respectively (Shen *et al.*, 1994). In the USA 16% of CMV-seropositive STD clinic attendees, and in Australia 6% of all patients attending a gynecological clinic for dysplasia, shed CMV by virus isolation (Embil *et al.*, 1985; Collier *et al.*, 1995; Mathijs *et al.*, 1997). In Western countries there is also considerable variability, depending upon age at first intercourse, number of recent partners, and use of barrier contraception. There was a significant association with other STDs, such as

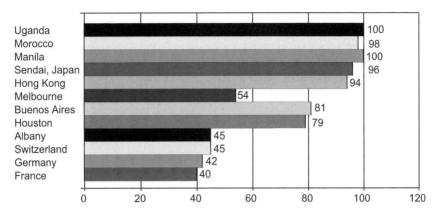

FIGURE 1.4 Worldwide seroprevalence for cytomegalovirus. Data from Ho et al., 1990

gonorrhea (Mindel and Sutherland, 1984; Collier *et al.*, 1990). In homosexual men CMV seropositivity was associated with increasing age, lifetime numbers of sexual partners and anal-receptive intercourse (Collier *et al.*, 1987; Mindel and Sutherland, 1984). A high rate of seropositivity (over 90%) has been documented in Western countries, associated with the shedding of CMV in semen in 20–30% of men for up to 22 months after seroconversion. Cytomegalovirus infection was associated with an alteration in T-lymphocyte subsets and ratio. The risk of CMV infection increased in patients coinfected with HIV, with higher rates of shedding and a greater likelihood of CMV disease, especially retinitis and gastrointestinal and neurological disease. Seminal shedding of CMV increases with decreasing CD4 counts in patients coinfected with HIV and CMV (Krieger *et al.*, 1995). Nationality and social class were also associated with CMV among heterosexuals. Gonorrhea was associated with CMV in both heterosexuals and homosexuals (Mindel and Sutherland, 1984). The seroprevalence among a cohort of adolescents in Cincinnati, USA, was higher at 62% than in a cohort of 16-year-old girls in Sweden (45%) (Anderson-Ellstrom *et al.*, 1995; Rosenthal *et al.*, 1997).

HUMAN PAPILLOMAVIRUS INFECTION

Human papillomaviruses are classified using DNA typing, as culture and serological assays are unsuitable. Based on differences detected by DNA hybridization, 'new' types are defined as having a less than 50% genomic DNA homology compared to previously accepted prototypes (Coggin and zur Hausen, 1979). Currently consideration is being given to reclassification on the basis of biological and pathogenic potential, related largely to particular regions of the HPV genome (E6/E7 and L1) (de Villiers, 1992). There are at least 100 human papillomaviruses, and more than 30 known to affect the genital tract. Further classification has been made according to the oncogenic potential of these viruses. For example, HPV types 6 and 11 are usually associated with condylomata acuminata (genital warts), whereas HPV types 16 and 18 are generally associated with flat warts of the cervix and cervical intraepithelial neoplasia and cervical carcinoma. HPV types 16 and 18 have also been associated with other genital and perigenital tumors, including penile, vulval and anal cancer (Wieland and Pfizer, 1997). The HPV types associated with genital lesions and genital tumors are summarized in Table 1.1).

The association of genital warts and sexual activity has been recognized for centuries. However, it was only with the suggestion of an epidemiological link between HPV type 16 and genital tract tumors that more definitive studies have been carried out. Using molecular techniques to study patients with genital warts has shown that the majority of these (well over 90%) are due to types 6 and 11. However, other types, including 2, 16, 30, 40, 44, 45, 54, 55 and 61, have also been found (Wieland and Pfizer, 1997). Subclinical lesions also occur, as evidenced by the fact that many patients who apparently have their warts apparently successfully treated recur, and that by using molecular techniques evidence of HPV infection can be demonstrated in normal-looking areas adjacent to clinical warts, and also in the partners of patients with warts (Sonnex, 1995).

Incidence data are available for condylomata acuminata, mainly from STD clinics in Europe and North America. These studies show that in the USA the incidence of condylomata acuminata increased eightfold (13–106/100 000) between the early 1950s and the late 1970s. Similar incidences have been noted in other developed countries. The reported prevalence of genital warts varies between 0.6 and 13%. These variations depend upon age, ethnicity, sexual orientation and other sexual factors, such as the numbers of lifetime sexual partners and condom use. Genital warts are now the commonest STD diagnosed from STD clinics in the UK (Government Statistical Service) and Australia (Donovan *et al.*, 1992), and the situation in other developed countries is likely to be similar. Over recent years there have been a large number of studies conducted using molecular techniques, in particular PCR, to detect subclinical

TABLE 1.1 Sexually transmitted diseases distributed throughout the world

HIV/AIDS

Genital herpes

Papillomavirus

Hepatitis B, C, A, E

Cytomegalovirus

Chlamydia trachomatis

Syphilis

Gonorrhea

HHV-8/KSHV

HTLV1

Tropical: chancroid, donovanosis, Lgv, Trichomoniasis

HPV. These have shown that in sexually active women HPV can be detected in 4–55% (Melkert *et al.*, 1993; Nishikawa *et al.*, 1991; Schiffman *et al.*, 1993; Kjaer *et al.*, 1993; ter-Meulen *et al.*, 1992; Coll-Seck *et al.*, 1994; van Doornum *et al.*, 1992; Engels *et al.*, 1992; Czeglédy *et al.*, 1992; Fairley *et al.*, 1992; Pao *et al.*, 1990; Critchlow *et al.*, 1992; Laga *et al.*, 1992; Kreiss *et al.*, 1992; Vermund *et al.*, 1991; Feingold *et al.*, 1990; St Louis *et al.*, 1993; Spinillo *et al.*, 1992); in 0–86% of heterosexual males (Wikström *et al.*, 1991; Critchlow *et al.*, 1992; Kataoka *et al.*, 1991; van Doornum *et al.*, 1992; Green *et al.*, 1991; Mandal *et al.*, 1991; Omar *et al.*, 1991; Nieminen *et al.*, 1991; Ostrow *et al.*, 1986; Grussendorf-Conen *et al.*, 1986); and in 0.8–78% of homosexual males (Kiviat *et al.*, 1993; Caussy *et al.*, 1990; Critchlow *et al.*, 1992; Melbye *et al.*, 1990; Palefsky *et al.*, 1990). However, these data are difficult to interpret and compare, as the source of the populations varied (including gynecology, STD and family planning clinics), the age varied from 18 to 53 years, and many studies did not report sexually related factors, in particular the number of partners, partners with warts, or condom usage. In addition, the studies varied in terms of which HPV type was looked for, and also which method was used to detect it. Despite these reservations, these studies suggest that subclinical HPV infections are extremely common in all parts of the world where studies have been done. HPV- and HIV-impaired cell-mediated immunity (CMI) is associated with the persistence of genital and perigenital HPV infection, including warts, and an increased incidence of cervical and anal squamous cell carcinoma. This has been reported in patients with neoplasia, in transplant recipients and in those infected with HIV (Melbye *et al.*, 1990; Palefsky *et al.*, 1990; de Villiers, 1992). Numerous cross-sectional studies have shown that women and men infected with HIV have an increased prevalence of clinical and subclinical (detection using molecular techniques) HPV infection. Using non-PCR methods (dot-blot hybridization, Southern blot and in situ hybridization), HPV was reported from 4–25% of women without HIV and 37–53% of those with HIV. However, using PCR-based methods the differences were more obvious, with a prevalence of 21–49% in HIV-negative women and 56–74% in HIV-positive women (ter-Meulen *et al.*, 1992; Coll-Seck *et al.*, 1994; Laga *et al.*, 1992; Kreiss *et al.*, 1992; Vermund *et al.*, 1991; Feingold *et al.*, 1990; St Louis *et al.*, 1993; Spinillo *et al.*, 1992). Similar differences have been noted in homosexual males (HIV-negative 6–78%, compared to 26–92% HIV-positive) (Kiviat *et al.*, 1993; Caussy *et al.*, 1990; Critchlow *et al.*, 1992; Kiviat *et al.*, 1993; Melbye *et al.*, 1990; Palefsky *et al.*, 1990). Anecdotal reports also suggest that HPV infections appear to be more persistent and perhaps less sensitive to therapy in HIV-positive individuals, particular when CD4 lymphocyte counts are low, and that patients with HPV-associated cervical or anal dysplasia progress more rapidly (Palefsky *et al.*, 1990). However, long-term follow-up studies have yet to be reported.

HEPATITIS B

Hepatitis B is one of the commonest and most widespread viral infections of humans. Furthermore, it is the major viral cause of cancer (hepatocellular

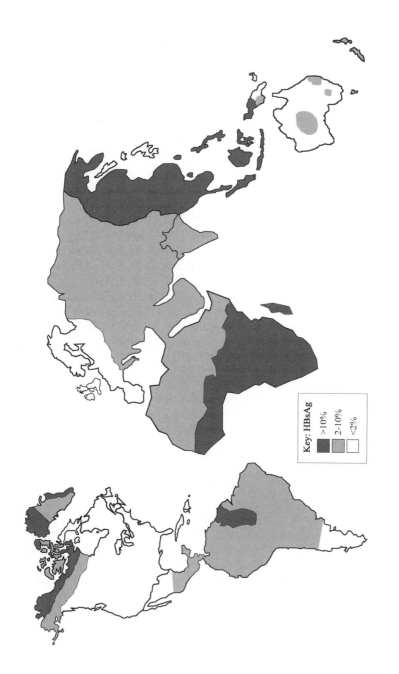

■ FIGURE 1.5 Worldwide distribution of hepatitis B carriers.

Key: HBsAg

>10%

2-10%

<2%

carcinoma), and also of cirrhosis, particularly in Asia and Africa. Infectious hepatitis B virions are found in almost all biological fluids, at high concentrations in the blood, menstrual fluid, semen, vaginal secretions and saliva; at low concentrations in the urine, but usually not in feces unless contaminated by blood. Therefore, the methods of transmission of hepatitis B virus are via the blood (including all forms of parenteral contact, such as injecting drug use and nosocomial transmission in hospitals), vertically (usually at parturition), and sexually. Hepatitis B surface antigen (HBsAg) has been found in breast milk, but there is no convincing evidence for transmission by this route (Alter *et al.*, 1986; Karayiannis *et al.*, 1985).

The prevalence of hepatitis B as measured by serum HBsAg is highly variable throughout the world (Figure 1.5), being at its lowest (<0.5%) in countries with high socioeconomic standards, including the USA, Australia, the UK and Northern Europe. The highest prevalences have been recorded in the Pacific Islands (20–40%), followed by Southeast Asia, China, Africa, Australian Aborigines and the Inuit of Canada (5–15%) (Glasgow *et al.*, 1997). Intermediate levels (1–5%) have been found in Mediterranean countries, Eastern Europe, India, Mexico and South America (Maupas and Melnick, 1981). The highest levels of endemicity of hepatitis B are almost certainly maintained by vertical transmission and, in some countries, horizontal transmission in childhood. The marked decrease in hepatitis B prevalence after immunization of neonates in Asia, especially Taiwan, supports vertical transmission as the predominant mode of spread in these countries (Beasley *et al.*, 1983). Intrafamily transmission through wounds, or shared implements such as razors and toothbrushes, is common among the families of hepatitis B carriers.

Sexual transmission occurs after acute infection and during chronic carriage, with a high likelihood of eventual transmission to the spouse (Alter *et al.*, 1986; Berris *et al.*, 1973; Sagliocca *et al.*, 1997). Male to female transmission is more efficient than female to male. A number of studies have now shown that there is a greater risk of sexual transmission of hepatitis B among homosexual men than heterosexuals in Western communities of low endemicity. It is highly efficient, with a high proportion of homosexual men exhibiting evidence of exposure to hepatitis B (30–60% are anti-HBc positive) and approximately 5% being chronic carriers (Fulford *et al.*, 1973; Juarez-Figueroa *et al.*, 1997; Schreeder *et al.*, 1982). Prostitutes also demonstrate a high rate of infection. In the United States, sexual transmission of hepatitis B increased during the 1980s while other forms of transmission decreased (CDC, 1988). Like other forms of transmission, sexual transmission is more efficient in patients with higher concentrations of infectious virions in blood, demonstrated by the presence of circulating HBsAg or HBV-DNA. In most other developed countries (eg the UK, Australia) increased frequencies of hepatitis B were observed among patients attending STD clinics (Fulford *et al.*, 1973). However, in some Western populations, such as the Midlands of the UK, sexual transmission of hepatitis B appears to be very low and of little consequence for immunization policies (El-Dalil *et al.*, 1997).

The risk of hepatitis B acquisition (measured by anti-HBc or HBsAg) increases with the increased numbers of lifetime sexual partners and the presence of other sexually transmitted diseases, such as syphilis and HIV. In

Western countries of low endemicity heterosexual acquisition of hepatitis B has been assessed by examining patients attending STD clinics and distinguishing the likelihood of sexual acquisition from other risk factors, such as intravenous drug use and intrafamily transmission, by association with sexually transmitted diseases and numbers of lifetime partners. There is a strong association with numbers of lifetime partners and the presence of syphilis, but usually not non-ulcerating STDs, suggesting that genital ulceration may be a risk for hepatitis B transmission as well as for HIV (Corona et al., 1996). Similar studies have recently been conducted in areas of intermediate or high hepatitis B endemicity, including East Africa, Central and South America, Asia (Singapore and India) and the Mediterranean (Italy) (Edmunds et al., 1996; Heng et al., 1995; Hon et al., 1993; Jacobs et al., 1997; Juarez-Figueroa et al., 1997; Scott-Wright et al., 1997; Corona et al., 1996). Despite the dominance of vertical transmission in most of these regions, a contribution of heterosexual transmission (and in some cases homosexual transmission) can be identified by strong association with other STDs, such as syphilis, and also numbers of lifetime partners. Again, syphilis as an ulcerating STD emerges as a strong risk factor for the acquisition of hepatitis B (Heng et al., 1995; Kaur et al., 1996; Jacobs et al., 1997; Scott-Wright et al., 1997).

Mathematical models have been developed to estimate the transmission dynamics of hepatitis B in the UK, with separate models being used for homosexual and heterosexual transmission (Edmunds et al., 1996; Williams et al., 1996). These are useful for evaluating the need for immunization, but may be hampered by confounding issues of compliance. However, many so-called Western communities of low endemicity now consist of populations of mixed ethnicity with different rates of hepatitis B infection. Hence in Australia, the UK, Canada and the USA, the mixing of Asian, Mediterranean and Pacific Islander populations within urban centers, and high infection rates within Aboriginal populations in Australia and South America, raises new levels of complexity in estimating the contribution of sexual transmission to hepatitis B prevalence in the community as a whole. In such communities each of these ethnic groups needs to be considered independently.

HEPATITIS DELTA

Hepatitis delta is a virus whose transmission depends upon the superinfection of hepatitis B carriers or coinfection of susceptible persons with hepatitis B and delta viruses. In Western countries the most common mode of transmission is injecting drug use, but it can also be transmitted sexually, although much less efficiently than hepatitis B. The hepatitis delta antibody prevalence among homosexual men is higher than in control populations and is not attributable to other risk behaviors such as injecting drug use (DeCock et al., 1998; Solomon et al., 1988). The distribution of hepatitis delta throughout the world is not uniform. Regions of high seroprevalence include the Mediterranean, the Middle East, West Africa, the Amazon basin of South America, Central Asia and some Pacific Islands. The modes of transmission that maintain such high seroprevalence are not clear (Torres et al., 1996).

HEPATITIS C

In the past decade there has been considerable debate in the literature as to whether hepatitis C is sexually transmitted. In support of sexual transmission, higher seroprevalences of hepatitis C have been reported in STD clinic attendees, those with a higher number of lifetime sexual partners or other sexually transmitted diseases, prostitutes and homosexual men (Melbye *et al.*, 1990; Rall and Dienstag, 1995; Dienstag, 1997). However, alternative risk factors, including injecting drug use, can account for many of these apparent cases of sexual transmission (Bresters *et al.*, 1993; Brettler *et al.*, 1992; Conry-Cantilena *et al.*, 1996; Osmond *et al* 1993). Anecdotal cases of sexual transmission have been reported, and sequence homology of a common hepatitis C viral strain in sexual partners without other risk factors has been demonstrated (Thomas *et al.*, 1995; Peano *et al.*, 1992). However, studies of stable monogamous sexual partners of patients with chronic hepatitis C usually failed to demonstrate an infection risk, especially compared to hepatitis B (Gordon *et al.*, 1992; Meisel *et al.*, 1995). The exception was a Japanese study of spouses of hepatitis C carriers showing a progressively increasing risk of transmission throughout married life, a result which is difficult to explain (Akahane *et al.*, 1994). Furthermore, the prevalence of hepatitis C infection in groups at high risk of other sexually transmitted diseases is much lower than with hepatitis B (30%) or HIV (10%–15%) (Eyster *et al.*, 1991). The seroprevalence of hepatitis C is also much more uniform throughout the world than for hepatitis B (van der Poel, 1994). Therefore, it is not surprising that there are few data on variations in patterns of sexual transmission around the world, ie compared to parenteral transmission, especially in injecting drug users, sexual transmission is relatively unimportant on a global scale.

HEPATITIS A

Like hepatitis B and delta, hepatitis A shows a non-uniform distribution of infection throughout the world, as measured by age-specific acquisition and seroprevalence. Although transmitted mainly by the fecal–oral route in food and water, and under other conditions of poor hygiene, it has recently been shown to be an important sexually transmitted disease among homosexual men. Since 1980, periodic epidemics of hepatitis A among homosexual men have occurred, especially as seroprevalence fell and the number of those susceptible increased (Christenson *et al.*, 1982; Coutinho *et al.*, 1983; Leentvaar-Kuijpers *et al.*, 1995). In 1991 a large epidemic occurred simultaneously in large US, Canadian, European and Australian cities (CDC, 1992; Mulhall *et al.*, 1995). The risk of infection has been clearly related to oroanal and digital–anal intercourse in case control studies. Some of these epidemics have spread to the rest of the community (Henning *et al.*, 1995; Katz *et al.*, 1997). In addition, low-level endemic infection may also continue in some Western communities (Katz *et al.*, 1997). However, random seroprevalence surveys of homosexual and heterosexual men in low- and intermediate-endemicity populations (Brighton, UK, and Madrid, Spain) between epidemics do not always show such differences, indicating that other risk factors may dominate in these populations, at least at the time of sampling (Ballesteros *et al.*, 1996; Nandwani *et al.*, 1994).

OTHER VIRUSES

There are several other viruses that are also spread heterosexually, although this is not often the dominant mode of transmission. Molluscum contagiosum is an example of a sexually spread poxvirus, which causes a minor irritative skin disease where the lesions are increased in size, density and propensity, to recur in patients with AIDS (Brown *et al.*, 1981). It occurs worldwide. The mode of transmission of human herpes virus type 8 (HHV8/KSHV), the possible cause of Kaposi's sarcoma, is still being clarified, but is likely to be sexual (Simpson *et al.*, 1996). Epstein–Barr virus and other hepatitis viruses (G) may also be spread sexually but, if so, this is not the dominant mode of transmission (Sixby *et al.*, 1986). Little is known of the worldwide variability in sexual transmission of these viruses.

CHLAMYDIA TRACHOMATIS

The epidemiology of genital chlamydial infection is less well documented than that for syphilis and gonorrhea. There are several reasons for this, including the lack of cheap reliable tests, the observation that most females (80%) and males (50%) are asymptomatic, and finally, that many men who present with urethral discharge are diagnosed as having NGU without confirmatory tests being performed. Most studies suggest that chlamydia is more common than gonorrhea, with an estimated 88 million adult cases worldwide in 1995 (Gerbase *et al.*, 1998). In countries where reasonable surveillance data are available (Europe, North America and Australia), chlamydia has emerged as the leading bacterial STD (Quinn *et al.*, 1996). Unlike gonorrhea and syphilis, chlamydia continued to increase throughout the 1980s. However, recent national control programs in some countries, in particular Sweden, and local control programs in some cities in Canada, the USA and Australia, have been very successful, and in these circumstances the incidence of chlamydia infection has gradually decreased (Garland *et al.*, 1993; Kihlström *et al.*, 1994; Crowley *et al.*, 1994; Hillis *et al.*, 1994; Patrick and Bowie 1994).

The World Health Organization has estimated that the incidence of chlamydial infection is virtually twice that of gonorrhea in developed countries. For example, in 1995 the estimated number of new cases of gonorrhea in women in North America was 830 000, compared to 1 640 000 cases of chlamydia (Gerbase *et al.*, 1998). In contrast, in developing countries chlamydial infection occurs with a similar frequency to gonorrhea. For example, in subSaharan Africa the estimated number of new cases of gonorrhea in 1995 was 7 300 000, and for chlamydia it was 6 960 000 (Gerbase *et al.*, 1998). The reasons for this difference may relate to more active surveillance and case finding for gonorrhea in developed countries.

The true prevalence and incidence of chlamydia has been difficult to establish. Unlike gonorrhea, chlamydia has been difficult to identify in the laboratory. The organism cannot be grown on conventional culture medium, but can be grown in cell culture. Other methods for identification have included direct immunofluorescence, enzyme immunoassay, and more recently polymerase chain reaction (PCR) and ligase chain reaction (LCR). The introduction of enzyme immunoassays and direct fluorescence made laboratory diagnosis accessible in many parts of the world. However, direct fluorescence is extremely labor intensive and requires an expert technician, and enzyme immunoassays have poor sensitivity and specificity. Culture, EIAs and direct fluorescence all require a urethral or cervical specimen, further limiting the usefulness of these tests. The molecular techniques, including PCR and LCR, have excellent specificity and sensitivity, and have the additional advantage that they can be performed on urine in both males and females. Unfortunately, they require technical expertise and are expensive, and consequently their availability in developing countries is limited.

SYPHILIS

In the early part of the 20th century syphilis was endemic throughout Europe, North America, and probably most of Africa and Asia, and the extent of the problem became apparent with the development of serological tests. The problem was exacerbated by social breakdown and huge population movements during the First World War (Adler, 1980). Following the First World War, the availability of clinical services and social stabilization led to a reduction in the incidence of syphilis. However, this was short-lived and syphilis incidence increased rapidly during the Second World War (Aral and Holmes, 1990). After the Second World War, syphilis again declined rapidly as a consequence of the availability of clinical services, and in particular the introduction of penicillin (Adler, 1980). In the 1960s and 1970s there was an increase in syphilis in many parts of the developed world, particularly in males, largely as a consequence of homosexual sex (Mindel *et al.*, 1987; MMWR, 1984). Most of this increase was seen in primary, secondary and early latent syphilis, whereas late syphilis and congenital syphilis have continued to decrease in most parts of the developed world.

The 1980s saw the widespread promotion of safer sex messages as a consequence of the HIV epidemic, and this led to a reduction in the incidence of infectious syphilis in many parts of Europe, North America and Australia (Adler, 1980) (Figure 1.6). In Western Europe, Australia and some parts of North America the incidence of syphilis has continued to decrease during the 1990s (Adler, 1980). Unfortunately, Eastern Europe and some inner-city areas in North America have seen an alarming rise in syphilis in recent years (MMWR, 1984; Tichonova *et al.*, 1997). In several of the newly independent states of the former Soviet Union, syphilis has increased 15–30 times, from 5–50/100 000 in 1990 to 120–170/100 000 in 1994 (WHO, 1996). A considerable part of this increase has probably occurred as a result of social breakdown in the former USSR, resulting in poverty, unemployment and large-scale migration. In addition, the availability of quality health care has declined and the process of partner notification has all but evaporated. Among other factors that may be

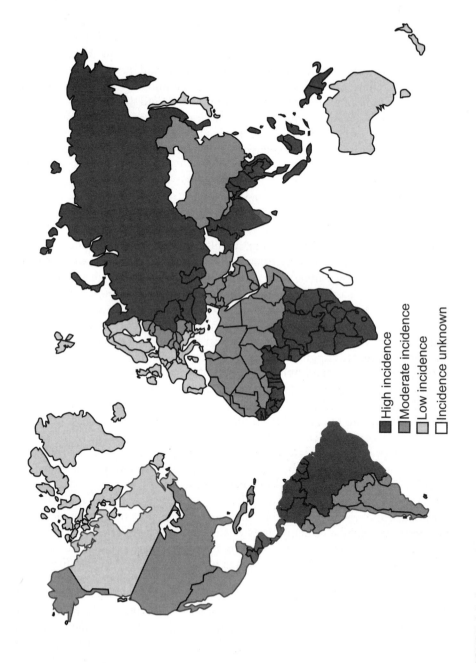

FIGURE 1.6 Geographic distribution of syphilis

High incidence

Moderate incidence

Low incidence

Incidence unknown

important are the rise in prostitution, the availability of sexually oriented materials, including pornography, and rapid and profound changes in sexual mores and behaviors. The situation in North America has been quite different. The incidence of syphilis began to increase in the United States among heterosexuals in the mid-1980s, from 13.7 cases/100 000 in 1981 to 184 cases/100 000 in 1989 (CDC, 1993). Most of this increase was seen in minority groups, particularly in inner cities in the East and the South. Much was associated with crack cocaine use, particularly when the drug was smoked (Dunn and Rolfs, 1991). A consequence of the epidemic was a dramatic increase in neonatal syphilis, from 158 cases in 1983 to 7219 cases in 1990. Fortunately, over the last 5 years the incidence has again begun to decrease. However, an estimated 130 000 new infections still occur each year (CDC, 1996).

The situation in the developing world is different. Syphilis remains a major burden in many developing countries, for both adults and children (Meheus and Schryver, 1989). For example, in Ethiopia syphilis seroprevalence was 12.8% in 1995 among blood donors (Rahlenbeck et al., 1997), and a study in South Africa among antenatal patients showed a seroprevalence of 6.5–9.0% (Wilkinson et al., 1997). WHO estimates in 1995 (in adults aged 15–49) for subSaharan Africa were 3.14% in males and 3.89% in females (Gerbase et al., 1998). In Asia, syphilis remains a common sexually transmitted disease, particularly in some of the larger cities, where commercial sex work appears to be a major source of infections (Kunawarak et al., 1995). In South and Southeast Asia in 1995, WHO estimates in adults aged 15–49 put the prevalence at 1.42% for males and 1.77% for females (Gerbase et al., 1998). Seroprevalence studies in parts of the world where endemic treponematoses (mainly yaws) are or were common may be difficult to unravel. Worldwide, the World Health Organization has estimated that up to 12 million new cases of syphilis in adults occur each year (Gerbase et al., 1998).

GONORRHEA

The trends in gonorrhea this century in many industrialized countries have partially mirrored those for syphilis, with peaks following both world wars and also in the late 1970s. However, there are some notable differences comparing gonorrhea and syphilis, and also in the trends in different countries. In the United Kingdom the number of new cases attending STD clinics peaked in 1977, at 41 542 males and 24 421 females (Adler, 1980) and has declined each year until 1994–95, when a small increase was noted (Government Statistical Service, 1996) (Figure 1.7). Similar trends have been seen in other industrialized countries, including Sweden, Australia, New Zealand, France and Switzerland. In most industrialized countries the annual incidence of gonorrhea is below 20/100 000 (Gerbase et al., 1998). In the USA, although gonorrhea has also declined from a peak in 1995, the incidence (165.1/100 000 in males and 149.5/100 000 in females) is considerably higher than in most other industrialized countries (CDC, 1996). There is also a huge difference in incidence rates across the USA, largely as a consequence of very high rates in inner cities, mainly in Afro-Americans. In 1995 the incidence was 1279.5/100 000 in black non-hispanic

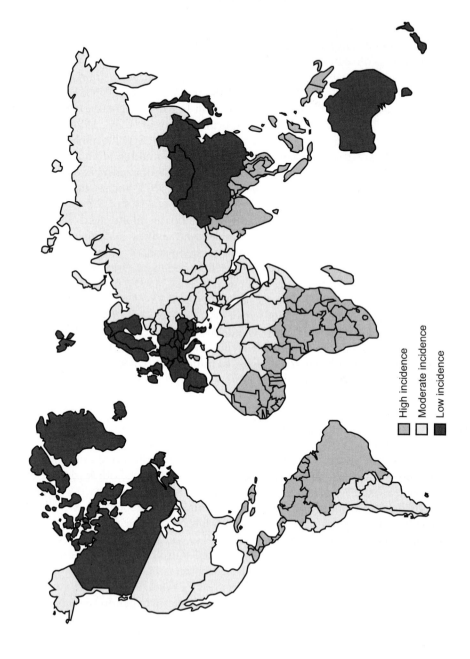

FIGURE 1.7 Geographic distribution of gonorrhea.

High incidence

Moderate incidence

Low incidence

males, compared to 21.1/100 000 in whites, and 913.1/100 000 in non-hispanic black females compared to 36.8/100 000 in white females. Much of this disparity can be explained on the basis of poverty and lack of access to medical care (CDC, 1996). Two recent studies from the United Kingdom have shown that in southeast London and in Leeds, being a member of a black ethnic group was associated with a higher risk of acquiring gonorrhea, even after controlling for socioeconomic status (Low et al., 1997; Lacey et al., 1997). The reasons why different ethnic groups have different risks for acquiring STDs are unclear, but may relate to sexual attitudes and behaviors (including ideas on condom use), choice of partners and patterns of sexual mixing. In the developing world, reliable figures for the incidence of sexually transferred diseases are scarce. However, there is some information from prevalence studies suggesting that gonorrhea in many developing countries is extremely common (Meheus and Schryver, 1989). In subSaharan Africa the estimated prevalence in 1995 in adults aged 15–49 was 1.98% in males and 2.80% in females, and WHO estimated that the number of new cases of gonorrhea in developing countries in 1995 was over 50 million (Gerbase et al., 1998).

THE TROPICAL SEXUALLY TRANSMITTED DISEASES

There are three STDs which are largely confined to the tropics: chancroid, lymphogranuloma venereum and Donovanosis. Chancroid is caused by the Gram-negative bacillus *Haemophilus ducreyi*, and is considered to be the commonest cause of genital ulcers in developing countries, particularly in subSaharan Africa. The global incidence of chancroid is estimated at 7 million (Figure 1.8). Outbreaks have occurred in Europe and North America, mostly in association with commercial sex, particularly in poor communities.

In the developing world, particularly in subSaharan Africa, genital ulcer disease (which is usually considered to be mainly due to chancroid) has been shown to be a major contributor to the HIV epidemic, and recent study has shown that controlling STDs, including genital ulcers, leads to a reduction in the acquisition of HIV (Grosskurth et al., 1995).

Lymphogranuloma venereum (LGV) is caused by *Chlamydia trachomatis* serovars L1, 2 and 3. The genital ulcers associated with lymphogranuloma venereum are generally transient, and these organisms are not considered to be a major cause of genital ulcers worldwide. However, the long-term consequences of LGV, in particular lymphatic obstruction and elephantiasis, are a major cause of morbidity in some tropical countries. Donovanosis, also known as granuloma inguinalae, is caused by an organism called *Calymomatobacterium granulomatis*. This infection has a well circumscribed geographic location, occurring in Papua New Guinea, Southeast India, the Carribean and northern Australia, and it has recently re-emerged in Southern Africa (Figure 1.9). The disease causes progressive destruction of the genitals and adjacent structures, but is amenable to treatment with many antibiotics. It is considered to be an important factor in the spread of HIV in those geographic areas where it occurs (Hart, 1997). Occasional cases of these diseases are seen in other countries in travelers or immigrants from endemic countries.

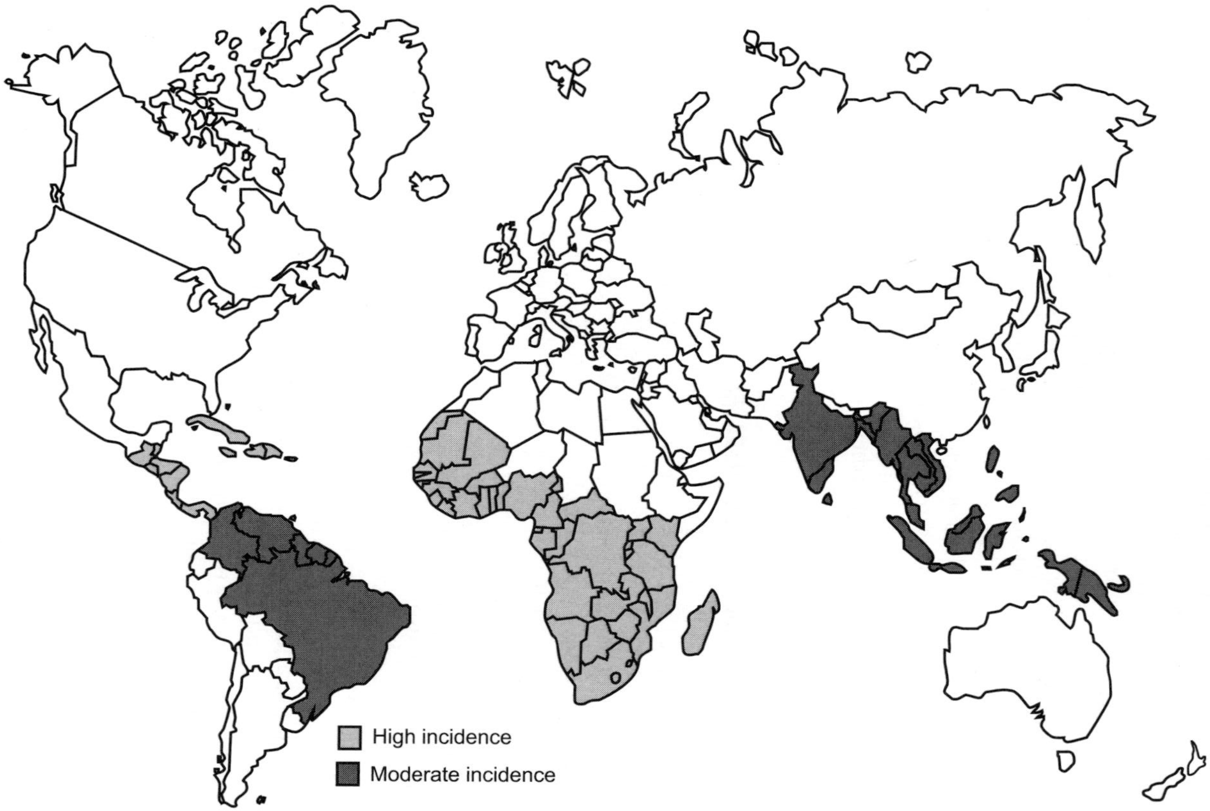

FIGURE I.8 Geographic distribution of chancroid.

High incidence

Moderate incidence

FIGURE 1.9 Geographic distribution of Donovanosis.

TRICHOMONIASIS

Trichomoniasis is caused by a flagellate parasite called *Trichomonas vaginalis*, which is non-invasive and causes vaginal discharge. Trichomoniasis is probably the commonest sexually transmitted infection worldwide, with 170 million cases, and is considered to be a major factor in the transmission of HIV in developing countries (Daly *et al.*, 1994, Grosskurth *et al.*, 1995).

ACKNOWLEDGEMENTS

The authors wish to thank Ms Brenda Wilson for typing the manuscript.

REFERENCES

Adler, M.W. (1980). Sexually transmitted diseases. The terrible peril: a historical perspective on the venereal diseases. *Br. Med. J.* **281**: 206–211

Akahane, Y., Kojima, M., Sugai, M. *et al.* (1994). Hapatitis C virus infection in spouses of patients with type C chronic liver disease. *Ann. Intern. Med.* **120**: 748–752

Albert, J., Franzen, L., Jansson, M. *et al.* (1992) Ugandan HIV-1 V3 loop sequences closely related to the U.S./European consensus. *Virology* **190**: 674–681

Alter, M.J., Ahtone, J., Weisfuse, I., Starko, K., Vacalis, T.D. and Maynard, J.E. (1986). Hepatitis B virus infection among heterosexuals. *JAMA* **256**: 1307–1310

Andersson-Ellstrom, A., Svennerholm, B. and Forssman, L. (1995). Prevalence of antibodies to herpes simplex virus types 1 and 2, Epstein–Barr virus and cytomegalovirus in teenage girls. *Scand. J. Infect. Dis.* **27**(4): 315–318

Aral, S.O. and Holmes, K.K. (1990). Epidemiology of sexual behavior and sexually transmitted diseases. *Int. J. Cancer* **46**: 214–219

Ashley, R.L., Militoni, J., Lee F., Nahmias, A. and Corey, L. (1988). Comparison of Western blot (immunoblot) and glycoprotein G-specific immunodot enzyme assay for detecting antibodies to herpes simplex virus types 1 and 2 in human sera. *J. Clin. Microbiol.* **26**(4): 662–667

Ballesteros, J., Dal-Re, R., Gonzalez, A. and del Romero, J. (1996). Are homosexual males a risk group for hepatitis A infection in intermediate endemicity areas? *Epidemiol. Infect.* **117**(1): 145–148

Barin, F., M'boup, S., Denis, F. *et al.* (1985). Serological evidence for virus related to simian T-lymphotropic retrovirus type III in residents of West Africa. *Lancet* **2**: 1387–1389

Barre-Sinoussi, F., Chermann, J.-C., Rey, F. *et al.* (1983). Isolation of T-lymphotropic retrovirus from a patient at risk for acquired immune deficiency syndrome (AIDS). *Science* **220**: 868

Bassett, I., Donovan, B., Bodsworth, N.J. *et al.* (1994). Herpes simplex virus type 2 infection of multi-partnered heterosexual men in Sydney. *Med. J. Aust.* **160**: 697–700

Beasley, R.P., Hwang, L.Y., Lee, G.C.Y. *et al.* (1983). Prevention of perinatally transmitted hepatitis B virus infections with hepatitis B immune globulin and hepatitis B vaccine. *Lancet* **2**: 1099–1102

Berris, B., Wrobel, D.M., Sinclair, J.C. and Feinman, S.V. (1973) Hepatitis B antigen in families of blood donors. *Ann. Intern. Med.* **79**: 690–693

Bobkov, A., Garaev, M.M., Rxhaninova, A. *et al.* (1994). Molecular epidemiology of HIV-1 in the former Soviet Union: Analysis of env V3 sequences and their correlation with epidemiology data. *AIDS* **8**: 619–624

Boucher, F.D., Yasukawa, L.L., Bronzan, R.N., Hensleigh, P.A., Arvin, A.M. and Prober, C.G. (1990). A prospective evaluation of primary genital herpes simplex virus type 2 infections acquired during pregnancy. *Pediatr. Infect. Dis. J.* **9**: 499–504

Brabin, L., Brabin, B.J., Doherty, R.R. *et al.* (1989). Patterns of migration indicate sexual transmission of HTLV-1 infection in non-pregnant women in Papua New Guinea. *Int. J. Cancer* **44**(1): 59–62

Bresters, D., Mauser-Bunschoten, E.P., Reesink, H.W. *et al.* (1993). Sexual transmission of hepatitis C virus. *Lancet* **342**: 210–211

Brettler, D.B., Mannuccio, P.M., Gringeri, A. *et al.* (1992). The low risk of hepatitis C virus transmission among sexual partners of hepatitis C-infected hemophilic males: an international, multicenter study. *Blood* **80**: 540–543

Brienig, M.K., Kingsley, L.A., Armstrong, J.A., Freeman, D.J. and Ho, M. (1990). Epidemiology of genital herpes in Pittsburgh: serological, sexual and racial corelates of apparent and inapparent herpes simplex infections. *J. Infect. Dis.* **162**: 299–309

Britt, W.J. and Alford, C.A. (1996). Cytomegalovirus. In: Fields Virology, 3rd edn. Eds. Fields, B.N., Knipe, D.M., Howley, P.M. *et al.* Lippincott-Raven, Philadephia

Brown, S.T., Nalley, J.F. and Kraus, S.T. (1981). Molluscum contagiosum. *Sex. Transm. Dis.* **8**: 227–234

Brown, Z.A., Selke, S., Zeh, J. *et al.* (1997). The acquisition of herpes simplex virus during pregnancy. *N. Engl. J. Med.* **337**(8): 509–515

Campodonico, M., Janssens, W., Heyndrickx, L. *et al.* (1996). HIV type 1 subtypes in Argentina and genetic heterogeneity of the V3 region. *AIDS Res. Hum. Retroviruses* **12**: 79–81

Cann, A.J. and Chen, I.S.Y. (1996). Human T-cell leukemia viruses types I and II. In: Fields Virology. Eds Fields, B.N., Knipe, D.M., Howley, P.M. *et al.* Lippincott-Raven, Philadelphia

Cates, W. and Hinman, A.R. (1991). Sexually transmitted diseases in the 1990's. *N. Engl. J. Med.* **325**: 1368–1370

Caussy, D., Goedert, J.J., Palefsky, J. *et al.* (1990). Interaction of human immunodeficiency and papilloma viruses: association with anal epithelial abnormality in homosexual men. *Int. J. Cancer* **46**: 214–19

Centers for Disease Control (1996). Sexually transmitted disease surveillance 1995. Division of STD/HIV Prevention September 1996

Centers for Disease Control (1993). Sexually transmitted disease surveillance 1992. Division of STD/HIV Prevention July 1993

Centres for Disease Control (1992). Hepatitis A among homosexual men – United States, Canada and Australia. *MMWR* **41**: 155, 161–164

Centers for Disease Control (1988a). Leads from the MMWR. Changing patterns of groups at high risk for hepatitis B in the United States. *JAMA* **260**(6): 761

Centers for Disease Control (1988b). Relationship of syphilis to drug use and prostitution: Connecticut and Philadelphia, PA. *MMWR* **37**: 755–764

Centers for Disease Control (1981). Kaposi's sacoma and Pneumocystis pneumonia among homosexual men – New York and California. *MMWR* **30**: 305–308

Chaix, M.-L., Chappey, C., Couillin, I., Rozenbaum, W., Levy, J.-P. and Saragosti, S. (1993). Diversity of the V3 region of HIV in Paris, France. *AIDS* **7**: 1199–1204

Christenson, B., Bottiger, M., Svensson, A. and Jeansson, S. (1992). A 15-year surveillance study of antibodies to herpes simplex virus types 1 and 2 in a cohort of young girls. *J. Infection* **25**(2): 147–54

Christenson, B., Brostrom, C.H., Bottiger, M. *et al.* (1982). An epidemic outbreak of hepatitis A among homosexual men in Stockholm. *Am. J. Epidemiol.* **116**: 599–607

Chu, S.Y. and Curran, J.W. (1997). Epidemiology of human immunodeficiency virus infection in the United States. In: DeVita, V.T. Jr., Hellman, S., Rosenberg, S.A. eds. AIDS. Etiology Diagnosis, Treatment and Prevention, 4th edn. Philadelphia: Lippincott-Raven, 137–165

Clavel, F., Mansinho, K., Chamaret, S. *et al.* (1987). Human immunodeficiency virus type 2 infection associated with AIDS in West Africa. *N Engl J Med* **316**: 1180–1185

Clewley, J.P., Arnold, C., Barlow, K.L., Grant, P.R. and Parry, J.V. (1996). diverse HIV-1 genetic subtypes in UK. *Lancet* **347**: 1487

Coggin, J.R. and zur Hausen, H. (1979). Workshop on papillomaviruses and cancer. *Cancer Res.* **39**: 545–6

Collier, A.C., Handsfield, H.H., Ashley, R. *et al.* (1995). Cervical but not urinary excretion of cytomegalovirus is related to sexual activity and contraceptive practices in sexually active women. *J. Infect. Dis.* **171**(1): 33–38

Collier, A.C., Handsfield, H.H., Roberts, P.L. *et al.* (1990). Cytomegalovirus infection in women attending a sexually transmitted disease clinic. *J. Infect. Dis.* **162**: 46–51

Collier, A.C., Meyers, J.D., Corey, L., Murphy, V.L., Roberts, P.L. and Handsfield, H.H. (1987) Cytomegalovirus infection in homosexual men. Relationship to sexual practices, antibody to human immunodeficiency virus, and cell-mediated immunity. *Am. J. Med.* **82**(3 Spec no): 593–601

Coll-Seck, A., Faye, M.A., Critchlow, C.W. *et al.* (1994). Cervical intraepithelial neoplasia and human papillomavirus infection among Senegalese women seropositive for HIV-1 or HIV-2 or seronegative for HIV. *Int. J. Sex. Transm. Dis. AIDS* **5**: 189–193

Conry-Cantilena, C., VanRaden, M., Gibble, J. *et al.* (1996). Routes of infection, viremia, and liver disease in blood donors found to have hepatitis C virus infection. *N. Engl. J. Med.* **334**: 1691–1696

Corey, L. (1990). Genital herpes in sexually transmitted diseases and aetiologic agents. Holmes, K.K., March, P., Sparling, P.S. *et al.* (eds) McGraw Hill, New York

Corona, R., Caprilli, F., Giglio, A. *et al.* (1996). Risk factors for hepatitis B virus infection among heterosexuals attending a sexually transmitted diseases clinic in Italy: role of genital ulcerative diseases. *J. Med. Virol.* **48**(3): 262–266

Coutinho, R.A., Albrecht-Van Lent, P., Lelie, N. *et al.* (1983). Prevalence and incidence of hepatitis A among male homosexuals. *Br. Med. J.* **287**: 1743–1745

Cowan, F.M., Johnson, A.M., Ashley, R., Corey, L. and Mindel, A. (1996). Relationship between antibodies to herpes simplex virus (HSV) and symptoms of HSV infection. *J. Infect. Dis.* **174**(3): 470–5

Critchlow, C.W., Holmes, K.K., Wood, R. *et al.* (1992). Association of HIV and anal HPV infection among homosexual men presenting to a public health department HIV antibody screening clinic. *Arch. Intern. Med.* **152**: 1673–1676

Crowley, T., Horner, P.J. and Nelki, J. (1994). Screening associated with reduced infection rates. *Br. Med. J.* **308**: 716–717

Cunningham, A.L., Lee, F.K., Ho, D.W.T. *et al.* (1993). Herpes simplex virus type 2 antibody in patients attending antenatal or STD clinics. *Med. J. Aust.* **158**: 525–528

Cunningham A.L., Turner R.R., Miller A.C., Para M.F. and Merigan T.C. (1985). Evolution of recurrent herpes simplex lesions. An immunohistologic study. *J. Clin. Invest.* **75**: 226–33

Curran, J.W., Lawrence, D.N., Jaffe, H. *et al.* (1984). Acquired immunodeficiency syndrome (AIDS) associated with transfusions. *N Engl J Med.* **310**: 69

Czeglédy, J., Rogo, K.O., Evander, M. and Wadell, G. (1992). High-risk human papillomavirus types in cytologically normal cervical scrapes from Kenya. *Med. Microbiol. Immunol.* **180**: 321–6

Davis, K.C., Horsburgh, C.R. Jr., Hasiba, U. *et al.* (1983). Acquired immunodeficiency syndrome in a patient with hemophilia. *Ann Intern Med.* **98**: 284

DeCock, K.M., Niland, J.C., Lu H.P. *et al.* (1988). Experience with human immunodeficiency virus infections in patients with hepatitis B virus and hepatitis delta virus infections in Los Angeles. *Am. J. Epidemiol.* **127**: 1250–1260

Decosas, J. and Adrien, A. (1997). Migration and HIV. *AIDS* **11**(Suppl A): S77–S84

Delaporte, E., Janssens, W., Peeters, M. *et al.* (1996). Epidemiology and molecular characteristics of HIV infection in Gabon, 1986–1994. *AIDS* **10**: 903–910

de Villiers, E-M. (1992). Laboratory techniques in the investigation of human papillomavirus infection. Donovan, B., Minichiello, V., Hart, G. Australia. Chapter in: STDs in Asia and the Pacific. Eds: Brown, T., Chan, R., Murgrotitchian, D., Mulhall, B.P., Plummer, D., Sittitrai, W. *Genitourinary Med.* **68**: 50–4

Dickover, R., Garraty, E., Herman, S. *et al.* (1996). Identification of levels of maternal HIV-1 RNA associated with risk of perinatal transmission. *J. Am. Med. Assoc.* **275**: 599–605

Dienstag, J.L. (1997). Sexual and perinatal transmission of hepatitis C. *Hepatology* **26**(3 Suppl): 66S-70S

Donovan, B., Minichiello, V. and Hart, G. (1992). Australia. Chapter in: STDs in Asia and the Pacific. Eds: Brown T., Chan R., Murgrotitchian D., Mulhall B.P., Plummer, D., Sittitrai W. *Genitourinary Med.* **68**: 50–54

Downie, J.C., Dwyer, D.E., Kazazi, F. *et al.* (1992). Identification of infection of an Australian resident with the human immunodeficiency virus type 2 (HIV-2). *Med. J. Aust.* **157**: 415–417

Dumitrescu, O., Kalish, M.L., Kliks, S.C., Bandea, C.I. and Levy, J.A. (1994). Characterization of human immunodeficiency virus type 1 isolates from children in Romania: Identification of a new envelope subtype. *J. Infect. Dis.* **169**: 281–288

Dunn, R.A. and Rolfs, R.T. (1991). The resurgence of syphilis in the United States. *Curr. Opin. Infect. Dis.* **4**: 3–11

Dwyer, D.E., Ge, Y.C., Wang, B. *et al.* (1997). first human immunodeficiency virus type 1 sequences in the V3 region, neg and upr genes from Papua New Guinea. *AIDS Res. Hum. Retroviruses* **13**: 625–627

Dwyer, D., Ge, Y., Bolton, W., Wang, B., Cunningham, A.L. and Saksena, N. (1996). Subtype B isolates of human immunodeficiency virus type 1 detected in Australia. *Ann. Acad. Med. Singapore* **25**: 188–191

Eberhardt-Phillips, J., Dickson, N., Paul, C. *et al.* (1998). Herpes Simples type 2 infection in a cohort ages 21 years. *Sex. Transm. Infect.* **74**: 216–218

Edmunds, W.J., Medley, G.F. and Nokes, D.J. (1996). The transmission dynamics and control of hepatitis B virus in the Gambia. *Stat. Med.* **15**(20): 2215–2233

El-Dalil, A.A., Jayaweera, D.T., Walzman, M. *et al.* (1997). Hepatitis B markers in heterosexual patients attending two genitourinary medicine clinics in the West Midlands. *Genitourinary Med.* **73**(2): 127–130

Embil, J.A., Garner, J.B., Pereira, L.J., White, F.M. and Manuel, F.R. (1985). Association of cytomegalovirus and herpes simplex virus infections of the cervix in four clinic populations. *Sex. Transm. Dis.* **12**(4): 224–228

Engels, H., Nyongo, A., Temmerman, M. *et al.* (1992). Cervical cancer screening and detection of genital HPV-infection and chlamydial infection by PCR in different groups of Kenyan women. *Ann. Soc. Belge Med. Trop.* **72**: 53–62

Eyster, M.E., Alter, J.J., Aledort, L.M., Quan, S., Hatzakis, A. and Goedert, J.J. (1991). Heterosexual co-transmission of hepatitis C virus (HCV) and human immunodeficiency virus (HIV). *Ann. Intern. Med.* **115**: 764–768

Fairley, C.K., Chen, S., Tabrizi, S.N., Leeton, K., Quinn, M.A. and Garland, S.M. (1992). The absence of genital human papillomavirus DNA in virginal women. *Int. J. Sex. Transm. Dis. AIDS* **3**: 414–417

Feingold, A.R., Vermund, S.H., Burk, R.D. *et al.* (1990). Cervical cytologic abnormalities and papillomavirus in women infected with human immunodeficiency virus. *J. of AIDS* **3**: 896–903

Field, P.R., Ho, D.W.T., Irving, W.L., Isaacs, D. and Cunningham, A.L. (1993). The reliability of serological tests for the diagnosis of genital herpes: A critique. *Pathology* **25**: 175–179

Fleming, D.T., McQuillan, G.M., Johnson, R.E. *et al.* (1997). Herpes simplex virus type 2 in the United States, 1976 to 1994. *N. Engl. J. Med.* **337**(16): 1105–1111

Forsgren, M., Skoog, E., Jeansson, S., Olofsson, S. and Giesecke, J. (1994). Prevalence of antibodies to herpes simplex virus in prpegnant women in Stockholm in 1969, 1983 and 1989: implications for STD epidemiology. *Int. J. STD AIDS* **592**: 113–116

Forsgren, M., Sterner, G., Anzen, B. and Enocksson, E., (1990). Management of women at term with prenancy complicated by herpes simplex. *Scand. J. Infect. Dis.* (Suppl), **77**: 58–66

Fulford, K.W.M., Dane, D.S., Catterall, R.D., Woof, R. and Denning, J.V. (1973). Australia antigen and antibody among patients attending clinic for sexually transmitted diseases. *Lancet* **1**: 1470–1473

Gallo, R.C., Salahuddin, S.Z., Popovic, M. *et al.* (1984). Frequent detection and isolation of cyto-pathic retroviruses (HTLV-III) from patients with AIDS and at risk for AIDS. *Science* **224**: 500

Gao, F., Robertson, D.I., Morrision, S.G. *et al.* (1996). The heterosexual human immunodeficiency virus type 1 epidemic in Thailand is caused by an intersubtype (A/E) recombinant of African origin. *J. Virol.* **70**: 7013–7029

Gao, F., Yue, L., Robertson, D.L. *et al.* (1994). Genetic diversity of human immunodeficiency virus type 2: Evidence of distinct sequence subtypes with differences in virus biology. *J. Virol.* **68**: 7433–7447

Garland, S.M., Gertig, D.M. and McInnes, J.A. (1993). Genital *Chlamydia trachomatis* infection in Australia. *Med. J. Aust.* **159**: 90–96

Gerbase, A.C., Rowley, J.T., Heymann, D.H.L., Berkley, S.F.B. and Piot, P. (1998). Global prevalence and incidence estimates of selected curable STDs. *Genitourinary Med.* WHO Supplement (in press)

Gerberding, J.L. (1995). Management of occupational exposures to blood-borne viruses. *N. Engl. J. Med* (332): 444–451

Gibson, J.J., Hornung, C.A., Alexander, G.R., Lee, F.K., Potts, W.A. and Nahmias, A.J. (1990). A cross-sectional study of herpes simplex virus types 1 and 2 in college students: occurrence and determinants of infection. *J. Infect. Dis.* **162**: 306–312

Glasgow, K.W., Schabas, R., Williams, D.C., Wallace, E. and Nalezyty, L.A. (1997). A Population-based hepatitis B seroprevalence and risk factor study in a northern Ontario town. *Can. J. Pub. Health* **88**(2): 87–90

Gordon, S.C., Patel, A.H., Kulesza, G.W., Barnes, R.E. and Silverman, A.L. (1992). Lack of evidence for the heterosexual transmission of hepatitis C. *Am. J. Gastroenterol.* **87**: 1849–1851

Gottlieb, M.S., Schroff, R., Schanker, H.M. *et al* (1981). *Pneumocystis carinii* pneumonia and mucosal candidiasis in previously healthy homosexual men: evidence of a new acquired cellular immunodeficiency. *N. Engl. J. Med* **305**: 1425

Government Statistical Service. (1996). Sexually transmitted diseases, England 1995: new cases seen at NHS genito-urinary medicine clinics. No. 14. London: HMSO, 1–15

Green, J., Monteiro, E., Bolton, V.N., Sanders, P. and Gibson, P.E. (1991). Detection of human papillomavirus DNA by PCR in semen from patients with and without penile warts. *Genito-urinary Med.* **67**: 207–210

Grez, M., Dietrich, U., Balfe, P. *et al.* (1994). Genetic analysis of human immunodeficiency virus type 1 and 2 (HIV-1 and HIV-2) mixed infections in India reveals a recent spread of HIV-1 and HIV-2 from a single ancestor for each of these viruses. *J. Virol.* **68**: 2161–2168

Grussendorf-Conen, E.I., de Villiers, E.-M. and Gissman, L. (1986) Human papillomavirus genomes in penile smears of healthy men. *Lancet*, ii, 1092 [letter]

Handsfield, H.H., Plummer, F.A., Simonsen, J.N. *et al.* (1991). Cofactors in male-female sexual transmission of human immunodeficiency virus type 1. *J. Infect. Dis.* **163**: 233–239

Hart, G. (1997). State-of-the-art clinical article: Donavanosis. *Clin. Infect. Dis.* **25**: 24–32

Heng, B.H., Goh, K.T., Chan, R., Chew, S.K., Doraisingham, S. and Quek, G.H. (1995). Prevalence of hepatitis B virus (HBV) infection in Singapore men with sexually transmitted diseases and HIV infection: role of sexual transmission in a city state with intermediate HBV endemicity. *J. Epidemiol. Commun. Health* **49**(3): 309–313

Henning, K.J., Bell, E., Braun, J. and Barker, N.D. (1995). A community-wide outbreak of hepatitis A: risk factors for infection among homosexual and bisexual men. *Am. J. Med.* **99**(2): 132–136

Hersh, B.S., Popovici, F., Apetrei, R.C. *et al.* (1991). Acquired immunodeficiency syndrome in Romania. *Lancet* **338**: 645–649

Heyndrickx, L., Janssens, W., Alary, M. *et al.* (1996). Genetic variability of HIV type 1 in Benin. *AIDS Res. Hum. Retroviruses* **12**: 1495–1497

Hillis, S., Nakashima, A., Amsterdam, L. *et al.* (1994). *Chlamydia trachomatis* infections: trends in prevalence, incidence, incidence of recurrent infection, pelvic inflammatory disease and ectopic pregnancy following a comprehensive program. In: Chlamydial Infections. Eds: Orfila, J., Byrne, G.I., Chernesky, M.A. *et al.* Bologna: Societa Editrice Esculapio, 67–70

Hinuma, Y., Komoda, H., Chosa, T. *et al.* (1982). antibodies to adult T-cell leukemia virus associated antigen (ATLA) in sera from patients with ATL and controls in Japan: a nation-wide sero-epidemiology study. *Int. J. Cancer* **29**: 631–635

Ho, M. (1990). Epidemiology of cytomegalovirus infections. *Rev. Infect. Dis.* **12**(7): S701–S710

Ho, D.W.T., Field, P.R., Irving, W.L., Packham, D.R. and Cunningham, A.L. (1993). Western blot ananlysis for diagnosis of HSV-2 primary and recurrent genital infections. *J. Clin. Microbiol.* **31**: 3157–3164

Ho, D.W.T., Field, P.R., Sjogren-Jansson, E., Jeansson, S. and Cunningham, A.L. (1992). Indirect ELISA for the detection of HSV-2 specific IgG and IgM antibodies with glycoprotein G (gG-2). *J. Virol. Meth.* **36**: 249–264

Holmberg, S.D., Horsburgh, C.R., Ward, J.W. and Jaffe, H.W. (1989). biological factors in the sexual transmission of human immunodeficiency virus. *J. Infect. Dis.* **160**: 116–125

Holmberg, S.D., Stewart, J.A., Gerber, A.R. *et al.* (1988). Prior herpes simplex virus type 2 infection as a risk factor for HIV infection. *JAMA* **259**(7): 1048–1050

Hou, M.C., Wu, J.C., Kuo, B.T. *et al.* (1993). Heterosexual transmission as the most common route of acute hepatitis B infection among adults in Taiwan – the importance of extending vaccination to susceptible adults. *J. Infect. Dis.* **167**: 938–941

Jacobs, B., Mayaud, P., Changalucha, J. *et al.* (1997). Sexual transmission of hepatitis B in Mwanza, Tanzania. *Sex. Transm. Dis.* **24**(3): 121–126

Jaffe, H.W., Francis, D.P., McLane, M.F. *et al.* (1984). Transfusion-associated AIDS: serologic evidence of human T-cell leukemia virus infection of donors. *Science* **223**: 1309

Joag, S.V., Stephens, E.B. and Narayan, O. (1996). Lentiviruses. In: Fields, B.N., Knipe, D.M., Howley, P.M. eds. *Fields Virology*, 3rd edn. Philadelphia: Lippincott-Raven, 1977–1996. (Vol 2)

Johnson, R.E., Nahmias, A.J., Magder, L.S., Lee, F.K., Brooks, C.A. and Snowden, C.B. (1989). A seroepidemiologic survey of the prevalence of herpes simplex virus type 2 infection in the United States. *N. Engl. J. Med.* **321**: 7–12

Juarez-Figueroa, L.A., Uribe-Salas, F.J., Conde-Glez, C.J. *et al.* (1997). Hepatitis B markers in men seeking human immunodeficiency virus antibody testing in Mexico City. *Sex. Transm. Dis.* **24**(4): 211–217

Kajiyama, W., Kashiwagi, S., Ikematsu, H., Hayashi, J., Nomura, H. and Okochi, K. (1986). Intrafamilial transmission of adult T cell leukemia virus. *J. Infect. Dis.* **154**: 851–857

Kalish, M.L., Luo, C.-C., Raktham, S. *et al.* (1995). The evolving molecular epidemiology of HIV-1 envelope subtypes in injecting drug users in Bangkok, Thailand: Implications for HIV vaccine trials. *AIDS* **9**: 851–857

Kalish, M.L., Luo, C.-C., Weniger, B.G. *et al.* (1994). Early HIV type 1 strains in Thailand were not responsible for the current epidemic. *AIDS Res. Hum. Retroviruses* **10**: 1573–1575

Karayiannis, P., Novick, D.M., Lok, A.S., Fowler, M.J., Monjardino, J. and Thomas, H.C. (1985). Hepatitis B virus DNA in saliva, urine, and seminal fluid of carriers of hepatitis B e antigen. *Br. Med. J. (Clin. Res.)* **290**: 1853–1855

Kataoka, A., Claesson, U., Hansson, B.G., Eriksson, M. and Lindh, E. (1991). Human papilloma-virus infection of the male diagnosed by Southern-blot hybridization and polymerase chain reaction: comparison between urethra samples and penile biopsy samples. *J. Med. Virol.* **33**: 159–164

Katz, M.H., Hsu, L., Wong, E., Liska, S., Anderson, L. and Janssen, R.S. (1997). Seroprevalence of and risk factors for hepatitis A infection among young homosexual and bisexual men. *J. Infect. Dis.* **175**(5): 1225–1229

Kaur, U., Sahni, S.P., Bambery, P. *et al.* (1996). Sexual behaviour, drug use and hepatitis B infection in Chandigarh students. *Natl. Med. J. India* **9**(4): 156–159

Kihlström, E. and Danielsson, D. (1994). Advances in biology, management and prevention of infections caused by *Chlamydia trachomatis* and *Neisseria gonorrhoeae*. *Curr. Opin. Infect. Dis.* **7**: 25–33

Kiviat, N., Rompalo, A., Bowden, R. *et al.* (1990). Anal human papillomavirus infection among human immunodeficiency virus-seropositive and -seronegative men. *J. Infect. Dis.* **162**: 358–361

Kiviat, N.B., Critchlow, C.W., Holmes, K.K. *et al.* (1993). Association of anal dysplasia and human papillomavirus with immuno-suppression and HIV infection among homosexual men. *AIDS* **7**: 43–49

Kitayaporn, D., Uneklabh, C., Weniger, B.G. *et al.* (1994). HIV-1 incidence determined retro-spectively among drug users in Bangkok, Thailand. *AIDS* **8**: 1443–1450

Kjaer, S.K., de Villiers, E.-M., Çaglayan, H. *et al.* (1993). Human papillomavirus, *Herpes simple* virus and other potential risk factors for cervical caner in a high-risk area (Greenland) and a low-risk area (Denmark) – a second look. *Br. J. Cancer* **67**: 830–837

Kondo, K., Xu, J. and Mocarski, E.S. (1996). Human cytomegalovirus latent gene expression in granulocyte–macrophage progenitors in culture and in seropositive individuals. *Proc. Natl. Acad. Sci. USA* **93**(20): 11137–11142

Koutsky, L.A., Stevens, C.E., Holmes, K.K. *et al.* (1992). Underdiagnosis of genital herpes by current clinical and viral isolation procedures. *N. Engl. J. Med.* **326**: 1533–1539

Krieger, J.N., Coombs, R.W., Collier, A.C. Ross, S.O., Speck, C. and Corey, L. (1995). Seminal shedding of human immunodeficiency virus type 1 and human cytomegalovirus: evidence for different immunologic controls. *J. Infect. Dis.* **171**(4): 1018–1022

Kreiss, J.K., Kiviat, N.B., Plummer, F.A. *et al.* (1992). Human immunodeficiency virus, human papillomavirus, and cervical intraepithelial neoplasia in Nairobi prostitutes. *Sex. Transm. Dis.* **19**: 54–59

Kunawararak, P., Beyrer, C., Natpratan, C. *et al.* (1995). The epidemiology of HIV and syphilis among male commercial sex workers in northern Thailand. *AIDS* **9**(5): 517–521

Lacey, C.J.N., Merrick, D.W., Bensley, D.C. and Fairley, I. (1997). Analysis of the sociodemography of gonorrhoea in Leeds, 1989–93. *Br. Med. J.* **314**: 1715–1718

Laga, M., Icenogle, J.P., Marsella, R. *et al.* (1992). Genital papillomavirus infection and cervical dysplasia – Opportunistic complications of HIV infection. *Int. J. of Cancer* **50**: 45–48

Lasky, M., Perret, J.-L., Peeters, M. *et al.* (1997). Presence of multiple non-B subtypes and divergent subtype B strains of HIV-1 in individuals infected after overseas deployment. *AIDS* **11**: 43–51

Lee, F.K., Coleman, R.M., Pereira L., Bailey, P.D., Tatsuno, M. and Nahmias, A.J. (1985). Detection of herpes simplex virus type 2 specific antibody with glycoprotein G. *J. Clin. Microbiol.* **22**: 641–644

Leentvaar-Kuijpers, A., Kool, J.L., Veugelers, P.J., Coutinho, R.A. and van Griensven, G.J. (1995). An outbreak of hepatitis A among homosexual men in Amsterdam, 1991–1993. *Int. J. Epidemiol.* **24**(1): 218–222

Lo, J.Y.C., Lim, W.W.L., Ho, D.W.T., Field, P.R. and Cunningham, A.L. (1998). Type-specific seroepidemiology of herpes simplex virus in Hong Kong. *Sex. Transm. Infect.* (in press)

Louwagie, J., Delwart, E.L., Mullins, J.I., McCutchan, F.E., Eddy, G. and Burke, D.S. (1994). Genetic analysis of HIV-1 isolates from Brazil reveals presence of two distinct genetic subtypes. *AIDS Res. Hum. Retroviruses* **10**: 561–567

Loveday, C. and Mercey, D. (1993). The prevalence of human retroviral infections in female patients attending a central London sexually trasmitted disease clinic: 1985–1990. *Genitourin Med.* **69**(1): 31–34

Low, N., Daker-White, G., Barlow, D. and Pozniak, A.L. (1997). Gonorrhoea in inner London: results of a cross sectional study. *Br. Med. J.* **314**: 1719–1723

Luzuriaga, K. and Sullivan, J.L. (1997). Transmission of the human immunodeficiency virus from mother to the fetus and infant. In: DeVita, B.T. Jr., Hellman, S., Rosenberg, S.A., eds. AIDS. Etiology, Diagnosis, Treatment and Prevention, 4th edn. Philadelphia: Lippincott-Raven, 167–173

Mandal, D., Haye, K.R., Ray, T.K., Goorney, B.P., Stanbridge, C.M. and Corbitt, G. (1991). Prevalence of occult human papillomavirus infection, determined by cytology and DNA hybridization, in heterosexual men attending a genitourinary medicine clinic. *Int. J. Sex. Transm. Dis. AIDS* **2**: 351–355

Marks, C., Tideman, R.L. and Mindel, A. (1997). Evaluation of sexual health services within Australia and New Zealand. *Med. J. Aust.* **166**(7): 348–352

Marlink, R. (1996). Lessons from the second AIDS virus, HIV-2. *AIDS* **10**: 689–699

Mastro, T.D., Satten, G.A., Nopkesorn, T., Sangkharomya, S. and Longini, I.M. (1994). Probability of female-to-male transmission of HIV-1 in Thailand. *Lancet* **343**: 204–207

Masur, H., Michelis, M.A., Greene, J.B. *et al.* (1981). An outbreak of community-acquired *Pneumocystis carinii* pneumonia: initial manifestation of cellular immune dysfunction. *N. Engl. J. Med.* **305**: 1431

Mathijs, J.M., Rawlinson, W.D., Jacobs, S. *et al.* (1991). Cellular localization of human cytomegalovirus reactivation in the cervix. *J. Infect. Dis.* **163**: 921–922

Maupas, P. and Melnick, J.L. (1981). Hepatitis B infection and primary liver cancer. *Prog. Med. Virol.* **27**: 1–5

Meheus, A. and De Schryver, A. (1989). Sexually transmitted diseases in developing countries. *Curr. Opin. Infect. Dis.* **2**: 25–30

Meisel, H., Reip, A., Faltus, B. *et al.* (1995). Transmission of hepatitis C virus to children and husbands by women infected with contaminated anti-D immunoglobulin. *Lancet* **345**: 1209–1211

Melbye, M., Palefsky, J., Gonzales, J. *et al.* (1990). Immune status as a determinant of human papillomavirus detection and its association with anal epithelial abnormalities. *Int. J. Cancer* **46**: 203–206

Melbye, M., Biggar, R.J., Wantzin, P., Krogsgaard, K., Ebbesen, P. and Becker, N.G. (1990). Sexual transmission of hepatitis C virus: cohort study (1981–9) among European homosexual men. *Br. Med. J.* **301**: 210–212

Melkert, P.W.J., Hopman, E., van den Brule, A. *et al.* (1993). Prevalence of HPV in cytomorphologically normal cervical smears, as determined by the polymerase chian reaction, is age-dependent. *Int. J. of Cancer* **53**: 919–923

Mertens, T. and Piot, P. (1997). Global aspects of human immunodeficiency virus epidemiology: General considerations. In: De Vita, V.T. Jr., Hellman, S., Rosenberg, S.A. eds. AIDS. Etiology, Diagnosis, Treatment and Prevention, 4th edn. Philadelphia: Lippencott-Raven, 103–118

Mertz, G.J., Benedetti, J., Ashley, R., Selke, S.A. and Corey, L. (1992). Risk factors for the sexual transmission of genital herpes. *Ann. Intern. Med.* **116**: 197–202

Mindel, A. and Sutherland, S. (1984). Antibodies to cytomegalovirus in homosexual and heterosexual men attending an STD clinic. *Br. J. Vener. Dis.* **60**(3): 189–192

Mindel, A., Tovey, S.J. and Williams, P. (1987). Primary and secondary syphilis, 20 years' experience. Epidemiology. *Genitourinary Med.* **63**: 361–364

Morbidity and Mortality Weekly Report. (1984). Syphilis – United States, 1983. **33**(30): 443

Mulhall, B.P., Hart, G. and Harcourt, C. (1995). Sexually transmitted diseases in Australia: a decade of change. Epidemiology and surveillance. *Ann. Acad. Med. Singapore* **24**(4): 569–578

Murphy, E., Korber, B., Georges-Courbot, M.-C. *et al.* (1993). Diversity of V3 region sequences of human immunodeficiency viruses type 1 from the Central African Republic. *AIDS Res. Hum. Retroviruses* **9**: 997–1006

Myers, G., Korber, B., Wain-Hobson, S. *et al.* (1995). Human Retroviruses and AIDS: A Compilation and Analysis of Nucleic Acid and Amino Acid Sequences. Los Alamos, New Mexico, (Laboratory LAN, ed)

Nahmias, A.J., Lee, F.K. and Beckman-Nahmias, S. (1990). Sero-epidemiological and -sociological patterns of herpes simplex virus infection in the world. *J. Infect. Dis.* (Suppl.) **69**: 19–36

Nandwani, R., Caswell, S., Boag, F., Lawrence, A.G. and Coleman, J.C. (1994). Hepatitis A seroprevalence in homosexual and heterosexual men. *Genitourinary Med* **70**(5): 325–328

Nieminen, P., Koskimies, A.I. and Paavonen, J. (1991). Human papillomavirus DNA is not transmitted by semen. *Int. J. Sex. Transm. Dis. AIDS* **2**: 207–208

Nishikawa, A., Fukushima, M., Shimada, M. *et al.* (1991). Relatively low prevalence of human papillomavirus 16, 18 and 33 DNA in the normal cervices of Japanese women shown by polymerase chain reaction. *J. J. of Cancer Res.* **82**: 532–538

O'Brien, T.R., George, J.R. and Holmberg, S.D. (1992). Human immunodeficiency virus type 2 infection in the United States. *J. Am. Med. Assoc.* **267**: 2775–2779

Omar, R., Choudhury, M., Fischer, J. and Ezpeleta, C. (1991). A 'Pap' test for men? Male urethral smears as screening tool for detecting subclinical human papillomavirus infection. *Urology* **37**: 110–115

Osmond, D.H., Padian, N.S., Sheppard, H.W., Glas, S., Shiboski, S.C. and Reingold, A. (1993). Risk factors for hepatitis C virus seropositivity in heterosexual couples. *JAMA* **269**: 361–365

Ostrow, R.S., Zachow, K.R., Niimura, M., Okagaki, T., Muller, S., Bender, M. and Faras, A.J. (1986). Detection of papillomavirus DNA in human semen. *Science* **231**: 731–733

Palefsky, J.M., Gonzales, J., Greenblat, R.M., Ahn, D.K. and Hollander, H. (1990). Anal intraepithelial neoplasia and anal papillomavirus infection among homosexual males with group IV HIV disease. *J. Am. Med. Assoc.* **263**: 2911–2916

Pao, C.C., Lin, C.-Y., Maa, J.-S., Lai, C.-H., Wu, S.-Y. and Soong, Y.-K. (1990). Detection of human papilloma-viruses in cervicovaginal cells using polymerase chain reaction. *J. Infect. Dis.* **161**: 113–115

Patrick, D.M. and Bowie. W.R. (1994). The epidemiology of *Chlamydia trachomatis* infection in British Columbia. In: Chlamydial Infections. Eds: Orfila, J., Byrne, G.I., Chernesky. M.A. *et al.* Bologna: Societa Editrice Esculapio, 32–35

Pavlakis, G.N. (1997). The molecular biology of human immunodeficiency virus type 1. In: DeVita, V.T. Jr., Hellman, S., Rosenberg, S.A. eds. AIDS. Etiology, Diagnosis, Treatment and Prevention, 4th edn. Philadelphia: Lippincott-Raven, 45–74

Peano, G.M., Genoglio, L.M., Menardi, G., Balbo, R., Marenchino, D. and Fenoglio, S. (1992). Heterosexual transmission of hepatitis C virus in family groups without risk factors. *Br. Med. J.* **305**: 1473–1474

Perkins, N.L., Coughlan, E.P., Franklin, R.A., Reid, M.R. and Taylor, J. (1996). Seroprevalence of herpes simplex virus type 2 antibodies in New Zealand sexual health clinic patients. *NZ Med. J.* **109**: 402–405

Piot, P., Quinn, T.C., Taelman, H. *et al.* (1984). Acquired immunodeficiency syndrome in a heterosexual population in Zaire. *Lancet* **2**: 65

Poon, M.C., Landay, A., Prasthofer, E.F. *et al.* (1983). Acquired immunodeficiency syndrome with *Pneumocystis carinii* pneumonia and *Mycobacterium avium-intracellulare* infection in a previously healthy patient with classic hemophilia: clinical, immunologic, and virologic findings. *Ann. Intern. Med.* **98**: 287

Quinn, T.C., Gaydos, C., Shepherd, M. *et al.* (1996). Epidemiologic and microbiologic correlates of *Chlamydia trachomatis* infection in sexual partnerships. *JAMA* **4**: 276(21):1737–1742

Rahlenbeck, S.I., Yohannes, G., Molla, K., Reifen, R. and Assefa, A. (1997). Infections with HIV, syphilis and hepatitis B in Ethiopia: a survey in blood donors. *Int. J. STD AIDS* **8**(4): 261–264

Rall, C.J.N. and Dienstag, J.L. (1995). Epidemiology of hepatitis C virus infection. *Semin. Gastrointest. Dis.* **6**: 3–12

Ratner, L., Haseltine, W., Patarca, R. *et al* (1985). Complete nucleotide sequence of the AIDS virus, HTLV-III. *Nature* **313**: 450–458

Robertson, D.L., Sharp, P.M., McCutchan, F.E. and Hahn, B.H. (1995). Recombination in HIV-1. *Nature* **374**: 124–126

Rosenblum, L.S., Hadler, S.C., Castro, K.G., Lieb, S. and Jaffe, H.W. (1990). Heterosexual transmission of hepatitis B in Belle Glade, Florida. *J. Infect. Dis.* **161**: 407–411

Rosenthal, S.L., Stanberry, L.R., Biro, F.M. *et al.* (1997). Seroprevalence of herpes simplex virus types 1 and 2 and cytomegalovirus in adolescents. *Clin. Infect. Dis.* **24**(2): 135–139

Sagliocca, L., Stroffolini, T., Amoroso, P. *et al.* (1997). Risk factors for acute hepatitis B: a case-control study. *J. Viral. Hepatol.* **4**(1): 63–66

Saksena, N. and Bolton, W. (1998). Dissemination and evolution of non-AIDS human retroviruses: The primate T-cell lymphoma/leukemia viruses types 1 and 2. Venereology, (In press)

Schiffman, M.H., Bauer, H.M., Hoover, R.N. *et al.* (1993). Epidemiologic evidence showing that human papillomavirus causes most cervical intraepithelial neoplasia. *J. Nat. Cancer Inst.* **85**: 958–964

Schreeder, M.T., Thompson, S.E., Hadler, S.C. *et al.* (1982). Hepatitis B in homosexual men: prevalence of infection and factors related to transmission. *J. Infect. Dis.* **146**: 7–15

Schreiber, G.B., Busch, M., Kleinman, S.H. and Korelitz, J.J. (1996). For the Retrovirus Epidemiology Donor Study. The risk of transfusion transmitted viral infections. *N. Engl. J. Med.* **334**: 1685–1690

Scott, G.B., Buck, B.E., Leterman, J.G. *et al.* (1984). Acquired immunodeficiency syndrome in infants. *N. Engl. J. Med.* **310**: 76

Scott-Wright, A., Hakre, S., Bryuan, J.P. *et al.* (1997). Hepatitis B virus, human immunodeficiency virus type-1, and syphilis among women attending prenatal clinics in Belize, Central America. *Am. J. Trop. Med. Hyg.* **56**(3): 285–290

Shen, C.Y., Chang, S.F., Lin H.J. *et al.* (1994). Cervical cytomegalovirus infection in prostitutes and in women attending a sexually transmitted disease clinic. *J. Med. Virol.* **43**(4): 362–366

Simpson, G.R., Schulz, T.F., Whitby, D. *et al.* (1996). Prevalence of Kaposi's sarcoma associated herpesvirus infection measured by antibodies to recombinant capsid protein and latent immunofluorescence antigen. *Lancet* **349**: 113–138

Sixby, J.W., Lemon, S.M. and Pagano, J.S. (1986). A second site for Epstein–Barr virus shedding: the uterine cervix. *Lancet* **2**: 122–124

Slobod, K.S., Rencher, S.D., Farmer, A., Smith, F.S. and Hurwitz, J.L. (1994). HIV type 1 envelope sequence diversity in inner city community. *AIDS Res. Hum. Retroviruses* **10**: 873–875

Smallman-Raynor, M. and Cliff, A. (1991). The spread of human immunodeficiency virus type 2 into Europe: a geographical analysis. *Int. J. Epidemiol.* **20**: 480–489

Solomon, R.E., Kaslow, R.A., Phaur, J.P. *et al.* (1988). Human immunodeficiency virus and hepatitis delta virus in homosexual men – a study of four cohorts. *Ann. Intern. Med.* **108**: 51–54

Sonnex, C. (1995). The clinical features of genital and perigenital human papilloma virus infection. In: Genital Warts. Human Papilloma Virus Infection. Ed: Mindel A., Edward Arnold, London

Spinillo, A., Tenti, P., Zappatore, R. *et al.* (1992). Prevalence, diagnosis and treatment of lower genital neoplasia in women with human immunodeficiency virus infection. *Eur. J. Obstet. Gynaecol. Reprod. Biol.* **43**: 235–241

St Louis, M.E., Icenogle, J.P., Manzila, T. *et al.* (1993). Genital types of papillomavirus in children of women with HIV-1 infection in Kinshasa, Zaire. *Int. J. Cancer* **54**: 181–184

Stuver, S.O., Tachibana, N., Okayama, A. *et al.* (1993). Heterosexual transmission of human T cell leukemia/lymphoma virus type I among married couples in southwestern Japan: an initial report from the Miyazaki cohort study. *J. Infect. Dis.* **197**: 57–65

Sullivan, M.T., Williams, A.E., Fang, C.T., Grandinetti, T., Poiesz, B.J. and Ehrlich, G.D. (1991). Transmission of human T-lymphotropic virus types I and II by blood transfusion: A retrospective study of recipients of blood components (1983 through 1988). *Arch. Intern. Med.* **151**: 2043–2048

ter-Meulen, J., Eberhardt, H.C., Luande, J. *et al.* (1992). Human papillomavirus (HPV). infection, HIV infection and cervical cancer in Tanzania, East Africa. *Int. J. Cancer* **51**: 515–521

Thomas, D.L., Zenilman, J.M., Alter, H.J. *et al.* (1995). Sexual transmission of hepatitis C virus among patients attending sexually transmitted disease clinics iin Baltimore – an analysis of 309 sex partnerships. *J. Infect. Dis.* **171**: 768–775

Tichonova, L., Borisenko, K., Ward, H., Meheus, A., Gromyko, A. and Renton, A. (1997). Epidemics of syphilis in the Russian Federation: trends, origins and priorities for control. *Lancet* **350**: 210–213

Tookey and Garland, S.M. (1990). Neonatal herpes simplex: Royal Women's Hospital 10-year experience with management guidelines for herpes in pregnancy. *Aust. NZ J Obstet. Gynaecol* **32**: 331–334

Torres, J.R. (1996). Hepatitis B and hepatitis delta virus infection in South America. *Gut* **38**(Suppl 2). S48–S55

Tsuji, Y., Doi, H., Yamabe, T., Ishimaru, T., Miyamoto, T. and Hino, S. (1990). Prevention of mother-to-child transmission of human T-lymphotropic virus type-1. *Paediatrics* **86**: 11–17

van Doornum, G.J.J., Hooykaas, C., Juffermans, L.H.J. *et al.* (1992). Prevalence of human papillomavirus infections among heterosexual men and women with multiple sexual partners. *J. Med. Virol.* **37**: 13–21

van der Poel, C.L. (1994). Hepatitis C virus epidemiology transmission and prevention. In: Reesink, H.W. ed. Hepatitis C Virus. Amsterdam: Karger, 137–163

Van Harmelen, J., Wood, R., Lambrick, M., Rybicki, E.P., Williamson, A.-L. and Williamson, C. (1997). An association between HIV-1 subtypes and mode of transmission in Cape Town, South Africa. *AIDS* **11**: 81–87

Vermund, S.H., Kelly, K.F., Klein, R.S., Feingold, A.R., Schreiber, K., Munk, G. and Burk, R.D. (1991). High risk of human papillomavirus infection and cervical squamous intraepithelial lesions among women with symptomatic human immunodeficiency virus infection. *Am. J. of Obst. Gynecol.* **165**: 392–400

Wain-Hobson, S., Sonigo, P., Danos, O., Cole, S. and Alizon, M. (1985). Nucleotide sequence of the AIDS virus, LAV. *Cell* **40**: 9–17

Wald, A., Corey, L., Cone, R., Hobson, A., Davis, G. and Zeh, J. (1997). Frequent genital herpes simplex virus 2 shedding in immunocompetent women. Effect of acyclovir treatment. *J. Clin. Invest.* **99**(5): 1092–1097

Weniger, B.G., Takebe, Y., Ou, C.Y. and Yamazaki, S. (1994). The molecular epidemiology of HIV in Asia. *AIDS* **8**: 13–28

Weniger, B.G., Limpakrnjanarat, K., Ungchusak, K. *et al.* (1991). The epidemiology of HIV infection and AIDS in Thailand. *AIDS* **5**(Suppl 2): S71–S85

Whitley, R.J. (1993) Neonatal herpes simplex virus infections. *J. Med. Virol.* (Suppl) **1**: 13–21

WHO. (1996). Epidemic of sexually transmitted diseases in Eastern Europe. Report on a WHO Meeting, Copenhagen, Denmark, pp 13–15

Wieland, U. and Pfizer, H. (1997). Papilloma virus in human pathology: Epidemiology, pathogenesis and oncogenic role. In: Human Papilloma Virus Infection: A Clinical Atlas. Eds: Gross, G.E., Barrasso, R. Berlin: Ullstein-Mosby, 1–18

Wikström, A., Lidbrink, P., Johansson, B. and von Krogh, G. (1991). Penile human papillomavirus carriage among men attending Swedish STD clinics. *Int. J. Sex. Transm. Dis. AIDS* **2**: 105–109

Wilkinson, D., Sach, M. and Connolly, C. (1997). Epidemiology of syphilis in pregnancy in rural South Africa: opportunities for control. *Trop. Med. Int. Health* **2**(1): 57–62

Williams, J.R., Nokes, D.J., Medley, G.F. and Anderson, R.M. (1996). The transmission dynamics of hepatitis B in the UK: a mathematical model for evaluating costs and effectiveness of immunization programmes. *Epidemiol. Infect.* **116**(1): 71–89

World Health Organisation (1997). Global AIDS surveillance. *Wkly Epidemiol. Rec.* **72**: 357–364

Zurita, S., Costa, C., Watts, D. *et al.* (1997). Prevalence of human retroviral infection in Quillabamba and Cuzco, Peru: a new endemic area for human T cell lymphotropic virus type 1. *Am. J. Trop. Med. Hyg.* **56**(5): 561–565

2

THE CHANGING EPIDEMIOLOGY OF SEXUALLY TRANSMITTED DISEASES: THE RUSSIAN EXPERIENCE

MIKHAIL GOMBERG,
VAGAN A. AKOVBIAN
AND
ANNA A. KOUBANOVA

Central Research Institute of Skin and Venereal Diseases, Moscow, Russia

HISTORICAL ASPECTS

The totalitarian system in the former USSR influenced every aspect of social activity, including an individual's sex life. Even this intimate part of human activity was confined within the strictest 'communist norms and standards'. To establish control over sexual behavior medical science, in particular venereology, was politicized and ideologized. According to Governmental Order N806, under socialism all venereal diseases were considered remnants of a capitalistic society, rather than a societal disease (Touranov and Daniushevski, 1951). Venereology, as a result, was supposed to fight not the diseases themselves, but the remains of capitalism. Acquiring a venereal disease was a misdemeanor that had no place in Soviet society. The words of the USSR's Minister of Justice, N. Krylenko, depict the authorities' attitude toward human sexual behavior. In 1934, after article N121, paragraph 1 in the Criminal Code of the Russian Federation outlawing homosexual contacts among men, was published, Krylenko said: '... after two decades of building socialism in the USSR there is no reason for anybody to be a homosexual. Those who insist on their homosexuality have to be considered counterrevolutionary elements and must be convicted for their anti-Soviet behavior' (Akovbian *et al.*, 1995a). At that time, being convicted of anti-Soviet activity carried a minimum sentence of 10 years in a gulag, or labor camp.

The system responsible for the struggle against sexually transmitted infections in Russia arose in 1918 out of the Medical Board of the NKVD (the predecessor of the KGB), where a special Central Commission was formed. At the end of 1918 this Commission was transferred to the Narcomzdrav of the Russian Federation (the predecessor of the Ministry of Health) (Akovbian and Prokhorenkov, 1995). Step by step, a unified hierarchical structured system was formed to fight venereal diseases. The structural unit of this system became a dispensary. The first dermatovenereological dispensary was organized in Russia in 1921 (Bourenkov, 1997).

In Russia the 'dispensary method' of managing venereal diseases encompassed not only medical aspects (diagnosis, treatment, partner notification, preventive measures etc.), but also included other measures that were believed to be necessary for 'social security'. Examples included the requirement that all patients with syphilis had to abstain from sexual contact for up to 5 years, and they were forbidden to marry. If a patient violated these rules he was punished according to the Russian Criminal Code (Victorov, 1980). According to articles N150 and 192 of the Code, and the Order of the Council of Ministers of the Russian Federation (24 January 1927), any person suspected of having a venereal disease was obliged to undergo a compulsory medical examination. If a person refused to notify all his sexual contacts he was subject to civil liability and fines.

The treatment schedules for syphilis in the USSR involved multiple injections of sodium benzyl penicillin (eight times a day) and, up to the beginning of the 1990s, included the use of bismuth salts. The treatment courses were intermittent and started in the hospital, where a patient had to stay for up to 1 month. After inpatient treatment the person underwent several other intermittent courses as an outpatient at the dispensary (Skripkin, 1980). If a patient was late or refused to come for the next treatment he was considered a criminal, and according to article N115 of the Criminal Code of the Russian Federation could be sued. Statistics indicate that each year more than 1000 people were convicted of violations of this article (Victorov, 1980). The average sentence was 1 year in prison or labor camp, where the patient was obliged to finish his treatment.

Thus, the Soviet system of struggle against STDs was extremely regimented. But, given the conditions of that era, when communist ideology influenced everything, such an aggressive approach was not surprising. The system also appeared to be extremely effective. Official statistics on venereal diseases became routinely available in Russia after World War II. Soon after the war a special order was issued by the Ministry of Health concerning the complete elimination of venereal diseases within 1–2 years, as there was no place for these diseases in the future communist society. After the outbreak of syphilis that followed the Second World War (115.6 cases per 100 000 people) the morbidity of the disease gradually decreased, reaching its nadir in 1963 (2.5 cases per 100 000 people). A slight rise was seen between 1973 and 1979 (26.5 cases per 100 000 people). This increase was explained at the time by the natural biological periods of activation of *Treponema pallidum* (12-year cycles of this activation were found; it was proposed that they corresponded to solar activity) (Akovbian *et al.*, 1995a). A new minimum was reached in 1989 (4.3 cases per 100 000 people).

It is well known what political changes took place in the USSR after 1985. At first, under *perestroika* there was little change in the dispensary and registration systems and in the reported cases of STDs.

STRUCTURE OF THE DERMATOVENEREOLOGICAL SERVICE IN POST-SOVIET RUSSIA

This service was organized according to territorial principles. In 1996–1997 there were 351 dermatovenereological dispensaries and 9221 dermatovenereologist specialists working in the country (0.6 per 10 000 people) (Ministry of Health, 1997). In 1997, there were 919 centralized diagnostic laboratories servicing dispensaries, including 299 serologic, 262 bacteriologic and 358 clinical–diagnostic laboratories. There were 31 678 beds in hospitals for inpatient management of dermatovenereological disorders (2.2 per 100 000 people). In addition to the dispensaries there were 61 departments of dermatology and venereology within medical universities, academies and schools, and seven departments of dermatology and venereology within academies and institutes of postgraduate education. Three scientific institutes of dermatology and venereology are working now in collaboration with these departments on scientific research in dermatology and venereology. The main challenge now for this 'medical army' is the effective control of STDs in a more open society.

STDS IN RUSSIA OVER THE PAST 4 YEARS

Before 1993 there were no official statistics concerning infections other than syphilis and gonorrhea. According to the new order N228 of the Russian Federation's Ministry of Health, seven other sexually transmitted infections (including bacterial vaginosis) were added to the compulsory registration list (Table 2.1) (Tikhonova, 1997). As may be seen from these data, all STDs except trichomoniasis and gonorrhea increased substantially between 1993 and 1996. Over this period, although there was a dramatic increase in syphilis cases (up 780% between 1993 and 1996), there was a paradoxical decline in cases of gonorrhea (down 38% between 1993 and 1996). This reduction may reflect a

TABLE 2.1   New cases of STDs in the Russian Federation, 1993–1996

Disease	Per 100 000 population				Dynamics
	1993	1994	1995	1996	1996:1993
Syphilis	33.9	85.8	177.0	264.6	+780%
Gonorrhea	230.9	204.6	173.5	139.0	− 38.3%
Trichomoniasis	327.9	335.0	343.9	341.5	+ 4.1%
Chlamydia	37.1	61.6	90.2	106.1	+290%
Ureaplasmosis	24.4	38.0	56.9	67.6	+280%
Bacterial vaginosis	45.0	102.2	139.3	153.0	+340%
Urogenital candidiasis	77.1	122.9	153.1	185.0	+240%
Genital herpes	8.5	7.4	8.8	10.8	+ 25.9%
Anogenital warts	17.8	20.3	20.9	22.6	+ 25.8%
Total					+1 282 700 cases

real decline in the number of gonococcal infections, as has been seen in developed countries. However, this seems unlikely given the substantial increase in most of the other reported STDs. It is also possible that this apparent decline is due to underreporting. Currently in Russia there are still remnants of the old reporting system. Consequently, regulations require the compulsory identification of patients and their sexual contacts. To avoid official registration a potential patient may forgo official medical services altogether and seek medical help elsewhere, through private practitioners or pharmacists. Alternatively, individuals with gonorrhea may forgo medical evaluation and rely on self-diagnosis and treatment. This is comparatively easy, especially in men, when a urethral discharge is present within a few days of sexual intercourse suggesting the diagnosis of gonorrhea. In this situation the individual may undertake self-treatment or ask for advice from his more experienced friends. In Russia, antibiotics are easily purchased in pharmacies or through non-licensed dealers without a prescription.

Cases of chlamydial genital infection have increased steadily in Russia since reporting began in 1993. Still, the number of reported cases is low relative to the number of cases of syphilis and gonorrhea. The possible reasons for this are presented in Table 2.2.

OUTBREAK OF SYPHILIS IN RUSSIA

The number of cases of syphilis was very low until 1991, when dramatic social changes began to sweep the country. In that year the number of syphilis cases began to double each year until 1995, when there were 177 registered cases per 100 000 people, 40 times more than in 1989 (Figure 2.1) (Tikhonova, personal communication). The distribution varied geographically (Figures 2.2 and 2.3), in rural areas the rate being about half that seen in cities. Suburban areas had intermediate rates of syphilis, with the number increasing with increasing proximity to the cities (Belyaev, 1996).

In some parts of Russia the situation with syphilis seems to be stabilizing, and in areas such as Moscow, St. Petersburg and Samara, a slight decline has been observed. Unfortunately, there are still some regions in Russia, such as Khakassia, Tuva and Sakhalin, where the number of cases is still growing. For

TABLE 2.2 Possible reasons for the apparent low number of chlamydial genital infections in Russia

No system of routine diagnostics for *C.trachomatis* in Russia

Microbiological tests to diagnose *C.trachomatis* are, as a rule, not free of charge. This undoubtedly leads to underutilization of these tests, and hence underdiagnosis of chlamydial infection

There is no screening for asymptomatic infection

Self-treatment may lead to underreporting. The most effective antibiotics, such as azithromycin, are widely available and may be easily purchased in a pharmacy without a prescription

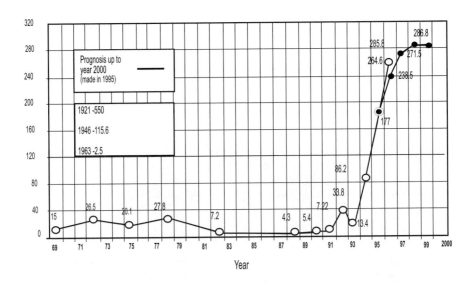

FIGURE 2.1 Rates of syphilis in the Russian Federation per 100 000 people.

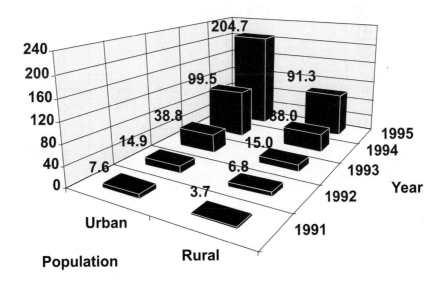

FIGURE 2.2 Notification rates of new cases of syphilis in urban and rural settings in the Russian Federation, 1991–1995 (Belyaev *et al.*, 1996).

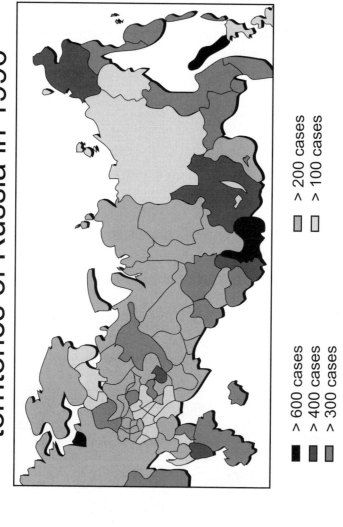

Syphilis morbidity distribution on some territories of Russia in 1996

> 600 cases
> 400 cases
> 300 cases

> 200 cases
> 100 cases

FIGURE 2.3 Distribution of syphilis cases in some of the territories of the Russian Federation in 1996 (per 100 000 people).

example, according to the preliminary data in some regions of Khakhassia, syphilis reached 902 cases per 100 000 people in 1997. The regions with increasing numbers of cases typically have a very low population densities and very high unemployment.

The number of cases of syphilis also varied according to age group and gender (Table 2.3) (Tikhonova *et al.*, 1997). Although about half of the cases in either gender was seen in the 20–29-year age group, the greatest increase was seen in females aged 15–17 years and males aged 18–19 years (126 and 99 times higher in 1996 than in 1988, respectively). In 1996, among males between the ages of 20 and 29 years, almost one in 100 had syphilis, whereas about one out of every 75 women aged 18–19 years was infected. One of the highest rates of syphilis was observed in the Moscow region, where, with a population of 6 million, one out of 64 females aged 18–19 years had syphilis (Tatiana Shouvalova, personal communication).

Sexual abuse has also been shown to be a factor in the spread of syphilis in Russia. A survey performed in the largest venereological hospital in Moscow reviewed the archival histories of 3890 female patients with syphilis from 1983 to 1995 (Loseva and Ibragimov, 1996). In 1983, 0.7% of all cases were acquired as a result of sexual abuse. In 1993, 6.0% of the cases resulted from sexual abuse, with 8.0% in 1994 and 7.0% in 1995. In 60% of the cases resulting from abuse, the diagnosis was not established until the recurrent secondary stage. This indicates that victims of sexual abuse are seeking medical help too late. The reasons for this may be psychological, as well as the fact that the majority of the patients were minors: more than half of the victims were under 18 years of age.

TABLE 2.3 Rates of syphilis cases by age and gender in the Russian Federation 1989–1996

Year	0–14 males	0–14 females	15–17 males	15–17 females	18–19 males	18–19 females	20–29 males	20–29 females	30–39 males	30–39 females	>40 males
1989	0.1	0.1	3.4	6.9	6.9	21.6	15.1	12.4	7.9	5.5	2.3
1990	0.1	0.1	2.7	8.8	8.8	24.3	19.4	10.0	9.0	6.6	2.5
1991	0.1	0.2	4.6	14.1	13.0	35.7	27.4	25.6	11.6	8.2	2.9
1992	0.3	0.4	9.0	31.5	26.8	72.8	52.8	49.6	19.5	14.1	4.9
1993	0.6	1.1	26.2	89.4	75.2	188.0	139.0	124.0	48.5	33.9	10.8
1994	1.4	3.4	65.3	217.0	187.0	407.0	351.0	311.0	125.0	89.2	27.4
1995	3.0	6.7	130.0	437.0	359.0	922.0	577.0	629.0	274.0	196.0	85.5
1996	5.9	10.6	189.0	605.0	507.0	1321.0	916.0	919.0	416.0	315.0	115.0
Ratio increase 1988–1996	99	106	90	176	99	80	68.0	79.0	48.0	53.0	43.0

Pregnant women are tested serologically in Russia up to three times during their pregnancy. Data from the Moscow region (Table 2.4) showed that syphilis in pregnant women increased from 0.01% of all pregnancies in 1989 to 0.60% in 1996, a 60-fold growth. In 1996 the Moscow region registered 769 pregnant women with syphilis. Of the 218 babies born to these women, 31 were diagnosed with congenital syphilis (Shouvalova, personal communication). Not surprisingly, the number of cases of congenital syphilis has increased throughout Russia, from 15 cases in 1990 to 470 in 1996. In 1990, those 15 cases accounted for 26% of all cases of syphilis in children aged 0–14 years, whereas the 470 cases in 1996 accounted for only 18% of all syphilis cases in children.

The dramatic increase in the number of cases of syphilis over the past 5 years indicates that Russia is experiencing an epidemic spread of the disease. The likely factors contributing to this are social, biomedical and political (Gromyko, 1996) (Table 2.5 and 2.6).

CHANGING CONDITIONS AND STRATEGIES FOR DERMATOVENEREOLOGICAL SERVICES IN RUSSIA

In order to control the outbreak of syphilis the Russian dermatovenereological service, which worked effectively in the past, has had to develop new approaches for controlling the spread of STDs. The differences between former and present-day conditions are summarized in Table 2.7. As discussed earlier, the strategy for controlling STDs under the Soviet system was based on compulsory screening and the use of legal sanctions to force patients to undergo treatment and to identify sexual contacts. In post-Soviet Russia certain elements of the old system still work, but on a different level. Approaches to STD control in the Soviet and post-Soviet eras are summarized in Table 2.8. One strategy that remains from Soviet times is the active identification of syphilis and gonorrhea cases: the approaches used for this are outlined in Table 2.9. As part of the active identification program, selected occupational groups (shown

TABLE 2.4 Syphilis cases in pregnant women in the Moscow Region

Year	Total number pregnancies	Deliveries	Cases of syphilis (% of all pregnancies)	Number of pregnant women with syphilis
1989	238 756	71 506	0.01	24
1990	215 108	64 788	0.02	43
1991	167 397	56 220	0.02	33
1992	148 427	47 563	0.04	59
1993	135 053	42 207	0.10	124
1994	145 912	43 697	0.20	300
1995	137 633	42 720	0.44	603
1996	126 039	42 255	0.60	769

TABLE 2.5 Social factors contributing to the epidemic spread of syphilis in Russia

Collapse of the old economic system, with resultant unemployment and poverty

Migration of population

Large income differentials

Introduction of prostitution and pornography

Increasing illegal drug use

Decrease in condom usage

Unfavorable changes in sexual behavior (earlier introduction to sexual activity and an increase in the number of sexual partners)

TABLE 2.6 Biomedical and political factors contributing to the epidemic spread of syphilis in Russia

Infected people remain contagious for long periods if not properly diagnosed and treated

Increasing use of poor-quality private medical care or self-diagnosis and treatment

Anonymous testing and treatment are not free of charge, thus potentially discouraging infected individuals from seeking medical care

Insufficient funds for free qualified medical care

Lack of a federal program against STDs

Lack of a unified system of primary prevention for HIV and STDs. Traditionally in Russia the AIDS service was completely separated from other STDs and had its own control programs

Inadequate legislation concerning STDs and prostitution

TABLE 2.7 Changing conditions for dermatovenereological service in Russia

Before 1991

Strict system of residential registration of every person in the country

Lack of unemployment

Prostitution and pornography were illegal and underground

Compulsory identification of a patient upon admission to medical care

Closed borders

After 1991

No strict system of residential registration and the appearance of millions of refugees and migrants throughout the country

Open borders

Unemployment and poverty

Widespread prostitution and pornography

Availability of anonymous treatment

TABLE 2.8 Approaches to STD control in the Soviet and Post-Soviet eras

Soviet era (before 1991)

Special syphilis departments at each dermatovenerological dispensary

Active identification of syphilis

Free of charge diagnosis, treatment and partner notification

System of contact tracing throughout the country

Compulsory screening for syphilis of all inpatients, pregnant women and occupational groups

Legal sanctions and collaboration with the police to force infected people to undergo treatment and identify sexual contacts

Compulsory inpatient treatment of syphilis

Follow-up for up to 5 years, during which a patient was obliged to be tested every 3 months and risked conviction if they had sexual contact

Programs of primary prevention were a formality on paper only

Post-Soviet era (after 1991)

Special syphilis departments at each dermatovenerological dispensary remain

Active identification of syphilis

The old system of free diagnosis and treatment still works, but only if the patient identifies himself; for anonymous testing and treatment payment is required

The system of contact tracing throughout the country became very difficult but much more confidential

The system of active identification of syphilis remains unchanged

Changes in the Criminal Code allow people to decline treatment and do not require identification of sexual contacts

Anonymous diagnosis and treatment is possible

Effective methods of outpatient management

The maximum time of serological follow-up of patients after treatment was reduced to 3 years

Programs of primary prevention are targeted to the high-risk groups

---

TABLE 2.9 Components of active identification of syphilis

Contact tracing (identifies one third of all notified cases of syphilis)

Serological testing of all inpatients

Serological testing of blood donors

Serological testing of pregnant women (up to three times during pregnancy)

Compulsory periodic testing of patients diagnosed with syphilis

Compulsory periodic testing of selected occupational groups

Occasional identification of syphilis by dermatologists and non-dermatologists in patients seeking medical help

Serological testing of patients after diagnosis of other STDs

Serological testing of closed populations (military, convicts etc.)

TABLE 2.10 Occupational groups for compulsory syphilis screening

Food retailers

Workers in catering facilities

Nursery workers

Obstetric and pediatric staff

Teachers and other workers of children's facilities

Hotel, hostel and swimming pool employees

Hairdressing, sauna and laundry workers

Taxi, intercity and interstate drivers

Others (at the discretion of the local authorities)

in Table 2.10) are required to undergo compulsory testing. Data from the Moscow region (presented in Table 2.11) show the number of cases of syphilis and gonorrhea detected by through such compulsory testing from 1985 to 1995 (Tikhonova *et al.*, 1997). Detection of syphilis was most common among food retailers (0.24% of all screened), nursery workers (0.12%) and workers in catering facilities (0.25%). In every occupational group the number of cases of syphilis exceeded the number of cases of gonorrhea. It is important to note that this system of compulsory screening appeared to be an effective means of combating venereal diseases in the years before *perestroika*. All the workers were registered with the police according to their place of residence, and were required to submit to regular medical examinations. In the event that syphilis or gonorrhea was caught at a late stage of infection, the administrative punishment fell upon the medical staff responsible. In addition, the law on mandatory employment existed, according to which every person had to work. If he refused he violated the criminal code. Today, this system is no longer effective. Less attention is paid to residential registration requirements, and people are often hired within these groups without the prerequisite medical examinations.

ROLE OF DEPARTMENTS OF ANONYMOUS TESTING AND TREATMENT IN CONTROLLING STDS

There are about 300 Departments of Anonymous Testing and Treatment (DATTs) in Russia, which provide qualified medical care for patients willing to pay. Unlike state-funded free medical clinics, DATTs are not required to identify the names of individuals who test positive for STDs. The effectiveness of these departments in syphilis recognition may be seen from Figure 2.4. Although these departments add only about 10% of all cases to the overall statistics, for the most part these are patients who probably would not come to the dispensary on their own. Furthermore, the number seeking medical help with early forms of syphilis is higher in this group.

The development of DATTs made it possible to change the strategy of syphilis management. Weekly injections of benzathine penicillin are effective

TABLE 2.11 Cases of syphilis and gonorrhea detected in the Moscow region by testing of selected occupational groups from 1985 to 1996

Year	Workers of catering facilities			Food retailers			Nursery workers			Obstetricians and pediatricians			Workers in boarding schools			Hairdressing, sauna and laundry workers			Staff of hotels, hostels and swimming pools			Total		
	n	Gn	Sy	n	Gn	Sy	n	Gn	Sy	n	Gn	Sy	n	Gn	Sy	n	Gn	Sy	n	Gn	Sy	n	Gn	Sy
1985	46 848	112	1	38 701	64	3	74 549	107	5	7244	7	–	5008	–	–	10 221	8	–	8339	4	–	190 912	302	9
1986	49 319	77	4	37 983	41	–	75 661	93	3	7362	10	1	5263	3	1	9741	8	1	7787	6	–	193 116	238	10
1987	46 707	77	4	37 983	41	–	75 661	93	3	7362	10	1	5352	4	–	10 494	11	–	8287	2	2	188 394	207	6
1988	46 511	45	2	36 356	36	2	75 155	56	–	10 505	8	–	5298	1	–	8602	12	–	7080	3	–	189 507	161	4
1989	47 024	38	1	34 248	28	1	77 865	47	1	8612	4	1	4820	1	–	7069	14	–	6780	3	–	186 418	135	4
1990	47 629	55	2	34 806	32	3	76 719	38	2	7989	1	1	4475	1	–	7663	10	–	7824	2	–	187 105	138	8
1991	41 472	40	1	31 064	32	1	72 743	39	4	6919	5	1	3909	1	1	6823	5	1	7269	4	–	170 199	126	8
1992	36 680	38	3	30 056	50	4	68 631	56	2	9208	8	–	4000	3	–	5519	5	–	5353	3	1	159 448	163	10
1993	26 805	52	7	30 287	52	9	61 194	41	13	8132	1	2	4628	1	–	4899	8	2	3896	1	2	139 841	156	35
1994	27 816	22	23	35 147	42	40	61 987	52	23	8411	6	1	5114	3	3	5418	9	1	4118	2	3	148 011	136	94
1995	28 500	22	33	43 310	42	79	63 789	41	52	8659	6	7	6462	2	2	5992	9	9	4571	4	3	161 283	126	186
1996	27 759	27	68	46 402	26	111	64 780	13	80	8848	–	9	7736	2	11	6637	4	13	4980	6	7	167 092	78	299

Gn, gonorrhea; Sy, syphilis.

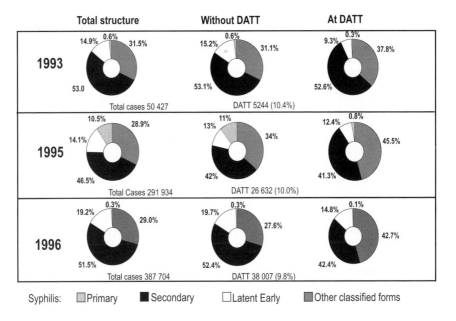

FIGURE 2.4 Detection of syphilis cases by the Departments of Anonymous Testing and Treatment (DATTs).

and do not require hospitalization. Given the epidemic that has occurred in Russia, the old system, with its compulsory hospitalization of all syphilis cases within 24 hours, would overcrowd the nation's hospitals. When outpatient management of syphilis became possible, there was great interest in developing other new effective methods, preferably oral, that could be used in patients allergic to penicillin, or with concomitant infections such as *Chlamydia trachomatis*. Russia has the world's greatest experience in the use of azithromycin (500 mg daily for 10 days) to treat syphilis. In many comparative trials with other antitreponemal antibiotics, azithromycin showed remarkable results.

CONCLUSIONS

A highly regimented society with widely accessible health care and strict punitive policies regarding sexually transmitted infections helped maintain a very low incidence of STDs in Soviet Russia. The political, social and economic upheaval that occurred in the late 1980s and early 1990s has resulted in a more open society, but with increases in unemployment, poverty, drug use, prostitution and tourism. These changes have contributed to a dramatic increase in STDs, particularly syphilis. As outlined in this chapter, strategic changes are occurring in the field of dermatovenereology in order to effectively deal with STDs in post-Soviet Russia.

REFERENCES

Akovbian, V.A., Tikhonova L.I., Mashkilleyson A.L. *et al.* (1995a). Syphilis in Russia: Historical experience, epidemiological analysis, prognosis. *STDs* [in Russian] **4**: 22–25

Akovbian, V.A., Prokhorenkov, V.I. (1995b). Sexually transmitted diseases: experience of the past and look in the future. *Vestnik dermatoilogii i venerologii* [in Russian] **3**: 16–19

Belyaev, Y.N. (1996). Notification rates of new cases of syphilis and gonorrhea in urban and rural population in Russia. Report from the Department of Sanitary-Epidemologic Surveillance, Ministry of Health of Russia

Bourenkov, S.P. (1977). Dispensaire. In: "Big Medical Encyclopedia [in Russian] 3rd edn." V.7, Moscow p 352

Gromyko, A.I. (1996). Epidemics of sexually transmitted diseases in Eastern Europe. *STDs* [in Russian] **6**: 22–25

Loseva, O.K., Ibragimov, R.A. (1996). Sexual abuse as a factor of spreading sexually transmitted diseases, *STDs* [Russian] **4**: 29–31

Ministry of Health of Russia (1997). Report of the Department of Statistics

Skripkin, Y.K. (1980). Treatment of syphilis. In: Shaposhnikov OK (ed). Venereal Diseases. Doctor's Manual. Moscow p 302–331

Tikhonova, L.I. (1997). STD morbidity and STD prevention in Russia. *STDs* [in Russian] **4**: 22–26

Tikhonova, L.I., Borisenko, K.K., Gromyko, A.I., Meheus, A., Renton, A. (1997). Epidemics of syphilis in the Russian Federation: trends, origins and priorities for control. *Lancet* **350**: 210–213

Touranov, N.M. and Daniushevski S.M. (eds) (1951). Handbook on organizing the struggle against venereal and contagious skin diseases [in Russian]. Moscow, Part IV, pp 260–261

Victorov, I.S. (1980). Criminal responsibility for distribution of sexually transmitted diseases [in Russian]. Saratov

3

THE GENITAL TRACT: DEVELOPMENTAL, ANATOMICAL AND MICROBIOLOGICAL FACTORS AFFECTING STD ACQUISITION AND STRATEGIES FOR PREVENTION

MORRIS D. COOPER

Department of Medical Microbiology and Immunology,
Southern Illinois School of Medicine, Springfield, IL, USA

INTRODUCTION

Any area of the body may be involved by sexually transmitted diseases (STDs), and many critical anatomical points may be important in some context of their diagnosis. There are two major differences in the anatomy and the examination between male and female patients. First, in females there is a genital tract and a separate urinary tract. In the male we find a single genitourinary tract, with the urethra serving as a common conduit for the excretory function of the urinary tract and the reproductive function of the delivery of semen. The second major difference is that the critical reproductive organs in the female are located in the pelvis, and are therefore less accessible and less easily examined. In order to appreciate the complexity of the human reproductive tract we need to examine the anatomy of these systems.

ANATOMY OF THE FEMALE GENITAL TRACT

The perineum is a diamond-shaped region bounded by the symphysis pubis anteriorly, the coccyx posteriorly and the ischial tuberosities laterally. The plane of the urogenital triangle is closed or occupied by a thick triangular membrane, the urogenital diaphragm. This closes the anterior floor of the perineum and defines the anterior wall of the ischiorectal fossa.

The region of the urogenital triangle contains two spaces, the superficial and deep perineal spaces. The superficial space is that lying between the urogenital diaphragm and the skin of the labia majora. The contents of the superficial perineal pouch include the greater vestibular glands (Bartholin's), the crura of the clitoris, the bulbs of the vestibule, which are composed of elongated erectile tissue, and the overlying superficial perineal muscles.

The deep perineal space is a potential space within the urogenital diaphragm, formed by the superior and the inferior fasciae (perineal membrane) of the deep transverse perineal muscle, which occupies the space. Together the muscle and the fascia form the urogenital diaphragm. The contents of the deep perineal space include the deep transfer perineal muscle and the circular muscle fibers surrounding the urethra and vagina, as these structure pierce the urogenital diaphragm.

Female urethra

The female urethra measures 3–4 cm in length from the bladder neck to the meatus in the anterior vestibule of the vagina. Proximally the mucosa comprises transitional epithelium, which gradually becomes stratified squamous as it courses distally. The lumen appears stellate in cross-section because of extensive longitudinal folding of the mucosa. Beneath the mucosa is the lamina propria, which is rich in vascular and neural plexuses. The muscular coat, similar to other body tubes, is composed of a double layer of smooth muscle, with the inner fibers circularly arranged and the outer layer disposed longitudinally. As the urethra traverses the urogenital diaphragm, circularly arranged striated muscle fibers form an external sphincter of the urethra. These fibers are innervated by the internal pudendal nerve, in contrast to the internal urethral sphincter at the bladder neck, which is innervated by the pelvic splanchnic nerve.

The entire length of the urethra has paraurethral glands running parallel to it, which are tuboalveolar outgrowths of the mucosa. Located in the lamina propria, these glands have their openings on the posterior and posterolateral walls of the urethra. At the distal end of the urethra there are usually two larger glands, commonly identified as Skene's glands, whose ducts are visible on the posterior wall. Both Skene's glands and the paraurethral glands are vulnerable to infection (Figure 3.1).

The Vagina

The vagina is a fibromuscular tube whose anterior and posterior walls are normally in contact with one another. There is a longitudinal ridge along the mucosal surface of both the anterior and posterior walls, from which secondary elevations called rugae extend laterally. The vaginal wall consists of three layers: (1) the mucous membrane, composed of stratified squamous non-keratinized epithelium and underlying lamina propria of connective tissue; (2) the muscular layer, composed of smooth muscle fibers disposed both longitudinally and circularly; and (3) the adventitia, a dense connective tissue that blends with the surrounding fascia.

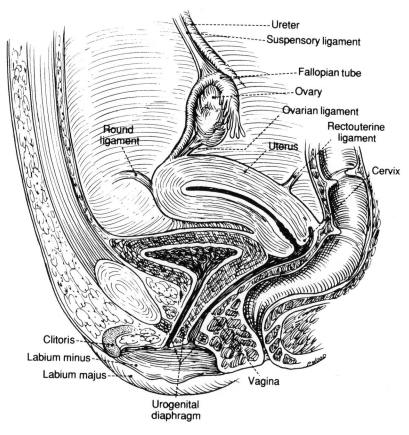

Ureter

Suspensory ligament

Fallopian tube

Ovary

Ovarian ligament

Rectouterine ligament

Round ligament

Uterus

Cervix

Clitoris

Labium minus

Labium majus

Vagina

Urogenital diaphragm

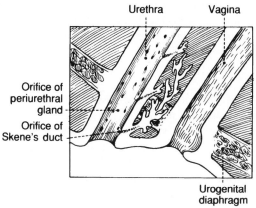

Urethra

Vagina

Orifice of periurethral gland

Orifice of Skene's duct

Urogenital diaphragm

FIGURE 3.1 Sagittal section of the female pelvis. Inset shows magnification of urethral glands.

There are no glands in the vaginal wall. During sexual stimulation the marked increase in fluid production in the vagina is believed to be due to transudation across the vaginal wall.

The stratified squamous epithelium of the adult vagina is several layers thick. The basal layer is a single layer of cylindrical cells with oval nuclei. Above this area are several layers of polyhedral cells that appear to be connected together much like those of the stratum spinosum of the epidermis. About these are several more layers of cells, which are more flattened in appearance and accumulate glycogen in their cytoplasm, the significance of which will be discussed below. They also exhibit keratohyalin granules intracellularly. This tendency toward keratinization, however, is not normally completed in the vaginal epithelium, and the surface cells always retain their nuclei.

The most superficial cells are desquamated into the vaginal lumen, where their intracellular glycogen is converted into lactic acid, probably by the bacteria normally resident in the vagina. The resulting acidity is believed to be important in protecting the female reproductive tract from infection by many pathogenic bacteria.

Estrogen stimulates the production of glycogen and maintains the thickness of the entire epithelium. Before puberty and after the menopause, when estrogen levels are relatively low, the epithelium thins and the pH is higher than in the reproductive years (neutral prior to puberty and 6.0 or higher after menopause). The thinness of the epithelium and the relatively high pH of the vaginal milieu are among the factors that are thought to render females in these groups more susceptible to vaginal infections.

The Uterus

The uterus is composed of two major components: (1) the expanded upper two-thirds of the organ, the body of the uterus, and (2) the cylindrical lower third, the cervix. The fundus is the rounded upper part of the body, superior to the points of entry of the uterine tubes. The isthmus is the short, slightly constricted zone between the body and the cervix. These two main components are different from one another in both structure and function.

The Cervix

The cervix consists primarily of dense collagenous connective tissue. Only about 15% of its substance is smooth muscle. In the isthmus, the uterine lumen narrows down to form the internal os. Below this point the lumen widens slightly to form the cervical canal (or endocervical canal). Finally, a constricted opening, the external os, at the lower end of the cervix, provides communication between the lumen of the cervix and that of the vagina.

Inside the cervix, the endocervical mucosa is arranged in a series of folds and ridges. A longitudinal ridge runs down the anterior wall and another down the posterior wall; from each of these, small folds run laterally. The resemblance of this arrangement to a tree trunk with upward-spreading branches has given rise to the term *arbor vitae uterina*, to describe the endocervical mucosa.

The part of the cervix that projects into the vagina (the *portio vaginalis*, or ectocervix) is covered by stratified squamous non-keratinizing epithelium. Usually in older women this type of epithelium extends for a very short distance into the cervical canal, where it forms a rather abrupt junction with the simple columnar epithelium lining the rest of the canal. The site of the squamo-columnar junction varies: it may occur higher up in the cervical canal, or the columnar epithelium may actually extend out beyond the external os, where it forms small patches known as physiologic eversion, or ectopy, on the vaginal surface of the cervix. Ectopy is usually present in adolescents and decreases during the third and fourth decades of life.

The mucosa contains large branched endocervical glands. In reality these are not true glands but are merely deep grooves or clefts (sometimes calls crypts), which serve to increase the surface area of the mucosa greatly. The epithelium of both the mucosal surface and the 'glands' is of the simple columnar type, in which almost all the cells are mucus secreting. A few ciliated cells are present. If the ducts of the glands become blocked, mucus accumulates inside them to form small lumps just under the surface (Nabothian cysts).

Unlike the mucous membrane of the body of the uterus, the endocervical mucosa does not slough off at menstruation. It does, however, respond to cyclic changes in the levels of ovarian hormones, estrogen and progesterone. It secretes up to 60 mg of mucus a day throughout much of the cycle, but near the time of ovulation (midcycle), when estrogen secretions reach a peak, the secretion rate increases 10-fold and the abundant clear mucus fills the cervical canal. It is then less viscous than at other times during the cycle, and is easily penetrated by spermatozoa.

The production of progesterone by the corpus luteum after ovulation (or during pregnancy) changes the quantity and properties of mucus produced: it becomes more viscous, less abundant, and much less penetrable by spermatozoa, acting as a plug to seal off the uterine cavity.

The Uterine Wall

The wall of the uterus is composed of three layers: (1) the endometrium, which is a glandular mucus membrane; (2) the myometrium, which is a smooth muscle layer; and (3) the serosa (Figures 3.2 and 3.3).

The Endometrium

The primary function of the endometrium is to provide a suitable site for the implantation and subsequent development of the embryo. It is a mucosa with a large population of glycogen-secreting glands and a highly vascularized environment. If there is no implantation most of the endometrium is sloughed, resulting in the menstrual flow, and is regenerated during the next menstrual cycle. This cyclic shedding and regeneration of the endometrium is controlled by the ovarian hormones estrogen and progesterone, the increase and decrease in these hormone levels regulating the growth and sloughing of the endometrium.

The thickness of the endometrium varies from 0.5 to 5 mm, depending on the stage of the menstrual cycle, being thickest a few days after ovulation, which

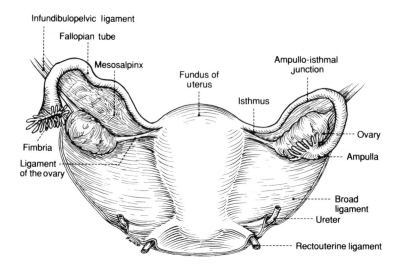

FIGURE 3.2 Posterior view of broad ligament and female reproductive organs.

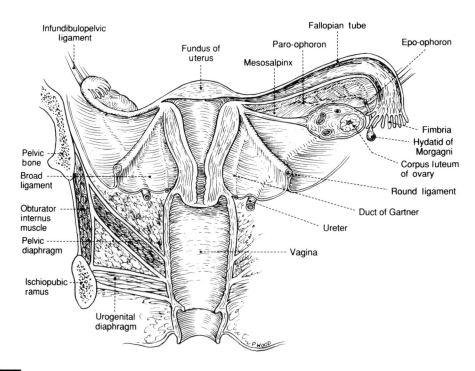

FIGURE 3.3 Coronal section of pelvis illustrating broad ligament, endopelvic space, pelvic diaphragm and urogenital diaphragm.

coincides with the time implantation would be expected. The endometrium consists of simple columnar epithelium and a highly cellular lamina propria (endometrial stroma), in which reside large numbers of tubular uterine glands. The epithelium is composed of both ciliated and non-ciliated (secretory) cells.

The endometrium is subdivided into a narrow, deeper layer next to the myometrium, the basalis; and a thicker more superficial layer, the functionalis, so called because this is the portion that is shed during menstruation.

The arteries that supply the endometrium play an important role in the onset of menstruation. Circumferentially oriented arteries in the myometrium give off numerous branches toward the endometrium. As they enter this area, the small basal arteries branch off to supply the basalis. As they enter the functionalis the arteries become highly contorted and are known as coiled or spiral arteries. Late in the menstrual cycle they spasmodically contract, which induces ischemia, necrosis, and an eventual sloughing of the functionalis.

During the menstrual cycle the endometrium passes through a number of phases. In the menstrual phase – approximately the first 4 days of a typical 28-day cycle – the functionalis is sloughed and the cellular debris and blood are discharged into the vagina. From day 5 until ovulation the endometrium is in the proliferative phase. The epithelium in the persisting portions of the uterine gland in the basalis grows and covers the denuded surface. Estrogen from the developing follicles in the ovary promotes rapid proliferation of the epithelium, glands and stroma. The endometrium may thicken by 2–3 mm at this time. Progesterone and estrogen from the corpus luteum stimulate the secretory phase (days 15–28), in which the epithelial cells begin to secrete. Accumulation of secretory products in the lumina of the glands, together with some edema of the stroma, causes the endometrium to increase further in thickness.

Later in the secretory phase the ovarian hormones decrease, and the changes that initiate menstruation occur. Intermittent constriction in the spiral arteries cause stasis of blood and ischemia of vessels and tissue in the area of supply. During the intervening periods of relaxation blood escapes from the weakened vessels, promoting menstrual hemorrhage.

The Myometrium

The myometrium consists of bundles of smooth muscle fibers separated by strands of connective tissue. Four layers of smooth muscle have been distinguished, but their boundaries are poorly defined owing to overlap between adjacent layers. In the innermost and outermost layers most of the muscle fibers are disposed longitudinally, whereas in the middle layers there are more circular fibers.

Estrogen is essential for the maintenance of normal size and function in myometrial smooth muscle cells.

The Serosa

The serosa is the peritoneal covering of the uterus; hence only the pelvic portion of the uterus has a serosa. The cervix has no serosa.

THE FALLOPIAN (UTERINE) TUBES

The uterine tubes are the structure in which fertilization and the early stages of embryo development take place. Therefore, the tubes must provide an environment in which a number of functions can occur. These include providing a suitable milieu for the transport of gametes to the midsections of the tubes for fertilization, and providing nutrients adequate for the embryo during early division. The transport mechanisms must be such that the embryo is delivered to the uterus at the appropriate time, in terms of its own development as well as the uterine lining's receptivity for implantation.

General Structure

The fallopian tubes are composed of four parts: (1) the infundibulum, the funnel-shaped end of the fallopian tube, which contains numerous fimbriae that surround the abdominal os; (2) the ampulla, which accounts for the majority of the length of the fallopian tube; (3) the isthmus, which is the narrow portion leading to the uterus; and (4) the interstitial portion, which extends the tube through the wall of the uterus and contains a small os connecting the cavities of the fallopian tube and the uterus. The mucous membrane and the muscular layer vary from one region to another.

The Infundibulum

After the rupture of the follicle and the release of the oocyte with its accompanying granulosa cells, the oocyte encounters the fimbriae of the infundibulum. Here, some of the granulosa cells are removed as it is transported into the ostium. The surface of the fimbriae is heavily ciliated and the ciliary motion is toward the ostium.

The epithelium of the fimbriae, as well as the entire fallopian tube, is responsive to ovarian hormones. As the estrogen concentration increases during the follicular phase of the menstrual cycle, the epithelium increases in height, reaching a maximum at midcycle. During the luteal phase cellular height decreases. There is little evidence for deciliation and reciliation during the menstrual cycle, but it is known that decreases in or withdrawal of estrogen results in deciliation. The fimbriae of the fallopian tubes in postmenopausal women are largely devoid of ciliated cells, whereas those from postmenopausal women on estrogen supplementation have large numbers of ciliated epithelial cells.

The Ampulla

The mucosa of the ampulla is arranged in an elaborate system of longitudinal folds, resulting in a series of fine channels between the folds of the lumen. In this area something less than 50% of the epithelial cells are ciliated, and those that are beat toward the isthmus.

Fertilization occurs in the proximal portion of the ampulla. There are potentially two mechanisms for transporting the cumulus mass to this area: ciliary motion and/or smooth muscle contractility.

In this area of the fallopian tube the role of the ciliary cells is rather clear, but that of the secretory cells is less so. Loss of cilia may be the result of infection, which can lead to poor gamete transport with the resultant effect being infertility or ectopic pregnancy. Microbial infections may have resulted in the occlusion of the fallopian tube, and persistent secretion may lead to the formation of hydrosalpinx.

The Isthmus

The folds encountered in the isthmus are simpler and lower. There is also a marked increase in the thickness of the muscle layer. The cilia of the epithelial cells beat toward the uterus. The isthmus has the capacity to transport spermatozoa distally toward the site of fertilization, and then later to conduct the developing embryo proximally. It is not known whether the isthmus controls the passage of cleaving embryos into the uterus.

BLOOD SUPPLY TO AND LYMPHATIC DRAINAGE OF THE FEMALE REPRODUCTIVE TRACT

Blood supply

The internal pudendal artery is the arterial trunk supplying blood to all of the perineal structure inferior to the pelvic diaphragm. It begins as a branch of the internal iliac, which is located subperitoneally in the lateral pelvis. It exits the bony pelvis, crosses the sacrospinous ligament and enters the ischiorectal fossa. At this point the artery, along with the internal pudendal vein and nerve, becomes enclosed by the obturator fascia, forming the pudendal canal. As the artery enters the pudendal canal it gives off an inferior rectal artery which supplies the anorectal junction. The remaining portion of the internal pudendal artery reaches the base of the urogenital diaphragm and gives off a series of perineal branches. These supply the contents of both superficial and deep perineal spaces, including the vagina, urethra and clitoris.

The venous drainage of both perineal triangles parallels the arterial supply. There is also a rich submucosal venous plexus in the distal vagina. Distension of these submucosal veins can produce vaginal or vulvar varices. The inferior rectal veins join the internal pudendal vein just as it leaves the ischiorectal fossa at the lesser sciatic foramen. Both the rectal and the vaginal submucosal plexuses penetrate the pelvic diaphragm, to communicate with the endopelvic space. Here, vaginal veins may anastomose with uterine veins and inferior rectal veins with middle rectal veins.

The blood supply of the uterus and the upper vagina is via a single arterial trunk, the internal iliac artery. This arises from the division of the common iliac artery at the junction of the sacrum and the ilium. Descending in the lateral pelvis subperitoneally it gives off a series of visceral branches, including rectal, uterine and vesical. These course medially to enter the endopelvic space at the base of the broad ligament. Before reaching the isthmus of the uterus, the uterine artery crosses superior to the ureter and gives branches to the vaginal fornix and cervix. Turning superiorly in the parametrial space of the broad ligament, a series of arterial branches is given to the body of the uterus until the artery anastomoses with the ovarian artery at the uterotubal junction.

The uterine vein is usually plexiform, coursing laterally in the base of the broad ligament before reaching the lateral pelvic wall. Here the plexus of veins forms a series of tributaries entering the internal iliac vein, which in turn empties into the inferior vena cava. Other veins in the endopelvic space include middle rectal veins draining the rectum.

The normal route of rectal venous flow is into the internal iliac vein. During pregnancy the fetus may partially occlude the inferior vena cava when the woman is recumbent, increasing venous resistance and diminishing pelvic venous flow into the inferior vena cava. Because the middle rectal veins also communicate with the superior rectal branches of the inferior mesenteric vein, there is the potential for pelvic blood to ascend via the portal circulation. None of the pelvic veins contains valves, which allows blood to take the path of least resistance. Middle rectal veins also communicate with inferior rectal veins; these are tributaries of the internal pudendal vein, which drains into the iliac veins before entering the inferior vena cava. Increased blood flow in these vessels, particularly in the last trimester of pregnancy, is a well known cause of hemorrhoids.

The ovarian arteries arise as lateral branches from the abdominal aorta, descend in the retroperitoneal space, cross the ala of the sacrum and enter the suspensory ligament of the ovary. As the ovarian artery enters the lateral edge of the broad ligament it courses medially between the two layers of the ligament, giving branches to the ovary and uterine tube.

The venous drainage of the structures in the superior part of the broad ligament is via the ovarian vein, which parallels the ovarian artery as the vein ascends in the retroperitoneal space. On the right side of the ovarian vein is a tributary of the inferior vena cava, whereas on the left side it drains into the left renal vein.

Lymphatic drainage

As a general rule the lymphatic drainage follows the blood supply of a region. However, the lymphatic drainage of the perineum differs in this respect because there is a dual pathway. Deep lymphatics course upward, following the pudendal vein, draining the deep parts of both the urogenital and the anal triangles. However, superficial lymphatics from the skin overlying the vulvar and anal areas course to the medial thigh, where they communicate with superficial inguinal lymph nodes. Adenopathy of the superficial inguinal nodes is well known in many vulvar and anal infections, as well as in carcinoma of these regions.

A plexus of uterine lymphatics parallels the course of uterine veins, entering regional lymph nodes along the internal iliac artery. From these nodes lymph trunks ascend to para-aortic nodes in the retroperitoneum.

Afferent lymphatics from the ovarian and fallopian tube accompany ovarian vessels to para-aortic lymph nodes in the retroperitoneum. The fundus of the uterus is drained in part by this same route, but also sends lymphatic vessels anteriorly, paralleling the course of the round ligaments of the uterus. This bilateral course carries afferent lymphatics to inguinal lymph nodes on both sides of the pelvis.

ANATOMY OF THE MALE GENITAL TRACT

Genitalia

The Penis

There are two parts of the penis: the base, which is attached to the pubis, and the pendular portions. Underlying the penile skin there are cavernous erectile bodies and the paired corpora cavernosa, which are primarily concerned with erection, and the corpus spongiosum which contains the urethra. These erectile bodies are separate structures at the base of the penis but are bound by fascia along its shaft. The corpora cavernosa are cylindrical bodies in the shaft region but taper markedly at the base, where they attach to the pubic ramus and perineal membrane. The corpus spongiosum has three parts: beginning at the perineum there is the bulb of the penis, the spongy portion, and the glans at the tip of the penis (Figure 3.4).

The base and proximal portion of the penile shaft are covered by the muscles. The paired ischiocavernosus muscles overlie the crura and corpus cavernosa. Another pair of muscles, the bulbospongiosus, overlies the corpus spongiosum.

Urethra and Glans

The urethra is divided into bulbous, spongy and glandular portions. The bulbous and spongy parts are lined by a pseudostratified columnar epithelium, except at the tip of the penis, termed the fossa navicularis, which is lined by stratified squamous epithelium. The epithelium contains small acini of mucous

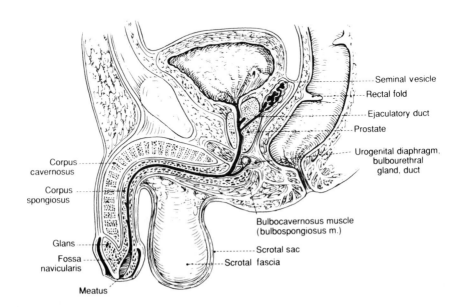

Corpus cavernosus
Corpus spongiosus
Glans
Fossa navicularis
Meatus

Seminal vesicle
Rectal fold
Ejaculatory duct
Prostate
Urogenital diaphragm, bulbourethral gland, duct
Bulbocavernosus muscle (bulbospongiosus m.)
Scrotal sac
Scrotal fascia

FIGURE 3.4 Sagittal section of pelvis and male reproductive system.

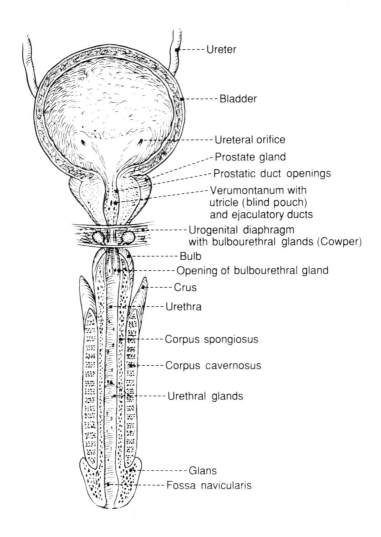

Ureter

Bladder

Ureteral orifice

Prostate gland

Prostatic duct openings

Verumontanum with
utricle (blind pouch)
and ejaculatory ducts

Urogenital diaphragm
with bulbourethral glands (Cowper)

Bulb

Opening of bulbourethral gland

Crus

Urethra

Corpus spongiosus

Corpus cavernosus

Urethral glands

Glans

Fossa navicularis

FIGURE 3.5 Coronal section of penis and urethra viewed anteriorly.

cells (glands of Littré), as well as mucosal and submucosal glands, termed urethral or periurethral glands (Figure 3.5). These glands can become infected and form abscesses.

On the superior surface of the corona of the glans penis, as well as on the undersurface near the frenulum, there are sebaceous glands, the glands of Tyson. These secrete a white cheesy type of material which, with desquamating epithelial cells, forms the smegma, a substance that accumulates between the prepuce and glands in uncircumcised men.

Scrotum

The scrotum may have small sebaceous cysts which may be multiple and, on occasion, become quite large or develop infections. The scrotum has two compartments which are divided in the midline. Each side is the mirror image of the other. The testis is the most anterior intrascrotal structure and must be examined carefully. The second most important structure is the epididymis, which lies immediately posterior to the testis.

Testis

The testis has two main functions: it produces sperm and it secretes male hormones. Sperm production takes place in the seminiferous tubules, whereas the production of testosterone, the major male hormone, takes place in the tissue located between the tubules. Each testis contains approximately 400–600 seminiferous tubules. Individual tubules are up to 70 cm in length and are coiled along most of this in order to be accommodated in a fascial compartment of the testis. These compartments are extensions of the outer fibrous capsule of the testis, the tunica albuginea. The seminiferous tubules join to form the rete testis, which is the connection to the excretory duct system. The lining of the seminiferous tubules contains two main types of cells, the developing sperm cells and the Sertoli cells, which support and presumably 'nurse' the sperm cells during the development process. Sperm are continuously produced in the testis from puberty to senility, following an orderly sequence of events. In the testis this process takes about 64 days. However, when they leave the testis the sperm cells are immature and unable to fertilize an egg.

Excretory Ducts

The excretory ducts transport sperm from the testis to the end of the male reproductive tract. They are composed of five components: the efferent ducts, the epididymis, the vas, the ejaculatory duct and the urethra.

GENITAL TRACT INFECTIONS

There are significant differences in male and female anatomy, as well as the reproductive physiology that accounts for more risk factors for STDs and their complications in women than in men. These differences also account for greater difficulty in the diagnosis of sexually transmitted infections in women, and further contributes to the higher risk of complications in women and an increased risk of transmission in the community.

In heterosexual men gonococcal and chlamydial infections, are usually limited to the anterior regions of the urethra. The progression of such infections into the posterior urethra and genital adnexae has not been well studied. Urethral discharge is a classic symptom of these diseases and is easily recognized by both patient and physician. However, the absence of urethral discharge does not exclude these diagnoses.

In women, several STD pathogens, including *Neisseria gonorrhoeae*, *Chlamydia trachomatis* and herpes simplex virus (HSV), have a predilection for the epithelial cells of the urethra, cervix and rectum simultaneously. This produces a wide range of symptoms and a variety of differential diagnoses. Localized infection can produce symptoms which are poorly localized and are easy to ascribe to the involvement of a contiguous site. For example, infections of either the vagina or the cervix can produce abnormal vaginal discharge. Finally, a lack of appreciation of the differing clinical signs of specific genital infection in women can often be attributed to inadequate inspection of the genitalia because of poor clinical skills, lack of speculums in developing countries, or reluctance to perform the inspection (Holmes, 1990).

Historically there has been a sense of mystery or lack of admission as to the true etiology of urogenital tract infections. Many terms, such as 'non-specific genital tract infection' and 'lower genital tract infection', complicate clinical diagnosis and therapy. However, over the last 25 years, and specifically in the last 5 years, our ability to accurately and rapidly diagnose sexually transmitted diseases has improved dramatically (Burczak *et al.*, 1994; Cone *et al.*, 1994 which should lead to improved management strategies and control of several STDs.

In a university study of women with urethral syndrome who had sterile bladder urine, the etiology was found to be related to the presence or absence of pyuria. Infection with *C. trachomatis* was demonstrated in 62% of women with pyuria and 6% of women without (Stamm *et al.*, 1980). In this report, pyuria was defined as fewer than eight leukocytes per ml of uncentrifuged midstream urine. *N. gonorrhoeae* was not found to be associated with the urethral syndrome in this population of university students. However, in indigent female populations attending emergency rooms, gonococcal infection was found to be significantly correlated with dysuria, accounting for 61% of the cases studied. *C. trachomatis* is probably the leading cause of urethral syndrome in female contacts of men with non-gonococcal urethritis (Paavonen *et al.*, 1986), whereas both *N. gonorrhoeae* and *C. trachomatis* may be important among women who are at high risk of gonorrhea. HSV infections, though not commonly found among women with urethral syndrome, can clearly produce urethritis, dysuria and pyuria in young sexually active women, and can be mistaken for bacterial cystitis if the vulvar lesions are not prominent.

Vaginal Infections

Age-dependent Differences in Vaginal Anatomy, Physiology and Flora

The normal anatomy, physiology and microbial etiology of the vagina are age-dependent (Cruickshank and Sharman, 1934). There are also obvious age-dependent differences in the source of vaginal infections. These factors account for different etiologies of vaginitis in neonates, infants, prepubertal girls and pre- and postmenopausal adults (Table 3.1).

During the first month of life the neonatal vagina is still under the influence of maternal estrogen, and is lined by stratified squamous epithelium. Following this period the vaginal epithelium is composed of cuboidal cells, and the pH of

TABLE 3.1 Etiology of lower urogenital tract sexually transmitted diseases in adult premenopausal women.

Site involved	Usual STD microbial etiology	Idiopathic
Urethritis	N. gonorrhoeae	
	C. trachomatis	
	HSV	
Vulvitis	HSV?-HPV	Vulvar papillomatosis
		Vulvar vestibulitis
		Essential vulvodynia
Endocervicitis	C. trachomatis	
	N. gonorrhoeae	
	HSV	
Ectocervicitis	HSV	

the healthy vagina remains at 7.0 until puberty. With the onset of puberty, and again under the influence of estrogen, the epithelium is composed of stratified squamous epithelial cells that contain glycogen. The normal flora of the vaginal vault contains *Lactobacillus acidophilus*, a facultative anerobe that produces H_2O_2, which produces lactic acid from the glycogen, and the pH falls to between 4.0 and 4.5 in normal adult females. The resultant fall in pH, as well as the production of hydrogen peroxide, is important in regulating the vaginal flora (Eschenbach *et al.*, 1989). The relative concentration of vaginal microbes has not been extensively determined in either neonates, infants or prepubertal girls, but the flora often is composed of diphtheroids, *Staphylococcus epidermiditis*, γ-hemolytic streptococci, lactobacilli and coliforms (Hammerschlag *et al.*, 1978; Hardy, 1941). Lactobacilli dominate the vaginal flora of normal postpubertal women (Eschenbach *et al.*, 1989).

Vaginitis in Premenarchal Girls

The vaginal squamous epithelium of the neonate is resistant to perinatally transmitted *N. gonorrhoeae* and *C. trachomatis* but is susceptible to perinatally transmitted vaginal candidiasis (Holmes, 1990). As the neonate grows into infancy the cuboidal vaginal epithelium becomes susceptible to the bacterial sexually transmitted diseases of *N. gonorrhoeae* and *C. trachomatis*, but becomes resistant to candidiasis. Vaginal infection by *N. gonorrhoeae* is rated among infants, and with few exceptions (Gutman and Holmes, 1989) is thought to represent postnatal acquisition, usually from abusive situations.

In older premenarchal girls the etiology of vaginal symptoms is correlated with the status of puberty and the presence or absence of signs of abnormal vaginal discharge. In one study of premenarchal patients with suspected vaginitis and age-matched controls (Hammerschlag *et al.*, 1978) a microbial pathogen was isolated from 53% of patients with abnormal vaginal discharge, none with only suspected vaginitis without discharge, and one of the control subjects. *N. gonorrhoeae* was isolated from one-third of abnormal vaginal discharges in prepubertal girls, whereas *Candida albicans* was isolated only from those premenarchal girls who were considered to be pubertal based on breast growth (Tanner stages II, III or IV). *C. trachomatis* has not been implicated as a common cause of prepubertal vaginitis. This agent was isolated from the vagina of about one-quarter of girls with gonococcal vaginitis, and appeared to be responsible for postgonococcal vaginitis following penicillin treatment in such cases (Rettig and Nelson, 1981). The occurrence of *C. trachomatis* infections in prepubertal sexually abused girls has been documented in several studies (Fuster and Neinstein, 1987).

Vaginal Infections in Adult Women

Vaginal infections are among the most common problems in clinical medicine, occurring in about 28% of women attending STD clinics (CDC, 1979). The most common types of vaginal infections are trichomoniasis, candidiasis and bacterial vaginosis. In postmenopausal women vaginal symptoms are very prevalent and the usual forms of infection need to be considered and differentiated from vaginal atrophy (Galask and Larsen, 1981, Osborne *et al.*, 1979). As estrogen supply wanes, the vaginal epithelium becomes thin and lubrication occurs less often with sexual arousal, the introitus narrows and the depth of the vagina decreases. These atrophic changes have been found to be negatively correlated with frequency of intercourse and with circulating concentrations of gonadotropins and androgens in postmenopausal women (Leiblum *et al.*, 1983).

Cervicitis

There are two types of cervicitis, *endocervicitis* (also known as mucopurulent cervicitis) and *ectocervicitis*. The etiology of endocervicitis usually includes the sexually transmitted pathogens *C. trachomatis*, *N. gonorrhoeae* and herpes simplex virus. HSV can also be associated with ectocervicitis.

Endocervicitis

Endocervicitis, together with subclinical urethral infection in both sexes, represents the tip of the iceberg of infections caused by *N. gonorrhoeae* and *C. trachomatis*. Control of these infections is largely via clinical diagnosis of urethritis in the male and the treatment of the female sex partners of such men. Endocervicitis produces symptoms less frequently than male urethritis, and the symptoms (e.g. vaginal discharge) are less distinctive than those of urethritis. Careful assessment of the clinical signs of mucopurulent cervicitis and appropriate detection of subclinical infection in the female are of paramount importance in the control of gonococcal and chlamydial infection.

Infection of the cervix is a reservoir for both sexual as well as perinatal transmission of pathogenic microorganisms, and can lead to at least two types of complications in the female: (1) ascending intraluminal spread of pathogenic organisms from the cervix, which can produce endometritis and salpingitis; and (2) ascending infection during pregnancy, which can result in chorioamnionitis, premature delivery and puerperal infection.

There is a lack of widely recognized signs of cervical inflammation, which leads to a confusing set of nomenclatures. This results in part from changes that occur in the cervix over the reproductive period and during the menstrual cycle (Goldacre *et al.*, 1978; Pixley, 1976; Singer, 1975), and from the difficulty in differentiating normal ectopic columnar epithelium from endocervicitis. This is complicated by the fact that cervical ectopy appears to be correlated with cervical infection by *C. trachomatis* (Arya *et al.*, 1981; Harrison *et al.*, 1985; Tait *et al.*, 1980). Ectopy presents in a majority of younger teenage girls, and decreases with increasing age. There is also a positive correlation with oral contraceptive use (Holmes, 1990).

Mucopurulent Cervicitis

Mucopurulent cervicitis is usually defined as the presence of a yellow exudate on a white cotton-tipped swab taken from a specimen of endocervical secretions, and is usually correlated with the isolation of *C. trachomatis*. These mucopurulent secretions were present in 62% of women with cervical chlamydial infection and in 12% of those with no cervical pathogen. Although mucopurulent cervicitis has been correlated with endocervical infection in most if not all studies, the sensitivity, specificity and predictive value have varied in different settings (Keur *et al.*, 1988; Schafer *et al.*, 1984; Swinker *et al.*, 1988). The cause of mucopurulent cervicitis in the absence of proven gonococcal, chlamydial or HSV infection remains uncertain. Various genital pathogens have been implicated, and it is likely that false negative cultures for chlamydia account for some cases (Paavouen, 1979; Paavouen *et al.*, 1986). In addition, it seems possible that oral contraceptive use and cervical ectopy *per se* may be associated with endocervical inflammation.

Ectocervicitis

Colposcopic examination of female STD clinic patients has shown that cervical HSV infection is highly correlated with cervical ulcers or necrotic lesions, and both *C. trachomatis* and cytomegalovirus (CMV) infection of the cervix are correlated with the colposcopic features of immature metaplasia. Cervical ulcerations or necrotic lesions were detected by colposcopy in 65% of women with positive cervical cultures for HSV, and in 11.5% of negative cultures for HSV.

In patients with endocervicitis due to *C. trachomatis* cervical biopsy may reveal intraepithelial inclusions, which are located in columnar or metaplastic cells. Such inclusions are best seen with immunofluorescence or immunoperoxidase staining, and contain typical elementary and reticulate bodies of *C. trachomatis*. The majority of cervical biopsies from patients with chlamydia-positive endocervicitis show superficial focal endocervical microulcerations, reactive endocervical cellular changes, stromal and epithelial cellular edema and dilation, and proliferation of subepithelial capillaries and stromal inflammatory infiltration, predominantly by plasma cells.

There is little recent information on gonococcal cervical infections or either the initial or recurrent episodes of HSV cervicitis. However, HSV infections differ from gonococcal and chlamydial infections in causing deep, necrotic ulcerations with stromal infiltrations, predominantly by lymphocytes.

Vaginal Discharge Associated with Infection or Cervicitis

In the majority of adolescent females columnar epithelium lies in an exposed position on the ectocervix at the onset of menarche. The prevalence of ectopy gradually declines through young adulthood. The term ectropion has been used to describe the parulous parous cervix, which opens as the blades of a vaginal speculum are spread to expose the endocervix. Ectopy not associated with visible evidence of mucopurulent exudate or with colposcopic epithelial abnormalities is a normal finding. Recurrent genital herpes involving the cervix alone produces lesions of the endocervix and the ectocervical squamous epithelium.

Not infrequently cervicitis coexists with vaginal infection, particularly with bacterial vaginosis or trichomonal vaginitis. Excluding the neonatal period, gonococcal vaginitis is the most common form of gonorrhea in children. In contrast to adults, in prepubertal girls the non-estrogenized alkaline vaginal mucosa may be colonized and infected with N. gonorrhoeae. Gonococcal vaginitis is usually a mild disease, perhaps because it is restricted to the superficial mucosa. Prepubertal vaginitis has been attributed not only to N. gonorrhoeae, but also to numerous other infectious agents. The incidence of gonorrhea, genitourinary chlamydia and syphilis in several studies of children known to have been sexually abused have been summarized, and gonococcal vaginitis is the most frequently recognized of these causes.

Upper Genital Tract Infection in Womn

Pelvic inflammatory disease (PID) is a clinical syndrome caused by the ascending spread of microorganisms from the lower to the upper genital tract. Usually it is associated with abdominal pain of recent onset, but with other symptoms from various organ systems. Abdominal pain is usually a hallmark for a clinical diagnosis of PID, or for considering laparoscopy to verify suspected acute salpingitis (Jacobson *et al.*, 1969).

Endometritis and Endometrial Infection

Endometritis was present in 47% of women who had mucopurulent cervicitis without symptoms of PID. Approximately 65% of individuals who were culture positive for cervical *C. trachomatis* had endometritis. Sweet *et al.* (1983) reported persistence of *C. trachomatis* in the endometrium of women treated for PID. *C. trachomatis* has also been recovered from the endometrium of infertile women. However, from individual histories it was unlikely that the chlamydial infection was of recent onset. In contrast, there are several studies in which *C. trachomatis* was not recovered from the cervix (Gump *et al.*, 1983), or only a single recovery in 185 endometrial specimens. Sellors *et al.* (1988) was unable to isolate *C. trachomatis* from either the cervix or the fallopian tubes of 52 women with tubal factor infertility. Therefore, chlamydial relationships to cervical, endometrial and tubal infection must be confirmed not only with rigorous microbiological investigations, but also with histopathological studies of these sites.

Salpingitis and Tubal Infection

C. trachomatis has been reported in about 15% of infertile women who had no clinical or laparoscopic evidence of acute salpingitis. Patton (1989) was unable to isolate *C. trachomatis* from the fallopian tubes of 60 infertile women with distal tubal scarring. Anestad *et al.* (1987) also reported that in 67 infertile women with tubal adhesion or occlusions *C. trachomatis* was not recovered from the cervix or fallopian tubes of any of them.

The relationship of tubal infertility to previous infection with *N. gonorrhoeae* has not been well studied. Tjiam *et al.* (1985) identified gonococcal pili antibodies in 61% of a group of infertile women and in 25% of women without scarred fallopian tubes. Less than a third of women with gonococcal pili antibodies had chlamydial antibodies, and half of these lacked a history of PID.

Ectopic pregnancy is an important sequela of salpingitis. It is estimated that half of all ectopic pregnancies can be attributed to previous PID (Weström *et al.*, 1981). Svensson *et al.* (1985) studied a group of 112 women with ectopic pregnancies: 65% had serum antibodies to *C. trachomatis* and 31% of these had no past history of PID. These studies suggest that a significant proportion of women with ectopic pregnancy, particularly those who have no other risk factors, may have chlamydial infection, and that many of these had atypical rather than typical PID.

MICROBIAL FACTORS INVOLVED IN STDs

Chlamydia trachomatis

The initial step in the infectious process caused by *C. trachomatis* involves attachment to human epithelial cells (Figure 3.6). This is essential in the pathogenesis of the infection because these obligate intracellular bacterial pathogens must be internalized by the host cell in order to convert to the metabolically active reticulate bodies which in turn, grow and divide within a membrane-bound environment. During this growth the reticulate body parasitizes the host cell for certain amino acids and ATP (Hatch *et al.*, 1982).

Attachment mechanisms have been examined by a number of investigators, but a singular high-affinity adhesin–receptor interaction has not been established. This may be attributed to: (a) the type of host cell infected; (b) different species of *Chlamydia*; (c) separate biovars of *C. trachomatis*; and (d) methods of *in vitro* inoculation. The chlamydial major outer membrane protein (MOMP), which constitutes nearly 60% of the total protein in the envelope, contributes to adherence by an electrostatic mechanism. Recently, a heparan sulphate-mediated attachment mechanism has been proposed for the lymphogranuloma venereum (LGV) biovar of *C. trachomatis* (Zhang and Stephens, 1992). However, the trachoma biovar is dominated by an adherence mechanism other than the glycosaminoglycan (GAG)-mediated mechanism for initial attachment to host cells. Although a GAG-dependent mechanism is required for mediating infectivity mutually exclusive of the mode of attachment, the dominance of GAG-dependent binding for the trachoma biovar justifies efforts to characterize trachoma biovar adhesin molecules other than the GAG ligand (Chen and

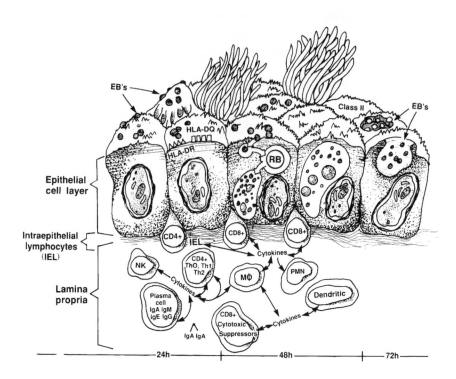

FIGURE 3.6 Immunobiology of chlamydial infection of the human fallopian tube mucosa. Elementary bodies (EBs) attach to the microvilli of epithelial cells prior to invasion of the epithelium. In some cases where EB aggregates are large, a local cytotoxicity is seen. The chlamydia have a normal developmental cycle in the epithelial cells but do penetrate the basolateral portion of the cell. The EBs cause a lesion in the luminal surface of the infected epithelial cells and are released where they are capable of infecting other cells.

Stephens, 1997; Danilition *et al.*, 1990; Joseph and Bose 1991; Lundemose *et al.*, 1993; Raulston *et al.*, 1993; Su *et al.*, 1990). Other investigators, using primary hormone-responsive endometrial epithelial cells, have shown that a receptor-mediated endocytic route exists which involves clathrin-coated pits. This implicates a specific adhesin–receptor interaction for attachment. There is also some evidence that a heat shock protein may be involved in adherence. There have been reports that antibody against the chlamydial hsp 70 homolog protects against chlamydial salpingitis in infected patients (Brunham *et al.*, 1987; Danilition *et al.*, 1990). Certain antibodies against chlamydial hsp70 neutralize infectivity *in vitro*, and antigenic reactivity against whole elementary bodies (EBs) in microimmunofluorescent and enzyme-linked immunosorbent assays implicate surface-accessible epitopes (Danilition *et al.*, 1990). Therefore, the envelope locality and its potential role in attachment of the chlamydial hsp70 constitutes an intensive area of investigation. Thus, it appears that the chlamydia have developed multiple methods of attachment to ensure successful propagation within the host cell.

The question of chlamydial toxicity to either the host or to cells *in vitro* is still unresolved. Toxicity has been demonstrated in *C. trachomatis* biovar LGV, trachoma and mouse, as well as numerous strains of *C. psittaci* (Bell and Theobald, 1962; Moulder, 1991; Rake and Jones, 1944; Tavenne *et al.*, 1964). The toxicity is associated with EBs and not RBs (reticulate bodies), and the toxic dose is large, i.e. 10^7–10^8 EBs per mouse. The multiplication-independent toxicity of chlamydia for host cells *in vitro* was first demonstrated with mouse peritoneal macrophages. An inoculum of the order of 10–100 EBs of *C. trachomatis* and *C. psittaci* per host cell damages macrophages within 2–10 hours, depending on the multiplicity of the chlamydia strain inoculated. Visual evidence of damage consists of rounding up, vacuolization and, by 20–24 hours, detachment from the substrate and lysis. Trypan blue and other vital dyes are no longer excluded and lysosomal and cytoplasmic enzymes are released in the medium, as well as the uptake of labeled amino acids being depressed (Moulder, 1991). The chlamydial cell wall contains a lipopolysacccharide that in many respects resembles the endotoxin of host-independent Gram-negative bacteria. It has been suggested that the toxic death of mice and cultured cells might be caused by massive contact with chlamydial endotoxin. However, toxic doses of *C. psittaci* kill both endotoxin-responsive and endotoxin-non-responsive mice at the same time after injection, and macrophages from non-responsive mice are only slightly more sensitive to EB toxicity *in vitro*. In addition, chlamydial lipopolysaccharide engenders genus-specific antibody, whereas antibody neutralization of multiplication-independent toxicity is strain specific, both in mice and in cell culture. The chlamydial toxin is probably an artifact. The large doses of chlamydia needed for immediate damage to mice or cultured cells are unlikely to exist outside the laboratory. Only if the natural cell targets of chlamydia are exquisitely sensitive to immediate toxic injury is there much chance of the phenomenon playing any role in the pathogenesis of chlamydial infection.

Much of the microbiological study regarding chlamydia has as its ultimate goal the production of a vaccine for the prevention of *C. trachomatis* infection. The feasibility of this is based upon the results achieved during the 1960s and 1970s using whole chlamydial EBs to immunize humans against trachoma. The best of these trials confidentially established that parenteral immunization of young children at high risk for trachoma significantly reduced the incidence (Britigan *et al.*, 1985). Parallel work in non-human primates allowed more precise evaluation of vaccine-induced protection (Grayston and Wang, 1978), and under these conditions EB vaccines prevented both infection and disease. However, the protection was restricted to the homologous chlamydial strain, and heterologous challenge in vaccinated animals led to a more severe disease. During the 1980s molecular elucidation of the antigenic structure of the chlamydial EB has generated a renewed interest in the design of an effective vaccine to prevent *C. trachomatis* infection.

Candidate antigens for a subunit vaccine have been based upon their identification as targets for antibody-mediated neutralization. However, it is important to note that antibody-mediated neutralization has not been demonstrated to reflect the immune effector mechanism in human immunity to chlamydial infection, and thus this criterion carries with it some uncertainty.

The use of molecules which are targets for neutralization restricts consideration to EB-specific surface-exposed antigens. Three candidate antigens are targets for neutralizing antibodies: chlamydial MOMP, hsp70 protein, and the MIP (macrophage infectivity potentiator) protein.

The chlamydial hsp70 was initially identified as a minor neutralization antigen of the organism. This is a bifunctional protein whose N terminus contains an ATPase domain and whose C terminus contains a peptide-binding domain. Two EB surface-exposed antigenic sites for the hsp70 have been identified: one resides in the putative 'ATPase domain' and the other in the 'peptide-binding domain'. The conserved nature of the sequence of the hsp70 among *C. trachomatis* strains makes it an attractive vaccine choice; however, its limited immunogenicity because of its few T-cell sites is an important obstacle to its use as a vaccine antigen (Zhong and Brunham, 1992).

Recently, a chlamydial gene was cloned for a 29 kDa protein which had substantial sequence identity to the *Legionella pneumophilia* MIP gene. The MIP is necessary for intracellular survival, and genetic studies have proved that it is an essential virulence factor. Chlamydial MIP was inhibitable with FK506 and rapamycin, and possessed peptidyl-prolyl *cis/trans* isomerase activity (Hundemose *et al.*, 1993). Antibody to chlamydial MIP neutralized the organism in cell culture (Hundemose *et al.*, 1992).

MOMP is a major vaccine candidate antigen: more is known about its genetic structure and immunochemistry than any other chlamydia antigen. It is the major variable surface protein of *C. trachomatis* (Caldwell *et al.*, 1991, 1992). The MOMP gene, termed *omp1* or *ompA*, has been cloned in its entirety from at least seven different *C. trachomatis* immunotypes (Stephens *et al.*, 1986, 1987). Comparative analysis of *omp1* DNA sequences shows that nucleotide substitution occurs throughout the open reading frame, but that flanking regulatory sequences are highly conserved (Stephens *et al.*, 1987). Despite the informative nature of the genetic and immunochemical data, we do not know the three-dimensional structure of MOMP, nor its topology within the outer membrane.

Current approaches to designing a MOMP subunit vaccine have been based on using oligopeptides (Qu *et al.*, 1993; Su *et al.*, 1992; Zhong *et al.*, 1993) or cloned recombinant fragments of MOMP as non-replicating antigens (Conlan *et al.*, 1990; Toye *et al.*, 1990), or live recombinant fragments such as *Salmonella* (Hayes *et al.*, 1991), polio virus (Caldwell, 1993; Murdin *et al.*, 1993) or phage (Zhong *et al.*, 1994) that deliver an amplifiable MOMP epitope to the host.

A remarkable and unexplained feature about the immunoepidemiology of *C. trachomatis* infection is the tight correlation between immunotypes and disease syndromes (Grayston and Wang, 1978). Trachoma is almost entirely due to infection with immunotypes A, B/Ba and C. Thus, a vaccine to prevent trachoma based on MOMP sequences from these three immunotypes is sensible and is a reasonable objective for a first-generation vaccine. However, given the fact that MOMP is a highly polymorphic protein, a vaccine based on MOMP will certainly be faced with difficulties because of antigenic variations in the MOMP. The tempo for genetic change in MOMP will be one of the determinants of vaccine effectiveness. Currently, it is assumed that MOMP polymorphism is relatively limited in extent and stable through time and space. However, molecular analysis of *omp1* sequences from individuals with trachoma (Dean *et al.*, 1992; Hayes *et al.*, 1992) or sexually transmitted chlamydial infections (Yang *et al.*, 1993) demonstrates unexpected sequence polymorphism.

The challenges to designing an effective vaccine for the prevention of C. *trachomatis* infection are summarized as follows: (1) the three-dimensional structure of MOMP and its topologic relationship to the cell envelope must be determined; (2) a genetic system for gene transfer in C. *trachomatis* must be established; (3) the epidemiology and genetic basis for *omp1* variation needs to be elucidated; and (4) the relationship of *in vitro* antibody-mediated neutralization to C. *trachomatis* immunity needs to be established (Brunham, 1994). Because a vaccine based on MOMP may prove impractical, other antigens such as hsp70 and MIP (and others) need to be evaluated. Also, studies of pathogenesis have helped thinking about the design of a chlamydial vaccine. Most remarkable are the observations that the protein product of a single gene, hsp60, is centrally involved in immune responses related to chlamydia-induced inflammation and scarring (Morrison *et al.*, 1989; Toye *et al.*, 1993). At present, the absence of a genetic system for chlamydia prevents pursuit of this line of inquiry.

Neisseria gonorrhoeae

Neisseria gonorrhoeae is the etiological agent of the sexually transmitted disease gonorrhea. Humans are the exclusive host of N. *gonorrhoeae*; some animal models have been described, but their usefulness in studying pathogenesis has been limited by their lack of similarity to human mucosal disease. For example, it has been impossible to establish relevant genital infections in any animal, including several primate species (McGee *et al.*, 1990). Thus, it has not been possible to study relevant mucosal immune responses to N. *gonorrhoeae* in experimental animals.

The most critical interaction between the host and N. *gonorrhoeae* occurs at the mucosal surface: the gonococcus must attach to the epithelial cell surface and multiply. Further, gonococci must evade non-specific defense mechanisms, as well as any specific defenses developed from prior infections. The epithelial cells of various mucosae appear to have common receptors for the gonococci, as clinical infections have been reported in the eye, the pharynx, the rectum and the urogenital tract. Particularly in the urogenital tract these organisms may invade the urethra and/or the endocervix, and remain for extended periods of time without initiating clinical symptoms. The mechanisms for the sequestering of these gonococci are unknown, but these sequestered isolates have unique nutritional requirements, an extreme sensitivity to antimicrobial agents, and limited ability to activate either the complement cascade or neutrophil chemoattractant factors (Britigan *et al.*, 1985).

Infection of the human mucosa by N. *gonorrhoeae* involves attachment, invasion, multiplication and transport to subepithelial tissue (Figure 3.7). In the following section these steps will be explored, with emphasis on the mechanisms that may be involved in protecting the host from infection.

The initial step in the infectious process is the attachment of the gonococci via its ligand(s) to a receptor on the microbial epithelial cell surface. The initial observations in this area were made by Kellog et al (1968), who determined that only the gonococcal types 1 and 2 (pili expressing), when used to infect male

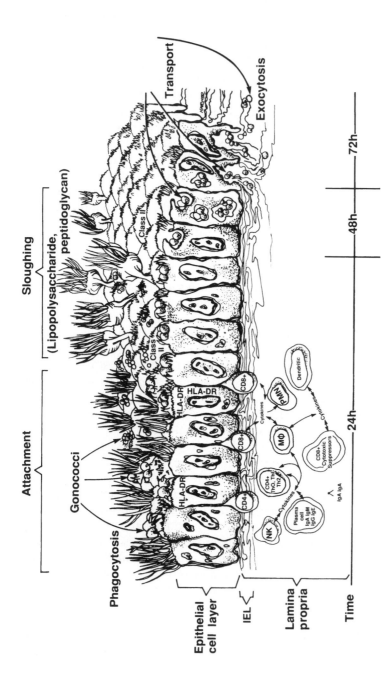

FIGURE 3.7 Immunobiology of gonococcal infection of human fallopian tube mucosa. Gonococci attach to the microvilli of non-ciliated epithelial cells and are internalized by a phagocytic mechanism that results in a membrane-bound vesicle. This process takes approximately 24 hours. At the same time, the gonococci liberated from their surface two molecules (lipopolysaccharide and peptidoglycan) that are toxic to the ciliated epithelial cells, and slough as a result of this action. Gonococci continue to migrate through the epithelial cell, and by 72 hours after infection they have begun an orderly exocytosis from the epithelial cell. At this point in the infection the gonococci are present in the lamina propria, where they can encounter a variety of immunocompetent cells to initiate both B- and T-cell responses, as well as the increased production of cytokines (i.e. **TNF-α**).

volunteers, were able to establish a urethritis. This suggests that pili are involved in the initial attachment to epithelial cells. Subsequent studies have demonstrated that other surface-expressed ligands (opacity proteins (Opa+), a 37 kDa protein) (Bessen and Gotschlich, 1986; Connell *et al.*, 1988) have also been implicated in attachment to host cells.

Both the Opa proteins and pili are interesting as attachment ligands. The Opa proteins comprise a group of at least 11 closely related outer membrane proteins, most of which give an opaque appearance to colonies. In some cases the Opa proteins appear to bind directly to some cells (Bessen and Gotschlich, 1986). Usually only a few are expressed at any given time. These proteins are highly variable and 'switching' between them occurs at a rate of 10^{-3} per cell per generation. Opa expression is controlled by variations in length of a five-nucleotide DNA repeat $(CTCTT)_n$ within the signal sequence-encoding region of the *Opa* gene. Variations in the numbers of the $(CTCTT)_n$ result in translational frame shifts, and therefore control expression of the protein at the level of translation (Murphy *et al.*, 1989; So *et al.*, 1985).

Pili, which are composed of pilin subunits, also undergo high frequencies of phase and antigenic variation. These variations are due to recombination between incomplete, variant 'silent' piling DNA cassettes and piling structural genes, as well as to other transcriptional control mechanisms (Haas and Meyer, 1986; Hagblom *et al.*, 1985). These high-frequency antigenic variations provide a mechanism for the organism to escape from immune surveillance. In addition these variants serve as specific ligands for different cell receptors (Cohen and Sparling, 1992). Expression of pili is strongly associated with increased adherence but the region of pilin involved in cell adherence to host cells has yet to be clarified (Robinson *et al.*, 1989; Rothbard *et al.*, 1988; Schoolnik *et al.*, 1984; Swanson, 1973; Ward *et al.*, 1974).

A number of laboratories have demonstrated that gonococci can invade epithelial cells *in vitro* (Cooper *et al.*, 1984; Makino *et al.*, 1991; McGree *et al.*, 1989, 1988; Weel *et al.*, 1991). Outer membrane proteins that may aid in or facilitate epithelial cell invasion include outer membrane protein (Por) and several of the Opa proteins. Differential expression of these proteins in various segments of the urogenital tract (urethra, endocervix, fallopian tube) may explain the unique anatomical localization of infection. Changes in phenotype between *in vitro* and *in vivo* may occur; the importance of this phenomenon to vaccine development is obvious. For example, Jerse *et al.* (1994) and Cohen *et al.* (1994) have demonstrated that experimental intraurethral human infection with Opa-negative gonococci leads to *in vivo* changes in Opa expression. The Opa-negative gonococci all expressed Opa, and in some subjects as many as three different Opa phenotypes were isolated. Further, there was no detectable difference in infectivity by an Opa-negative variant and one expressing an Opa protein (OpaF) that was highly represented in reisolates from the patients. These results indicate that there is a strong selection for expression of one or more Opa proteins *in vivo*, but that no single protein is preferentially expressed during early infection in the urethra. Similarly, Schneider *et al.* (1991) studied lipo-oligosaccharide (LOS) variation in human volunteers challenged intraurethrally with gonococci, and demonstrated an evolution of the LOS phenotype during the course of infection. This change reflected, in part, sialylation of the LOS by cytidine monophosphate-N-acetylneuraminic acid (CMP-NANA). This change provides gonococci with an unstable form of serum resistance that may be required for infection. The gonococcus also has a high-frequency variation ($\sim 10^{-3}$) in the core sugars of LOS (Apicella et al., 1987). The mechanisms for this variable expression are unknown. In conclusion, N. gonorrhoeae has a number of cell surface molecules that are important in interactions with host cells, and which could be targets for host immune response. The instability of these structures makes the development of a protective response difficult, if not impossible.

Mucosal Antibody Responses

Host responses to gonococcal infection at the mucosal level have not been extensively studied. Modern immunological methods are just beginning to be applied to the understanding of molecular immunological events occurring at the mucosal level. There are a number of both non-specific and specific factors that are associated with mucosal infection by gonococci which may play larger or smaller roles in providing protection. Inflammation associated with this infection is seen as a failure of a combination of host defense mechanisms and the direct cytotoxicity of certain surface-expressed molecules of the gonococcus.

Human serum non-specifically contains antibodies which are bactericidal for many gonococci. This is true even from individuals who have no prior history of gonococcal infection. These natural antibodies are generally directed against a component present in LOS, but can be directed against a variety of other gonococcal antigens (Densen et al., 1987; Ison et al., 1986; Lammel et al., 1985). However, individuals who have previously been exposed to gonococci possess antibodies to multiple antigens both in serum and mucosal secretions (Cohen and Sparling, 1992). Antigonococcal IgG, IgM and IgA have all been detected at mucosal surfaces, but IgA is the predominant antibody. Both sIgA$_1$ and sIgA$_2$ are present and produced locally by mucosal plasma cells. The precise function of these sIgA antibodies in providing antigonococcal immunity is unknown, but it does not appear to be opsonization. Despite the fact the gonococci produce an IgA$_1$ protease, this does not appear to be a virulence factor, as genetically constructed organisms lacking IgA$_1$ protease attach to, invade and damage human fallopian tube tissue with the same efficiency as do those with the IgA$_1$ protease (Cooper et al., 1984).

Gonococcal Interaction with Phagocytic Cells

There is little doubt that gonococci interact with phagocytic cells. Urethral exudates from experimentally infected males contain neutrophils, some of which contain intracellular gonococci. Not all neutrophils in these exudates contain intracellular organisms and not all organisms isolated from such exudates are intracellular. Phagocytes have different microbicidal mechanisms for use under anerobic and aerobic conditions. Both anerobic and aerobic organisms have been recovered from vaginal secretions (Bartlett et al., 1977) and the male urethra (Stamey, 1980). Phagocytes generate O_2^-, H_2O_2 and other reactive oxygen intermediates, which (in concert with microbicidal proteins) might be expected to kill gonococci. However, gonococcal adaptation is probably an important part of resistance in vivo. Gonococci exposed to neutrophils increase their production of catalase, which facilitates resistance to neutrophils. Likewise, the use of neutrophil lactate by gonococci allows sufficient oxygen consumption to inhibit neutrophil generation of O_2^-, which could ultimately lead to an anerobic environment. Neutrophil-derived nitric oxide could replace oxygen as a terminal electron acceptor, which is required for bacterial growth. Inadequate attachment of Opa$^-$ gonococci and/or gonococci expressing variant Opa proteins could lead to extracellular survival; piling and/or other gonococcal factors could also inhibit phagocytosis directly. The formation of a unique 20 kDa protein after phagocytosis could permit intracellular survival. Translocation of gonococcal Por protein to the phagocyte could inhibit neutrophil deregulation, also leading to intracellular survival of gonococci.

Cell-mediated Immunity (CMI), Cytokine Production and Tissue Injury

Another important aspect of the inflammatory response is the contribution of cell-mediated immunity, as characterized by the interaction between lymphocytes and macrophages and the generation of a variety of cytokines. The female reproductive tract contains all the elements required fully to respond immunologically to bacterial invasion. The mucosal epithelial cell surface contains class II molecules. In particular, HLA-DR and HLA-DQ have been expressed on the surface of human fallopian tube epithelial cells (Cooper *et al.*, 1990). The presence of MHC antigens was not restricted to any particular anatomical region of the fallopian tube. There is uniform expression of HLA-DR along the luminal surfaces of the epithelial cells. In the confluent areas there was more prominent staining on the tips of the mucosal villi than in the crypts. HLA-DQ was expressed less uniformly and was located only at the luminal surfaces, and not intracellularly or along the basolateral borders. These studies suggest these epithelial cells could serve as antigen-presenting cells to CD4$^+$T cells.

To determine whether the mucosal epithelial cells that expressed class II could transport an exogenous antigen to underlying immunocompetent cells, dual labeling experiments were performed. Following incubation of fallopian tube organ cultures with rhodamine-labeled LOS, the LOS was present in class II expressing cells as well as in association with immunocompetent cells in the lamina propria. The cells so labeled also appear to secrete interleukin 1 (IL-1), a further prerequisite to their functioning as antigen-presenting cells. Whereas the LOS downregulated the expression of class II, particularly that of HLA-DR, interferon-γ stimulated enhanced expression of various HLA antigens on the surface of epithelial cells. These results suggest that fallopian tube epithelial cells may be important in the mucosal immune response as antigen-presenting cells. The enhanced expression of HLA class I and class II antigens could increase the activation of both CD4$^+$ and CD8$^+$ intraepithelial lymphocytes. The stimulation of CD4$^+$T cells in the presence of antigen and IL-1 could result in activation of delayed-type hypersensitivity and/or B cells. Conversely, if CD8$^+$ lymphocytes are activated, this could lead to either non-specific or antigen-specific suppression of the response, which might explain why mucosal immunity to gonococcal infection is often impaired.

McGee *et al.* (1992) proposed the theory that some of the damage seen in LOS-damaged fallopian tubes is partly due to the action of tumor necrosis factor (TNF). Exogenously applied recombinant human TNF damaged the mucosa in concentrations as low as 1–10 µg ml^{-1}. This cytotoxic activity could be blocked by anti-TNF antibody. Additional studies have shown that there is a time-dependent increase in the concentration of TNF in gonococcal-infected fallopian tube organ cultures, compared to concentrations in uninfected controls. It can be speculated that some of the damage that leads to scarring following *in vivo* infection may be due to the host response triggered by microbial activities.

Grifo *et al.* (1988) studied the level of interferon-γ (IFN) in patients with pelvic inflammatory disease. About 28% of the cases were of gonococcal origin, and 88% of these had increased levels of circulating IFN-γ. The occurrence of detectable levels of IFN-γ in the peripheral circulation of these patients suggests that elevated interferon levels might exist at the site of infection. High local concentrations of interferon-γ in the fallopian tube induced the expression of class II antigens on the cell membrane of epithelial cells and macrophages (Grifo *et al.*, 1988). There are several hypothetical mechanisms by which this can exacerbate the pathogenesis of the disease. In one, high levels of class II antigens on cell surfaces selectively activate suppressor T cells by a process known as the autologous mixed lymphocyte reaction (Glimcher *et al.*, 1981). Such a reaction results in the elaboration of soluble suppressor factors which in time down-regulate cell-mediated immune responses. A second hypothesis would postulate that class II antigen expression on epithelial cells causes them to lose 'self' status and renders them susceptible to attack by autologous T cells. This results in the appearance of cytotoxic T cells specifically reactive to class II-expressing cells. Whether IFN-γ stimulates the inhibitory or the stimulatory immune response pathway depends on local concentrations of IFN and the numbers of class II-positive cells.

Effective vaccination against gonococcal infection has been an elusive goal. Until host mechanisms that might protect against infection are better characterized and understood, an effective vaccine that targets the genital tract for protection may continue to be out of reach. Recent evidence suggest that mucosal immune responses, particularly local (vaginal) IgG antibody levels, may mirror those measured in serum (Bouvet *et al.*, 1994). Therefore, a consideration of the data available on serum-based immune mechanisms may shed light on protection against mucosal pathogens such as *N. gonorrhoeae*. In particular, efficacy of past vaccines targeted to prevent infection acquired by the mucosal route have shown a correlation between efficacy and levels of IgG antibody in serum that result from vaccination (Robbins *et al.*, 1995).

A successful vaccine candidate against gonorrhea may require one or more of the following elements: the generation of serum or mucosal antibodies that either facilitate complement-mediated killing of the organism, and/or enhance phagocytosis and microbial killing by polymorphonuclear leukocytes; the ability to stimulate an immune response that blocks the attachment of the gonococcus to host tissues; and the evocation of cell-mediated defenses that prevent infection or modify fallopian tube damage. With the intention of satisfying one or more of these criteria, at least two gonococcal vaccine candidates have been tested in humans over the past 20 years (Boslego *et al.*, 1991; Tramont, 1989). *N. gonorrhoeae* are obligate human pathogens, and therefore the most relevant testing of vaccine candidates can only be performed in humans. Nevertheless, certain useful predictions have been gleaned using animal models of infection. The chimpanzee model first developed by Lucas *et al.* (1971) mimics uncomplicated gonococcal infection in humans with respect to incubation period, certain clinical manifestations, sexual transmission, local cellular response in the exudate, *in vivo* culture characteristics of the infected material, and systemic immune responses. Arko *et al.* (1976) inoculated male chimpanzees parenterally with a formalinized whole-cell vaccine and showed strain-related resistance to gonococcal challenge. Buchanan and Arko (1977) used purified outer membrane (prepared from the same strains used in the chimpanzees noted above) to immunize guinea pigs, and again showed strain-specific immunity in a chamber model.

In order to assess LOS as a vaccine candidate its antibody response was studied. Antibody against LOS has been shown to have several important functions: complement activation, bactericidal activity (Apicella *et al.*, 1986; Rice *et al.*, 1977, 1996) and opsonic activity (Densen *et al.*, 1987; Ross and Densen, 1985). Although these properties make LOS an excellent candidate, considerable LOS heterogeneity is displayed by the gonococcus. Certain other limitations preclude the use of LOS as a vaccine antigen. First, the toxicity of the lipid A moiety of LOS limits its potential use as a vaccine immunogen. Second, purification of oligosaccharide (OS) from LOS may modify its antigenicity (Yamasaki *et al.*, 1988) and may result in a T cell-independent saccharide antigen that may be poorly immunogenic (Kayhty *et al.*, 1984; Moiser *et al.*, 1977). Alternative strategies to the use of pure saccharide vaccine may include conjugation to a protein carrier and the production of anti-idiotype monoclonal antibodies (Mabs), which may act as function 'molecular mimics' (Tramont *et al.*, 1981).

Buchanan *et al.* (1973) showed that infected patients made antibodies against purified pili, and Tramont *et al.* (1981) showed that antibody generated by the vaccination of volunteers with a gonococcal pilus could block the attachment of gonococci to human epithelial cells *in vitro*. In a small trial, Brinton *et al.* (1982) reported that a parenteral gonococcal pilus vaccine was protective in a human challenge model when the homologous strain was used as the challenge organism. A large randomized placebo-controlled, double-blinded efficacy trial of purified gonococcal pilus vaccine composed of a single pilus type was tested in 3250 volunteers (US military personnel stationed in the Republic of South Korea) (Boslego *et al.*, 1991). Vaccinees developed a sustained ELISA antibody response to homologous and heterologous pili, but the titers against heterologous antigen were only 40% as high as the homologous titers. Local antibodies measured from semen were also seen against homologous and heterologous pili. However, there were no increases in antibody titers that inhibited gonococcal attachment *in vitro*, and this vaccine failed to protect men against gonococcal urethritis in an open field trial. In designing this or a subsequent vaccine trial, a graded risk of acquiring gonorrhea was not considered in susceptibility to gonorrhea. Although prevaccination serum antibody (IgG and IgA) levels against gonococcal pili initially were low in the military personnel, failure to consider antibody status against other common cross-reacting antigens may have undermined the effect of the vaccine. A placebo/control trial using a human challenge might have facilitated this analysis.

In a recent American trial that took place in 1985 (Tramont, 1989), a placebo/control human challenge trial was performed. Sixty-three male volunteers either were immunized with a vaccine prepared from the outer membranes of a single strain of *N. gonorrhoeae* or were given a placebo. These men were challenged intraurethrally with viable organisms 2–4 weeks after completing the vaccination course. No significant difference in infection after challenge was observed in the two groups, but resistance to infection was high: 40% in vaccinees and 32% in placebo recipients. The goal for this outer membrane-derived vaccine preparation was enrichment of the Por protein. This study emphasizes that the use of a placebo group and stratification for pre-existing immunity will be important considerations in the future design of gonococcal vaccine trials that involve vaccine candidates to which there may already be partial immunity.

Experimental challenge of humans to evaluate vaccine candidates offers several advantages. These are: (1) the ability, in some cases, to use genetically constructed deletional mutants, deficient in the vaccine-targeted epitope(s), to establish the connection between virulence and the epitope; (2) the ability to assess naturally acquired immunity (placebo recipients), defined in terms of the vaccine candidate in protecting against disease; and (3) the ability to stratify participants on the basis of pre-existing immunity (placebo recipients) and postimmunization responses (vaccine recipients), while challenging participants with different-sized inocula to assess the quantitative role of immunity (Rice *et al.*, 1996).

Genital Herpes Infection

Herpes viruses (types 1 and 2) are acquired by direct contact of the mucosal surface with the viral particle. The fundamental principle of pathogenesis is the propensity of the virus to replicate at the mucosal surface. With viral replication at the site of infection, either an intact virion or, more simply, the naked capsid enters the nerve termini and is carried by retrograde axonal flow to the dorsal root ganglia where, after a round of viral replication, latency is established. After latency is established, reactivation can occur with a proper provocative stimulus and the virus becomes evident at mucocutaneous sites, appearing as skin vesicles or mucosal ulcers. From the site of infection the virus can disseminate in either the immunocompromised host or the neonate. Thus, the local immune response should play a decisive role in the prevention of infection, particularly recurrent infections. However, the herpes viruses have a unique ability to become latent by residing in immunoprivileged sites. They are therefore sheltered from the local immune response until there is viral reactivation. Thus, the initial and subsequent secondary immune responses, particularly those initiated during recurrence, are poorly understood. The clinical manifestations of primary and recurrent genital herpes have been described (Corey *et al.*, 1983, 1985; Corey and Spear, 1986a,b).

The local humoral and cellular immune responses are relevant to the modulation and neutralization of infectious viral particles. This is particularly true of the protective effect of local immunity against viruses that invade the human respiratory and gastrointestinal tracts. In these sites there is local expression of a cellular immunity and the production of secretory antibodies, particularly IgA. The local cell-mediated immune response to HSV includes local cell infiltration, local cytokine production and changes in the levels of various immunoregulatory molecules. As herpes infection is normally restricted to the epidermis, in recurrent cutaneous lesions dermal and epidermal cellular infiltrates present in and around herpetic lesions. In the infiltrates of cutaneous lesions (Cunningham *et al.*, 1985) there is a preponderance of CD4$^+$ lymphocytes in all stages, and the CD4/CD8 ratio is approximately 1–4 times greater than that seen in peripheral blood. Cells with morphological and ultrastructural similarities to those of large granular lymphocytes (LGLs) have been seen in herpetic lesions (Heng *et al.*, 1989), and are probably similar to those seen in mucosal surfaces. There is evidence that (HLA) class II expression is upregulated in cutaneous infection, and this may augment the class II restricted CD4$^+$ lymphocyte-mediated specific CMI response. There have been a number of attempts to correlate non-specific CMI responses (NK (natural killer) ADCC (antibody-dependent cellular cytotoxicity) and IFN-α) or specific CMI (lymphoproliferation, cytotoxicity or IFN-γ) with HSV disease. These associations have included incidence, severity and/or timing of the disease. However, these studies have looked mainly at systemic manifestations of these responses, and almost nothing is known about what occurred locally in regard to these CMI responses. Because genital HSV infections are characterized by epithelial cell replication, intracellular growth and cell to cell spread, successful control of the infection will probably involve both systemic and mucosal immunity. The understanding of the anti-HSV-2 CMI at the mucosal level is lacking.

Humoral mucosal immunity to genital herpes has been best studied in cervical secretions. Secretory antibodies to HSV-2 have been demonstrated by several techniques, including neutralization, radioimmunoassay (RIA), enzyme-link immunoassay (EIA) and indirect fluorescent assay (IFA) (Ashley and Koelle, 1992). Ashley *et al.*, (1994) demonstrated protein-specific antibodies to HSV-2 in the cervicovaginal secretions of women with primary genital HSV-2. Using Western blot analysis cervical IgG and IgA responses to the HSV-2 proteins VP5, gB and gD were detected in most patients within 2 weeks of onset. Cervical IgM to most HSV-2 proteins appeared within 6–10 days. Cervical IgG and IgA persisted for weeks, but the IgM response waned. This phenomenon was associated with isotype switching in the local compartment converting IgM B cells to IgA-committed B cells (Strober and Harriman, 1991). Cervical IgM responses were not detected when symptomatic recurrences developed. It was also observed that women with oral HSV-1 infections have cervical antibodies to HSV-1, and anamnestic responses to HSV proteins have been documented when such women acquired genital HSV-2 infection. Thus, it appears that there are a series of complex antibody responses in the human female genital tract induced in response to HSV infections. The association of mucosal antibody with clearance of HSV-2 from the cervix raises the possibility that site-specific immunity or mucosal immune stimulation by oral vaccines may be effective. However, until the contribution of mucosal immunity in natural genital HSV infection is understood, the impact of oral vaccines for this infection remains problematic.

This brief overview of the immune responses induced by the three most common causes of sexually transmitted diseases indicates that this field of study is in its infancy. Although little is known about the mucosal aspects of these three diseases, even less information exists concerning the mucosal defense mechanisms induced by other sexually transmitted infections of the female genital tract, including syphilis, chancroid and cytomegalovirus.

The increasing number of women being infected by the human immuno-deficiency virus through heterosexual intercourse suggests that understanding the local responses is a critical challenge for future research. Hopefully, this challenge will be met and the goal of eradicating sexually transmitted diseases can be achieved.

VACCINE CONCEPTS FOR THE FUTURE: DNA VACCINES

DNA vaccination has been studied with increasing intensity as a potentially successful technology for the future ever since the initial demonstration that direct transfection *in vivo* with DNA could be used to express foreign proteins and thereby induce protective immune responses (Donnelly *et al.*, 1997a).

DNA vaccines represent a novel means of expressing antigens *in vivo* for the generation of both humoral and cellular immune responses. They have elicited protective immunity in a number of preclinical models of diseases (Donnelly *et al.*, 1997b). DNA vaccines employ genes encoding proteins of pathogens or tumors, rather than using the proteins themselves, a live replica-ting vector or an attenuated version of the pathogen itself. DNA vaccines consist of a bacterial plasmid with a strong viral promoter, the gene of interest, and a polyadenylation/transcriptional termination sequence. The plasmid is grown in bacteria (*Escherichia coli*), purified, dissolved in a saline solution, and then simply injected into the host. The DNA plasmid is taken up by host cells, where the encoded protein is made. The plasmid is made without an origin of repli-cation that is functional in eukaryotic cells: such plasmids neither replicate in the mammalian host nor integrate within the chromosomal DNA of the animal. The role of antigen-presenting cells in the immune responses that ensue upon expression of the encoded foreign protein have been studied for intramuscular immunization, where the muscle cells are the primary protein-expressing cell type, but not for other routes of immunization, such as that employing a 'gene gun' to propel gold beads coated with DNA into the epidermis.

Protective immunity induced by a DNA vaccine was first demonstrated in a mouse model of influenza. DNA encoding influenza nucleoprotein (NP) induced high-titer antibodies and cytotoxic T lymphocytes (CTL) and cross-strain protection against a lethal challenge with the virus (Ulmer *et al.*, 1993). Anti-NP antibodies were detected by an ELISA and endpoint titers of $> 10^6$ were measured, indicating that DNA vaccination is a very effective means of gen-erating polyclonal antibodies without the need for purified protein. Both primary CTL (non-specifically stimulated with concanavalin A and IL-2) and antigen-restimulated CTL (restimulated with peptide-pulsed or virus-infected syngeneic spleen cells) were detected *in vitro*. The cross-strain protection was conferred by CMI responses, as adoptive transfer of anti-NP antibodies into naive mice did not protect them from challenge (Ulmer *et al.*, 1993), and depletion of T cells *in vivo* prior to challenge abrogated NP DNA-induced protection. Humoral and CMI responses to NP are long lived and have been detected 2 years after immunization of mice with NP DNA.

In other areas of infectious disease, immune responses to antigens encoded by and delivered as plasmid DNAs have been raised in a variety of species (e.g. chickens, mice, ferrets, cattle and non-human primates) to a variety of antigens: influenza virus hemagglutinin, matrix protein and nucleoprotein, human immunodeficiency virus 1 (HIV-1) gp120 and gp160, bovine herpes virus gIV, rabies virus surface glycoprotein, hepatitis B virus (HBV) surface antigen, hepatitis C virus (HCV) core antigen, malarial circumsporzoite protein, *Mycobacterium tuberculosis* heat shock protein 65 and *Mycoplasma pulmonis* antigens. Moreover, in different models of viral diseases inoculation with plasmid DNA has been found to be protective.

Vaccines are also being looked at for use as contraceptives, as well as preventation against sexually transmitted diseases. There is little information on the induction of protective immune responses in the genital tract mucosa, making progress towards a vaccine goal very slow. Livingston *et al.*, (1998) tested the *in vivo* transfection of a model DNA-based antigen delivered by gene gun technology to induce an antibody response detectable in vaginal secretions. Female rats were immunized with plasmids encoding human growth hormone (HGH) under the control of a cytomegalovirus promoter by way of vaginal mucosa, Peyer's patch and/or abdominal skin routes. HGH was found at all three sites, showing that transfection *in vivo* with cytomegalovirus/HGH was successful. The level of plasmid in the vaginal tissues was very similar to that found in the skin. The rats immunized vaginally maintained vaginal immunoglobulin A and G antibody titers to HGH for a minimum of 14 weeks. Those immunized by other routes showed inconsistencies, with no significant vaginal antibody titers after 6 weeks. This study did show that DNA-based immunization by the gene gun method may be effective in inducing local immunity in the female genital tract.

The question becomes, why is gene vaccination so effective at inducing long-lived and potent immune responses? Potential answers to this question are provided by considering some attributes of an effective vaccine: (1) it provides an immunogenic antigen; (2) it is processed via major histocompatibility complex (MHC) classes I and II; (3) it stimulates immunological memory; and (4) it contains an adjuvant.

In studies involving a murine model of HSV infections by Kuklin *et al.* (1997), plasmid DNA encoding glycoprotein B(gB) of HSV-1 was used to evaluate both systemic and distal mucosal immunity. These investigators used an intranasal (i.n.) route of administration and compared it to an intramuscular (i.m.) delivery route. Their results demonstrated that i.n. administration of DNA induces both systemic and mucosal immunity at the genital and enteric levels. However, the levels of immunity induced were less than ideal and did not approach those resulting from live virus infection. As regards systemic immunity, the i.m. route also proved superior. However, i.m. immunization was less effective than i.n. in inducting mucosal IgA. One aim of mucosal administration is to exploit the well-known common mucosal defense mechanisms and induce barrier levels of immunity at multiple mucosal surfaces. This becomes an important concept for preventing invasion by pathogens that enter the body via mucosa and cause damage at the mucosal site.

REFERENCES

Ånestad, G., Lunde, O., Moen, M., and Dalaker, K. (1987). Infertility and chlamydial infection. *Fertil. Steril.* **48**: 787–790

Apicella, M.A., Shero, M., Jarvis, G.A., Griffiss, J.M., Mandrell, R.E., and Schneider, H. (1987). Phenotypic variation in epitope expression of *Neisseria gonorrhoeae* lipooligosaccharide. *Infect. Immun.* **55**: 1755–1761

Apicella, M.A., Westerink, M.A., Morse, S.A., Schneider, H., Rice, P.A., and Griffiss, J.M. (1986). Bactericidal antibody response of normal human serum to the lipooligosaccharide of *Neisseria gonorrhoeae*. *J. Infect. Dis.* **153**: 520–526

Arko, R.J., Duncan, W.P., Brown, W.J., Peacock, W.L., and Tomizawa, T. (1976). Immunity in infection with *Neisseria gonorrhoeae*: duration and serological response in the chimpanzee. *J. Infect. Dis.* **133**: 441–447

Arya, O.P., Mallinson, H., and Goddard, A.D. (1981). Epidemiological and clinical correlates of chlamydial infection of the cervix. *Br. J. Vener. Dis.* **57**: 118–124

Ashley, R.L., Corey, L., Dalessio, J. *et al.* (1994). Protein-specific cervical antibody responses to primary genital herpes simplex virus type 2 infections. *J. Infect. Dis.* **170**: 20–26

Ashley, R., and Koelle, D.M. (1992). In: 'Sexually Transmitted Diseases', (ed T.C. Quinn), Immune responses to genital herpes infection, pp. 201–238, Raven Press, New York

Bartlett, J.G., Onderdonk, A.B., Drude, E. *et al.* (1977). Quantitative bacteriology of the vaginal flora. *J. Infect. Dis.* **136**: 271–277

Bell, S.D., Jr. and Theobald, B. (1962). Differentiation of trachoma strains on the basis of immunization against toxic death in mice. *Ann. N.Y. Acad. Sci.* **98**: 337–346

Bessen, D., and Gotschlich, E.C. (1986). Interactions of gonococci with HeLa cells: attachment and the role of protein II. *Infect. Immun.* **54**: 154–160

Boslego, J.W., Tramont, E.C., Chung, R.C. *et al.* (1991). Efficacy trial of a parenteral gonococcal pilus vaccine in men. *Vaccine* **3**: 154–162

Bouvet, J-P., Belec, L., Pires, R., and Pillot, J. (1994). Immunoglobulin G antibodies in human vaginal secretions after parenteral vaccination. *Infect. Immun.* **62**: 3957–3961

Brinton, C.C., Jr., Wood, S.W., Brown, A. *et al.* (1982). In: 'Seminars in Infectious Diseases Vol. 4, Bacterial Vaccines' (ed, J.B. Robbins, J.H. Hill and J.C. Sadoff), The development of neisserial pilus vaccine for gonorrhea and meningococcal meningitis, pp. 140–150, Thieme-Stratton, New York

Britigan, B.E., Cohen, M.S., and Sparling,. P.F. (1985). Gonococcal infection: a model of molecular pathogenesis. *N. Engl. J. Med.* **312**: 1683–1694

Brunham, R.C. (1994). In: 'Chlamydial Infections' (ed. J. Orfilia, G.I. Byrne, M.A. Chernesky, J.T. *et al.*) Vaccine Design for the Prevention of *Chlamydia trachomatis* infection, pp. 73–82, Societa Editrice Escalapio, Bologna

Brunham, R.C., Peeling, R., Maclean, I., McDowell, H., Perrson, K., Osser, S. (1987). Postabortal *Chlamydia trachomatis* salpingitis: correlating risk with antigen-specific serological responses and with neutralization. *J. Infect. Dis.* **155**: 749–755

Buchanan, T.M., and Arko, R.J. (1977). Immunity to gonococcal infection induced by vaccination with isolated outer membranes of *Neisseria gonorrhoeae* in guinea pigs. *J. Infect. Dis.* **135**: 879–887

Buchanan, T.M., Swanson, J., Holmes, K.K., Kraus, S.J., and Gotschlich, E.C. (1973). Quantitative determination of antibody to gonococcal pili. Changes in antibody levels with gonococcal infection. *J. Clin. Invest.* **52**: 2896–2909

Burczak, J.D., Chernesky, M.A., Tomazic-Allen, S.T. *et al.* (1994). In: 'Chlamydial Infections', (ed. J. Orfilia, G.I. Byrne, M.A., Chernesky, J.T. *et al.*), Application of LCR to the detection of *Chlamydia trachomatis* in urogenital specimens from men and women, pp. 322–325, *Societa Editrice Escalapio*, Bologna

Caldwell, H.D. (1993). A polio virus hybrid expressing a neutralization epitope from the major outer membrane protein of *Chlamydia trachomatis* is highly immunogenic. *Infect. Immun.* **61**: 4406–4414

Caldwell, H.D., Kromhout, J. and Schachter, J. (1981). Purification and partial characterization of the major outer membrane protein of *Chlamydia trachomatis*. *Infect. Immun.* **31**: 1161–1176

Caldwell, H. and Schachter, J. (1982). Antigenic analysis of the major outer membrane protein of *Chlamydia* spp. *Infect. Immun.* **35**: 1024–1031

Centers for Disease Control. (1979). Non reported sexually transmitted diseases. *MMWR* **28**: 61

Chen, J.C.R. and Stephens, R.S. (1997). *Chlamydia trachomatis* glycosaminoglycan-dependent and independent attachment to eukaryotic cells. *Microbial. Pathol.* **22**: 23–30

Cohen, M. and Sparling, P.F. (1992). Mucosal Infection with *Neisseria gonorrhoeae*: Bacterial adaptation and mucosal defenses. *J. Clin. Invest.* **89**: 1699–1705

Cohen, M.S., Cannon, J.G., Jerse, A.E., Charniga, L., Isbey, S.F. and Whicker, L. (1994). Human experimentation with *Neisseria gonorrhoeae*: rationale, methods, and implications for the biology of infection and vaccine development. *J. Infect. Dis.* **169**: 532–537

Cone, R.W., Hobson, A.C., Brown, Z. *et al.* (1994). Frequent detection of genital herpes simplex virus DNA by polymerase chain reaction among pregnant women. *JAMA* **272**: 792–796

Conlan, J.W., Ferris, S., Clarke, I.N. and Ward, M.E. (1990). Isolation of recombinant fragments of the major outer-membrane protein of *Chlamydia trachomatis:* Their potential as subunit vaccines. *J. Gen. Microbiol.* **136**: 2013–2020

Connell, T.D., Black, W.J., Kawula, T.H. *et al.* (1988). Recombination among protein II genes of *Neisseria gonorrhoeae* generates new coding sequences and increases structural variability in the protein II family. *Mol. Microbiol.* **2**: 227–236

Cooper, M., Duncan-Moehling, S. and Moticka, E. In: 'Neisseria 1990', (ed. M., Achtman, P. Kohl, C. Marchal, G. Morelli, A. Seiler, B. Thiesen), Ability of cells of the human fallopian tube to act as antigen presenting cells, pp. 573–578, Walter de Gruyter, New York

Cooper, M.D., McGee, Z.A., Mulks, M.H., Koomey, J.M. and Hindman, T.L. (1984). Attachment to and invasion of human fallopian tube mucosa by an IgA1 protease-deficient mutant of *Neisseria gonorrhoeae* and its wild-type parent. *J. Infect. Dis.* **150**: 737–744

Corey, L. (1985). In: 'The Herpes Viruses,. Vol 4', (ed. B. Roizman and C. Lopez), The natural history of genital herpes simplex virus, pp. 1–31, Plenum Press, New York

Corey, L., Adams, H.G., Brown, Z.A. and Holmes, K.K. (1983). Genital herpes simplex virus infection: clinical manifestations, course and complications. *Ann. Intern. Med.* **98**: 958–972

Corey, L. and Spear, P. (1986a). Infections with herpes simplex viruses. *N. Engl. J. Med.* **314**: 686–691

Corey, L. and Spear, P.G. (1986b). Infections with herpes simplex viruses. *N. Engl. J. Med.* **314**: 749–757

Cruickshank, R. and Sharman, A. (1934). The biology of the vagina in the human subject. II: The bacterial flora and secretion of the vagina at various age-periods, and their relation to glycogen in the vaginal epithelium. *J. Obstet. Gynaecol. Br. Emp.* **41**: 208–226

Cunningham A.L., Turner, R.R., Miller, A.C., Para, M.F., and Merigan, T.C. (1985). Evolution of recurrent herpes simplex lesions: an immunohistologic study. *J. Clin. Invest.* **75**: 226–233

Danilition, S.L., Maclean, I.W., Peeling, R., Winston, S. and Brunham, R.C. (1990). The 75-kilodalton protein of *Chlamydia trachomatis:* a member of the hear shock protein 70 family? *Infect. Immun.* **58**: 189–196

Dean, D., Schachter, J., Dawson, C.R. and Stephens, R.S. (1992). Comparison of the major outer membrane protein variant sequence regions of B/Ba isolates: A molecular epidemiologic approach to *Chlamydia trachomatis* infections. *J. Infect. Dis.* **166**: 383–392

Densen, P., Gulati, S. and Rice, P.A. (1987). Specificity of antibodies against *Neisseria gonorrhoeae* that stimulate neutrophil chemotaxis. Role of antibodies directed against lipooligosaccharides. *J. Clin. Invest.* **80**: 78–87

Donnelly, J.J., Ulmer, J.B. and Liu, M.A. (1997a). DNA Vaccines. *Life Sci.* **60**: 163–172

Donnelly, J.J., Ulmer, J.B., Shiver, J.W. and Liu, M.A. (1997b). DNA Vaccines. *Ann. Rev. Immunol.* **15**: 617–648

Eschenbach D.A., Davick, P.R., Williams, B.L. *et al.* (1989). Prevalence of hydrogen peroxide-producing *Lactobacillus* species in normal women and women with bacterial vaginosis. *J. Clin. Microbiol.* **27**: 251–256

Fuster, C.D. and Neinstein, L.S. (1987). Vaginal *Chlamydia trachomatis* prevalence in sexually abused prepubertal girls. *Pediatrics* 79: 235–238

Galask, R.P. and Larsen, B. (1981). Identifying and treating genital tract infections in post-menopausal women. *Geriatrics* 36: 69–79

Glimcher, L.H., Longo, D.L., Green, I., and Schwartz, R.H. (1981). Murine syngeneic mixed lymphocyte response. I. Target antigens are self Ia molecules. *J. Exp. Med.* 154: 1652–1670

Goldacre, M., Loudon, J.N., Watt, B. *et al.* (1978). Epidemiology and clinical significance of cervical erosion in women attending a family planning clinic. *Br. Med. J.* 1: 748–750

Grayston, J.T. and Wang, S-P. (1978). The potential for vaccine against infection of the genital tract with *Chlamydia trachomatis*. *Sex. Transm. Dis.* 5: 73–77

Grifo, J.A., Jeremias, J., Ledgerm W.J. and Witkin, S.S. (1988). Interferon-y in the diagnosis and pathogenesis of pelvic inflammatory disease. *Am. J. Obstet. Gynecol.* 160: 26–31

Gump, D.W., Gibson, M. and Ashikaga, T. (1983). Evidence of prior pelvic inflammatory disease and its relationship to *Chlamydia trachomatis* antibody and intrauterine contraceptive device use in infertile women. *Am. J. Obstet. Gynecol.* 146: 153–159

Gutman, L.T. and Holmes, K.K. (1989). In 'Infectious Diseases of the Fetus and Newborn Infant, 3rd edn.' (ed. J.S. Remington and J.O. Klein). Gonococcal Infections, Saunders, Philadelphia

Haas, R. and Meyer, T.F. (1986). The repertoire of silent pilus genes in *Neisseria gonorrhoeae*: Evidence for gene conversion. *Cell* 44: 107–115

Hagblom, P., Segal, E., Billyard, E., and So, M. (1985). Intragenic recombination leads to pilus antigenic variation in *Neisseria gonorrhoeae*. *Nature* 315: 156–158

Hammerschlag, M.R., Alpart, S., Roshar, I. *et al.* (1978). Microbiology of the vagina in children: normal and potentially pathogenic organisms. *Pediatrics* 62: 57–62

Hardy, G.C. (1941). Vaginal flora in children. *Am. J. Dis. Child.* 62: 939–954

Harrison, H.R., Costin, M. and Meder, J.B. (1985). Cervical *Chlamydia trachomatis* infection in university women: Relationship to history, contraception, ectopy and cervicitis. *Am. J. Obstet. Gynecol.* 153: 244–251

Hatch, T.P., Al-Hossainy, E. and Silverman, J.A. (1982). Adenine nucleotide and lysine transport in *Chlamydia psittaci*. J. Bacteriol. 150, 662–670

Hayes, L.J., Bailey, R.L., Mabey, D.C.W. *et al.* (1992). Genotyping of *Chlamydia trachomatis* from a trachoma-endemic village in the Gambia by a nested polymerase chain reaction: Identification of strain variants. *J. Infect. Dis.* 166: 1173–1177

Hayes, L.J., Conlan, J.W., Everson, J.S., Ward, M.E. and Clarke, I.N. (1991). *Chlamydia trachomatis* major outer membrane protein epitopes expressed as fusions with LamB in an attenuated aroA strain of *Salmonella typhimurium*; their application as potential immunogens. *J. Gen. Microbiol.* 137: 1557–1564

Heng, M.C.Y., Allen, S.G., Heng, S.Y., Matsuyama, R. and Fazier, J. (1989). An electron microscopic study of the epidermal infiltrate in recurrent herpes simplex. *Clin. Exp. Dermatol.* 14: 199–202

Holmes, K.K. (1990). In: 'Sexually Transmitted Diseases. 2 ed' (ed. K.K. Holmes, P. Mårdh, P.F. Sparling, P.J. Wiesner, W. Cates, Jr., S.M. Lemon and W.E. Stamm), Lower genital tract infections in women: cystitis, urethritis, vulvovaginitis and cervicitis, pp. 527–545, McGraw Hill Information Services Company, New York

Ison, C.A., Hadfield, S.G., Bellinger, C.M., Dawson, S.G. and Glynn, A.A. (1986). The specificity of serum and local antibodies in female gonorrhea. *Clin. Exp. Immun.* 65: 198–205

Jacobson, L., and Weström, L. (1969). Objectivized diagnosis of pelvic inflammatory disease. *Am. J. Obstet. Gynecol.* 105: 1088–1098

Jerse, A.E., Cohen, M.S., Drown, P.M. *et al.* (1994). Multiple gonococcal opacity proteins are expressed during experimental urethral infection in the male. *J. Exp. Med.* 179: 911–920

Joseph, T.D., and Bose, S.K. (1991). A heat-labile protein of *Chlamydia trachomatis* binds to HeLa cells and inhibits the adherence of chlamydia. *Proc. Natl. Acad. Sci. USA* 88: 4054–4058

Kayhty, H., Karanko, V., Peltola, H. and Makela, P.H. (1984). Serum antibodies after vaccination with *Haemophilus influenzae* tube b capsular polysaccharide and responses to reimmunization: no evidence of immunologic tolerance or memory. *Pediatrics* 74: 857–865

Kellogg, D.S., Cohen, I.R., Norins, L.C., Schroeter, A.L. and Reising, C. (1968). *Neisseria gonorrhoeae* II. Colonial variation and pathogenicity during 35 months *in vitro*. *J. Bacteriol.* 96: 596–605

Kent, G.P., Harrison, H.R., Berman, S.M. and Kennlyside, R.A. (1988). Screening for *Chlamydia trachomatis* infection in a sexually transmitted disease clinic: Comparison of diagnostic tests with clinical and historical risk factors. *Sex. Transm. Dis.* **15**: 51–57

Kuklin, N., Daheshia, M., Karem, K., Manickan, E. and Rouse, B. (1997). Induction of mucosal immunity against herpes simplex virus by plasmid DNA immunization. *J. Virol.* **71**: 3138–3145

Lammel, C.J., Sweet, R.L., Rice, P.A. *et al.* (1985). Antibody–antigen specificity in the immune response to infection with *Neisseria gonorrhoeae*. *J. Infect. Dis.* **152**: 990–1001

Leiblum, S., Bachmaun, G., Kemmann, E., Colburn, D. and Swartzman, L. (1983). Vaginal atrophy in the postmenopausal women. *JAMA* **249**: 2195–2198

Livingston, J.B., Lu, S., Robinson, H. and Anderson, D.J. (1998). Immunization of the female genital tract with a DNA-based vaccine. *Infect. Immun.* **66**: 322–329

Lucas, C.T., Chandler, Jr., F., Martin, Jr., J.E. and Schmale, J.D. (1971). Transfer of gonococcal urethritis from man to chimpanzee. An animal model for gonorrhea. *JAMA* **216**: 1612–1614

Lundemose, A.G., Kay, J.E. and Pearce, J.H. (1993). *Chlamydia trachomatis* Mip-like protein has peptidyl-prolyl *cis-trans* isomerase activity that is inhibited by FK506 and rapamycin and is implicated in initiation of chlamydial infection. *Mol. Microbiol.* **7**: 777–783

Lundemose, A.G., Rouch, D.A., Birkelund, S., Christiansen, G. and Pearce, J.H. (1992). *Chlamydia trachomatis* Mip-like protein. *Mol. Microbiol.* **6**: 2539–2548

Makino, S., vanPutten, J.P. and Meyer, T.F. (1991). Phase variation of the opacity outer membrane protein controls invasion by *Neisseria gonorrhoeae* into human epithelial cells. *EMBO J.* **10**: 1307–1315

McGee, Z.A., Clemens, C.M., Jensen, R.L., Klein, J.J., Barley, L.R. and Gorby, G.L. (1992). Local induction of tumor necrosis factor as a molecular mechanism of mucosal damage by gonococci. *Microbial. Pathol.* **12**: 333–341

McGee, Z.A., Gregg, C.R., Johnson, A.P., Kalter, S.S. and Taylor-Robinson, D. (1990). The evolutionary watershed of susceptibility to gonococcal infection. *Microbial Pathol.* **9**: 131–139

McGee, Z.A., Gorby, G.L., Wyrick, P.B., Hodinka, R. and Hoffman L.H. (1988). Parasite-directed endocytosis. *Rev. Infect. Dis.* 10, S311–S316

McGee, Z.A. and Woods, Jr., M.L. (1987). Use of organ cultures in microbiological research. *Ann. Rev. Microbiol.* **41**: 294–300

Moiser, D.E., Zaldivar, N.M., Goldings, E., Mond, J., Scher, I. and Paul, W.E. (1977). Formation of antibody in the newborn mouse: study of T-cell independent antibody response. *J. Infect. Dis.* 136, S14–19

Morrison, R.P., Belland, R.J., Lyng, K. and Caldwell, H.D. (1989). Chlamydial disease pathogenesis: The 57 kD chlamydial hypersensitivity antigen is a stress response protein. *J. Exp. Med.* **170**: 1271–1283

Moulder, J.W. (1991). Interactions of chlamydiae and host cells *in vitro*. *Microbiol. Rev.* **55**: 143–190

Murdin, A.D., Su, H., Manning, D.S., Klein, M.H., Parnell, M.J. and Caldwell, H.D. (1993). A polio virus hybrid expressing a neutralization epitope from the major outer membrane protein of *Chlamydia trachomatis* is highly immunolgenic. *Infect. Immun.* **61**: 4406–4414

Murphy, G.L., Connell, T.D., Barritt, D.S., Koomey, M. and Cannon, J.G. (1989). Phase variation of gonococcal protein II: regulation of gene expression by slipped-strand mispairing of a repetitive DNA sequence. *Cell* **56**: 539–547

Osborne, N.G., Wright, R.C. and Grubin, L. (1979). Genital bacteriology: A comparative study of premenopausal women with post menopausal women. *Am. J. Obstet. Gynecol.* **135**: 195–198

Paavonen, J. (1979). *Chlamydia trachomatis* induced urethritis in female partners of men with non gonococcal urethritis. *Sex. Transm. Dis.* **6**: 69–71

Paavonen J., Critchlow, C.W., DeRouen, T. *et al.* (1986). Etiology of cervical inflammation. *Am. J. Obstet. Gynecol.* **154**: 556–564

Patton, D.L. (1989). Silent pelvic inflammatory disease: Morphological, physiological, and serological findings. *Am. J. Obstet. Gynecol.* **3**: 622–630

Pixley, E. (1976). Basic morphology of the prepubertal and youthful cervix: Topographic and histologic features. *J. Reprod. Med.* **16**: 221–230

Qu, Z., Cheng, X., de la Maza, L.M. and Peterson, E.M. (1993). Characterization of a neutralizing monoclonal antibody directed at variable domain I of the major outer membrane protein of *Chlamydia trachomatis* C- complex serovars. *Infect. Immun.* **61**: 1365–1370

Rake, G. and Jones, H.P. (1944). Studies on lymphogranuloma venereum. II. The association of specific toxins with agents of the lymphogranuloma-psittacosis group. *J. Exp. Med.* **79**: 463–485

Raulston, J.E., Davis, C.H., Schmiel, D.H., Morgan, M.W. and Wyrick, P.B. (1993). Molecular characterization and outer membrane association of a *Chlamydia trachomatis* protein related to the hsp70 family of proteins. *J. Biol. Chem.* **268**: 23139–23147

Rettig, P.J. and Nelson, J.D. (1981). Genital tract infection with *Chlamydia trachomatis* in prepubertal children. *J. Pediatrics* **99**: 206–210

Rice, P.A., Gulati, S., McQuillen, D.P. and Ram, S. (1996). In: 'Tenth International Pathogenic *Neisseria* Conference', (eds. W.D. Zollinger, C.E. Frasch, and C.D. Deal), Is there protective immunity to gonococcal disease?, pp. 3–8

Rice, P.A. and Kasper, D.L. (1977). Characterization of gonococcal antigens responsible for induction of bactericidal antibody in disseminated infection. *J. Clin. Invest.* **60**: 1149–1158

Robbins, J.B., Schneerson, R. and Szu, S.C. (1995). Perspective: Hypothesis: Serum IgG antibody is sufficient to confer protection against infectious diseases by inactivating the inoculum. *J. Infect. Dis.* **171**: 1387–1398

Robinson, E.J., Clemens, C.M., Schoolnik, G.K. and McGee, Z.A. (1989). Probing the surface of *Neisseria gonorrhoeae*: Immunoelectron microscopic studies to localize cyanogen bromide fragment 2 in gonococcal pili. *Mol. Microbiol.* **3**: 57–64

Ross, S.C. and Densen, P. (1985). Opsonophagocytosis of *Neisseria gonorrhoeae*: interaction of local and disseminated isolates with complement and neutrophils. *J. Infect. Dis.* **151**: 33–41

Rothbard, J.B., Fernandez, R., Wang, L., Teng, N.N. and Schoolnik, G.K. (1985). Antibodies to peptides corresponding to a conserved sequence of gonococcal pilins block bacterial adhesion. *Proc. Natl. Acad. Sci. USA* **82**: 915–919

Schafer, M.A., Beck, A., Blain, B. *et al.* (1984). *Chlamydia trachomatis*: Important relationships to race, contraception, lower genital tract infection, and Papanicolaou smear. *J. Pediatrics* **104**: 141–146

Schneider, H., Griffiss, J.M., Boslego, J.W., Hitchcock, P.J., Zahos, K.M. and Apicella, M.A. (1991). Expression of paragloboside-like lipooligosaccharides may be a necessary component of gonococcal pathogenesis in men. *J. Exp. Med.* **174**: 1601–1605

Schoolnik, G.K., Fernandez, R., Tai, J.Y., Rothbard, J., and Gotschlich, E.C. (1984). Gonococcal pili: primary structure and receptor binding domain. *J. Exp. Med.* **159**: 351–370

Sellors, J.W., Mahony, J.B., Chernesky, M.A., and Rath, D.J. (1988). Tubal factor infertility: An association with prior chlamydial infection and asymptomatic salpingitis. *Fertil. Steril.* **49**: 451–457

Singer, A. (1975). The uterine cervix from adolescence to menopause. *Br. J. Obstet. Gynaecol.* **82**: 81–99

So, M., Billyard, E., Deal, C. *et al.* (1985). Gonococcal pilus: genetics and structure. *Curr. Topics Microbiol. Immunol.* **118**: 13–28

Stamey, T.A. (1980). In 'Pathogenesis and treatment of urinary tract infections', Williams & Wilkins, Baltimore

Stamm, W.E., Wagner, K.F., Amsel, R. *et al.* (1980). Causes of acute urethral syndrome in women. *N. Engl. J. Med.* **303**: 409–415

Stephens, R.S., Mullenbach, G., Sanchez-Pescador, R. and Agabian, N. (1986). Sequence analysis of the major outer membrane protein gene from *Chlamydia trachomatis* serovar L^2. *J. Bacteriol.* **168**: 1277–1282

Stephens, R., Sanchez-Pescador, R., Wagar E.A., Inouye, C. and Urdea, M.S. (1987). Diversity of *Chlamydia trachomatis* major outer membrane protein genes. *J. Bacteriol.* **169**: 3879–3885

Strober, W. and Harriman, G.R. (1991). In: 'Mucosal Immunology I. Basic Principles' (ed R.P. MacDermott and C.O. Elson), The regulation of IgA B-cell differentiation, pp. 473–494, W.B. Saunders, Philadelphia

Su, H. and Caldwell, H.D. (1992). Immunogenicity of a chimeric peptide corresponding to T helper and B cell epitopes of the *Chlamydia trachomatis* major outer membrane protein. *J. Exp. Med.* **175**: 227–235

Su, H., Watkins, N.G., Zhang, Y.X. and Caldwell, H.D. (1990). *Chlamydia trachomatis*-host cell interactions: role of the chlamydial major outer membrane protein as an adhesin. *Infect. Immun.* **58**: 1017–1025

Svensson L., Mårdh, P.A., Ahlgren, M. and Nordanskjold, F. (1985). Ectopic pregnancy and antibodies to *Chlamydia trachomatis. Fertil. Steril.* **44**: 313–317

Swanson, J. (1973). Studies on gonococcal infection. IV. Pili: their role in attachment of gonococci to tissue culture cells. *J. Exp. Med.* **137**: 571–589

Sweet, R.L., Schachter, J. and Robbie, M.O. (1983). Failure of beta-lactam antibiotics to eradicate *Chlamydia trachomatis* in the endometrium despite apparent clinical cure of acute salpingitis. *JAMA* **250**: 2641–2645

Swinker, M.I., Young, S.A., Clevenger, R.L., Neely, J.L. and Palmer, J.E. (1988). Prevalence of *Chlamydia trachomatis* cervical infection in a college gynecology clinic: Relationship to other infections and clinical features. *Sex Transm. Dis.* **15**: 133–136

Tait, I.A., Rees, E., Hobson, D., Byng, R.E. and Tweedie, M.C. (1980). Chlamydial infection of the cervix in contacts of men with nongonococcal urethritis. *Br. J. Vener. Dis.* **56**: 37–45

Taverne, J., Blyth, W.A. and Reeve, P. (1964). Toxicity of the agents of trachomatis and inclusion conjunctivitis. *J. Gen. Microbiol.* **37**: 271–275

Tjiam, K.H., Zeilmaker, G.H., Alberda, A.T. *et al.* (1985). Prevalence of antibodies to *Chlamydia trachomatis, Neisseria gonorrhoeae,* and *Mycoplasmma hominis* in infertile women. *Genito-urinary Med.* **61**: 175–178

Toye, B., Laferriere, C., Claman, P., Jessamine, P. and Peeling, R. (1993). Association between antibody to the chlamydia heat-shock protein and tubal infertility. *J. Infect. Dis.* **68**: 1236–1240

Toye, B., Zhong, G., Peeling, R. and Brunham, R.C. (1990). Immunologic characterization of a cloned fragment containing the species-specific epitope from the major outer membrane protein of *Chlamydia trachomatis. Infect. Immun.* **58**: 3909–3913

Tramont, E.C. (1989). Gonococcal Vaccines. *Clin. Microbiol. Rev.* 2, S74–77

Tramont, E.C., Sadoff, J.C., Boslego, J.W. *et al.* (1981). Gonococcal pilus vaccine. Studies of antigenicity and inhibition of attachment. *J. Clin. Invest.* **68**: 881–888

Ulmer, J.B., Donnelly, J.J., Parker, S.E. *et al.* (1993). Heterologous protection against influenza by injection of DNA encoding a viral protein. *Science* **259**: 1745–1749

Ward, M.E., Watt, P.J. and Robertson, J.N. (1974). The human fallopian tube: a model for gonococcal infection. *J. Infect. Dis.* **129**: 650–659

Weel, J.F.L., Hopman C.T.P. and vanPutten J.P.M. (1991). Insitu expression and localization of *Neisseria gonorrhoeae* opacity proteins in infected epithelial cells apparent role of Opa proteins in cellular invasion. *J. Exp. Med.* **6**: 1395–1406

Weström, L., Bengtsson, L.P. and Mardh, P.A. (1981). Incidence, trends, and risks of ectopic pregnancy in a population of women. *Br. Med. J.* **282**: 15–18

Yamasaki, R., Schneider, H., Griffiss, J.M. and Mandrell, R. (1988). Epitope expression of gonococcal lipooligosaccharide (LOS). Importance of the lipoidal moiety for expression of an epitope that exists in the oligosaccharide moiety of LOS. *Mol. Immunol.* **25**: 799–809

Yang, C.L., Maclean, I. and Brunham, R.C. (1993). DNA sequence polymorphism of the *Chlamydia trachomatis* omp 1 gene. *J. Infect. Dis.* **168**: 1225–1230

Zhang, J.P. and Stephens, R.S. (1992). Mechanism of attachment of *Chlamydia trachomatis* to eukaryotic host cells. *Cell* **69**: 861–869

Zhong, G. and Brunham, R.C. (1992). Antigenic analysis of the chlamydial 75-kilodalton protein. *Infect. Immun.* **60**: 1221–1224

Zhong, G., Smith, G.P., Berry, J. and Brunham, R.C. (1994). Conformational mimicry of a chlamydial neutralization epitope on filamentous phage. *J. Biol Chem.*, 269, 24183–24188

Zhong, G., Toth, I., Reid, R. and Brunham, R.C. (1993). Immunogenicity evaluation of a lipidic amino acid-based synthetic peptide vaccine for *Chlamydia trachomatis. J. Immunol.* **151**: 3728–3736

4
MUCOSAL IMMUNITY OF THE GENITAL TRACT

DAVID I. BERNSTEIN
AND
GREGG N. MILLIGAN

Division of Infectious Diseases,
University of Cincinnati College of Medicine,
Cincinnati, Ohio, USA

INTRODUCTION

The sexually transmitted diseases discussed in this book – indeed most infections – are initiated at mucosal surfaces. Therefore, protection of the mucosa plays a pivotal role in preventing viral, bacterial and fungal infections. The host has developed unique immune mechanisms to protect these surfaces, most notably the presence of secretory IgA antibody. It is now well accepted that the immune system can be divided into systemic and secretory components, although there is interaction between the two. The concept of mucosal immunity grew from the recognition that antibody isotypes in secretions differed from those found in the blood, IgA, most commonly dimeric IgA, being the most abundant isotype in secretions. Further research led to the concept of a common mucosal immune system whereby the induction of immune responses at one mucosal surface led to the distribution of immune cells to other mucosal surfaces (reviewed in Staats and McGhee, 1996; Phillips-Quagliata and Lamm, 1994; Mestecky, 1987). The most intensely studied surface is that of the gastrointestinal tract, whereas the least studied surface is the reproductive tract and, as will be discussed, much needs to be learned so that we can efficiently induce both humoral and cell-mediated immunity to protect this surface.

This chapter will provide an overview of the mucosal immune system and its various components. The induction of secretory IgA and its transport to mucosal surfaces, as well as T-cell mediated immunity, will be discussed. The differences between the genital tract as an inductive site for immunity versus its role as an effector site for protection of the male and female genital tracts will be emphasized. The genital tract is unique among mucosal surfaces in that it must respond to foreign proteins and protect the reproductive system from invading organisms, while (at least at selected times) not destroying other foreign cells, specifically spermatocytes and the developing fetus. This may in part account for why the genital tract is not a good inductive site, and may explain some of the unique endocrine/immunologic interactions that are observed here. The genital tract also differs from the intestinal and respiratory mucosal surfaces in the way antigens are presented, i.e. the lack of antigen-transporting epithelial or M cells. Further, although secretory IgA is present, IgG, and not IgA, is the predominant antibody isotype in the cervicovaginal secretions.

The lymphoid tissues of mucosal surfaces contain more B and T cells than any other tissue in the body (Mestecky, 1987), perhaps because they must protect such a large surface area. There are several hundred square meters of mucosal surface, compared to $1.8\,m^2$ of skin. It has been estimated that there are 10^{10} lymphocytes within $1\,m$ of the intestine (Brandtzaeg and Baklien, 1976). T-helper cells are the largest cellular component of the mucosal immune system and, as discussed later, play a central role in regulating the mucosal immune response. The major immunoglobin at most mucosal surfaces in higher mammals is IgA, comprising more than 60% of the total immunoglobulin produced. The majority of this is dimeric or secretory IgA that is produced locally in the lamina propria of the gastrointestinal, upper respiratory and genitourinary tracts, as opposed to the monomeric IgA that predominates in serum. The best-studied mucosal immune compartment is that of the gastro-intestinal (GI) tract, which has been termed GALT (gastrointestinal-associated lymphoid tissue). There has been a great interest in using this system, by way of oral immunization, to induce protection of not only the GI tract but also of other mucosal surfaces through the common mucosal immune system. The BALT, or bronchial-associated lymphoid tissue, has also been evaluated and used as an inductive site for protection of the respiratory system, and also has a means to provide protection for other mucosal surfaces, including the genital tract, as discussed later in this chapter.

The following sections will provide a brief overview of the components of the mucosal immune system, followed by a discussion of the female and male genital tract immune system. Non-specific, innate immune mechanisms that also might be involved in protection of the genital mucosa, such as the role of mucus, pH and defensins and protegrins (Lehrer and Ganz, 1996), will not be discussed.

OVERVIEW OF MUCOSAL IMMUNE SYSTEM

Antigen Sampling

Immune responses are normally initiated by the delivery of antigen into organized lymphoid tissue, where the cellular and cytokine milieu induces the activation, proliferation and differentiation of immune effector cells. The mucosal epithelium serves as an effective barrier that restricts the large number of antigens exposed to mucosal surfaces from having direct access to lymphoid cells. As a result, mechanisms have evolved that allow the host to effectively sample antigens present at the mucosal surface. Different strategies of antigen sampling are employed which are dependent primarily on the type of mucosal epithelia present (Neutra *et al.*, 1996). At mucosal sites such as the intestine, which are covered by a single layer of epithelial cells sealed by tight junctions, directional transport of antigens into lymphoid aggregates occurs at unique regions via specialized epithelial cells, referred to as microfold or M cells. At mucosal sites covered by stratified or pseudostratified epithelium, such as the pharynx, vagina, urethra and esophagus, antigen is endocytosed primarily by dendritic cells, Langerhans' cells or macrophages, which then transport the antigen to regional lymph nodes (Figure 4.1A).

The best-studied example of antigen sampling through M cells occurs at the intestinal lumen and involves directional transcytosis of antigen across the

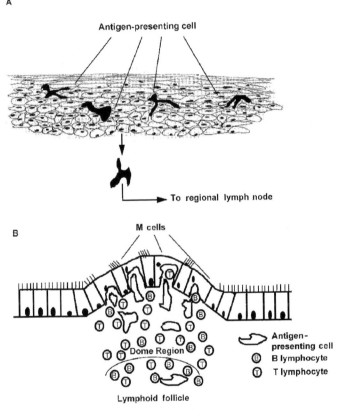

FIGURE 4.1 Antigen sampling across mucosal surfaces. A Uptake of antigen across stratified or pseudostratified epithelial surfaces. Antigen from invading microorganisms or which has passively diffused across the epithelia is taken up and processed by antigen-presenting cells. These cells then migrate to regional lymph nodes and present antigen to **B** and **T** lymphocytes. **B** Directional transcytosis of antigen across simple epithelial membranes. Antigen is endocytosed by **M** cells and is transported to the basolateral membrane. Following exocytosis, antigen is presumably taken up by antigen-presenting cells in the dome region, and subsequently moved into the lymphoid follicle and presented to **B** and **T** lymphocytes.

epithelial barrier and into lymphoid follicles by M cells. Antigen is prohibited from passively diffusing through the intestinal epithelium by physical exclusion, owing to the thick mucin-like glycocalyx which covers the densely packed microvilli found on the apical surface of enterocytes. Antigen sampling at the intestinal lumen is further restricted by tight junctions between these cells (Neutra *et al.*, 1996). By contrast, M cells that cover the mucosal-associated lymphoid tissue of the gut are covered by a less dense glycocalyx, and their apical surface is not modified by microvilli. Instead, these cells possess a well defined endocytic domain which allows for endocytosis of macromolecules, particles and microorganisms. Macromolecules and microorganisms which adhere to the M-cell apical membrane may be phagocytosed or endocytosed primarily via clathrin-coated pits (Neutra *et al.*, 1987; Figure 4.1B). Non-adherent antigens may be taken up by fluid-phase pinocytosis (Owen, 1977; Bockman and Cooper, 1973). After uptake, the antigens are transported to the basolateral membrane with apparently little or no degradation. The basolateral membrane of M cells is deeply invaginated, forming large pockets in relatively close proximity to the apical membrane (Neutra *et al.*, 1996). Lymphocytes and macrophages have been demonstrated within these pockets, as well as in the dome structure that overlies the lymphoid follicle. Antigen is exocytosed into the pocket, where it is presumably taken up by antigen-presenting cells, processed, and presented to lymphocytes in the follicle. Activated T and B lymphocytes primed by this antigen then leave the follicle and home to the lamina propria regions of the intestinal mucosa, or to distal mucosal sites.

Immunoglobulins

In contrast to the immunoglobulin content of serum, approximately 80% of the immunoglobulin detected in secretions of mucosal effector sites is polymeric IgA (Kiyono *et al.*, 1992), which is elicited by environmental antigens at mucosal inductive sites. Further, 75–80% of the B cells present in Peyer's patches express surface IgA, whereas few B cells in peripheral lymph nodes express this isotype (Butcher *et al.*, 1982).

Although little is known about the relative importance of secretory antibody in protection against STDs, pathogen-specific IgA and IgG are secreted from plasma cells within the urogenital mucosa or transudated from the serum into the female genital tract (Lehner and Miller, 1996; Ashley *et al.*, 1994; Kozlowski *et al.*, 1997; Hordnes *et al.*, 1996; Belec *et al.*, 1995; Ogra and Ogra, 1973). Secretory IgA, the predominant antibody isotype at most mucosal surfaces, is ideally suited to its role in protection of the mucosal surface because it is resistant to proteolytic cleavage (Mogens and Russell, 1994; Underdown and Mestecky, 1994; Staats and McGhee, 1996). The primary function of IgA is thought to be prevention of adherence or surface neutralization of invading pathogens. However, recent evidence suggests that IgA may also be able to neutralize virus by forming intracellular complexes within epithelial cells, thereby preventing viral replication (Burns *et al.*, 1996; Mazanec *et al.*, 1995; Mazanec *et al.*, 1992).

Two Cα genes encoding two separate IgA subclasses are present in humans. These two subclasses are slightly different structurally and are differentially expressed in human tissue (Kett *et al.*, 1986; Brandtzaeg *et al.*, 1986). IgA1 has a

13 amino acid extension, which may confer a greater segmental flexibility on these molecules and could conceivably influence antigen binding (Pumphrey, 1986). The presence of the extended hinge region in IgA1 molecules renders them more susceptible to microbe-produced proteases. The human IgA subclasses differ also in the types of sugars comprising their carbohydrate side chains. Most notably, IgA1 contains N-acetylgalactosamine, which is absent in IgA2 (Wold *et al.*, 1990; Mestecky *et al.*, 1986; Baenziger and Kornfeld, 1974a; Baenziger and Kornfeld, 1974b). The majority of serum IgA and approximately 30–50% of IgA in secretions is comprised of IgA1. Moreover, the ratio of IgA1 to IgA2 antibody varies among mucosal sites. IgA1 predominates in the secretions of most mucosal tissues, except in the large intestine and the female genital tract (Kutteh *et al.*, 1988).

Secretory IgA effectively serves as a barrier to microorganisms at mucosal surfaces. Whereas serum IgA originates mainly from the bone marrow, IgA present in mucosal secretions is synthesized almost entirely locally. Approximately 80% of serum IgA exists in monomeric form, whereas IgA in mucosal secretions is composed primarily of disulfide-linked dimers and tetramers (Mestecky and McGhee, 1993). Associated with polymeric IgA (pIgA) of mucosal secretions are a J-chain polypeptide of approximately 15–16 kDa and secretory component (SC), a glycoprotein of approximately 100 kDa (Figure 4.2), which are covalently linked by disulfide bands to the Cα region of IgA. The predominance of secretory IgA over other immunoglobulins in mucosal secretions reflects the specific transport of pIgA by an SC-mediated mechanism (Brandtzaeg, 1995).

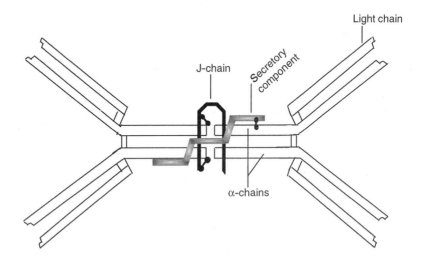

FIGURE 4.2 Structure of secretory IgA. IgA monomers are linked in a tail-to-tail configuration and are covalently associated with two other polypeptides. J-chain is linked to the penultimate cysteine residues on the carboxy termini of the two α chains of a single IgA monomer. Secretory component is linked by disulfide bonds to the Cα2 domain of the other IgA monomer.

J-chain is added to the IgA molecule before secretion by plasma cells (Moldoveanu et al., 1984), and is thought to enhance the polymerization of IgA monomers in an end-to-end configuration (Svehag and Bloth, 1970; Bastian et al., 1992; Abel and Grey, 1968). Results of studies with J-chain deficient mice suggest that IgA polymerization can occur in the absence of J-chain, but that it may be less efficient, or perhaps the polymers may be less stable than in wild-type mice (Hendrickson et al., 1995). J-chain association with IgA monomers may result in a conformation of the resulting pIgA, which can interact with the polymeric immunoglobulin receptor (pIgR) found on the basolateral plasma membrane of secretory epithelial cells. Initially, pIgA and pIgR associate non-covalently, followed by endocytosis and transcytosis of the ligand–receptor complexes to the apical membrane. Polymeric IgA becomes covalently linked to the pIgR by disulfide linkage of the Cα2 domain and the fifth domain of the pIgR (Mestecky and McGhee, 1987) prior to external translocation. Cleavage of the transmembrane pIgR at the luminal membrane releases secretory IgA, which is composed of pIgA and covalently associated SC. Although the majority of pIgA probably reaches mucosal secretions in this manner, studies with J-chain deficient mice suggest that alternate mechanisms for transport of IgA may exist. Similar levels of IgA were detected in mucosal and glandular secretions of J-chain deficient and wild-type animals; however, the mucosal IgA of J-chain deficient mice was not associated with SC, suggesting that secretion apparently occurred by mechanisms distinct from conventional pIgR-mediated transport (Hendrickson et al., 1996).

Whereas the role of secretory IgA is clearer for respiratory infections such as influenza (McNeal et al., 1994; Clements et al., 1986; Tomoda et al., 1995) and gastrointestinal infections such as rotavirus (Clements et al., 1986; Burns et al., 1996), the role of IgA in protection of the genital tract against STDs is uncertain, although shedding of herpes simplex virus (HSV) has been correlated to cervicovaginal IgA levels (Merriman et al., 1984). Other studies, however, suggest a less important role for local antibody in protection from chlamydia and HSV (Ramsey et al., 1988; Igietseme and Rank, 1991). For example, Johansson reported that γ-interferon receptor-deficient mice were susceptible to reinfection with Chlamydia trachomatis despite the presence of high levels of local IgA and IgG (Johansson et al., 1997), and others have correlated resistance to the levels of antigen-specific T cells in the genital mucosal tract of guinea pigs (Igietseme and Rank, 1991). Similarly, Milligan has reported that T cells, predominantly CD4[+] cells, are required for protection of the vaginal mucosa of HSV-immune mice following rechallenge (Milligan et al., 1998). Thus, although protection of the vaginal mucosa of immune mice was diminished by depletion of CD8[+] T cells in these studies, protection was more adversely affected by CD4[+] T-cell depletion. In fact, HSV-2 replicated in the vaginas of immune mice depleted of CD4[+] T cells with similar kinetics to naive mice for the first 5 days following vaginal rechallenge with HSV-2, despite the presence of high titers of vaginal and serum antibodies. Therefore, in these animal models protection from infection against these common STDs appears to be mediated more by CD4[+] cells and γ-interferon than by antibody.

B Cells

The majority (75–80%) of B cells present in the Peyer's patches of the intestine express surface IgA (Butcher *et al.*, 1982). Mechanisms have obviously evolved at this site and other mucosal immune inductive sites which preferentially skew the B-cell response towards the production of IgA. Given the protective effect of mucosal IgA and the desirability of inducing protection at mucosal surfaces, a great deal of effort has been expended in understanding the molecular mechanisms involved in the induction of mucosal IgA responses.

The cellular and cytokine milieu that comprises the microenvironment of the mucosal lymphoid follicles has been assumed to be responsible for the preferential development of IgA responses. It is possible that non-committed B cells are activated and switched preferentially to IgA expression in the presence of environmental antigens and IgA switch signals in the mucosal lymphoid follicles. Alternatively, it has been suggested that B cells that have taken the initial steps toward switching to IgA expression may preferentially home to the mucosal immune inductive sites (reviewed in Strober and Ehrhardt, 1994).

T cells influence B-cell activation and differentiation events through the release of cytokines or by cognate interactions. An early, essential role of T cells in the development of B-cell responses occurs by interaction of the CD40 ligand expressed on the surface of activated T cells, with the CD40 antigen expressed on the B-cell surface resulting in B-cell activation. CD40/CD40L interactions are responsible for the development of primary and secondary immune responses to thymus-dependent antigens and the generation of memory B cells (reviewed in Foy *et al.*, 1996). In addition, signaling via CD40 may prepare the B cell for switch differentiation (Aruffo *et al.*, 1993). Additional T-cell influences on the developing mucosal antibody responses have also been implicated in the preferential switch to IgA, and subsequently on the differentiation of IgA-committed B cells to IgA-secreting plasma cells. In this regard, Peyer's patch T cells have been shown to induce IgA isotype switch in LPS-activated IgM-expressing B cells *in vitro* to a much greater extent than splenic T cells (Kawanishi *et al.*, 1983). Thus, whereas the addition of splenic T-cell clones to LPS activated B cells resulted in a switch to predominantly IgG+ B cells, Peyer's patch T-cell clones induced a switch primarily to IgA expression. Presumably, this difference represented the release of soluble mediators or the induction of a membrane-bound component unique to Peyer's patch T cells. Although the presence of unique IgA switch T cells in Peyer's patch T cells is an intriguing possibility, other groups have been unable to demonstrate preferential skewing of an antibody response by these cells (Arny *et al.*, 1989; Al Maghazachi *et al.*, 1988a,b).

Cytokines released by T cells and other cellular constituents of organized lymphoid follicles have been shown to influence the nature and magnitude of the antibody response. Given the differences in isotype predominance in B-cell responses initiated at mucosal rather than systemic locations, it has been thought that local differences in cytokine profiles may influence the response to particular isotypes. In this regard, interleukin-4 (IL-4) has been shown to be a switch factor for IgG1 and IgE (Snapper and Paul, 1987; Coffman and Carty, 1986; Coffman *et al.*, 1986). Similarly, transforming growth factor B1 (TGFB1) has been found to be capable of switching B cells from IgM to IgA expression (Sonada *et al.*, 1989; Coffman *et al.*, 1989). Switching to IgA, and subsequent secretion of IgA antibody by terminally differentiated B cells, has been shown to be potentiated by other cytokines, including IL-2, IL-4, IL-5, IL-6 and IL-10. In particular, IL-5 has been shown to increase IgA production by IgA-expressing B cells (Sonada *et al.*, 1989; Kim and Kagnoff, 1990). Although IL-4 alone does not have this effect, it has been shown to augment the production of IgA in association with IL-5 (Eckmann *et al.*, 1992; Murray *et al.*, 1987). IL-6 has been shown to increase IgA secretion in IgA-expressing B cells (Beagley *et al.*, 1989), and IL-10 has been implicated in differentiation of IgA-expressing cells to IgA-secreting plasma cells (Defrance *et al.*, 1992).

Dendritic cells have long been known to play important roles in T-cell activation through their ability to present antigens efficiently (Steinman, 1991). Recent studies have demonstrated that dendritic cells also play an important role in the development of B-cell responses (Clark, 1997). Purified dendritic cells have been shown to directly stimulate CD40-activated memory B cells to proliferate and secrete IgA and IgG, and to stimulate naive B cells in the presence of IL-2 to produce IgM (Dubois *et al.*, 1997). Further studies by the same group demonstrated that, in the presence of TGFB1 and IL-10, human dendritic cells greatly enhanced isotype switching to IgA by CD40-activated naive B cells (Fayette *et al.*, 1997). These results raise the intriguing possibility that dendritic cells, in addition to T cells and cytokines, may provide a unique signal or enhance a weak signal involved in isotype switch to IgA. Interestingly, studies by Sangster *et al.* (1997) suggest that dendritic cells from the site of antigen deposition, rather than at the site of immune response induction, may be involved in preferential skewing of the developing B-cell response to particular isotypes.

T cells

T lymphocytes are found in abundance in immune inductive sites of the gastrointestinal tract. More than 95% of these cells express the α/β T-cell receptor, and include both CD4[+] and CD8[+] subsets. Approximately 60% are CD4[+] (Kiyono *et al.*, 1992), including T cells capable of mediating B-cell isotype switch to IgA (Strober and Sneller, 1988). In addition, murine T cells bearing Fcα receptors and capable of inducing B-cell differentiation to IgA production have been cloned from the Peyer's patch (Mestecky and McGhee, 1987).

Following antigen stimulation, T cells migrate from mucosal immune inductive sites to lamina propria regions of mucosal effector sites. Approximately 40–60% of lymphocytes in the lamina propria of the gastrointestinal tract are T cells (Brandtzaeg, 1984). Such cells are predominantly CD4[+], and most have been shown to secrete Th2-type cytokines, including IL-5 (Taguchi, *et al.*, 1990). T lymphocytes have also been detected in the intraepithelial spaces of the gastrointestinal tract; however, the majority of these cells are CD8[+] (Brandtzaeg *et al.*, 1988). Also, large numbers of murine intraepithelial T lymphocytes utilize the γ/δ T-cell receptor (Taguchi *et al.*, 1991; Mosley *et al.*, 1991). In contrast to the lamina propria T lymphocytes, approximately equal numbers of IL-5 and IFNγ-secreting cells were detected in intraepithelial lymphocyte populations, suggesting the presence of an approximately equal distribution of both Th1- and Th2-like CD4$^+$ T cells (Taguchi et al., 1990). Interestingly, IL-5-secreting CD8$^+$ T lymphocytes have also been detected among intraepithelial lymphocytes (Taguchi et al., 1990). CD8$^+$ cytotoxic T lymphocytes specific for intestinal viral pathogens have been demonstrated in Peyer's patches, as well as in the lamina propria and intraepithelial effector regions (London et al., 1986, 1989; Offit and Dudzik, 1989; Offit et al., 1991).

The vagina and cervix contain all the components necessary to develop a cell-mediated immune response. These tissues contain epithelial Langerhans' cells, a dendritic cell that is considered a professional antigen-presenting cell, and T lymphocytes that express α/β or γ/δ TCR (Parr and Parr, 1994; Staats and McGhee, 1996; Parr and Parr, 1991; Neutra and Kraehenbuhl, 1996; Nandi and Allison, 1991). The presence and origin of T lymphocytes, as well as the induction of cellular immune responses in the vagina and cervix, have not been extensively studied and consequently are not well understood. Lymphoid follicles analogous to the Peyer's patches of the intestine have been detected only in the endocervix of humans. However, a population of intraepithelial T lymphocytes has been detected distributed uniformly along the length of the cervical epithelium (Fox, 1993). These intraepithelial T cells are composed primarily of the CD8$^+$ subset (Morris et al. 1983; Edwards and Morris, 1985; Peters, 1986; Roncalli et al., 1988), with CD8:CD4 ratios varying between 3:2 and 9:1. T cells have also been demonstrated in the stroma of the human cervical epithelia; however, the subset composition of these cells is not agreed upon and has been reported to reflect a predominance of CD8$^+$ cells (Tay et al., 1987; DiLoreto et al., 1989), a predominance of CD4$^+$ T cells (Roncalli et al., 1988), or a random distribution of both (Edwards and Morris, 1985). The data concerning T-cell populations in the human vagina are scarce, but demonstration of primarily CD8$^+$ T cells in the vaginal squamous epithelium has been reported (Edwards and Morris, 1985). In a report evaluating the vaginal mucosa of mice, the majority of T cells were CD4$^+$ (Fidel et al., 1996).

Like the local immunoglobulin levels of the urogenital tract, the cell-mediated immune responses also appear to be influenced by the menstrual cycle. Antigen-presenting cells in the uterus and vagina can be influenced by hormones (Wira and Rossoll, 1995; Prabhala and Wira, 1995), and the CD8$^+$ cytolysis that is found throughout the female reproductive tract also is hormonally mediated (White et al., 1997). In the uterine endometrium T-cell mediated cytolysis is present during the proliferative phase of the menstrual cycle but not during the subsequent postovulatory phase, suggesting that the high levels of estradiol and progesterone present during days 14–28 of the menstrual cycle downregulate CTL activity in the uterus. The lymphoid aggregates that are occasionally found in uterine endometrium are mostly CD8$^+$, and are also influenced by the menstrual cycle (Yeaman et al., 1997). This could affect the ability to resist infection but not reject an embryo. Cytolytic activity appears to persist in the cervix and vagina throughout the menstrual cycle.

Protection against chlamydia (Ramsey et al., 1988; Igietseme and Rank, 1991), HSV (Milligan et al., 1998; Milligan and Bernstein, 1997; McDermott et al., 1989), and perhaps candida (Fidel et al., 1995) seem most related to local T-cell mediated immunity. Following chlamydia infection in guinea pigs, the susceptibility to reinfection was directly proportional to the presence of antigen-specific T cells in the genital mucosa (Igietseme and Rank, 1991). Using a murine model of genital HSV infection, Milligan and Bernstein found that HSV-specific T cells, predominantly CD4$^+$ cells, were present in the genital lymph nodes (gLN) 4 days post intravaginal inoculation, but were not detected in the genital tract or spleen until later (Milligan and Bernstein, 1995a), suggesting that the cellular immune response originates in the gLN. The mechanisms by which these cells effect virus clearance is currently unclear, but apparently involves IFN-γ and antigen non-specific effector cells, such as macrophages and neutrophils (Milligan and Bernstein, 1997; Milligan, unpublished results). CD8$^+$ T cells are not required for virus clearance but can apparently function to clear virus in the absence of CD4$^+$ T cells (Milligan and Bernstein, 1997). Antigen-specific T cells appear to persist in the vaginal mucosa for at least 2 months, and appeared to provide protection following reinoculation.

Following local (vaginal) vaccination, T cells appear to be critical to the protection of the vaginal mucosa and dorsal root ganglia (the site of HSV latency). As shown in Figure 4.3, when mice that had been intravaginally infected with HSV-2 were rechallenged intravaginally with another HSV-2 strain 30 days later, the immunized mice were protected from the lethal

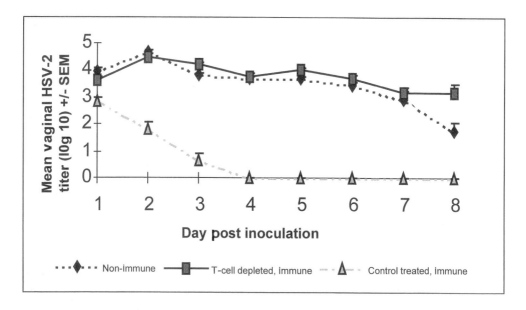

FIGURE 4.3 Protection of the vaginal mucosae of HSV-immune mice requires T cells. BALB/c mice were inoculated intravaginally with an attenuated (thymidine-kinase deficient) HSV-2 strain, HSV-2 tk-. After 30 days, one group of immune mice was depleted of T cells by treatment with a cocktail of monoclonal antibodies specific for Thy 1.2, CD4 and CD8, and a second group was treated with an isotype-matched control antibody. These mice and a group of age-matched non-immune mice were then challenged with a fully virulent strain of HSV-2. Vaginal virus titers were determined from vaginal swabs taken daily after challenge by plaque titration on Vero cell monolayers.

challenge and virus titers were reduced by approximately 90% at 24 hours after challenge. However, vaginal virus replication in immune mice that were depleted of T cells was indistinguishable from that in naive mice, in spite of the presence of high levels of local and serum antibody (Milligan *et al.*, 1998). Although the depletion of either CD4$^+$ or CD8$^+$ cells resulted in poorer protection of the vaginal mucosa, depletion of CD4$^+$ cells had a greater negative effect. Similarly, in immune mice depleted of T cells HSV-2 could be detected in the dorsal root ganglia following rechallenge. No infectious virus could be detected in control treated immune mice. In contrast to protection of the vaginal mucosa, however, either CD4$^+$ or CD8$^+$ cells were able to protect the ganglia (Milligan *et al.*, 1998).

Another recent finding is that only immunization via a mucosal route – in this case intranasal immunization with a recombinant adenovirus expressing an HSV glycoprotein – induced long-lasting memory T cells in the mucosa (Gallichan and Rosenthal, 1996a). In this study, the investigators compared intranasal to systemic immunization and found anti-HSV cytolytic T cells to be present in both mucosal and systemic compartments shortly after immunization. However, several months after immunization only those mice immunized by the intranasal route had anamnestic CTL activity in the mucosal tissues, whereas systemic immunization only induced long-lasting activity in the spleen. It is interesting to speculate that this could explain why the recent trial of a subunit HSV glycoprotein vaccine showed only short-term protection (L. Corey, personal communication).

In the guinea-pig model of chlamydial infection, it also appears that protection from reinfection correlates to the presence of antigen-specific T cells which are present at day 30 but greatly reduced by day 75 (Igietseme and Rank, 1991). Short-term T-cell mediated immunity has also been implicated in protection of the GI tract. Thus, in one study, CD8$^+$-mediated protection against reinfection was complete shortly after infection, but only partially protective later (Franco *et al.*, 1997). The duration of T-cell persistence at mucosal surfaces may therefore be implicated in protection (Ahmed and Gray, 1996).

FEMALE GENITAL TRACT

The uptake of protein antigens across the vaginal epithelium has been studied mainly in animal models. Proteins do not cross the vaginal epithelium well during estrus, when it is thickest (Parr and Parr, 1990). This may be so that the vaginal mucosa can serve as an effective barrier to antigen sampling during times of increased fertility, and prevent the induction of immune reactions to sperm or seminal proteins (Parr and Parr, 1994). However, the uptake of protein antigens across the vaginal epithelium has been demonstrated in a murine model during early pregnancy and at diestrus, when the epithelium is thinnest (Parr and Parr, 1990, 1991). Thus, the ability of protein to cross the vaginal epithelium is influenced by hormonal changes in the epithelial layer. Invasion of the vaginal epithelium by microorganisms can also be influenced by hormonal effects. In animal models of infections with sexually transmitted pathogens, mice in early pregnancy or in diestrus are more readily infected intravaginally with herpes simplex virus type 2 (Baker and Plotkin, 1978) or chlamydia (Tuffrey et al., 1986).

Langerhans' cells, a subset of dendritic antigen-presenting cells to T lymphocytes, present in the vaginal epithelium have been shown to capture luminal proteins (Parr and Parr, 1990). These cells have also been demonstrated in various epithelial layers of the human vagina and cervix (Edwards and Morris, 1985), and are thought to endocytose antigen and transport it to regional lymphoid aggregates or lymph nodes. In this regard, lymphoid aggregates are absent in the vagina but have been reported in the human oviduct and cervix (Otsuki, 1989; Edwards and Morris, 1985). However, the role of these sites in the initiation of immune responses is unclear. Mice lack organized lymphoid aggregates in the female reproductive tract (Parr and Parr, 1985, 1988), and immune responses to vaginally delivered proteins or intravaginally inoculated HSV-2 are thought to initiate in the regional lymph nodes (Milligan and Bernstein, 1995a).

Both IgA and IgG are present in the genital secretions of females and vary during the menstrual cycle, but IgG appears to be the predominant isotype (Quesnel et al., 1997; Kutteh et al., 1996; Parr and Parr, 1994). The amount and isotype of antibody differs at various sites in the female genital tract, suggesting differences in the distribution of ASC, SC, and transudation of serum immunoglobulins. Most of the IgA in genital secretions of women is secretory and is derived from local mucosal plasma cells, whereas IgG is derived both from local secretions and (mainly) serum transudation (Quesnel et al., 1997; Kutteh and Mestecky, 1994; Hocini et al., 1995). IgA appears to originate mainly from the uterus/endocervix as the result of active transport by SC through the uterine epithelium. In both mice and humans, removal of the uterus results in approximately a 90% reduction in IgA concentrations, but no change in IgG levels (Parr and Parr, 1990, 1994; Lubeck et al., 1994). The uterus contains SC and IgA plasma cells in the glandular epithelium and IgA in the glandular lumen (Kutteh et al., 1988). Interferon-γ increases the secretory component as well as IgA levels in uterine secretions (Prabhala and Wira, 1991).

The endocervix of women is lined by columnar epithelial cells and contains numerous glands with SC on the luminal and glandular epithelium, making it a likely candidate for IgA production (Kutteh et al., 1988, 1993; Kutteh and Mestecky, 1994; Parr and Parr, 1994; Wira et al., 1994). The ectocervix, however, is similar to the vagina and contains few glands, SC or IgA plasma cells. The mouse cervix is different from the human endocervix and, like the vagina, is lined with stratified squamous epithelium and contains only a few IgA plasma cells. SC and IgA plasma cells are also found in the isthmus and ampulla of the oviduct, suggesting that this might be another site for antibody production (Parr and Parr, 1990; Kutteh et al., 1988).

IgG antibody is mainly derived from transudation through the vaginal epithelium (Brandtzaeg et al., 1994; Mestecky, 1987; Parr and Parr, 1994; Underdown and Mestecky, 1994), although in one recent study, IgG-secreting cells, rather than IgA, represented the predominant antibody-secreting cell (ASC) in the cervix (Crowley-Nowick et al., 1995).

Studies of antibody levels have been hampered by the lack of a quantitative system for collecting samples. In a recent study, Quesnel et al. (1997) compared vaginal washes with two newer methods for collection: snow-strips, which are absorptive paper strips, and wicks. Both provided more accurate estimates of antibody secretions. Capture of undiluted specimens showed that IgA levels were highest in the endocervix and lowest in the vagina. This supports earlier findings showing that IgA-secreting plasma cells were most abundant in the endocervix (Kutteh and Mestecky, 1994). IgG levels did not vary significantly by site.

A unique aspect of the local genital tract immune response in females is that the levels of immunoglobulins and lymphocytes fluctuate as a result of hormonal changes (Parr and Parr, 1994; Wira and Sandoe, 1977; Wira et al., 1994). In animal models IgA levels are higher during estrus and proestrus, and are lowest at diestrus. This may be explained by the higher number of IgA plasma cells detected during proestrus and estrus (McDermott et al., 1980; Canning and Billington, 1983), and by the increased levels of SC at these times (Sullivan et al., 1983), SC being responsible for the transport of IgA into mucosal tissues, as discussed above. IgG levels are highest during proestrus and lowest during estrus, when the vaginal epithelium is thickest and could impede transudation from the serum (Kutteh et al., 1996; Parr and Parr, 1994). Like the levels of total immunoglobulin, pathogen-specific vaginal IgA levels following intranasal vaccination were also significantly higher during estrus than during diestrus or proestrus, whereas the IgG levels were highest during diestrus (Gallichan and Rosenthal, 1996b). The cyclic changes in SC and antibody levels have also been observed in humans (Quesnel et al., 1997; Suzuki et al., 1984; Sullivan et al., 1984; Sullivan and Wira, 1983).

Recently an Elispot assay was used to unambiguously determine the source of IgA and IgG antibody in the murine female genital tract following HSV infection (Milligan and Bernstein, 1995b). It was found that following vaginal HSV infection in mice, large numbers of HSV-specific IgG ASC were present in the gLN at day 7, and peaked between days 7 and 9 post infection. Lower numbers of IgA ASC were also present in the gLN on days 7–9, confirming the predominance of IgG in vaginal washes of these mice. Neither IgG nor IgA ASC could be detected in the genital mucosa following primary infection.

Following rechallenge of these animals, HSV-specific ASC were present as early as day 4 in the gLN, and again were predominantly IgG ASC, although the number of IgA ASC were increased compared to primary infection. Also, as opposed to primary infection, ASC were easily detected in the genital mucosa but appeared somewhat later than in the gLN. About equal numbers of IgG- and IgA-producing cells were detected. The quantity of IgG in the vaginal wash increased following rechallenge, but a much larger increase in IgA was seen, so that titers were approximately equal after rechallenge. In contrast, IgG antibodies remained the predominant isotype in the serum. Similar findings have been reported following intranasal immunization of mice with adenovirus expressing an HSV glycoprotein (Gallichan and Rosenthal, 1996b). In this report, following vaginal HSV challenge both IgG and IgA ASC were found in the gLN and genital mucosal tissue, but again IgG predominated. These studies suggest that genital humoral responses originate in the gLN following HSV infection, and that gLN serve as the primary source of the HSV-specific IgG- and IgA-secreting cells present in the genital mucosa after vaginal infection. A similar conclusion was made by others based on immunization with non-replicating antigens in the mouse pelvis (Thapar *et al.*, 1990).

MALE GENITAL TRACT

There are few studies investigating the male genital tract (reviewed in Parr and Parr, 1994; Fowler, 1993; Pudney and Anderson, 1993; Mestecky and Fultz, 1999). IgA and IgG are found in seminal fluid and appear to originate mostly from the prostate gland, and to a lesser extent from the seminal vesicles. Striking differences in the level of IgA and IgG have been reported (Mestecky and Fultz, 1999). Specific antibodies to chlamydia, HIV and gonococci have been detected in the semen (Boslego *et al.*, 1991; Eggert-Kruse *et al.*, 1996; Wolff *et al.*, 1992). Most of the antibody appears to be IgG, and is felt to be serum derived (Wolff *et al.*, 1992).

The majority of immunoglobulin in prostate fluid appears to be IgG, but IgA is also present and is mostly secretory (Fowler *et al.*, 1982; Fowler and Mariano, 1982). This is supported by reports of IgA plasma cells in the prostate (Bene *et al.*, 1988; Doble *et al.*, 1990; Fowler, 1993; Parr and Parr, 1994), along with higher levels of SC than in other male reproductive organs (Crowley-Nowick *et al.*, 1995). In rats SC can be found in the ejaculatory ducts and secretory ducts of accessory cells and urethral glands (Parr and Parr, 1989). As in the female genital tract, the levels of SC in the male genital tract are also hormonally mediated (Crowley-Nowick *et al.*, 1995).

Immune cells, including macrophages, neutrophils and lymphocytes, are found in seminal fluid but little is known about their origin (Harrison *et al.*, 1991; Wolff and Anderson, 1988; El-Demiry *et al.*, 1986; Parr and Parr, 1994). Mucosal lymphoid nodules are only rarely found (Parr and Parr, 1994).

IMMUNIZATION STRATEGIES TO INDUCE LOCAL PROTECTION

The understanding that systemic immunization may not provide optimum protection of mucosal surfaces, including the genital mucosa (Marthas *et al.*, 1992), led to the introduction of mucosally delivered vaccines. In spite of this recognition, however, only one of the commonly used vaccines, oral polio virus (OPV), is administered via a mucosal surface. This vaccine has been shown to provide better protection against intestinal polio virus replication than the systemic administration of inactivated polio virus (IPV) (Ogra, 1984). More recently, another orally administered vaccine for rotavirus infection, the most common cause of severe diarrhea in infants, has been shown to provide good protection against severe rotavirus disease (Bernstein *et al.*, 1995; Rennels *et al.*, 1996), and live attenuated influenza vaccines given intranasally have also provided significant protection (Edwards *et al.*, 1994; Gruber *et al.*, 1996; Belshe *et al.*, 1998). Further, injected polymeric anti-influenza IgA was shown to be transported into nasal secretions and provided protection against an influenza challenge in mice (Renegar and Small, 1991). These examples illustrate the utility of mucosal vaccines that induce protection through the well-characterized GALT and BALT systems against intestinal and respiratory pathogens. However, much less is known about inducing protection for the genital tract.

Preclinical approaches, including direct immunization of the genital tract and use of the common mucosal immune system through oral or intranasal immunization, have been explored. Other strategies have concentrated on inducing immunity of the gLN through directed immunization of the draining lymph nodes, or immunization into the pelvis. A recent report (Quiding-Jabrink *et al.*, 1997) suggests that the route of immunization can affect the expression of adhesion markers on antibody-secreting cells (ASC), and thereby influence the distribution of these cells. Most ASC induced by systemic immunization of volunteers expressed L-selectin, an adhesion molecule conferring specificity for peripheral lymph nodes, whereas following oral and rectal immunization ASC expressed α4 β7, an adhesion molecule with specificity for mucosal tissues.

Genital Tract Immunization

Direct immunization of the female genital tract with killed antigen appears to be a poor method of inducing local antibodies, although low levels of specific antibodies can be produced in uterine and/or vaginal secretions (Haneberg *et al.*, 1994; Thapar *et al.*, 1990; Ogra *et al.*, 1981; Parr *et al.*, 1988; Parr and Parr, 1994). As discussed above, this is partially explained by the lack of an organized lymphoreticular system in the genital tract, as opposed to the efficient organization found in the GALT and BALT systems. Part of the failure may also be explained by the inability of antigens to penetrate the luminal epithelium, followed by the rapid loss of antigen from the genital tract. It has been reported that even infection with an invasive replicating pathogen such as HSV is severely limited unless the mucus secretions and the epithelium are thinned by treatment with progesterone (Baker and Plotkin, 1978; Milligan and Bernstein, 1995). Similar findings have been reported for infections with chlamydia in mice (Tuffrey *et al.*, 1986). Therefore, interpretation of local genital immunization studies should be considered in light of the discussion above regarding the influence of the menstrual cycle on many aspects of the local induction of immunity (Wira and Prabhala, 1993).

In contrast to the poor induction of immune responses by non-replicating antigens, intravaginal inoculation of HSV-2 or chlamydia provides protection against reinoculation with that organism. However, it should be emphasized that immune induction by intravaginal inoculation is quite different from that induced at other mucosal surfaces. The response is primarily IgG and not IgA, and probably originates in the draining lymph nodes not directly associated with the mucosal surface (Milligan and Bernstein, 1995b).

Nevertheless, protective responses can be elicited by immunization with replicating organisms. Originally, McDermott showed that a prior vaginal infection with an attenuated HSV strain provided protection from a lethal rechallenge with a virulent strain (McDermott *et al.*, 1984). Further, he showed that protection could be provided by passive transfer of T cells from the draining lymph nodes, but not by splenic T cells or B cells (McDermott *et al.*, 1989). These observations were extended, as discussed above, to show that both

HSV-specific B (Milligan and Bernstein, 1995b) and T cells (Milligan and Bernstein, 1995a) are found in the genital mucosa shortly after challenge of HSV-immune mice. Further, it appears that protection of the vaginal mucosa correlates better to the presence of HSV specific CD4$^+$ T cells than antibody (Milligan et al., 1998).

The combination of vaginal immunization with immunization at another more potent mucosal inductive site, most commonly oral, has also been investigated. Vaginal–oral combinations have been used successfully to induce IgG and IgA antibodies and CTL responses (Bergmeier et al., 1995; Klavinskis et al., 1996).

Genital mucosal immunization in males is less well studied than in females. Topical urethral immunization with a recombinant simian immunodeficiency virus antigen, covalently linked to cholera toxin B subunit and augmented with oral immunization with the same antigen and killed cholera organisms, induced secretory IgA and IgG in urethral secretions and seminal fluid of non-human primates (Lehner et al., 1994b). In addition, antigen-specific T-cell responses were found in the gLN. In these studies only genital–oral immunization and not rectal–oral immunization, induced specific IgA and IgG antibodies in the urethra and seminal fluid, but both induced T- and B-cell responses in the regional lymph nodes (Lehner et al., 1994b).

Mucosal Immunization

The concept of a common mucosal immune system should allow immunization at one mucosal surface to induce protective immune responses at distal sites (McDermott and Bienenstock, 1979; Mestecky, 1987). There are a number of recent studies that have compared several vaccination routes, most commonly oral and nasal, as well as combinations of these routes with systemic boosts or vice versa. Oral immunization has several advantages, most notably the ease of administration. The main drawback of this route is that antigen must survive the enzymes and acidic environment of the upper gastrointestinal tract. Therefore, there has been great interest in the use of biodegradable microspheres for oral immunization (Morris et al., 1994; Challacombe et al., 1997). An early report took advantage of the use of live attenuated polio vaccine to show that oral administration could induce antibodies in the female genital tract (Ogra and Ogra, 1973). Specific IgA antibodies were also produced in the genital tract of mice by oral immunization with live influenza virus vaccine (Briese et al., 1987). Oral vaccination with live organisms has also been used to protect against vaginal challenge with HSV (Irie et al., 1992) and chlamydia (Nichols et al., 1978; Cui et al., 1991). Most recently, HIV neutralization antibodies and SIV antibody and T-cell responses have been induced by oral immunization (Kubota et al., 1997; Bukawa et al., 1995). Oral immunization may need to be combined with systemic immunization to provide the best protection, as was shown with a simian immunodeficiency model of HIV (Marx et al., 1993).

The type of T-cell response (Th1 vs. Th2) following oral immunization appears to be related to the immunogen, and can be influenced by adjuvants. For example, the administration of tetanus with cholera toxin as an adjuvant induces Th2 cells and cytokines (IL-4, IL-5 and IL-10), whereas administration with a recombinant *Salmonella* elicits a Th1-type response as well as Th2 (Yamamoto *et al.*, 1996). In another study the administration of oral but not systemic IL-12 was able to redirect a Th2 response to a protein given with cholera toxin to a Th1 response (Marinaro *et al.*, 1997). The other strategy that has received attention is the nasal route. Several recent studies have emphasized that this route is an induction site for mucosal (including genital tract), and systemic immune responses (Berquist *et al.*, 1997; Pal *et al.*, 1996; Mestecky and Fultz, 1999). Evaluation of this route in rodents, non-human primates, and humans has shown that nasal immunization induces an immune response not only in the local secretions but also in the female genital tract secretion (Mestecky and Fultz, 1999; Rudin *et al.*, 1998; Kozlowski *et al.*, 1999; Asanuma *et al.*, 1999). Immunization with attenuated viral vectors (Bernstein and Tepe, 1994; Muster *et al.*, 1995; Gallichan *et al.*, 1993) and DNA (Kuklin *et al.*, 1997) has been shown to induce vaginal antibody and/or protection. This strategy also offers an easy route of administration that has shown some success against respiratory pathogens. This is well illustrated by the recent success of the cold-adapted influenza vaccine given intranasally to children (Edwards *et al.*, 1994; Gruber *et al.*, 1996; Belshe *et al.*, 1998). Intranasal vaccines also induce antibody at other mucosal surfaces, including the genital tract. Vaginal IgA antibody and protection has been induced following intranasal immunization with recombinant adenovirus expressing an HSV glycoprotein in mice (Gallichan *et al.*, 1993). Similarly, recombinant vaccinia expressing an HSV glycoprotein has been used to examine the effect of route on protection using the guinea-pig model of genital herpes (Bernstein and Tepe, 1994). Results showed that intranasal immunization decreased vaginal virus replication, the severity of the primary disease, and subsequent recurrences, although protection was best when the vaccinia recombinant was given intravaginally. HIV-specific antibody has been induced by intranasal immunization with a chimeric influenza virus (Muster *et al.*, 1995). The use of mucosal adjuvants has also improved the response of both intranasal and intravaginal immunization to induce vaginal antibodies (DiTommaso *et al.*, 1996).

There is also interest in rectal immunization because of reports describing the presence of an organized lymphoreticular system in the lower GI tract (MacDonald and Spencer, 1994; Langman and Rowland, 1986; O'Leary and Sweeney, 1986), and successful immunization using this route (Lehner *et al.*, 1993; Hordnes *et al.*, 1995; Klavinskis *et al.*, 1996; Kozlowski *et al.*, 1999). In one report, intrarectal immunization induced the highest levels of IgA and IgG antibodies in rectal secretions but was least effective for inducing antibodies in the female genital tract (Kozlowski *et al.*, 1997, 1999). Only vaginal immunization induced both IgA and IgG in the cervix and vagina, but did not induce antibodies in the rectum. In another study the combination of oral and rectal immunization induced rectal secretory antibodies and cell-mediated immune responses in the gLN (Lehner *et al.*, 1993).

Non-mucosal Routes

There have been a few studies evaluating immunization strategies designed to deliver antigen to the lymph nodes draining the genital tract – so-called directed immunization (Klavinskis, *et al.*, 1996; Lehner *et al.*, 1994a, b, 1996, 1999) – as well as immunization of the presacral space (Thapar *et al.*, 1990). Targeted lymph node immunization induced secretory IgA and IgG antibody in the vaginal, male urethral, rectal and seminal fluids, as well as serum (Lehner *et al.*, 1994a). Directed immunization also elicited CTL responses, with high responses in gLN (Klavinskis *et al.*, 1996; Lehner *et al.*, 1996). These studies add support to the experiments discussed above, implicating the gLN as the source for antibody and T-cell responses in the genital tract.

SUMMARY

The genital mucosal immune system involves a complex interaction of cells, cytokines and antibodies that is regulated in part by hormonal influences. There is evidence that secretory IgA and local cell-mediated immune mechanisms contribute to the resolution of STDs, and to protection from infection. However, much more needs to be learned regarding local defense mechanisms and how to best induce protective responses in the genital tract. Many of the newest vaccine strategies, including DNA vaccines, vectored vaccines, disabled viral vaccines and mucosal delivery routes, are all being evaluated in preclinical and early clinical trials. Successful protection from STDs may depend on our ability to develop vaccine strategies that induce mucosal immune responses.

REFERENCES

Abel, C.A. and Grey, H.M. (1968). Studies on the structure of mouse gamma-A myeloma proteins. *Biochemistry* **7**: 2682–2688

Ahmed, R. and Gray, D. (1996). Immunological memory and protective immunity: understanding their relation. *Science* **272**: 54–60

Al Maghazachi, A. and Phillips-Quagliata, J.M. (1988a). Con A-propagated, auto-reactive T cell clones that secrete factors promoting high IgA responses. *Int. Arch. Allergy Appl. Immunol.* **86**: 147–156

Al Maghazachi, A. and Phillips-Quagliata, J.M. (1988b). Keyhole limpet hemocyanin-propagated Peyer's patch T cell clones that help IgA responses. *J. Immunol.* **140**: 3380–3388

Arny, M., Kelly-Hatfield, P., Lamm, M.E. *et al.* (1989). The function of T cells from different lymphoid organs in regulating the proportions of plasma cells expressing various isotypes. *Cell Immunol.* **89**: 95–112

Aruffo, A., Farrington, M., Hollenbaugh, D. *et al.* (1993). The CD40 ligand, gp39, is defective in activated T cells from patients with X-linked hyper IgM syndrome. *Cell* **72**: 291–300

Asanuma, H., Aizawa, C., Kurata, T., Tamura, S. (1988). IgA antibody-forming cell responses in the nasal-associated lymphoid tissue of mice vaccinated by intranasal, intravenous and/or sub-cutaneous administration. *Vaccine* **16**: 1257–62

Ashley, R.L., Corey, L., Dalessio, J. *et al.* (1994). Protein-specific cervical antibody responses to primary genital herpes simplex virus type 2 infections. *J. Infect. Dis.* **170**: 20–26

Baenziger, J. and Kornfeld, S. (1974a). Structure of the carbohydrate units of IgA1 immunoglobulin. I. Composition, glycopeptide isolation, and structure of the asparagine-linked oligosaccharide units. *J. Biol. Chem.* **249**: 7260–7269

Baenziger, J. and Kornfeld, S. (1974b). Structure of carbohydrate units of IgA1 immunoglobulin. II. Structure of the O-glycosidically linked oligosaccharide units. *J. Biol. Chem.* **249**: 7270–7281

Baker, D.A. and Plotkin, S.A. (1978). Enhancement of vaginal infection in mice by herpes simplex virus type 2 with progesterone. *Proc. Soc. Exp. Biol. Med.* **158**: 131–134

Bastian, A., Kratzin, H., Eckart, H. *et al.* (1992). Intra- and Interchain disulfide bridges of the human J chain in secretory immunoglobulin A. *Biol. Chem.* **373**: 1255–1263

Beagley, K.W., Block, S.L., Lee, F. *et al.* (1989). Interleukins and IgA synthesis. Human and murine interleukin 6 induce high rate IgA secretion in IgA-committed B cells. *J. Exp. Med.* **169**: 2133–2148

Belec, L., Tevi-Benissan, C., Lu, X.S. *et al.* (1995). Local synthesis of IgG antibodies to HIV within the female and male genital tracts during asymptomatic and pre-AIDS stages of HIV infection. *AIDS Res. Hum. Retroviruses* **11**: 719

Belshe, R.B., Mendelman, P.M., Treanor, J. *et al.* (1998). Efficacy of trivalent live attenuated intranasal influenza vaccine in children. *N. Engl. J. Med.* **338**: 1405–1412

Bene, M.C., Studer, A., Faure, G. (1988). Immunoglobulin-producing cells in human prostate. *Prostate* **12**: 113–117

Bergmeier, L.A., Tao, L., Gearing, A.J. *et al.* (1995). Induction of IgA and IgG antibodies in vaginal fluid, serum and saliva following immunization of genital and gut associated lymphoid tissue. *Adv. Exp. Med. Biol.* **371B**: 1567–1573

Bergquist, C., Johansson, E.L., Lagergard, T., Holmgren, J., Rudin, A. (1997). Intranasal vaccination of humans with recombinant cholera toxin B subunit induces systemic and local antibody responses in the upper respiratory tract and the vagina. *Infect. Immun.* **65**: 2676–84

Bernstein, D.I. and Tepe, E.R. (1994). Effect of route of immunization on HSV-2 vaccine efficacy. [Abstract] International Herpesvirus Workshop, Vancouver, British Columbia, Canada 1994

Bernstein, D.I., Glass, R.I., Rogers, G. *et al.* (1995). Evaluation of rhesus rotavirus monovalent and tetravalent reassortant vaccines in U.S. children. *J. Am. Med. Assoc.* **273**: 1191–1196

Bockman, D.E. and Cooper, M.D. (1973). Pinocytosis by epithelium associated with lymphoid follicles in the bursa of fabricius, appendix, and Peyer's patches. *Am. J. Anat.* **136**: 455–478

Boslego, J.W., Tramont, E.C., Chung, R.C. *et al.* (1991). Efficacy trial of a parenteral gonococcal pilus vaccine in men. *Vaccine* **9**: 154–162

Brandtzaeg, P., and Baklien, K. (1976). Immunohistochemical studies of the formation and epithelial transport of immunoglobulins in normal and diseased human intestinal mucosa. *Gastroenterology* **11**: 1–45

Brandtzaeg, P. (1984). Research in gastrointestinal immunology. State of arts. *Scand. J. Gastroenterol.* **114**: 137–156

Brandtzaeg, P., Kett, K., Rognum, T.O. *et al.* (1986). Distribution of mucosal IgA and IgG subclass-producing immunocytes and alterations in various disorders. *Monogr. Allergy* **20**: 179–194

Brandtzaeg, P., Sollid, L.M., Thrane, P.S. *et al.* (1988). Lymphoepithelial interactions in the mucosal immune system. *Gut* **29**: 1116–1130

Brandtzaeg, P., Rabinovich, R., Lamm, M.E. *et al.* (1994). In: 'Handbook of Mucosal Immunology' (ed. Ogra, P.L., Mestecky, J., Lamm, M.E., Strober, W., McGhee, J.R., Bienenstock, J.), Function of Mucosal Immunoglobulins. pp. 113–126. Academic Press, San Diego, CA

Brandtzaeg, P. (1995). Molecular and cellular aspects of the secretory immunoglobulin. *APMIS* **103**: 1–19

Briese, V., Pohl, W.D., Noack, K. *et al.* (1987). Influenza specific antibodies in the female genital tract of mice after oral administration of live influenza vaccine. *Arch. Gynaecol.* **240**: 153–157

Bukawa, H., Sekigawa, K., Hamajima, K. *et al.* (1995). Neutralization of HIV-1 by secretory IgA induced by oral immunization with a new macromolecular multicomponent peptide vaccine candidate. *Nature Med.* **1**: 681–685

Burns, J.W., Siadat-Pajouh, M., Krishnaney, A.A. *et al.* (1996). Protective effect of rotavirus VP6-specific IgA monoclonal antibodies that lack neutralizing activity. *Science* **272**: 104–107

Butcher, E.C., Rouse, R.V., Coffman, R.L. *et al.* (1982). Surface phenotype of Peyer's patch germinal center cells: Implications for the role of germinal centers in B cell differentiation. *J. Immunol.* **129**: 2698–2707

Canning, M.B. and Billington, W.D. (1983). Hormonal regulation of immunoglobulin and plasma cells in the mouse uterus. *J. Endocrinol.* **97**: 419–424

Challacombe, S.J., Rahman, D. and O'Hagan, D.T. (1997). Salivary, gut, vaginal and nasal antibody responses after oral immunization with biodegradable microparticles. *Vaccine* **15**: 169–175

Clark, E.A. (1997). Regulation of B lymphocytes by dendritic cells. *J. Exp. Med.* **185**: 801–803

Clements, M.L., Betts, R.F., Tierney, E.L. *et al.* (1986). Serum and nasal wash antibodies associated with resistance to experimental challenge with influenza A wild-type virus. *J. Clin. Microbiol.* **24**: 157–160

Coffman, R.L. and Carty, J. (1986). A T cell activity that enhances polyclonal IgE production and its inhibition by interferon-gamma. *J. Immunol.* **136**: 949–954

Coffman, R.L., Ohara, J., Bond, M.W. *et al.* (1986). B cell stimulatory factor-1 enhances the IgE response of lipopolysaccharide-activated B cells. *J. Immunol.* **136**: 4538–4541

Coffman, R.L., Lebman, D.A. and Schrader, B. (1989). Transforming growth factor β specifically enhances IgA production by lipopolysaccharide-stimulated murine B lymphocytes. *J. Exp. Med.* **170**: 1039–1044

Crowley-Nowick, P.A., Bell, M., Edwards, R.P. *et al.* (1995). Normal uterine cervix: characterization of isolated lymphocyte phenotypes and immunoglobulin secretion. *Am. J. Reprod. Immunol.* **34**: 241–247

Cui, Z.D., Tristram, D., LaScolae, L.J. *et al.* (1991). Induction of antibody response to chlamydia trachomatis in the genital tract by oral immunization. *Infect. Immun.* **59**: 1465–1469

Defrance, T., Vandervliet, B., Briere, F. *et al.* (1992). Interleukin-10 and transforming growth factor β cooperate to induce anti-CD40-activated naive human B cells to secrete immunoglobulin A. *J. Exp. Med.* **175**: 671–682

DiLoreto, C., Beltrami, C.A., DeNictolis, M. *et al.* (1989). Immunohistochemical characterization and distribution of Langerhans' cells in normal epithelium of the uterine cervix. *Basic Appl. Histochem.* **33**: 39–48

DiTommaso, A., Saletti, G., Pizza, M. *et al.* (1996). Induction of antigen-specific antibodies in vaginal secretions by using a non-toxic mutant of heat-labile enterotoxin as a mucosal adjuvant. *Infect. Immun.* **64**: 974–979

Doble, A., Walker, M.M., Harris, J.R.W. *et al.* (1990). Intraprostatic antibody deposition in chronic abacterial prostatitis. *Br. J. Urol.* **65**: 598–605

Dubois, B., Vandervliet, B., Fayette, J. *et al.* (1997). Dendritic cells enhance growth and differentiation of CD40-activated B lymphocytes. *J. Exp. Med.* **185**: 941–951

Eckmann, L., Morzycka-Wroblewska, E., Smith, J.R. *et al.* (1992). Cytokine-induced differentiation of IgA B cells: studies using an IgA expressing B-cell lymphoma. *Immunology* **76**: 235–241

Edwards, J.N.T. and Morris, H.B. (1985). Langerhan's cells and lymphocyte subsets in the female genital tract. *Br. J. Obstet. Gynaecol.* **92**: 974–982

Edwards, K.M., Dupont, W.D., Westrich, M.K. *et al.* (1994). A randomized controlled trial of cold-adapted and inactivated vaccines for the prevention of influenza A disease. *J. Infect. Dis.* **169**: 68–76

Eggert-Kruse, W., Buhlinger-Gopfarth, N., Rohr, G. *et al.* (1996). Antibodies to chlamydia trachomatics in semen and relationship with parameters of male fertility. *Hum. Reprod.* **11**: 1408–1417

El-Demiry, M.I.M., Young, H., Elton, R.A. *et al.* (1986). Leukocytes in the ejaculate from fertile and infertile men. *Br. J. Urol.* **58**: 715–720

Fayette, J., Dubois, B., Vandenabeek, S. *et al.* (1997). Human dendritic cells skew isotype switching of CD40-activated naive B cells towards IgA_1 and IgA_2. *J. Exp. Med.* **185**: 1909–1918

Fidel, P.L., Jr., Lynch, M.E., Conaway, D.H. *et al.* (1995). Mice immunized by primary vaginal *Candida albicans* infection develop acquired vaginal mucosal immunity. *Infect. Immun.* **63**: 47–553

Fidel, P.L., Jr., Wolf, N.A. and KuKuruga, M.A. (1996). T lymphocytes in the murine vaginal mucosa are phenotypically distinct from those in the periphery. *Infect. Immun.* **64**: 3793–3799

Fowler, J.E., Jr., Kaiser, D.L. and Mariano, M. (1982). Immunologic response of the prostate to bacteriuria and bacterial prostatitis: Part I. Immunoglobulin concentrations in prostatic fluid. *J. Urol.* **128**: 158–164

Fowler, J.E., Jr. and Mariano, M. (1982). Immunologic response of the prostate to bacteriuria and bacterial prostatitis: Part II. Antigen-specific immunoglobulin in prostatic fluid. *J. Urol.* **128**: 165–170

Fowler, J.E., Jr. (1993). In: 'Local Immunity in Reproductive Tract Tissues' (ed. Griffin, P.D., Johnson, P.M.), Immunity to Infection in the Male Reproductive Tract. pp.341–355. Oxford University Press, Oxford, UK

Fox, H. (1993). In: 'Local Immunity in Reproductive Tract Tissues' (ed. Griffin, P.D., Johnson, P.M.), Immunocompetent Cells in the Cervix and Vagina. pp.177–186. Oxford University Press, New York, NY

Foy, T.M., Aruffo, A., Bajorath, J. *et al.* (1996). Immune regulation by CD40 and its ligand GP39. *Ann. Rev. Immunol.* **14**: 591–618

Franco, M.A., Tin, C. and Greenberg, H.B. (1997). CD8[+] T cells can mediate almost complete short-term and partial long-term immunity to rotavirus in mice. *J. Virol.* **71**: 4165–4170

Gallichan, W.S., Johnson, D.C. and Graham, F.L. (1993). Mucosal immunity and protection after intranasal immunization with recombinant adenovirus expressing herpes simplex virus glycoprotein B. *J. Infect. Dis.* **168**: 622–629

Gallichan, W.S. and Rosenthal, K.L. (1996a). Effects of the estrous cycle on local humoral immune responses and protection of intranasally immunized female mice against herpes simplex virus type 2 infection in the genital tract. *Virology* **224**: 487–497

Gallichan, W.S. and Rosenthal, K.L. (1996b). Long-lived cytotoxic T lymphocyte memory in mucosal tissues after mucosal but not systemic immunization. *J. Exp. Med.* **184**: 1879–1890

Gruber, W.C., Belsche, R.B., King, K.C. *et al.* (1996). Evaluation of live attenuated influenza vaccines in children 6–18 months of age: safety, immunogenicity, and efficacy. *J. Infect. Dis.* **173**: 1313–1319

Haneberg, B., Kendall, D., Amerongen, H.M. *et al.* (1994). Induction of specific immunoglobulin A in the small intestine, colon-rectum and vagina measured by a new method for collection of secretions from local mucosal surfaces. *Infect. Immun.* **62**: 15–23

Harrison, P.E., Barratt, C.L.R., Robinson, A.J. *et al.* (1991). Detection of white blood cell populations in the ejaculates of fertile men. *J. Reprod. Immunol.* **19**: 95–98

Hendrickson, B.A., Conner, D.A., Ladd, D.J. *et al.* (1995). Altered hepatic transport of immunoglobulin A in mice lacking the J chain. *J. Exp. Med.* **182**: 1905–1911

Hendrickson, B.A., Rindisbacher, L., Corthesy, B. *et al.* (1996). Lack of association of secretory component with IgA in J chain-deficient mice. *J. Immunol.* **157**: 750–754

Hocini, H., Barra, A., Belec, L. *et al.* (1995). Systemic and secretory humoral immunity in the normal human vaginal tract. *J. Immunol.* **42**: 269

Hordnes, K., Digranes, A., Haugen, I.L. *et al.* (1995). Systemic and mucosal antibody responses to group B streptococci following immunization of the colonic-rectal mucosa. *J. Reprod. Immunol.* **28**: 247–262

Hordnes, K., Tynning, J., Kvam, A.I. *et al.* (1996). Colonization in the rectum and uterine cervix with group B streptococci may induce specific antibody responses in cervical secretions of pregnant women. *Infect. Immun.* **64**: 1643

Igietseme, J.U. and Rank, R.G. (1991). Susceptibility to reinfection after a primary chlamydial genital infection is associated with a decrease of antigen-specific T cells in the genital tract. *Infect. Immun.* **59**: 1346–1351

Irie, H., Harada, Y., Kataoka, M. *et al.* (1992). Efficacy of oral administration of live herpes simplex virus type 1 as a vaccine. *J. Virol.* **66**: 2428–2434

Johansson, M., Schon, K., Ward, M. *et al.* (1997). Genital tract infection with *Chlamydia trachomatis* fails to induce protective immunity in gamma interferon receptor-deficient mice despite a strong local immunoglobulin A response. *Infect. Immun.* **65**: 1032–1044

Kawanishi, H., Saltzman, L. and Strober, W. (1983). Mechanisms regulating IgA class-specific immunoglobulin production in murine gut-associated lymphoid tissues. I. T cells derived from Peyer's patches that switch sIgM B cells to SIgA B cells *in vitro*. *J. Exp. Med.* **157**: 437–450

Kett, K., Brandtzaeg, P., Radl, J. *et al.* (1986). Different subclass distribution of IgA-producing cells in human lymphoid organs and various secretory tissues. *J. Immunol.* **136**: 3631–3635

Kim, P.H. and Kagnoff, M.F. (1990). Transforming growth factor beta 1 increases IgA isotype switching at the clonal level. *J. Immunol.* **145**: 3773–3778

Kiyono, H., Bienenstock, J., McGhee, J.R. *et al.* (1992). The mucosal immune system: features of inductive and effector sites to consider in mucosal immunization and vaccine development. *Reg. Immunol.* **4**: 54–62

Klavinskis, L.S., Bergmeier, L.A., Gao, L. *et al.* (1996). Mucosal or targeted lymph node immunization of macaques with a particulate SIVp27 protein elicits virus-specific CTL in the genito-rectal mucosa and draining lymph nodes. *J. Immunol.* **157**: 2521–2527

Kozlowski, P.A., Cu-Uvin, S., Neutra, M.R. *et al.* (1997). Comparison of the oral, rectal, and vaginal immunization routes for induction of antibodies in rectal and genital tract secretions of women. *Infect. Immun.* (in press)

Kozlowski, P.A., Cu-Uvin, C., Neutra, M.R., Flanigan, T.P. (1999). Mucosal vaccination strategies for women. *J. Infect. Dis.* 179, S493–8

Kubota, M., Miller, C.J., Imaoka, K. *et al.* (1997). Oral immunization with simian immunodeficiency virus p55gag and cholera toxin elicits both mucosal IgA and systemic IgG immune responses in non-human primates. *J. Immunol.* 158: 5321–5329

Kuklin, N., Daheshia, M., Karem, K. *et al.* (1997). Induction of mucosal immunity against herpes simplex virus by plasmid DNA immunization. *J. Virol.* 71: 3138–3145

Kutteh, W.H., Hatch, K.D., Blackwell, R.E. *et al.* (1988). Secretory immune system of the female reproductive tract: I. Immunoglobulins and secretory component-containing cells. *Obstet. Gynecol.* 71: 56–60

Kutteh, W.H., Edwards, R.P., Menge, A.C. *et al.* (1993). In: 'Local Immunity in Reproductive Tract Tissues' (ed. Griffin, P.D., Johnson, P.M.), IgA Immunity in Female Reproductive Tract Secretions. pp.229 Oxford University Press, New Delhi, India

Kutteh, W.H. and Mestecky, J. (1994). Secretory immunity in the female reproductive tract. *Am. J. Reprod. Immunol.* 31: 40

Kutteh, W.H., Wester, R., Prince, S.J. *et al.* (1996). Variations in immunoglobulins and IgA subclasses of human uterine cervical secretions around the time of ovulation. *Clin. Immunol.* 104: 538

Langman, J.M. and Rowland, R. (1986). The number and distribution of lymphoid follicles in the human large intestine. *J. Anat.* 194: 189–194

Lehner, T., Brookes, R., Panagiotidi, C. *et al.* (1993). T- and B-cell functions and epitope expression in non-human primates immunized with simian immunodeficiency virus antigen by the rectal route. *Proc. Natl. Acad. Sci. USA* 90: 8638–8642

Lehner, T., Bergmeier, L.A., Tao, L. *et al.* (1994a). Targeted lymph node immunization with simian immunodeficiency virus p27 antigen to elicit genital, rectal, and urinary immune responses in nonhuman primates. *J. Immunol.* 153: 1858–1868

Lehner, T., Tao, L., Panagiotidi, C. *et al.* (1994b). Mucosal model of genital immunization in male rhesus macaques with a recombinant simian immunodeficiency virus p27 antigen. *J. Virol.* 68: 1624–1632

Lehner, T. and Miller, C.J. (1996). In: 'Mucosal Vaccines' (ed. Kiyono, H., Ogra, P.L., McGhee, J.R.), Rectal and Genital Immunization with SIV/HIV. pp.357–371. Academic Press, San Diego, CA

Lehner, T., Wang, Y., Cranage, M. *et al.* (1996). Protective mucosal immunity elicited by targeted iliac lymph node immunization with a subunit SIV envelope and core vaccine in macaques. *Nature Med.* 2: 767–775

Lehner, T., Wang, Y., Ping, L., Bergmeier, L. Mitchell, E., Cranage, M. *et al.* (1999). The effect of route of immunization on mucosal immunity and protection. *J. Infect. Dis.* 179, S489–92

Lehrer, R.I. and Ganz, T. (1996). Endogenous vertebrate antibiotics. Defensins, protegrins, and other cysteine-rich antimicrobial peptides. *Ann. N.Y. Acad. Sci.* 797: 228–239

London, S.D., Rubin, D.H. and Cebra, J.J. (1986). Gut mucosal immunization with reovirus serotype I/L stimulates virus-specific cytotoxic T cell precursors as well as IgA memory cells in Peyer's patches. *J. Exp. Med.* 165: 830–847

London, S.D., Cebra, J.J. and Rubin, D.H. (1989). Intraepithelial lymphocytes contain virus-specific, MHC-restricted cytotoxic cell precursors after gut mucosal immunization with reovirus serotype 1/Lang. *Reg. Immunol.* 2: 98–102

Lubeck, M.D., Natuk, R.J., Chengalvala, M. *et al.* (1994). Immunogenicity of recombinant adenovirus-human immunodeficiency virus vaccines in chimpanzees following intranasal administration. *AIDS Res. Hum. Retroviruses* 10: 1443–1449

MacDonald, T.T. and Spencer, J. (1994). In: 'Handbook of Mucosal Immunology' (ed. Ogra, P.L., Lamm, M.E., McGhee, J.R., Mestecky, J., Strober, W., Bienenstock, J.), *Gut-associated Lymphoid Tissue.* pp.415–424. Academic Press, San Diego

Marinaro, M., Boyaka, P.N., Finkelman, F.D. *et al.* (1997). Oral but not parenteral interleukin (IL)-12 redirects T helper 2 (Th2)-type responses to an oral vaccine without altering mucosal IgA responses. *J. Exp. Med.* 185: 415–427

Marthas, M., Miller, C.J., Sutjipto, S. *et al.* (1992). Efficacy of live-attenuated and whole-inactivated simian immunodeficiency virus vaccines against vaginal challenge with virulent SIV. *J. Med. Primatol.* **21**: 99–107

Marx, P.A., Compans, R.W., Gettie, A. *et al.* (1993). Protection against vaginal SIV transmission with microencapsulated vaccine. *Science* **260**: 1323–1327

Mazanec, M.B., Kaetzel, C.S., Lamm, M.E. *et al.* (1992). Intracellular neutralization of virus by immunoglobulin A antibodies. *Proc. Natl. Acad. Sci. USA* **89**: 6901–6905

Mazanec, M.B., Coudret, C.L. and Fletcher, D. (1995). Intracellular neutralization of influenza virus by immunoglobulin A anti-hemagglutinin monoclonal antibodies. *J. Virol.* **69**: 1339–1343

McDermott, M.R. and Bienenstock, J. (1979). Evidence for a common mucosal immunologic system. I. Migration of B immunoblasts into intestinal, respiratory and genital tissues. *J. Immunol.* **122**: 1892–1898

McDermott, M.R., Clark, D.A. and Bienenstock, J. (1980). Evidence for a common mucosal immunologic system. II. Influence of the estrous cycle on B immunoblast migration into genital and intestinal tissues. *J. Immunol.* **124**: 2536–2539

McDermott, M.R., Goldsmith, C.H., Rosenthal, K.L. *et al.* (1989). T lymphocytes in genital lymph nodes protect mice from intravaginal infection with herpes simplex virus type 2. *J. Infect. Dis.* **159**: 460–466

McDermott, N., Smiley, J., Leslie, P. *et al.* (1984). Immunity in the female genital tract after intravaginal vaccination of mice with an attenuated strain of herpes simplex virus type 2. *J. Virol.* **51**: 747–753

McNeal, M.M., Broome, R.L. and Ward, R.L. (1994). Active immunity against rotavirus infection in mice is correlated with viral replication and titers of serum rotavirus IgA following vaccination. *Virology* **204**: 642–650

Merriman, H., Woods, S., Winter, C. *et al.* (1984). Secretory IgA antibody in cervicovaginal secretions from women with genital infection due to herpes simplex virus. *J. Infect. Dis.* **149**: 505–510

Mestecky, J. and Fultz, P.N. (1999) Mucosal immune system for the human genital tract. *J. Infect. Dis.* **79**: S470–4

Mestecky, J., Russell, M.W., Jackson, S. *et al.* (1986). The human IgA system: A reassessment. *Clin. Immunol. Immunopathol.* **40**: 105–114

Mestecky, J. (1987). The common mucosal immune system and current strategies for induction of immune responses in external secretions. *J. Clin. Immunol.* **7**: 265–276

Mestecky, J. and McGhee, J.R. (1987). Immunoglobulin A (IgA): molecular and cellular interactions involved in IgA biosynthesis and immune response. *Adv. Immunol.* **40**: 153–245

Mestecky, J. and McGhee, J.R. (1993). In 'Reproductive Tract Tissues' (ed. Griffin, P.D., Johnson, P.M.), The Secretory IgA System in Local Immunity. pp.53–72. Oxford University Press, New York, NY

Milligan, G.N. and Bernstein, D.I. (1995a). Generation of humoral responses against herpes simplex virus type 2 (HSV-2) in the murine female genital tract. *Virology* **206**: 234–241

Milligan, G.N. and Bernstein, D.I. (1995b). Analysis of herpes simplex virus specific T cells in the murine female genital tract following genital infection with HSV-2. *Virology* **212**: 481–489

Milligan, G.N. and Bernstein, D.I. (1997). Interferon gamma enhances resolution of herpes simplex type 2 (HSV-2) infection of the murine genital tract. *Virology* **229**: 259–268

Milligan, G.N., Bernstein, D.I. and Bourne, N. (1998). T lymphocytes are required for protection of the vaginal mucosa and sensory ganglia of immune mice against reinfection with herpes simplex virus type 2. *J. Immunol.* **160**: 6093–6100

Mogens, K. and Russell, M.W. (1994). In 'Handbook of Mucosal Immunology' (ed. Ogra, P.L., Mestecky, J., Lamm, M.E., Strober, W., McGhee, J.R., Bienenstock, J.), Function of Mucosal Immunoglobulins. pp.127–137. Academic Press, San Diego, CA

Moldoveanu, Z., Egan, M.L. and Mestecky, J. (1984). Cellular origins of human polymeric and monomeric IgA: Intracellular and secreted forms of IgA. *J. Immunol.* **133**: 3156–3162

Morris, H.H.B., Gatter, K.C., Stein, H. *et al.* (1983). Langerhans' cells in human cervical epithelium: an immunohistological study. *J. Obstet. Gynecol.* **90**: 400–411

Morris, W., Steinhoff, M.C. and Russell, P.K. (1994). Potential of polymer microencapsulation technology for vaccine innovation. *Vaccine* **12**: 5–11

Mosley, R.L., Whetsell, M. and Klein, J.R. (1991). Proliferative properties of murine intestinal intraepithelial lymphocytes (IELs): IEL-expressing TCRαβ or TCRγ/δ are largely unresponsive to proliferative signals mediated via conventional stimulation of the CD3-TCR complex. *Int. Immunol.* 3: 563–569

Murray, P.D., McKenzie, D.T., Swain, S.L. *et al.* (1987). Interleukin 5 and interleukin 4 produced by Peyer's patch T cells selectively enhance immunoglobulin A expression. *J. Immunol.* 139: 2669–2674

Muster, T., Ferko, B., Klima, A. *et al.* (1995). Mucosal model of immunization against human immunodeficiency virus type 1 with a chimeric influenza virus. *J. Virol.* 69: 6678–6686

Nandi, D. and Allison, J.P. (1991). Phenotypic analysis and gamma delta-T cell receptor repertoire of murine T cells associated with the vaginal epithelium. *J. Immunol.* 147: 1773–1778

Neutra, M.R. and Kraehenbuhl, J.-P. (1996). In: 'Mucosal Vaccines' (ed. Kiyono, H., Ogra, P.L., McGhee, J.R.), Antigen Uptake by M Cells for Effective Mucosal Vaccines, pp.41–55. Academic Press, San Diego, CA

Neutra, M.R., Pringault, E. and Kraehenbuhl, J.-P. (1996). Antigen sampling across epithelial barriers and induction of mucosal immune responses. *Ann. Rev. Immunol.* 14: 275–300

Neutra, M.R., Phillips, T.L., Mayer, E.L. *et al.* (1987). Transport of membrane-bound macro-molecules by M cells in follicle-associated epithelium of rabbit Peyer's patch. *Cell Tissue Res.* 247: 537–546

Nichols, R.L., Murray, E.S. and Nisson, P.E. (1978). Use of enteric vaccines in protection against chlamydial infections of the genital tract and the eye of guinea pigs. *J. Infect. Dis.* 138: 742–746

O'Leary, A.D. and Sweeney, E.C. (1986). Lymphoglandular complexes of the colon: structure and distribution. *Histopathology* 10: 267–283

Offit, P.A. and Dudzik, K.I. (1989). Rotavirus-specific cytotoxic T lymphocytes appear at the intestinal mucosal surface after rotavirus infection. *J. Virol.* 63: 3507–3512

Offit, P.A., Cunningham, S.L. and Dudzik, K.I. (1991). Memory and distribution of virus-specific cytotoxic T lymphocytes (CTL) and CTL precursors after rotavirus infection. *J. Virol.* 65: 1318–1324

Ogra, P.L. and Ogra, S.S. (1973). Local antibody response to poliovaccine in the human female genital tract. *J. Reprod. Immunol.* 110: 1307–1311

Ogra, P.L., Yamanaka, T. and Losonsky, G.A. (1981). In 'Reproductive Immunology' (ed. Gleicher, N.), Local Immunologic Defenses in the Genital Tract, pp.381–394. Alan R. Liss, Inc. New York, NY

Ogra, P.L. (1984). Mucosal immune response to poliovirus vaccines in childhood. *Rev. Infect. Dis.* 6, S361

Otsuki, Y. (1989). Lymphatics and lymphoid tissue of the fallopian tube: Immunoelectronmicro-scopic study. *Anat. Rec.* 225: 288–289

Owen, R.L. (1977). Sequential uptake of horseradish peroxidase by lymphoid follicle epithelium of Peyer's patches in the normal unobstructed mouse intestine: an ultrastructural study. *Gastro-enterology* 72: 440–451

Pal, S., Peterson, E.M., de la Maza, L.M. (1996). Intransal immunization induces long-term protection in mice against a Chlamydia trachomatis genital challenge. *Infect. Immun.* 64: 5341–8

Parr, E.L. and Parr, M.B. (1985). Secretory immunoglobulin binding to bacteria in the mouse uterus after mating. *J. Reprod. Immunol.* 8: 71–82

Parr, E.L. and Parr, M.B. (1988). Binding of C3 to bacteria in the mouse uterus after mating. *J. Reprod. Immunol.* 12: 315–319

Parr, E.L., Parr, M.B. and Thapar, M. (1988). A comparison of specific antibody responses in mouse vaginal fluid after immunization by several routes. *J. Reprod. Immunol.* 14: 165–176

Parr, M.B. and Parr, E.L. (1989). Immunohistochemical localization of secretory component and immunoglobulin A in the urogenital tract of the male rodent. *J. Reprod. Fertil.* 85: 115–124

Parr, E.L. and Parr, M.B. (1990). A comparison of antibody titers in mouse uterine fluid after immunization by several routes, and the mucosal immune responses. *Annual Review of Immunology* 14: 275–300 effect of the uterus on antibody titers in vaginal fluid. *J. Reprod. Fertil.* 89: 619–625

Parr, M.B. and Parr, E.L. (1991). Langerhans' cells and T lymphocyte subsets in the murine vagina and cervix. *Biol. Reprod.* 44: 491–498

Parr, M.B. and Parr, E.L. (1994). In: 'Handbook of Mucosal Immunology' (ed. Ogra, P.L., Mestecky, J., Lamm, M.E., Strober, W., McGhee, J.R., Bienenstock, J.), Mucosal immunity in the female and male reproductive tracts. pp.677–689. Academic Press, San Diego, CA

Peters, W.M. (1986). Nature of 'basal' and 'reserve' cells in oviductal and cervical epithelium in man. *J. Clin. Pathol.* **39**: 306–312

Phillips-Quagliata, J.M. and Lamm, M.E. (1994). In: 'Handbook of Mucosal Immunology' (ed. Ogra, P.L., Strober, W., Mestecky, J., McGhee, J.R., Lamm, M.E., Bienenstock, J.), Lymphocyte Homing to Mucosal Effector Sites. pp.225–239. Academic Press, San Diego, CA

Prabhala, R.H. and Wira, C.R. (1991). Cytokine regulation of the mucosal immune system: *In vivo* stimulation by interferon-γ of secretory component and immunoglobulin A in uterine secretions and proliferation of lymphocytes from spleen. *Endocrinology* **129**: 2915–2923

Prabhala, R.H. and Wira, C.R. (1995). Sex hormone and IL-6 regulation of antigen presentation in the female reproductive tract mucosal tissues. *J. Immunol.* **155**: 5566–5573

Pudney, J.A. and Anderson, D.J. (1993). In: 'Local Immunity in Reproductive Tract Tissues.' (ed. Griffin, P.D., Johnson, P.M.), Organization of Immunocompetent Cells and Their Function in the Male Reproductive Tract. pp.131–145. Oxford University Press, Oxford, UK

Pumphrey, R.S.H. (1986). Computer models of the human immunoglobulins. Binding sites and molecular interactions. *Immunol. Today* **7**: 206–211

Quesnel, A., Cu-Uvin, S., Murphy, D. *et al.* (1997). Comparative analysis of methods for collection and measurement of immunoglobulins in cervical and vaginal secretions of women. *J. Immunol. Meth.* **202**: 153–161

Quiding-Jabrink, M., Nordstrom, I., Granstrom, G. *et al.* (1997). Differential expression of tissue-specific adhesion molecules on human circulating antibody-forming cells after systemic, enteric, and nasal immunizations. A molecular basis for the compartmentalization of effector B cell responses. *J. Clin. Invest.* **99**: 1281–1286

Ramsey, K.H., Soderberg, L.S. and Rank, R.G. (1988). Resolution of chlamydial genital infection in B-cell-deficient mice and immunity to reinfection. *Infect. Immun.* **56**: 1320–1325

Renegar, K.B. and Small, P.A., Jr. (1991). Passive transfer of local immunity to influenza virus infection by IgA antibody. *J. Immunol.* **146**: 1972–1978

Rennels, M.B., Glass, R.I., Dennehy, P.H. *et al.* (1996). Safety and efficacy of high dose rhesus-human reassortant rotavirus vaccines: Report of the National Multicenter Trial. *Pediatrics* **97**: 7–13

Roncalli, M., Sideri, M., Gie, P. *et al.* (1988). Immunophenotypic analysis of the transformation zone of human cervix. *Lab. Invest.* **58**: 141–149

Rudin, A., Johansson, E.L., Bergquist, C., Holmgren, J. (1998). Differential kinetics and distribution of antibodies in serum and nasal and vaginal secretions after nasal and oral vaccination of humans. *Infect. Immun.* **66**: 3390–6

Sangster, M.Y., Mo, X.Y., Sealy, R. *et al.* (1997). Matching antibody class with pathogen type and portal of entry. Cognate mechanisms regulate local isotype expression patterns in lymph nodes draining the respiratory tract of mice inoculated with respiratory viruses, according to virus replication competence and site of inoculation. *J. Immunol.* **159**: 1893–1902

Snapper, C.M. and Paul, W.E. (1987). Brell stimulatory factor-1 (interleukin 4) prepares resting B murine B cells to secrete IgG1 upon subsequent stimulation with bacterial lipopolysaccharide. *J. Immunol.* **139**: 10–17

Sonada, E., Matsumoto, R., Hitoshi, Y. *et al.* (1989). Transforming growth factor β induces IgA production and acts additively with Interleukin 5 for IgA production. *J. Exp. Med.* **170**: 1415

Staats, H.F. and McGhee, J.R. (1996). In 'Mucosal Vaccines' (ed. Kiyono, H., Ogra, P.L., McGhee, J.R.), Application of Basic Principles of Mucosal Immunity to Vaccine Development. pp.17–39. Academic Press, San Diego, CA

Steinman, R.M. (1991). The dendritic cell system and its role in immunogenicity. *Ann. Rev. Immunol.* **9**: 271–296

Strober, W. and Sneller, M.C. (1988). Cellular and molecular events accompanying IgA B cell differentiation. *Monogr. Allergy* **24**: 181–190

Strober, W. and Ehrhardt, R.O. (1994). In 'Handbook of Mucosal Immunology' (ed. Ogra, P.L., Mestecky, J., Lamm, M.E., Strober, W., McGhee, J.R., Bienenstock, J.), Regulation of IgA B Cell Development. pp.159–176. Academic Press, San Diego, CA

Sullivan, D.A., Underdown, B.J. and Wira, C.R. (1983). Steroid hormone regulation of free secretory component in the rat uterus. *Immunology* **49**: 379–386

Sullivan, D.A. and Wira, C.R. (1983). Variations in free secretory component levels in mucosal secretions of the rat. *J. Immunol.* **130**: 1330–1335

Sullivan, D.A., Richardson, G.S., MacLaughlin, D.T. *et al.* (1984). Variations in the levels of secretory component in human uterine fluid during the menstrual cycle. *J. Steroid Biochem.* **20**: 509–513

Suzuki, M., Ogawa, M., Tamada, T. *et al.* (1984). Immunohistochemical localization of secretory component and IgA in human endometrium in relation to menstrual cycle. *Acta Histochem. Cytochem.* **17**: 223–229

Svehag, S.E. and Bloth, B. (1970). Ultrastructure of secretory and high-polymer serum immunoglobulin A of human and rabbit origin. *Science* **168**: 847–849

Taguchi, T., McGhee, J.R., Coffman, R.L. *et al.* (1990). Analysis of Th1 and Th2 cells in murine gut-associated tissues. Frequencies of CD4$^+$ and CD8$^+$ T cells that secrete IFN-γ and IL-5. *J. Immunol.* **145**: 68–77

Taguchi, T., Aicher, W.K., Fujihashi, K. *et al.* (1991). Novel function for intestinal intraepithelial lymphocytes. Murine CD3+, $\gamma\delta$ TCR$^+$ T cells produce IFNγ and IL-5. *J. Immunol.* **147**: 3736–3744

Tay, S.K., Jenkins, D., Maddox, P. *et al.* (1987). Lymphocyte phenotypes in cervical intraepithelial neoplasia and human papillomavirus infection. *Br. J. Obstet. Gynaecol.* **94**: 16–21

Thapar, M., Parr, E.L. and Parr, M.B. (1990). Secretory immune responses in mouse vaginal fluid after pelvic, parenteral or vaginal immunization. *Immunology* **70**: 121–125

Tomoda, T., Morita, H., Kurashige, T. *et al.* (1995). Prevention of influenza by the intranasal administration of cold-recombinant, live-attenuated influenza virus vaccine: importance of interferon-gamma production and local IgA response. *Vaccine* **13**: 185–190

Tuffrey, M., Falder, P., Gale, J. *et al.* (1986). Salpingitis in mice induced by human strains of *Chlamydia trachomatis*. *Br. J. Exp. Pathol.* **67**: 605–616

Underdown, B.J. and Mestecky, J. (1994). In 'Handbook of Mucosal Immunology' (ed. Ogra, P.L., Mestecky, J., Lamm, M.E., Strober, W., McGhee, J.R., Bienenstock, J.), Mucosal Immunoglobulins. pp.79–97. Academic Press, San Diego, CA

White, H.D., Crassi, K.M., Givan, A.L. *et al.* (1997). CD3$^+$ CD8$^+$ CTL activity within the human female reproductive tract: influence of stage of the menstrual cycle and menopause. *J. Immunol.* **158**: 3017–3027

Wira, C.R. and Sandoe, C.P. (1977). Sex steroid hormone regulation of IgG and IgA in rat uterine secretions. *Nature* **268**: 534–536

Wira, C.R. and Prabhala, R.H. (1993). In: 'Local Immunity in Reproductive Tract Tissues' (ed. Griffin, P.D., Johnson, P.M.), The Female Reproductive Tract is an Inductive Site for Immune Responses: Effect of Oestradiol and Antigen on Antibody and Secretory Component Levels in Uterine and Cervico-vaginal Secretions Following Various Routes of Immunization. pp.271–293. Oxford University Press, Oxford, UK

Wira, C.R., Richardson, J. and Prabhala, R. (1994). In: 'Handbook of Mucosal Immunology' (ed. Ogra, P.L., Mestecky, J., Lamm, M.E., Strober, W., McGhee, J.R., Bienenstock, J.), Endocrine Regulation of Mucosal Immunity: Effect of Sex Hormones and Cytokines on the Afferent and Efferent Arms of the Immune System in the Female Reproductive Tract. pp.705–718. Academic Press, San Diego, CA

Wira, C.R. and Rossoll, R.M. (1995). Antigen-presenting cells in the female reproductive tract: influence of the estrous cycle on antigen presentation by uterine epithelial and stromal cells. *Endocrinology* **136**: 4526–4534

Wold, A.E., Mestecky, J., Tomana, M. *et al.* (1990). Secretory immunoglobulin A carries oligosaccharide receptors for *Escherichia coli* type 1 fimbrial lectin. *Infect. Immun.* **58**: 3073–3077

Wolff, H. and Anderson, D.J. (1988). Immunohistologic characterization and quantitation of leukocyte subpopulations in human semen. *Fertil. Steril.* **49**: 497–504

Wolff, H., Mayer, K., Seage, G. *et al.* (1992). A comparison of HIV-1 antibody classes, titers and specificities in paired semen and blood samples from HIV-1 seropositive men. *J. AIDS* **5**: 65–69

Yamamoto, M., Vancott, J.L., Okahashi, N. *et al.* (1996). The role of Th1 and Th2 cells for mucosal IgA responses. *Ann. N. Y. Acad. Sci.* **778**: 64–71

Yeaman, G.R., Guyre, P.M., Fanger, M.W. *et al.* (1997). Unique CD8$^+$ T cell-rich lymphoid aggregates in human uterine endometrium. *J. Leuk. Biol.* **61**: 427–435

II

NON-SPECIFIC STRATEGIES FOR CONTROL

5

BEHAVIORAL AND PSYCHOLOGICAL FACTORS ASSOCIATED WITH STD RISK

SUSAN L. ROSENTHAL*
SHEILA S. COHEN[†]
AND
FRANK M. BIRO*

** Division of Adolescent Medicine,
Children's Hospital Medical Center,
and University of Cincinnati College of Medicine,
Cincinnati, Ohio, USA*

[†] *Central Psychiatric Clinic and Department of Pediatrics,
University of Cincinnati College of Medicine,
Cincinnati, Ohio, USA.*

INTRODUCTION

Reducing the frequency of risk behaviors and increasing health-promoting behaviors are key aspects in strategies designed to prevent sexually transmitted disease (STD) acquisition among sexually active individuals. All aspects of STD prevention have a behavioral component, even those designed to provide biological protection, such as vaccines. For example, vaccines do not provide protection if people are not immunized, as is evidenced by hepatitis B vaccine experience. Another example of the interplay between behavioral factors and biologic outcome is the association of an early initiation of intercourse with biological risk. Early initiation is also related to other behavioral risks, such as multiple partners, and may occur because of sociocultural factors such as a history of sexual abuse.

Risk factors are directly associated with acquiring the infection (i.e. causal factors). Risk markers are those variables associated with acquiring the infection but which do not directly cause the infection. Biological, behavioral and sociocultural risk factors/markers interact with each other and lead to infection. This chapter describes behavioral risk factors and markers, the influences on risk (both psychological and behavioral), and the psychological response to acquisition, together with a discussion of the approaches to prevention.

RISK FACTORS/MARKERS

Onset of Puberty

Puberty increases the adolescent's awareness of her or his body and sexual feelings and interests (Crockett and Petersen, 1993). Puberty has behavioral effects which are mediated through the influence of hormonal changes, and through the differential ways a person views him/herself once secondary sexual characteristics have developed. However, the timing of puberty relative to one's peers has a great impact on STD risk. Early-maturing girls are at greater risk for early initiation of intercourse (Brooks-Gunn and Paikoff, 1993), and an early onset of sexual activity is associated with an increase in behavioral and biological risks. Women who have their sexual debut between 10 and 14 years of age are significantly more likely to have multiple partners, more likely to report having sex with high-risk partners (i.e. intravenous drug-users or HIV-infected men), and more likely to have a history of an STD (Greenberg *et al.*, 1992).

Female Susceptibility

Within any given age group females have greater rates of STDs than males and suffer more profound consequences, such as involuntary infertility and cervical cancer (Institute of Medicine, 1997). Women's increased biological susceptibility may be due to the characteristics of female anatomy which make them especially sensitive to certain STDs (Cates, 1990). As noted above, a younger age at first intercourse in females is associated with greater risk of acquiring multiple episodes of an STD (Rosenthal *et al.*, 1994a). For up to 3 or 4 years following menarche, the columnar epithelium extends to the outer surface of the cervix. This leads to greater risk of *Chlamydia trachomatis* infection, since chlamydia binds to and infects the columnar epithelium (Harrison *et al.*, 1985).

Role of Asymptomatic Infections

Asymptomatic infection does not prompt affected individuals to seek health care and possibly take precautions to avoid transmitting their disease. Asymptomatic infection is very common: for example, among college women who tested positive for chlamydia, 79% had no symptoms of disease (Keim *et al.*, 1992). For viral infection, such as human papillomavirus (HPV), more than 90% of those infected may have not clinical signs or symptoms of infection (Moscicki, 1990). The consequence of delaying treatment varies according to the type of infectious process and the length of time the pathogen is infectious. For example, for bacterial infections the interaction of an asymptomatic infection with frequent partner switching (a risk factor for chlamydia) may lead to significant spread of the infection. For viral pathogens, which may be intermittently infectious, the frequency of partner changes may be less important, and even in stable monogamous relationships there may be risk. Asymptomatic infections make a large contribution to the high rates of STD transmission owing to the increased risk of infecting a subsequent contact (Corey, 1994; Potterat *et al.*, 1987).

Sociocultural Context

The rates for most STDs are greater in sexually active adolescents than in adults (Biro, 1992), and among both adolescents and adults there are subgroups that represent core groups for STD transmission. Factors such as poverty, low educational attainment and less power in one's immediate social world, as well as within society, are associated with risk. This is related not only to higher rates of risky behavior, but also to the high prevalence of disease in these partner pools, as well as lack of access to appropriate medical care. Homeless adolescents and adults, runaway adolescents, adolescents in detention and commercial sex workers all have high rates of sexual risk-taking and STDs (Breakey *et al.*, 1989; DiClemente *et al.*, 1991; Rosenblum *et al.*, 1992; Sugerman, 1991). In addition, children and adolescents who have been sexually abused are at risk for STD acquisition (Zierler *et al.*, 1991), which may occur as a direct result of the abuse (Siegel *et al.*, 1995). In addition, individuals with a past experience of abuse are vulnerable to future poor sexual decision-making. Lastly, another group at risk for STD acquisition is neonates born to high-risk women (Peloquin and Davidson, 1988).

PSYCHOLOGICAL/BEHAVIORAL INFLUENCES

Knowledge and Perception of Risk

Perceived risk is a critical variable in social psychological theories (health belief model, theory of reasoned action, protection motivation theory) addressing the reasons individuals either adopt, or fail to adopt, risk-reducing behaviors (Rosenthal *et al.*, 1994b). In order to protect oneself against STDs and manage one's risk, one needs accurate information and an understanding of risk. Knowledge is not sufficient, as even with complete knowledge and an accurate perception of risk an individual may not have the behavioral skills or motivation to implement protective behavior. For example, if a young woman perceives that requesting condom use will lead to sexual aggression, her perception of her risk of STD acquisition may not influence condom use behavior (Rosenthal and Cohen, 1994).

Both adults and adolescents often have inaccurate information regarding STDs (American Social Health Association, 1996; MacDonald *et al.*, 1990). Adults underestimate the prevalence of STDs and can name few other than HIV (American Social Health Association, 1996). This underestimation may contribute to difficulties in assessing one's own risk. In addition, engaging in high-risk behaviors without experiencing negative consequences (i.e. having sex with multiple partners without protection, and subsequently testing negative for STD acquisition) may result in one's perceiving one's personal susceptibility as low. Also, college students may underestimate the risk for STD associated with serial monogamy (Hernandez and Smith, 1990). Such perceptions of low personal risk may foster a sense of invulnerability and create little motivation to adopt protective measures. People may focus their concern on HIV, leading them to inadequately consider the risk of non-HIV STDs (MacDonald *et al.*, 1994). In addition, if individuals are primarily focused on their risk of pregnancy, condom use may decrease with the adoption of hormonal methods of birth control.

An important aspect of knowledge about STDs is the recognition that many infections are asymptomatic and are identified only through screening (Staat *et al.*, in press). Experience with an STD may not improve the ability to understand symptoms. Adolescents who have had an STD were more accurate in their knowledge regarding concrete symptoms and treatment of gonorrhea, but did not develop an accurate understanding of the potential for asymptomatic infections (Biro *et al.*, 1994). When people do not understand the concept of asymptomatic infection they may fail to accurately understand the risks of transmission, and may believe that they can tell whether a partner is infectious (Institute of Medicine, 1997).

In addition to the concept of asymptomatic infection, individuals may not recognize symptoms. They may attribute symptoms to the wrong cause, or may assume that the symptoms will remit spontaneously (Hook *et al.*, 1997; Staat *et al.*, in press). Any of these factors may lead to a failure to seek treatment. For example, although herpes simplex virus (HSV) infection may be inapparent in 71–89% of adolescents and college students (Rosenthal *et al.*, 1997a), such infection may not be truly asymptomatic. It is possible to teach adult women to recognize previously inapparent symptoms of herpes simplex infection (Langenberg *et al.*, 1989).

Personality Variables

Personality is considered to be those aspects of an individual that represent enduring dispositional traits. Although personality may be difficult to change, and may therefore, not be an appropriate target of interventions, understanding the role of personality in sexual decision-making may aid prevention planning (Jaccard and Wilson, 1991). For example, it has been suggested that heterosexuals who were more extroverted had increased STD acquisition (Hart, 1977), possibly because extroversion is associated with riskier sexual behaviors (Eysenck and Wilson, 1979). Among adolescent girls, those who were either impulsive or prone to anxiety scored higher on overt sexual behaviors (Muram *et al.*, 1991).

Another important aspect of personality is the perception that one can control destiny rather than it being controlled by others, by fate or by luck. This may be because one does not believe one should control reproductive health (i.e. it is the other gender's responsibility), or because one does not believe one has the skills to implement protective measures. Not all groups of high-risk individuals believe control over reproductive health is a desirable and expected outcome, and this belief may be related to lack of health-promoting behaviors (Millstein and Moscicki, 1995; St. Lawrence, 1993; Visher, 1986).

Partner Relationships

Sexual behaviors take place in the context of a relationship and although sexual intercourse is an intimate behavior, it does not always occur in a relationship of mutual respect, exclusivity or honest communication. Relying on a partner's report of past behavior may not provide a good assessment of risk, as couples' reports of their sexual histories show only fair agreement, which further deteriorates with time (Ellish *et al.*, 1996). The nature of the relationship influences condom use. Key relationship characteristics related to condom use include whether the relationship is perceived as casual, the level of trust in the relationship, and, within committed relationships, whether condoms were used at the onset of the relationship (Jadack *et al.*, 1997; Lear, 1995; Pilkington *et al.*, 1994; Reisen and Poppen, 1995). For more information about the determinants of barrier methods, see Chapter 6.

An important aspect of a relationship is the real or perceived power (or lack thereof) of each partner. Women may find it more difficult than men to implement protective measures because of the real or perceived power imbalance between men and women (Elias and Heise, 1994). Although women can insist on condom use during intercourse and refuse to have sexual intercourse if they are not used, ultimately their male partners control condom usage. For some women, fear of sexual coercion may prevent their insistence on condom use (Rosenthal and Cohen, 1994). An unequal sense of power in relationships and a concern about control have encouraged the development of female methods of protection. The female condom is already available, and topical microbicides, more appropriately called chemoprophylaxis, are under development. However, women prefer the male condom to the female condom, and being in control may not be the most popular feature of the female condom (Eldridge *et al.*, 1995; Gollub *et al.*, 1995).

Substance Use

The relationship between substance use and STD acquisition is extremely complex. People who use and abuse substances are more likely to have STDs, and population trends of STDs follow trends for substance use (Institute of Medicine, 1997; Marx *et al.*, 1991). Substance use is related to early sexual initiation, to having riskier as well as greater numbers of partners, and lack of condom use. Substance use has been demonstrated to be related to risk of STDs among adults and adolescents, and, for homosexual men, to a relapse to previous unsafe sexual behaviors (Siegel *et al.*, 1989; Stall *et al.*, 1986). Causal links between substance use and high-risk sexual behaviors have been suggested. These include impaired judgment associated with intoxication, underlying personality styles, including sensation seeking or problem behavior syndrome, or social situations associated with health-compromising behavior. However, it is often difficult to distinguish causality from association. To establish that substance use caused, rather than was associated with, STD risk, one would have to establish that the same individual who routinely used condoms, behaved differently (i.e. did not use condoms) when under the influence of substances (Fortenberry, 1995).

Societal Influences

American society has difficulty discussing sexuality, which contributes to the risk of STDs. Historically, sexuality was viewed from a moralistic perspective, and STDs were considered the result of 'sinful' behavior. Many patients had difficulty obtaining care for STDs, both from their private physicians and from hospitals. Society continued to have a moralistic, disapproving attitude toward the early victims of AIDS, through identifying AIDS with disenfranchised groups such as homosexuals and intravenous drug users. These negative attributes hinder communication that fosters health-promoting behaviors (Institute of Medicine, 1997).

For adolescents, social discomfort with sexuality may be even more pervasive. A primary source of information is parents, yet communication between parents and teens regarding sexuality may not take place at all, given parental discomfort or lack of knowledge or skills. In addition, parents who deny the possibility that their teen is or will become sexually active are unlikely to have conversations about healthy sexuality (Institute of Medicine, 1997). Parents are not the only ones who may feel uncomfortable discussing sexuality with teenagers: teenagers describe health-care providers as not providing them with the opportunity to discuss issues of importance to them, perhaps because the care providers are uncomfortable about doing so (Ammerman *et al.*, 1992; Marks *et al.*, 1983).

BEHAVIORAL ASPECTS OF BIOMEDICAL APPROACHES

Health-care Access and Utilization

For both adolescents and adults, access to health care is necessary in order for an individual to obtain appropriate diagnosis and treatment of STDs. Approximately 25% of adolescents and young adults considered at high risk for STD acquisition do not have health-care coverage (Institute of Medicine, 1997). This is important because uninsured individuals delay seeking care for health problems longer than those who have private insurance or Medicaid coverage (Donelan *et al.*, 1996; Freeman *et al.*, 1987). The national trend toward managed care for both privately insured and Medicaid populations, and subsequent changes in health-care delivery and financing, may have a significant impact on the delivery of public health services, including STD-related services. The move to managed care provides opportunities such as a focus on public health, the ability to support highly trained medical staff, and the ability to develop sophisticated information systems to track epidemiological trends. Potential risks in the move toward managed care include systems of financial disincentives, the lack of confidentiality for adults and minors, and the complexities of treating partners. Other aspects that affect health-care utilization include past experiences with the system, language skills, and cultural beliefs and personal priorities (Amaro and Gornemann, 1991; Institute of Medicine, 1997).

There are some locations of care services that may focus on prevention and treatment of STDs; these include some school-based clinics, adolescent clinics, and public STD clinics. Public STD clinics historically have been a major source of care for men, and in larger cities they may be overwhelmed by patient numbers. Women are often served in community clinics such as Planned Parenthood. The most effective programs of the future will probably be those that bring the public and private sectors together in joint efforts. An example of such a program for adolescents, the California Partnership for Adolescent Chlamydia Prevention, was described in the Institute of Medicine report (1997). This statewide program brings together the private, public, community and academic agencies in an effort to address policy issues and coordinate clinical care.

Vaccination Programs

An important biomedical component of STD prevention and control will be vaccines and immunotherapy. However, in order for these programs to be effective, behavioral strategies will have to be implemented to foster vaccine acceptance. Barriers to implementation include the providers' behavior, patients' lack of access and their beliefs, and cost limitations. Vaccines are not universally accepted by care providers and patients, even for non-sexually transmitted infections (Fedson, 1994; Siegel et al., 1994; Wood et al., 1995). Children whose living situations are temporary, or whose families are more mobile, are less likely to be immunized (Barrett and Ramsay, 1993). For those in health-care systems, vaccination can be facilitated by taking advantage of opportunities; however, this assumes that the patient is in a health-care system with adequate record-keeping.

The only sexually transmitted infection for which a vaccine is already available is hepatitis B. Initial efforts using a high-risk strategy were not successful in significantly reducing the rate of infection. It appears that once a universal strategy is implemented, the rates of vaccination increase. One immunization problem for high-risk populations is that completion of the series involves three injections, although some immunity may be provided with only a partial series. Another difficulty is the cost, but programs such as Vaccines for Children have made the vaccine free for all children under 19 years of age. Successful strategies for the vaccination of adolescents have been described (Kollar et al., 1994).

Health beliefs (e.g. efficacy, severity, safety) play a role in the acceptance of and compliance with vaccination, as they do with other aspects of STD prevention (Rosenthal et al., 1995c; Rosenthal et al., 1999). In addition, social norms play an important role, thus a universal vaccination strategy may be critical (Rosenthal et al., 1997b; Rosenthal et al., 1999). Reducing the stigma associated with being vaccinated for a disease for which one mechanism of transmission is sexual behavior is important. Finally, it appears that vaccination acceptance may be part of a general health-promoting lifestyle (i.e. seatbelt use, having cholesterol levels checked, and lower alcohol use) (Zimet et al., 1997).

PSYCHOLOGICAL/BEHAVIORAL RESPONSES TO ACQUISITION

Partner Notification and Follow-Up

Partner notification is a critical part of STD prevention programs. It is important not only to prevent the spread of infection to other partners, but also to prevent the index patient from being reinfected (Potterat et al., 1987). The most common methods of partner notification are patient (self) referral, active contact tracing, and conditional referral (professionals obtain partners' names, and index patients are allowed time to notify partners). For adolescents, provider referral may be more successful (Oh *et al.*, 1996). Patient referral requires that the individual is in a situation in which it is safe to tell the partner and that they understand the public health implications should they no longer be involved with that partner (Rosenthal *et al.*, 1995a). Patient assistance techniques to help facilitate the referral, such as a phone call, may be useful (Oxman *et al.*, 1994).

Psychological Reactions to STD

Understanding the psychological sequelae of STDs is important because it may provide the knowledge necessary to prevent psychological morbidity and future episodes. Studies have primarily used a cross-sectional design, which limits the ability to determine whether STD acquisition modifies attitudes and behavior, or whether behaviors and attitudes modify the risk. For example, adolescents who had a less negative perception of STDs were more likely to have multiple STD episodes (Rosenthal *et al.*, 1997b). STD acquisition has the potential to be perceived as a medical or an interpersonal event: this may affect how the person copes with an STD. It is possible that those individuals most concerned about their health may respond differently from those most concerned about the implications for their relationship. In general, it appears common to respond to STD acquisition by wishing it had never happened or avoiding reminders of it, a strategy not necessarily perceived as helpful (Rosenthal and Biro, 1991; Rosenthal *et al.*, 1995b).

There has been little research examining potential interventions to reduce the psychological or future medical impact of STD acquisition. Most of the studies have focused on herpes simplex virus (HSV) infection, and the outcome of subsequent recurrences. For example, one study found that group psychotherapy using cognitive restructuring decreased HSV rates over a 3-month follow-up period (McLarnon and Kaloupek, 1988). In addition, persons with higher levels of social support (Manne and Sandler, 1984), particularly herpes-specific social support (Vanderplate *et al.*, 1988), have lower recurrences of HSV. However, the psychological impact of having frequent recurrences may have been underestimated, and assumptions of causality regarding psychological morbidity or recurrences need to be made cautiously (Mindel, 1993).

PREVENTION APPROACHES

All efforts to promote STD risk reduction should be based on sound principles of behavioral change, be accessible and culturally appropriate, and be evaluated with sound research design. Interventions can occur at several levels, including the individual, the couple or society. Similarly, programs can have a variety of targeted outcomes, such as changing knowledge and attitudes, behavior, or STD prevalence rates. It has been helpful to use social psychological theories of behavior change to formulate intervention development. These theories (Health Belief Model, Theory of Reasoned Action, Protective Motivation Theory) differ in their details, but include as a common theme understanding the influences that determine why an individual does or does not perform a given behavior (Biro and Rosenthal, 1995; Rosenthal *et al.*, 1994b). To evaluate the likelihood that a person will engage in a protective behavior, one needs to determine his or her intentions. Intention is a culmination of the person's evaluation of the risks and benefits, an evaluation of the values or expectations of others, and the import-ance of complying with those expectations. These theories typically assume that the behavior in question is under volitional control. Unfortunately, the beha-vioral targets of STD reduction often are not under an individual's direct control, particularly for women. The most commonly used barrier method remains the male condom, and disenfranchised groups (i.e. those engaging in survival sex) may not have complete control of partner selection or partner acceptance of condoms. In addition, one must understand the cultural influences on attitudes. It may be difficult for those from a different culture to establish the correct language to assess the constructs in question. For example, 'dating' may not be a word used to describe the early stages of romantic relationships for inner-city adolescents. Also, programs must incorporate the developmental or cognitive level of the individual client, particularly when serving adolescents or cognitively impaired adults. Individuals may have the same facts but understand them at different levels (Levitt *et al.*, 1991). Although these theories do not depend directly on behavioral theory (behavior as determined by the antecedents and consequences), past experience influences future experience. However, the fact that there is no direct correlation between engaging in a behavior and acquisition of a STD may lead to inaccurate risk perceptions.

Sex Education in Schools

STD prevention begins with providing clear messages regarding sexuality before an individual becomes sexually experienced. Despite considerable concern about the impact of sex education in schools, there is no evidence that sex education or school-based clinics promote early engagement in sexual behavior. In fact, these services can provide important access to those adolescents who have less available parents (Adelman *et al.*, 1993). However, there is no uniformity regarding training of those that teach these programs and the content of the programs. For example, only 16% of teachers report having received training, and only 37% teach how to use condoms appropriately (CDC, 1996). Characteristics of school programs that are effective include a focus on reducing sexual risk-taking, the use of social learning theory, the incorporation of experiential activities to teach accurate information about risks and ways to avoid unprotected intercourse, recognition of social and media influences, reinforcement of group norms against unprotected intercourse, and modeling and practice in communication skills (Kirby *et al.*, 1994).

Social Marketing

Social marketing principles are increasingly a part of public health interventions, such as heart disease and teenage pregnancy. Social marketing has been confused with other approaches that use the media, such as public communication campaigns. Public communications focus on cognitive aspects or changing knowledge. Social marketing strategies are not single commercials, but rather integrated approaches that take place for a predetermined period on a large scale (Smith, 1991).

The contribution of social marketing is in the attention to integrating comprehensive planning with systematic audience research. It is the application of commercial marketing tools to foster change with a social purpose. The relevant aspects influencing marketing are: product, place, price and promotion (Smith, 1991). The product can be an idea, commodity, behavior or service which is being packaged to meet consumer needs. Unfortunately, our efforts to reduce STD acquisition and its sequelae do not lead to an easily manipulated and altered product. Place is the location in which the product or concept is offered. For STD reduction this might include addressing issues of access to care and the availability of resources. Price is the personal or economic cost involved. For example, the 'price of condom use' for a woman might be fear of the partner breaking up with her, or perceived social norms. Finally, promotion involves the interpersonal systems in which the consumer might interact. Social marketing, as with theories of social psychology, are only helpful to the extent that the research accurately determines consumer perceptions (Smith, 1991).

CONCLUSION

The prevention and treatment of STDs is a highly complex behavioral and psychological phenomenon. Successful programs must be multifaceted and tailored to the different needs of targeted groups, and occur at all levels of prevention (Rosenthal et al., 1994b). Primary prevention programs focus on reducing the number of individuals who acquire an STD. Primary prevention includes encouraging adolescents to delay the initiation of intercourse, fostering healthy sexual decision-making, including the use of barrier methods, and the acceptance of vaccines. Secondary prevention focuses on the management of current infections and the prevention of partner and neonate transmission. This includes access to quality medical care and partner notification strategies. Tertiary prevention is focused on preventing medical and psychological sequelae, and builds on secondary preventive efforts. Prompt diagnosis and treatment reduce both the likelihood that the infection will be transmitted to another individual and the likelihood that the infection will lead to medical sequelae. STDs remain a significant public health problem, and all three aspects of prevention must be implemented to minimize the risk to the index patient and their partners, and of vertical transmission to infants.

REFERENCES

Adelman, H.S., Barker, L.A. and Nelson, P. (1993). A study of a school-based clinic: Who uses it and who doesn't? *J. Clin. Child Psychol.* **22**: 52–59

Amaro, H. and Gornemann, I. (1991). In: 'Research Issues in Human Behavior and Sexually Transmitted Diseases in the AIDS Era' (eds. J.N. Wasserheit, S.O. Aral, and K.K. Holmes), Health Care Utilization for Sexually Transmitted Diseases: Influence of Patient and Provider Characteristics, pp. 140–160, American Society for Microbiology, Washington DC

Ammerman, S.D., Perelli, E., Adler, N., Irwin, C.A. (1992). Do adolescents understand what physicians say about sexuality and health? *Clin. Pediatr.* **31**: 590–595

American Social Health Association (ASHA) (1996). Teenagers know more than adults about STDs, but knowledge among both groups is low. *STD News*, 3, 1,5

Barrett, G. and Ramsay, M. (1993). Improving uptake of immunization: Mobile children miss out. *Br. Med. J.* **307**: 681–682

Biro, F.M. (1992). Adolescents and Sexually Transmitted Diseases. Maternal and Child Health Technical Information Bulletin. Washington DC: National Center for Education in Maternal and Child Health in cooperation with the Maternal and Child Health Bureau, Health Resources and Services Administration, Public Health Service, U.S. Department of Health and Human Services

Biro, F.M. and Rosenthal, S.L. (1995). Adolescents and sexually transmitted diseases: Diagnosis, developmental issues, and prevention. *J. Pediatr. Health Care* **9**: 256–262

Biro, F.M., Rosenthal, S.L. and Stanberry, L.S. (1994). Knowledge of gonorrhea in adolescent females with a history of STD. *Clin. Pediatr.* **33**: 601–605

Breakey, W.R., Fischer, P.J., Kramer, M. *et al.* (1989). Health and mental health problems of homeless men and women in Baltimore. *JAMA* **262**: 1352–1357

Brooks-Gunn, J. and Paikoff, R.L. (1993). In: 'Promoting the Health of Adolescents: New Directions for the Twenty-First Century' (eds. S.G. Millstein, A.C. Petersen, and E.O. Nightingale), Sex is a Gamble, Kissing is a Game: Adolescent sexuality and health promotion, pp. 180–208. Oxford University Press, New York

CDC. (1996). School-based HIV-prevention education – United States, 1994. *MMWR*, **45**: 760–765

Cates, W. (1990). The epidemiology and control of sexually transmitted diseases in adolescents. *Adolesc. Med. State Art Rev.* **1**: 409–427

Corey, L. (1994). The current trend in genital herpes: Progress in prevention. *Sex. Transm. Dis.* 21(suppl), s38–s44

Crockett, L.J. and Petersen, A.C. (1993). In: 'Promoting the Health of Adolescents: New Directions for the Twenty-First Century' (eds. S.G. Millstein, A.C. Petersen, and E.O. Nightingale), Adolescent Development: Health Risks and Opportunities for Health Promotion. pp. 13–37. Oxford University Press, New York

DiClemente, R.J., Lanier, M.M., Horan, P.F. and Lodica, M. (1991). Comparison of AIDS knowledge, attitudes, and behaviors among incarcerated adolescents and a public school sample in San Francisco. *Am. J. Public Health* **81**: 628–630

Donelan, K., Blendon, R.J., Hill, C.A. *et al.* (1996). Whatever happened to the health insurance crisis in the United States? Voices from a national survey. *JAMA* **276**: 1346–1350

Eldridge, G.D., St. Lawrence, J.S., Little, C.E., Shelby, M.C. and Brasfield, T.L. (1995). Barriers to condom use and barrier method preferences among low-income African-American women. *Women's Health* **23**: 73–89

Elias, C.J. and Heise L.L. (1994). Challenges for the development of female-controlled vaginal microbicides. *AIDS* **8**: 1–9

Ellish, N.J., Weisman, C.S., Celentano, D. and Zenilman, J.M. (1996). Reliability of partner reports of sexual history in a heterosexual population at a sexually transmitted diseases clinic. *Sex. Transm. Dis.* **23**: 446–452

Eysenck, H.J. and Wilson, G. (1979). 'The Psychology of Sex.' Dent, London

Fedson, D.S. (1994). Adult immunization: Summary of the National Vaccine Advisory Committee Report. *JAMA* **272**: 1133–1137

Fortenberry, J.D. (1995). Adolescent substance use and sexually transmitted diseases risk: A review. *J. Adolesc. Health* **16**: 304–308

Freeman, H.E., Blendon, R.J., Aiken, L.H., Sudman, S., Mullinix, C.F. and Corey, C.R. (1987). Americans report on their access to health care. *Health Affairs* **6**: 6–18

Gollub, E.L., Stein, Z. and El-Sadr, W. (1995). Short-term acceptability of the female condom among staff and patients at a New York City hospital. *Fam. Plann. Perspect.* **27**: 155–158

Greenberg, J., Magder, L. and Aral, S. (1992). Age at first coitus: A marker for risky sexual behavior in women. *Sex. Transm. Dis.* **19**: 331–334

Harrison, H.R., Phil, D., Costin, M. *et al.* (1985). Cervical *Chlamydia trachomatis* infection in university women: Relationship to history, contraception, ectopy, and cervicitis. *Am. J. Obstet. Gynecol.* **153**: 244–251

Hart, G. (1977). 'Sexual Maladjustment and Disease: An Introduction to Modern Venereology.' Nelson-Hall, Chicago

Hernandez, J.T. and Smith, F.J. (1990). Inconsistencies and misperceptions putting college students at risk of HIV infection. *J. Adolesc. Health Care* **11**: 295–297

Hook, E.W., Richey, C.M., Leone, P. *et al.* (1997). Delayed presentation to clinics for sexually transmitted diseases by symptomatic patients: A potential contributor to continuing STD morbidity. *Sex. Transm. Dis.* **24**: 443–448

Institute of Medicine (U.S.) Committee on Prevention and Control of Sexually Transmitted Diseases. (1997). 'The Hidden Epidemic: Confronting Sexually Transmitted Diseases' (eds, T.R. Eng, and W.T. Butler). National Academy Press, Washington DC

Jaccard, J. and Wilson, T. (1991). In: 'Research Issues in Human Behavior and Sexually Transmitted Diseases in the AIDS Era' (eds. J.N. Wasserheit, S.O. Aral, and K.K. Holmes), Personality Factors Influencing Risk Behaviors, pp. 177–200, American Society for Microbiology, Washington DC

Jadack, R.A., Fresia, A., Rompalo, A.M. and Zenilman, J. (1997). Reasons for not using condoms of clients at urban sexually transmitted disease clinics. *Sex. Transm. Dis.* **24**: 402–408

Keim, J., Woodard, P. and Anderson, M.K. (1992). Screening for *Chlamydia trachomatis* in college women on routine gynecological exams. *J. Am. Coll. Health* **41**: 17–19, 22–23

Kirby, D., Short, L., Collins, J. *et al.* (1994). School-based programs to reduce sexual risk behaviors: A review of effectiveness. *Public Health Rep.* **109**: 339–360

Kollar, L.M., Rosenthal, S.L. and Biro, F.M. (1994). Hepatitis B vaccine series compliance in adolescents. *Pediatr. Infect. Dis. J.* **13**: 1006–1008

Koutsky, L.A., Ashley, R.L., Holmes, K.K. *et al.* (1990). The frequency of unrecognized type 2 herpes simplex virus infection among women: Implications for the control of genital herpes. *Sex. Transm. Dis.* **17**: 90–94

Langenberg, A., Benedetti, J.K., Jenkins, J. *et al.* (1989). Development of clinically recognizable genital lesions among women previously identified as having 'asymptomatic' herpes simplex virus type 2 infection. *Ann. Int. Med.* **110**: 882–887

Lear D. (1995). Sexual communication in the age of AIDS: The construction of risk and trust among young adults. *Soc. Sci. Med.* **41**: 1311–1323

Levitt, M.Z., Selman, R.L. and Richmond, J.B. (1991). The psychosocial foundations of early adolescents' high-risk behavior: Implications for research and practice. *J. Res. Adolesc.* **1**(4): 349–378

MacDonald, N.E., Fisher, W.A., Wells, G.A., Doherty, J.A. and Bowie, W.R. (1994). Canadian street youth: Correlates of sexual risk-taking activity. *Pediatr. Infect. Dis. J.* **13**: 690–697

MacDonald, N.E., Wells, G.A., Fisher, W.A. *et al.* (1990). High-risk STD/HIV behavior among college students. *JAMA* **263**: 3155–3159

Manne, S. and Sandler, I. (1984). Coping and adjustment to genital herpes. *J. Behav. Med.* **7**: 391–410

Marx, R., Aral, S.O., Rolfs, R.T., Sterk, C.E. and Kahn, J.G. (1991). Crack, sex, and STD. *Sex. Transm. Dis.* **18**: 92–101

Marks, A., Malizio, J., Hoch, J., Brody R. and Fisher M. (1983). Assessment of health needs and willingness to utilize health care resources of adolescents in a suburban population. *J. Pediatr.* **102**: 456–460

McLarnon, L.D. and Kaloupek, D.G. (1988). Psychological investigation of genital herpes recurrence: Prospective assessment and cognitive-behavioral intervention for a chronic physical disorder. *Health Psychol.* **7**: 231–249

Millstein, S.G. and Moscicki, A-B. (1995). Sexually-transmitted disease in female adolescents: Effects of psychosocial factors and high risk behaviors. *J. Adolesc. Health* **17**: 83–90

Mindel, A. (1993). Long-term clinical and psychological management of genital herpes. *J. Med. Virol.* Suppl. 1, 39–44

Moscicki, A-B. (1990). In: 'Adolescent Medicine: State of the Art Reviews: AIDS and Other Sexually Transmitted Diseases' (eds. M. Schydlower, M-A. Shafer), Genital Human Papillomavirus Infections, pp. 451–469, Hanley & Belfus, Philadelphia, PA

Muram, D., Rosenthal, T.L., Tolley, E.A., Peeler, M.M. and Dorko, B. (1991). Race and personality traits affect high school senior girls' sexual reports. *J. Sex Educ. Therapy* **17**: 231–243

Oh, M.K., Boker, J.R., Genuardi, F.J., Cloud, G.A., Reynolds, J. and Hodgens, J.B. (1996). Sexual contact tracing outcome in adolescent chlamydial and gonococcal cervicitis cases. *J. Adolesc. Health* 18: 4–9

Oxman, A.D., Scott, E.A.F., Sellors, J.W. *et al.* (1994). Partner notification for sexually transmitted diseases: An overview of the evidence. *Can. J. Public Health* 85, S41–S47

Peloquin, L.J. and Davidson, P.W. (1988). In: 'Handbook of Pediatric Psychology' (ed. D.K. Routh), Psychological Sequelae of Pediatric Infectious Diseases, pp. 222–257, Guilford Press, New York

Pilkington, C.J., Kern, W. and Indest, D. (1994). Is safer sex necessary with a 'safe' partner? Condom use and romantic feelings. *J. Sex. Res.* 31: 203–210

Potterat, J.J., Duke, R.L. and Rothenberg, R.B. (1987). Disease transmission by heterosexual men with gonorrhea: An empiric estimate. *Sex. Transm. Dis.* 14: 107–110

Reisen, C.A. and Poppen, P.J. (1995). College women and condom use: Importance of partner relationship. *J. Applied Soc. Psychol.* 25: 1485–1498

Rosenblum, L., Darrow, W., Witte, J. *et al.* (1992). Sexual practices in the transmission of hepatitis B virus and prevalence of hepatitis delta virus infection in female prostitutes in the United States. *JAMA* 267: 2477–2481

Rosenthal, S.L. and Biro, F.M. (1991). A preliminary investigation of psychological impact of sexually transmitted diseases in adolescent females. *Adolesc. Pediatr. Gynecol.* 4: 198–201

Rosenthal, S.L. and Cohen, S.S. (1994). Primary prevention of sexually transmitted disease: Self-efficacy in the context of sexual coercion. *Adolesc. Pediatr. Gynecol.* 7: 63–68

Rosenthal, S.L., Biro, F.M., Succop, P.A., Cohen, S.S. and Stanberry, L.R. (1994a). Age of first intercourse and risk of sexually transmitted disease. *Adolesc. Pediatr. Gynecol.* 7: 210–213

Rosenthal, S.L., Cohen S.S. and Biro, F.M. (1994b). In: 'Risk, Resilience and Prevention: Promoting the Well Being of All Children' (ed. R.J. Simeonsson), Sexually Transmitted Diseases: A Paradigm for Risk Taking Among Teens, pp. 239–264. Paul H. Brookes, Baltimore

Rosenthal, S.L., Baker, J.G., Biro, F.M. and Stanberry, L.R. (1995a). Secondary prevention of STD transmission during adolescence: Partner notification. *Adolesc. Pediatr. Gynecol.* 8: 183–187

Rosenthal, S.L., Biro, F.M., Cohen S.S., Succop, P.A. and Stanberry, L.R. (1995b). Strategies for coping with sexually transmitted diseases by adolescent females. *Adolescence* 30: 655–666

Rosenthal, S.L., Kottenhahn, R.K., Biro, F.M. and Succop, P.A. (1995c). Hepatitis B vaccine acceptance among adolescents and their parents. *J. Adolesc. Health* 17: 248–254

Rosenthal, S.L., Stanberry, L.R., Biro, F.M. *et al.* (1997a). Seroprevalence of herpes simples virus type 1 and type 2 and cytomegalovirus in adolescents. *Clin. Infect. Dis.* 24: 135–139

Rosenthal, S.L., Biro, F.M., Succop, P.A., Bernstein, D.I. and Stanberry, L.R. (1997b). Impact of demographics, sexual history, and psychological functioning on the acquisition of STDs in adolescents. *Adolescence* 32: 757–769

Rosenthal, S.L., Lewis, L.M., Succop, P.A., Bernstein, D.I. and Stanberry, L.R. (1999). College students' attitudes regarding vaccination to prevent genital herpes. *Sex. Transm. Dis.* 26: 438–443

St. Lawrence, J.S. (1993). African-American adolescents' knowledge, health-related attitudes, sexual behavior, and contraceptive decisions: Implications for the prevention of adolescent HIV infection. *J. Consult. Clin. Psychol.* 61: 104–112

Siegel, R.M., Schubert, C.J., Myers, P.A. and Shapiro, R.A. (1995). The prevalence of sexually transmitted diseases in children and adolescents evaluated for sexual abuse in Cincinnati: Rationale for limited STD testing in prepubertal girls. *Pediatrics* 96: 1090–1094

Siegel, R.M., Baker, R.C., Kotagal, U.R. and Balistreri, W.F. (1994). Hepatitis B vaccine use in Cincinnati: A community's response to the AAP recommendation of universal hepatitis B immunization. *J. Natl. Med. Assoc.* 86: 444–448

Siegel, K., Mesagno, F.P., Chen, J-Y. and Christ, G. (1989). Factors distinguishing homosexual males practicing risky and safer sex. *Soc. Sci. Med.* 28: 561–569

Smith, W.A. (1991). In: 'Research Issues in Human Behavior and Sexually Transmitted Diseases in the AIDS Era' (eds. J.N. Wasserheit, S.O. Aral and K.K. Holmes), Organizing Large-Scale Interventions for Sexually Transmitted Disease Prevention, pp. 219–242, American Society for Microbiology, Washington DC

Staat, M.A., Rosenthal, S.L., Biro, F.M. and Stanberry, L.R. (1999). In: 'STDS in Adolescents: Challenges for the 21st Century' (ed. P.J. Hitchcock, R. Boruch, B. Flay, *et al.*), Sexually Transmitted Pathogens and Diseases, Oxford University Press, New York. In press

Stall, R., McKusick, L., Wiley, J., Coates, T.J. and Ostrow, D.G. (1986). Alcohol and drug use during sexual activity and compliance with safe sex guidelines for AIDS: The AIDS Behavioral Research Project. *Health Educ. Q.* **13**: 359–371

Sugerman, S.T. (1991). Acquired immunodeficiency syndrome and adolescents: Knowledge, attitudes, and behaviors of runaway and homeless youths. *Am. J. Dis. Child.* **145**: 431–436

VanderPlate, C., Aral, S.O. and Magder, L. (1988). The relationship among genital herpes simplex virus, stress, and social support. *Health Psychol.* **7**: 159–168

Visher, S. (1986). The relationship of locus of control and contraception use in the adolescent population. *J. Adolesc. Health Care* **7**: 183–186

Wood, D., Pereyra, M., Halfon, N., Hamlin, J. and Grabowsky, M. (1995). Vaccination levels in Los Angeles public health centers: The contribution of missed opportunities to vaccinate and other factors. *Am. J. Public Health* **85**: 850–853

Zierler, S., Feingold, L., Laufer, D., Velentgas, P., Kantrowiz-Gordon, I. and Mayer, K. (1991). Adult survivors of childhood sexual abuse and subsequent risk of HIV infection. *Am J of Public Health*, 81, 572–575

Zimet, G.D., Fortenberry, J.D., Fife, K.H., Tyring, S.K., Herne, K. and Douglas, J.M. (1997). Acceptability of genital herpes immunization: The role of health beliefs and health behaviors. *Sex. Transm. Dis.* **24**: 555–560

6
PHYSICAL BARRIER METHODS: ACCEPTANCE, USE AND EFFECTIVENESS

JOSEPH KELAGHAN

Contraception and Reproductive Health Branch,
Center for Population Research,
National Institute of Child Health and Human Development;
now at National Institutes of Health,
National Cancer Institute, Bethesda,
Maryland, USA

INTRODUCTION

Condoms are the only devices available that provide substantial protection against the most common sexually transmitted diseases. They offer the additional advantage of being contraceptives, as STDs occur in sexually active individuals among whom contraception is largely desired. Condoms are an adjunct to and not a substitute for other safer sexual practices. Their consistent and correct use lowers risk beyond that achieved by limiting the number of one's sexual partners, reducing or eliminating the use of drugs and alcohol associated with sex, improving communication with partners, and other measures. Condoms are particularly useful for reducing risk that is beyond the knowledge or control of the user (most often a partner's risky behavior). The demonstrated low degree of condom failure (i.e. breaking or completely slipping off the penis during intercourse) ensures excellent risk reduction to consistent users who are at moderate to low risk or who have adequately lowered their risk in other ways.

The earliest years of the AIDS epidemic led to recommendations by the Surgeon General of the United States (Koop, 1988) and the Centers for Disease Control (1988) for the use of male condoms for personal protection from infection. One of the stated goals of the United States Public Health Service, as described in Healthy People 2000[1] is to double the use of male condoms (based on the 417 million sold in 1988) as a step towards reducing the incidence of gonorrhea and chlamydia infection. Mosher and Pratt (1990) reported that the use of male condoms had indeed increased among the American public by 1988. The female condom was approved in 1993 by the FDA for use as a barrier

---

[1] *Healthy People 2000.* US Dept of Health and Human Services DHHS (PHS) 91–50212

contraceptive intended to prevent STDs. This new condom is not yet widely used, but future developments in marketing, promotion and user preference will determine its role in protecting against STDs and unintended pregnancy. Although the maximum protection against STDs afforded by condoms has not been consistently demonstrated in clinical trials, recommendations for their use can justifiably be based on laboratory studies and studies documenting the low breakage and slippage rates of condoms in actual use. Some have suggested that spermicide, used either alone or as an adjunct to condoms, might also provide protection, although the degree of protection is not as well demonstrated as that achieved by male or female condoms.

The vast majority of male condoms are made from latex, although those made from sheep intestinal membrane comprise a small part of the market, and synthetic condoms made from polyurethane have recently become available. Most information relevant to STD prevention pertains to latex condoms, which the FDA allows to carry a label claiming a protective effect against transmission of HIV and some other STDs. Natural-membrane condoms do not carry such a claim because of concern that the pores in this natural material might allow the transmission of viruses, although transfer of infectious particles through this material during use has not been demonstrated. Because of the limited data available, polyurethane condoms carry a label stating that use for protection from STDs is limited to those with an allergy to latex.

USE OF BARRIER METHODS

The most recent cycle of the United States National Survey of Family Growth was conducted in 1995 (Abma *et al.*, 1997). This consisted of interviews with a national sample of women aged 15–44 years in the civilian, non-institutionalized population. Fully 20.4% of sexually active contraceptive users reported using male condoms; the use of diaphragms and spermicidal preparations was 2% or less, and only 0.1% of women aged 20–24 (and fewer women in other age groups) reported use of the female condom. Approximately two-thirds of condom users reported consistently using these devices, whereas 20% reported their use during less than 20% of acts of intercourse. Although reported barrier use was more common among younger, never-married women and those with more education and higher incomes, consistent use was more common among older, married or previously married white women with more education and higher incomes.

EVIDENCE THAT CONDOMS PROTECT AGAINST SEXUALLY TRANSMITTED DISEASE

Nature of the Evidence

The ability of both male and female condoms to interfere with the transmission of some sexually transmitted organisms is demonstrated by *in vitro* studies documenting the characteristics of the materials, and by *in vivo* studies documenting a low rate of breakage or complete slippage during intercourse.

Although demonstrating a reduced acquisition of disease in users is important, it is much more difficult, and such studies have largely not shown the degree of protection suggested by the simpler and more readily interpretable studies mentioned above. Studies utilizing disease as an endpoint depend on the subject's continued compliance with condom use for the period of study (usually 6 months), accurate reporting of non-use, and the investigator's knowledge regarding each subject's risk of exposure to sexually transmitted organisms. Clinical trials are useful for demonstrating that prevention does occur, for identifying characteristics associated with successful use, and for evaluating the anticipated benefits of a program to promote condom use. However, predicting the protection provided when the device is used is more accurately established by demonstrating that sexually transmissible organisms do not pass through the condom, and measuring rates at which these products break or slip off the penis during use.

In vitro Studies

Carey *et al.* (1992) and Reitmeijer *et al.* (1988) have demonstrated that HIV and particles of similar size do not permeate intact latex condoms. Similarly, intact latex condoms do not allow the transmission of herpes simplex virus (Conant *et al.*, 1984), cytomegalovirus (Katznelson *et al.*, 1984) or hepatitis B virus (Minuk *et al.*, 1987). If latex is impermeable to viruses, it is almost certainly impermeable to the much larger organisms responsible for the most common STDs. Indeed, Barlow (1977) and Judson *et al.* (1989) have demonstrated the impermeability of latex to organisms causing gonorrhea and chlamydia, respectively. The *in vitro* studies provide assurance that a properly used condom provides excellent protection against a wide variety of STDs if it does not break or slip during use and expose one partner to the potentially infected fluids of the other.

Slippage and Breakage Studies

Both retrospective and prospective studies have measured condom breakage and slippage. The former are easier to conduct but suffer from potential inaccuracies in reporting. Retrospective surveys collect information from volunteer subjects concerning experience with condoms during a specified period of time (frequently 1 month, 6 months or 1 year) prior to interview. As respondents can only estimate the total number of condoms used during the period of recall, and are inclined to selectively remember a condom break (and perhaps to report mistakenly one that occurred prior to the period of requested recall), retrospective surveys provide less than exact quantitation. Such studies have reported breakage of 0.2% (Albert *et al.*, 1995) to 7.3% (Richters *et al.*, 1993) of condoms used. The source of the sample appears to influence the reported breakage rate. A national survey of American men (Grady and Tanfer, 1994) reported breakage in 1.9% of condoms used by 1226 men in the 6 months prior to interview. Richters *et al.* (1995) reported breakage in 4.9% of condoms used in the previous year by 108 men recruited at an STD clinic. Albert *et al.* (1991), however, found that women studied at a family planning clinic reported that only 0.8% of condoms used broke in the previous year. Retrospective studies of gay men using condoms for anal intercourse have found breakage rates ranging from 3.0% (Thompson *et al.*, 1993), reported in the year prior to interview among a population of gay men in New York City, to 5–9% (Judson *et al.* 1989) of condoms used in the previous 6 months by gay men recruited at an STD clinic. Studies including information on slippage have reported that 1.0–4.4% of condoms slipped off the penis during intercourse, suggesting that failure of the condom to prevent exposure to semen might be as high as 10%, although most studies find total failure rates in the range of 3–5%.

Prospective studies of condom breakage recruit subjects who agree to report their experience with the subsequent specified number of condoms (usually 6–12) or to report on condoms used while in the study (usually 1–12 months). A small study of 40 subjects using up to 10 condoms each (Gøtzsche and Hørding, 1988) detected a breakage rate of 5%. Steiner *et al.* (1993, 1994a) reported two prospective trials: the earlier study detected a breakage rate of 4.4% among 260 couples each asked to use four condoms from new lots, and the later study found a breakage rate of 3.7% for 268 couples asked to use six new condoms each. Although both these studies included some condoms that were known to be damaged from long storage in adverse conditions, the results reported above are limited to the use of new condoms. Zekeng *et al.* (1993), investigating 273 women in Cameroon for up to 1 year with monthly follow-up visits as part of a study of HIV infection, reported that 4.8% of condoms used broke. Sparrow and Lavill (1994) reported a breakage rate of 5.6% among 540 subjects using 3754 condoms during 1 month. Three studies of commercial sex workers in brothels allowed investigators to examine used condoms, in addition to interviewing subjects. Albert *et al.* (1995) found no breaks in 353 condoms used by 41 sex workers over 3 days. Rugpao *et al.* (1993) reported breakage in 5.9% of 5559 condoms examined in 30 brothels in Thailand during a 3-month period. In another study of seven brothels in the same part of Thailand, Rugpao *et al.* (1997) reported breakage in only 1.8% of sex acts involving a single condom during a 3-month period. They attributed the decline in breakage rate between the two studies to increased experience with condom use, subsequent to an educational campaign. Overall, prospective studies have reported breakage rates from 0.0% to 5.9%, and slippage rates from 0.1% to 6.6%, values in agreement with those from retrospective studies.

Non-latex Male Condoms and Female Condoms

A polyurethane male condom was introduced to the US market in 1994. Polyurethane condoms can be used with oil-based lubricants, have a longer shelf-life, and are an acceptable alternative for those who have developed a latex allergy. Perhaps the greatest benefit from existing and future non-latex condoms is the expansion of choices for those considering the use of condoms. Although much less research exists for these condoms than for latex condoms, Voeller *et al.* (1991) demonstrated that the polyurethane material prevents transmission of particles the size of the smallest sexually transmitted virus. A randomized clinical trial[2] comparing an experimental polyurethane condom with a latex condom found that 4.3% of the polyurethane condoms broke during use, and that 4.5% slipped off the penis. Although both these rates are in the range described for latex condoms, they are significantly higher than the corresponding rates for the latex condom used as a comparison (0.6% and 1.2% for breakage and slippage, respectively). Although further research will increase the precision of the breakage and slippage rates for non-latex condoms, the limited evidence available suggests that they provide good protection against STDs.

In vitro evidence also documents the ability of the polyurethane female condom to interfere with the transmission of sexually transmitted organisms (Drew *et al.*, 1990), and Soper *et al.* (1993) demonstrated protection against reinfection with *Trichomonas vaginalis* among women using the female condom consistently during a 45-day period. Research is ongoing to further clarify the slippage and breakage rates, to identify characteristics associated with consistent and correct use of this device, to elucidate problems experienced by women using this method, and to gauge the protection afforded by this method against STDs. The limited research concerning this device suggests that it provides good protection for women who find that they can use it consistently.

Clinical and Observational Studies

Although the physical properties of condom materials and the *in vivo* slippage and breakage studies indicate that male condoms provide excellent protection against STDs, most epidemiologic studies and clinical trials have demonstrated lower levels of protection than anticipated. These findings are quite variable. Two observational studies of HIV infection demonstrated excellent protection. De Vincenzi (1994) found no infections among female partners of HIV-positive men using condoms consistently during an average follow-up period of 20 months, and Hart (1984) detected no gonorrhea among men in a war environment who consistently used condoms. Most of the remaining studies have demonstrated more modest reductions. Pemberton *et al.* (1972) reported a reduction of 49% in urethral gonorrhea among male condom users at an STD clinic compared to non-users, and Barlow (1977) found a 75% reduction in gonorrhea among men at another STD clinic. Austin *et al.* (1984) reported a modest reduction of 13% in cervical gonorrhea among women at an STD clinic reporting the use of condoms, and Kelaghan (1982) reported a 40% reduction in risk of pelvic inflammatory disease among hospitalized women who reported condom use in the 3 months before interview. The discrepancy between slippage and breakage studies and studies with disease as an endpoint implies that inconsistent use by participants accounts for the variable results. Studies by Wulfert and Wan (1993) and by Ku *et al.* (1994), as well as data from the 1995 National Survey of Family Growth (Abma *et al.*, 1997) and the CDC (Ku *et al.*, 1994), indicate that many condom users do not use the product at every act of

---

[2] Study of the efficacy, acceptability and safety of a non-latex (polyurethane) male condom; Final Report to National Institute of Child Health and Human Development 1997; Contract N01-HD-1–3109.

intercourse. As many observational studies and clinical trials have not differentiated between consistent and inconsistent users, the studies have not
measured the effectiveness of the product. They do, however, demonstrate
the inadequacy of using condoms less than consistently.

Assessment of Condom Effectiveness

The possibility that a user might be exposed to the potentially infectious fluids of
a partner in 5–10% of condom uses is sobering, but the converse is also true: 90–
95% of the time condoms protect the user from contact with potentially
infectious fluids. Put another way, the correct use of a condom decreases the
user's risk by at least 90% each time one is used. Furthermore, these rates overestimate the rate of clinically relevant failure, as they include breakage occurring
at any time, from opening the condom wrapper to removing it after intercourse.
Although breaks occurring before intercourse or after withdrawal are important
to report and might reflect the quality of the condom, they might also reflect user
characteristics, such as manicure, or techniques in opening the condom package,
and such breaks do not lead to exposure if they are noted and a replacement is
used. Steiner *et al.* (1994b) have proposed reporting breakage and slippage as
clinical and non-clinical, clinical breaks and slips being those that occur during
intercourse and have the potential to cause an adverse event. Despite the substantial data suggesting that the correct use of condoms provides a high degree of
protection, such protection is not absolute. Perlman *et al.* (1990) advised caution
for those in situations where the risk of infection is great and the infection is
serious (most commonly an HIV-negative partner of someone who is HIV-
positive).

Other factors suggest that condoms might provide even greater protection
than indicated in the slippage and breakage studies. Both the retrospective and
the prospective studies identified characteristics associated with condom failure,
and many of these are remediable. The age of the condom is a major predictor of
breakage: Condoms more than 2 years old should not be used, and users should
check the date on the wrapper and dispose of those that are out-of-date.
Increased experience with condoms is consistently associated with lower rates
of slippage and breakage, indicating that competence with condom use can be
achieved. This observation is confirmed by several studies reporting very low
breakage rates for condoms used by commercial sex workers. Some studies find
that not living with the partner, fewer years of education, and increased numbers
of partners are associated with an increase in condom breakage or slippage.
Some of these characteristics are probably directly associated with failure (for
example, living with a partner could facilitate the development of routines
conducive to proper condom use), but some are probably surrogate markers for
other, unstudied and less easily measured traits. A review article by Silverman
and Gross (1997) indicates that breakage during anal intercourse is usually
reported to be greater than during vaginal intercourse, but the number of
observations is small and the causes are not well elucidated. The use of lubricants
has been variably reported to increase (Sparrow and Lavill, 1994) or decrease
(Gabbay and Gibbs, 1996) the rates of condom breakage, although Steiner *et al.*
(1994a) found a protection against breakage only when a water-based lubricant

was used with old condoms. Although oil-based lubricants should not be used with a latex condom because of demonstrated deterioration in the latex (Voeller *et al.*, 1989), no *in vivo* study has identified the use of these lubricants as a cause of breakage.

Before leaving the assessment of condom efficacy, it is important to note that the major reason why condoms fail is not breakage or slippage, but non-use. The protection provided by condoms against STDs (or pregnancy) is limited to the times they are used. It is useful to compare exposure to a partner's potentially infectious secretions for someone using condoms consistently and someone using condoms during 70% of acts of intercourse. If both users have six acts of intercourse a month, the former will possibly experience one exposure in a 4-month period (based on a breakage/slippage rate of 5% per use). During the same period the inconsistent user will experience seven exposures from not using a condom, and possibly one exposure from slippage or breakage. For someone not using condoms consistently, increasing the frequency of use is much more important than concerns about mechanical failure.

SPERMICIDES

Hicks *et al.* (1985), Jennings and Clegg (1993) and Bourinbaiar *et al.* (1994) have all demonstrated that spermicides containing nonoxynol-9 (N-9) inactivate HIV and other sexually transmitted viruses (Singh *et al.*, 1976). Studies by Bolch and Warren (1973) and by Singh *et al.* (1972) have shown that N-9 also inhibits the growth of *Neisseria gonorrheae*; the data are not as clear regarding inactivation of *Chlamydia trachomatis* (Kelly *et al.*, 1985; Ehret and Judson, 1988), but Kappus and Quinn (1986) propose that N-9 might offer protection in use by preventing infection while not inactivating the organism. Although the physical properties of condoms, together with results of the slippage and breakage studies, provide assurance that their consistent use provides better protection than is demonstrated in observational studies, the assurance is not as certain for spermicides. Despite the fact that spermicides containing N-9 inactivate all sexually transmitted viruses and bacteria studied at very low concentrations and in a very short time, critical factors such as distribution in the vagina, minimal effective dose and final concentration during coitus potentially influence the effectiveness of these products even during correct and consistent use. These factors are difficult to evaluate and remain largely undetermined. Further caution arises from the study of Weir *et al.* (1995), who observed that the use of N-9 was associated with an increased incidence of genital ulcers, and from the study of Roddy *et al.* (1993), who reported detectable genital irritation in women using high doses of N-9. Observational studies of the effect of N-9 on the acquisition of STDs have shown variable results, largely being less encouraging than similar studies of condoms. Jick *et al.* (1982) discovered an 87% decrease in gonorrhea among spermicide users at a health maintenance organization, whereas Louv *et al.* (1988) found only a 24% decrease in gonorrhea and a 26% decrease in chlamydia among spermicide users in a randomized clinical trial at an STD clinic. Rosenberg *et al.* (1987) calculated a 33% decline in gonorrhea and a 69% decline in chlamydia in a randomized clinical trial of the N-9 sponge in a population of commercial sex workers. It cannot be as readily assumed, as for condoms, that the less than perfect efficacy of spermicides in preventing sexually transmitted diseases is largely a matter of compliance. Recommendations for the use of these products for protection against STDs will have to depend on demonstration in clinical trials to a greater degree than for condoms.

Because of the uncertainty surrounding the degree of protection afforded by the use of spermicidal preparations, they are usually recommended as an adjunct to the consistent use of condoms. No clinical evidence is likely to demonstrate the effectiveness of this combined regimen for two major reasons: (1) the correct use of condoms reduces the risk of STDs to levels so low that it would be statistically difficult to detect a further benefit from the addition of spermicide; and (2) observation among those who select themselves to be inconsistent users of condoms is subject to selection bias, both for those who choose not to use the physical barrier and for those who choose to use the spermicide. Nevertheless, it is reasonable to recommend the dual use of barriers so long as the recommendation does not interfere with the message to use condoms. Spermicides are sometimes recommended as a second line of protection for those situations in which a physical barrier is not used. Although such a recommendation might be counterproductive if it discourages condom use, this strategy is reasonable as long as the user understands that the supporting evidence is limited.

CONCLUSION

Although condoms are not perfect protection against STDs, they are the only devices available that do provide protection, and that protection is substantial. Their utility is proven by a low degree of failure (i.e. breaking or completely slipping off the penis during intercourse) and is supported by clinical and epidemiologic studies demonstrating a reduction in disease incidence among users. The available evidence assures excellent reduction in risk to consistent users who are at moderate to low risk, or who have adequately lowered their risk in other ways.

REFERENCES

Abma J., Chandra, A., Mosher, W., Peterson, L. and Piccinino, L. (1997). Fertility, family planning, and women's health: New data from the 1995 National Survey of Family Growth. National Center for Health Statistics. *Vital Health Stat.* 23(19)(8)

Albert, A.E., Hatcher, R.A. and Graves, W. (1991). Condom use and breakage among women in a municipal hospital family planning clinic. *Contraception* 43(2): 167–176

Albert, A.E., Warner, D.L., Hatcher, R.A., Trussell, J. and Bennett C (1995). Condom use among female commercial sex workers in Nevada's legal brothels. *Am. J. Pub. Health* 85(11): 1514–1520

Austin, H., Louv, W.C., Alexander, W.J. (1984). A case-control study of spermicides and gonorrhea. *J. Am. Med. Assoc.* 251: 2822

Barlow, D. (1977). The condom and gonorrhoea. *Lancet* ii: 811

Bolch, O.H. and Warren, J.C. (1973). In vitro effects of Emko on *Neisseria gonorrhoeae* and *Trichomonas vaginalis*. *Am. J. Obstet. Gynecol.* 11: 1145–1148

Bourinbaiar, A.S. and Lee-Huang, S. (1994). Comparative in vitro study of contraceptive agents with anti-HIV activity: gramicidin, nonoxynol-9, and gossypol. *Contraception* **49**: 131–137

Carey, R.F., Herman, W.A., Retta, S.M., Rinaldi, J.E., Herman, B.A. and Athey, T.W. (1992). Effectiveness of latex condoms as a berrier to human immunodeficiency virus-sized particles under conditions of simulated use. *Sex. Transm. Dis.* **19**(4): 230–234

CDC (1988). Condoms for prevention of sexually transmitted diseases. *MMWR* **37**(9): 133–137

Conant, M.A., Spicer, D.W. and Smith, C.D. (1984). Herpes simplex virus transmission: condom studies. *Sex. Transm. Dis.* **11**(2): 94–95

de Vincenzi, I. (1994). A longitudinal study of human immunodeficiency virus transmission by heterosexual partners. European Study Group on Heterosexual Transmission of HIV. *N. Engl. J. Med.* **331**(6): 341–346

Drew, W.L., Blair, M., Miner, R.C. and Conant, M. (1990). Evaluation of the virus permeability of a new condom for women. *Sex. Transm. Dis.* **17**: 110–112

Ehret, J.M. and Judson, F.N. (1988). Activity of nonoxynol-9 against *Chlamydia trachomatis*. *Sex. Transm. Dis.* **15**: 156–157

Gabbay, M. and Gibbs, A. (1996). Does additional lubrication reduce condom failure? *Contraception* **53**: 155–158

Gøtzsche, P.C. and Hørding, M. (1988). Condoms to prevent HIV transmission do not imply truly safe sex. *Scand. J. Infect. Dis.* **20**: 233–234

Grady, W.R. and Tanfer, K. (1994). Condom breakage and slippage among men in the United States. *Fam. Plann. Perspect.* **26**(3): 107–112

Hart, G. (1984). Factors influencing venereal infection in a war environment. *Br. J. Vener. Dis.* **50**: 68–72

Hicks, D.R., Martin, L.S., Getchell, J.P. *et al.* (1985). Inactivation of HTLV-III/LAV-infected cultures of normal human lymphocytes by nonoxynol-9 in vitro (letter). *Lancet* 1985; **2**(8469–70): 1422–1423

Jennings, R. and Clegg, A. (1993). The inhibitory effect of spermicidal agents on replication of HSV-2 and HIV-1 in vitro. *J. Antimicrob. Chemother.* **32**(1): 71–82

Jick, H., Hannan, M.T., Stergachis, A., Heidrich, F., Perera, D.R. and Rothman, K.J. (1982). Vaginal spermicides and gonorrhea. *JAMA* **248**(13): 1619–1621

Judson, F.N., Ehret, J.M., Bodin, G.F., Levin, M.J. and Rietmeijer, C.A.M. (1989). In vitro evaluations of condoms with and without nonoxynol 9 as physical and chemical barriers against *Chlamydia trachomatis*, herpes simplex virus type 2, and human immunodeficiency virus. *Sex. Transm. Dis.* **16**(2): 51–56

Kappus, E.W. and Quinn, T.C. (1986). The spermicide nonoxynol-9 does not inhibit *Chlamydia trachomatis* in vitro. *Sex. Transm. Dis.* **13**(3): 134–137

Katznelson, S., Drew, W.L. and Mintz, L. (1984). Efficacy of the condom as a barrier to the transmission of cytomegalovirus. *J. Infect. Dis.* **150**(1): 155–157

Kelaghan, J., Rubin, G.L., Ory, H.W. and Layde, P.M. (1982). Barrier-method contraceptives and pelvic inflammatory disease. *JAMA* **248**: 184–187

Kelly, J.P., Reynolds, R.B., Stagno, S., Louv, S.C. and James, W. (1985). In vitro activity of the spermicide nonoxynol-9 against *Chlamydia trachomatis*. *Antimicrob. Agents Chemother.* **27**(5): 760–762

Koop, C.E. (1988). Understanding AIDS – a message from the Surgeon General. HHS Publication No. (CDC) HHS-88-8404

Ku, L., Sonenstein, F.L., Pleck, J.H. (1994). The dynamics of young men's condom use during and across relationships. *Fam. Plann. Perspect.* **26**: 246–251

Louv, W.C., Austin, H., Alexander, W.J., Stagno, S. and Cheeks, J. (1988). A clinical trial of nonoxynol-9 for prevention gonococcal and chlamydial infections. *J. Infect. Dis.* **158**(3): 518–522

Minuk, G.Y., Bohme, C.E., Bowen, T.J. *et al.* (1987). Efficacy of commercial condoms in the prevention of hepatitis B virus infection. *Gastroenterology* **93**: 710–714

Mosher, W.D. and Pratt, W.F. (1990). Contraceptive use in the United States, 1973–88. Advance Data from Vital and Health Statistics of the National Center for Health Statistics 182

Pemberton, J., McCann, J.S., Mahony, J.D.G., MacKenzie, G., Dougan, H. and Hay, I. (1972). Sociomedical characteristics of patients attending a VD clinic and the circumstances of infection. *Br. J. Vener. Dis.* **48**: 391–396

Perlman, J.A., Kelaghan, J., Wolf, P.H., Baldwin, W., Coulson, A. and Novello, A. (1990). HIV risk difference between condom users and nonusers among U.S. heterosexual women. *J AIDS* **3**: 155–165

Reitmeijer, C.A.M., Krebs, J.W., Feorino, P.M. and Judson, F.N. (1988). Condoms as physical and chemical barriers against human immunodeficiency virus. *JAMA* **259**: 1851–1853

Richters, J., Donovan, B., Gerofi, J. (1993). How often do condoms break or slip off in use? *Int. J. STD AIDS* **4**: 90–94

Richters, J., Gerofi, J. and Donovan, B. (1995). Why do condoms break or slip off in use? An exploratory study. *Int. J. STD AIDS* **6**: 11–18

Roddy, R.E., Cordero, M., Cordero, C., Fortney, J.A. (1993). A dosing study of nonoxynol-9 and genital irritation. *Int. J. STD AIDS* **4**(3): 165–170

Rosenberg, M.J., Rojanapithayakorn, W., Feldblum, P.J. and Higgins, J.E. (1987). Effect of the contraceptive sponge on chlamydial infection, gonorrhea, and candidiasis. *JAMA* **257**(17): 2308–2312

Rugpao, S., Pruithithada, N., Yutabootr, Y., Prasertwitayakij, W. and Tovanabutra, S. (1993). Condom breakage during commercial sex in Chiang Mai, Thailand. *Contraception* **48**: 537–547

Rugpao, S., Beyrer, C., Tovanabutra, S. *et al.* (1997). Multiple condom use and decreased condom breakage and slippage in Thailand. *J AIDS Hum. Retroviruses* **14**: 169–173

Silverman, B.G. and Gross, T.P. (1997). Use and effectiveness of condoms during anal intercourse: a review. *Sex. Transm. Dis.* **24**: 11–17

Singh, B., Cutler, J.C. and Utidjian, H.M. (1972). Studies in development of a vaginal preparation providing both prophylaxis against venereal disease, other genital infections and contraception. II. In vitro effect of contraceptive and non-contraceptive preparations on *Treponema pallidum* and *Neisseria gonorrhoeae*. *Br. J. Vener. Dis.* **48**: 57–64

Singh, B., Postic, B. and Cutler, J.C. (1976). Virucidal effect of certain chemical contraceptives on type 2 herpesvirus. *Am. J. Obstet. Gynecol.* **125**: 422–425

Soper, D.E., Shoupe, D., Shangold, G.A., Shangold, M.M., Gutmann, J. and Mercer, L. (1993). Prevention of vaginal trichomoniasis by compliant use of the female condom. *Sex. Transm. Dis.* **20**(3), 137–139

Sparrow, M.J. and Lavill, K. (1994). Breakage and slippage of condoms in family planning clients. *Contraception* **50**: 117–129

Steiner, M., Piedrahita, C., Glover, L. and Joanis, C. (1993). Can condom users likely to experience condom failure be identified? *Fam. Plann. Perspect.* **25**: 220–226

Steiner, M., Piedrahita, C., Glover, L., Joanis, C., Spruyt, A. and Foldesy, R. (1994a). The impact of lubricants on latex condoms during vaginal intercourse. *Int. J. STD AIDS* **5**: 29–36

Steiner, M., Trussell, J., Glover, L., Joanis, C., Spruyt, A. and Dorflinger, L. (1994b). Standardized protocols for condom breakage and slippage trials: a proposal. *Am. J. Pub. Health* **84**: 1897–1900

Thompson, J.L.P., Yager, T.J. and Martin, J.L. (1993). Estimated condom failure and frequency of condom use among gay men. *Am. J. Pub. Health* **83**: 1409–1413

Voeller, B., Coulson, A., Bernstein, G.S. and Nakamura, R. (1989). Mineral oil lubricants cause rapid deterioration of latex condoms. *Contraception* **39**(1): 95–101

Voeller, B., Coulter, S.L. and Mayhan, K.G. (1991). Gas, dye and viral transport through poly-urethane condoms. *JAMA* **266**: 2986

Weir, S.S., Roddy, R.E., Zekeng, L. and Feldblum, P.J. (1995). Nonoxynol-9 use, genital ulcers, and HIV infection in a cohort of sex workers. *Genitourinary Med* **71**(2): 78–81

Wulfert, E. and Wan, K.W. (1993). Condom use: a self-efficacy model. *Health Psychol* **12**(5): 346–353

Zekeng, L., Feldblum, P.J., Oliver, R.M., Kaptue, L. (1993). Barrier contraceptive use and HIV infection among high-risk women in Cameroon. *AIDS* **7**: 725–731

7
TOPICAL MICROBICIDES

PENELOPE J. HITCHCOCK

Sexually Transmitted Diseases Branch,
National Institute of Allergy and Infectious Diseases,
Rockville, Maryland, USA

INTRODUCTION

The human immunodeficiency virus (HIV) pandemic has focused attention on sexually transmitted diseases (STDs), both because HIV infection is a fatal STD and because other STDs are risk factors for the sexual transmission of HIV. Current global estimates indicate that over 40 million people are infected with HIV, the cause of acquired immune deficiency syndrome (AIDS), and the majority of these infections were acquired through sexual intercourse. Unless effective preventive measures to stop sexual transmission are implemented, the number of HIV infections will continue to grow.

Separate from the HIV epidemic, STDs cause significant morbidity and mortality and contribute greatly to increasing health-care costs. In the United States in 1998, an estimated 15 million new cases of STDs occurred, 64% of which were in people under 24 years old, including 3 million teenagers. In 1998 cost estimates associated with STDs alone exceeded $8 bn, excluding the cost of HIV infections attributable to STDs.

Furthermore, STDs disproportionately affect females, fetuses and newborns. Gonococcal and chlamydial infections cause pelvic inflammatory disease, infertility and ectopic pregnancy. Several common STDs adversely affect pregnancy and can result in spontaneous abortion, stillbirth, chorioamnionitis, and premature rupture of membranes, preterm delivery or postpartum endometritis. Neonatal infections include gonococcal and chlamydial conjunctivitis, which may lead to blindness; chlamydial pneumonia, which may lead to chronic respiratory disease; congenital syphilis; and herpes encephalitis. Genital warts can occasionally cause recurrent respiratory papillomatosis in children. Moreover, 'high-risk' human papillomavirus infections cause cervical cancer, one of the most common cancers in women throughout the world (Holmes *et al.*, 1999).

It is now clear that the risk of becoming infected or infecting others with HIV is substantially increased if one has an STD, such as chancroid, genital herpes, syphilis, trichomoniasis, gonorrhea or chlamydial infection. Over 75 studies examining the role of STDs in HIV transmission have been conducted. In 15, STD effects could be assessed independently of sexual behavior effects; both ulcerative and non-ulcerative STDs increased the risk of HIV transmission. Although the individual risk of HIV transmission associated with genital ulcer diseases appears to be higher than the discharge diseases (up to 10-fold compared to three- to fivefold), the high prevalence of discharge diseases results in a much higher population attributable risk. Recent studies have implicated bacterial vaginosis as a risk factor for HIV infection. This is of particular concern as up to 50% of women may suffer from these diseases. Furthermore, data from over 80 reports on the natural history of STDs in HIV infected people suggest that, at a community level, HIV infection may increase the prevalence of some STDs (e.g. genital ulcers). If coinfection with HIV prolongs or augments the infectiousness of individuals with STDs, and if the same STDs increase the risk of HIV acquisition, these infections may greatly amplify one another. This 'epidemiological synergy' may be fueling the explosive growth of the HIV pandemic in some populations (Wasserheit, 1991).

RATIONALE

Based on the recommendations from numerous international conferences, a consensus has emerged that safe, effective, female-controlled chemical barriers that will block transmission are needed to prevent sexually transmitted HIV infection, as well as other STDs. Currently available mechanical and chemical barriers have many limitations. Although the male condom, if used consistently and correctly, is a very effective barrier against the transmission of HIV and gonorrhea, it has a limited spectrum of efficacy for the other STDs. Importantly, it requires the active cooperation of the male partner and therefore cannot be independently implemented at the discretion of the female. Although the female condom only requires partner consent, virtually nothing is known about its efficacy in preventing bacterial and viral STDs. Spermicides have *in vitro* activity against most sexually transmitted pathogens, including HIV; however, well designed clinical studies testing whether spermicides prevent STDs/HIV infection have not provided evidence to support this claim. Furthermore, several studies have revealed that spermicides can cause mucosal erosions and ulcers; a recent study revealed increased shedding of HIV in association with spermicide use in HIV in infected women.

In addition to the limited efficacy of these existing methods, there are many situations in which personal, social or cultural barriers interfere with a woman's ability to successfully negotiate and implement the use of barriers that could reduce the risk of infection. Specifically, the need for a method that can be implemented by women is grounded in the high prevalence of non-consensual sex, sex without condom use, and risky behaviors that occur without partner knowledge. Just as oral contraceptives dramatically enhanced the ability of women to avoid unwanted pregnancy, effective, female-controlled topical microbicides are urgently needed to enhance the ability of women to avoid sexually transmitted infections. Furthermore, chemical barriers that inactivate pathogens in vaginal/cervical secretions, as well as in the ejaculate could reduce female-to-male as well as male-to-female transmission (Elias and Heise, 1993; Global Program on AIDS, 1993; Hitchcock *et al.*, 1994).

Topical microbicides are here defined as preparations for intravaginal or intrarectal use that are microbicidal (virucidal and/or bactericidal); such products will prevent sexually transmitted infections. The ideal microbicide should be *invisible* (colorless, odorless, tasteless), *innocuous* (non-toxic) and *inexpensive*; it should also have a long shelf-life, be easy to store, fast-acting for an appropriate duration, effective pre and post coitus, available without a prescription, and 'safe' for use at least once or twice daily.

Candidate classes of microbicidal compounds include (but are not limited to) detergents, defensins and other antimicrobial peptides, antibodies, organometallic complexes, acidic buffers, carbohydrates, sulfonated and non-sulfonated polymers, chemokine blockers, biological dyes, lipids and fatty acids, and reverse transcriptase inhibitors.

TWO TYPES OF ACTIVITY

Ideally, topical microbicides would not be inherently spermicidal but could be formulated with or without spermicidal activity; in fact, microbiocides that do not kill or harm sperm may be inherently safer. Non-contraceptive microbicides would be extremely useful for women who wish to become pregnant, or for those who use one of the many safe, effective methods for contraception that either have no protective effect against infection or, arguably, exacerbate the risk of infection (e.g. surgical sterilization, IUD use). Indeed, a person's contraceptive choices may change over a lifetime; however, no matter what an individual's current contraceptive preference, if they are sexually active they will desire/require protection from sexually transmitted infections (Stone and Hitchcock, 1994).

RESEARCH STRATEGY

The development of safe, effective topical microbicides involves a broad-based research agenda that spans basic biomedical and behavioral research, product development and clinical/behavioral evaluation. There is a pressing need to evaluate the currently available spermicides and determine whether or not their laboratory-documented microbicidal activity predicts clinical efficacy. Numerous studies evaluating these products have been/are being undertaken and have provided an opportunity to develop methods for studying product safety and efficacy (*vis-à-vis* prevention of infection). In parallel, studies on the fundamental aspects of inactivating infectious agents in the vagina and the rectum are being conducted. Product development research that includes formulation and preclinical efficacy and toxicity is also in progress, with the overall objective of identifying new candidates for clinical trials.

BASIC BIOMEDICAL RESEARCH

Arguably the most productive approach to identifying safe, effective strategies for blocking the early steps in the infectious process is based on collaborative, multidisciplinary efforts in microbiology, immunology, reproductive biology, reproductive toxicology and cell biology.

Early Steps in Infectious Processes

Studies delineating the chronology and biology of the early steps in the host–pathogen interaction include, but are not limited to: primary and secondary molecular interactions between the target cells and the pathogens; inflammation and altered kinetics and infectious dose; and cell-free and cell-associated transmission of infection.

Preclinical Microbicide Evaluation

Microbicides are evaluated using *in vitro* systems, *ex vivo* systems (using primary and established human cells lines and organ cultures) and animal models. The systems have been used to measure the bactericidal/virucidal activity and toxicity of currently available spermicidal products, new topical microbicides and inactive ingredients (carriers) in existing or new products.

Biology of the Reproductive Tract

Exogenous products should not disrupt the natural defense mechanisms of the vaginal ecosystem. To that end, there has been a focused effort to identify and characterize the 'endogenous' anatomical, physiological, hormonal, immunological and microbiological factors of the female reproductive tract that play a role in resistance and susceptibility to infection. These include, but are not limited to, vaginal pH, mucus, estrogen and other hormones, cervical ectopy and normal flora, including lactobacilli.

Reproductive Toxicology

Studies on the characterization of the spermicidal, teratogenic, mutagenic or carcinogenic properties of topical microbicides must also be conducted before a microbicide is approved for human use. Animal models have been developed that measure vaginal–cervical irritation/toxicity/inflammation, or effects on sperm or embryogenesis; these have been adopted from the contraceptive development field. The traditional rabbit model for measuring local effects has been used for studying spermicidal contraceptives; it is currently recommended for measuring the local toxicity of candidate microbicides.

PRODUCT DEVELOPMENT

The evaluation of putative topical microbicides is pursued in a stepwise fashion using model systems for efficacy, model systems for toxicity, and formulation development. Once the active ingredient has been formulated, efficacy and toxicity studies are repeated. The first measure of a potential microbicide is made in laboratory models. The simplest of questions is asked first: What are the virucidal/bactericidal properties of the active ingredient in the test tube or the tissue culture system? If the activity of the active ingredient is high, a variety of strains and cell lines are studied to fully characterize the compound. Spermicidal activity and toxicity for normal vaginal flora (*Lactobacillus* spp.) and yeast are measured, as is the toxicity for rabbit vaginal epithelium.

Once the assessment of the active ingredient is completed, promising compounds are formulated, using good manufacturing practices, into gels or other vehicles that would be suitable for use in the vagina or rectum; the studies are then repeated. If the formulated product is active, models may be used to assess its toxicity and efficacy in the vaginas of mice, guinea pigs, cats or primates, depending upon the spectrum of activity predicted for the product. A number of other tests are required of the clinical formulation. The assessment of the physicochemical properties of the formulated compound includes solubility, stability (real and accelerated time) and bioavailability. A wide range of toxicity studies are conducted, including *in vitro* mutagenesis, systemic absorption and systemic toxicity. Concurrent with phase I/II clinical testing, other studies are conducted in mammals, including oral toxicity, local irritation in male rabbits, hypersensitivity and phototoxicity, segment II reproductive toxicity, carcinogenicity and segment III reproductive toxicology.

Classes of New Products

The spermicides available in the United States and Europe are detergents. Active ingredients found in over-the-counter spermicides include nonoxynol-9, octoxynol-9 (approved in the US and Europe), benzalkonium chloride and menfegol (approved only in Europe). The spermicidal activity of these detergents is based on disruption of the lipid-rich envelope of sperm. Given that the cell membrane of the sperm is fundamentally the same as other eukaryotic membranes, the therapeutic index (the ratio of the concentration required for spermicidal function and the concentration that causes non-specific damage of other cells) is low. However, since the inception of the research effort to develop topical microbicides many new compounds have been identified, some representing novel, more specific strategies for inactivating infectious agents. The following overview of new products is meant to provide examples and is by no means comprehensive.

Detergents

The general mechanism of action of detergents is disruption of lipid bilayers. In order to increase the safety profile, other types of detergents have been examined for virucidal and microbicidal activity, as well as activity against the host cell membranes. Considering that HIV and genital herpes virus, herpes simplex virus, are enveloped viruses, identifying non-toxic virucidal detergents is difficult. On the other hand, the lipid membranes of bacteria are sufficiently different from host membranes that non-toxic microbicidal detergents have been identified. Increased molecular weight, achieved by altering chain length, results in increased microbicidal activity and decreased cytotoxicity. Recently, alkyl sulfates have been shown to have virucidal activity against human papillomavirus. Interestingly, octoxynol-9, one of the detergents currently approved for use as an active ingredient in over-the-counter spermicides, is relatively nontoxic for eukaryotic cells. Examples of other 'new age' detergents include C31G, sodium dodecyl sulfate, and chlorhexidine: some detergents are also being examined in combination.

Defensins

Within the last 10 years a new class of antimicrobial peptides has been identified. These are broadly distributed molecules and have been found in many animal species, including humans. From an evolutionary perspective these molecules represent acute-phase reactants that offer a first line of defense against infection. Currently two types of defensins are being studied for their potential as microbicides: the maganins and the protegrins. These are steroidal and linear molecules, respectively; they disrupt membranes by forming voltage-sensitive channels. The association with membranes is thought to be based on protein–protein interactions, and thereby achieves a high degree of specificity.

Acid Buffering Compounds

This class of products is based on the characteristics of the normal vaginal ecosystem. The pH of the healthy vagina is in the acidic range, except when semen is present; the buffering capacity of semen ejaculate increases the pH to 7–8, an alkalinity which is required to maintain viable sperm. Laboratory studies have demonstrated that acidic pH is detrimental to HIV and other sexually transmitted pathogens. These powerful buffering agents are able to maintain a pH of approximately 4, even in the presence of semen.

Carbohydrate Polymers

Modified and unmodified carbohydrate polymers may act as microbicides. Some interactions are reversible (static) and others are irreversible (cidal). Examples include carrageenan, heparin, sulfonated azo dyes and a naphthalene sulfonated polymer.

Lactobacillus Products

This approach reflects the importance of normal flora in the resistance to infection. Several preparations have been developed which contain viable *Lactobacillus* spp. These products provide an exogenous source of bacteria to recolonize the vagina with hydrogen peroxidase-producing lactobacilli. These lactobacilli can inactivate HIV *in vitro*.

Lipids and Fatty Acids

These molecules are found in breast milk and bile (cholic acid), and are active ingredients in currently available antifungal preparations. Antibacterial and antiviral activity varies depending upon the particular lipid or fatty acid molecule. These molecules probably act as surfactants and disrupt membrane integrity.

HIV-specific Molecules

Examples include chemokine receptor-blocking agents, anti-gp 120 proteins (e.g. cyanovirin-N), and reverse transcriptase inhibitors. These are likely to be HIV-specific approaches. The first class of molecule functions by selective inhibition of HIV entry into cells, targeting either chemokine receptor function or gp 120 function. The reverse transcriptase inhibitors interfere with secondary steps in the infectious process. Arguably these must permeate the host cell membrane in order to have an effect.

Other Approaches

Organo-metallic compounds, quarternary ammonium salts and β-lacto-globulins are examples of other classes of compounds currently being evaluated for microbicidal potential. Monoclonal antibodies are also being evaluated. Arguably, a cocktail of neutralizing antibodies could be used to provide passive immunity to a number of sexually transmitted pathogens. The ability to produce large quantities of functional antibodies in plant cells (plantibodies) may make this approach economically feasible.

Products for Rectal Use

Virtually all pathogens that can be transmitted through vaginal intercourse can also be transmitted through anal intercourse. Importantly, it is easier to transmit HIV by anal intercourse, both because of the CD4-like receptors on rectal epithelial cells and because of the trauma to the epithelial surface that is likely to result in direct inoculation into the bloodstream. Furthermore, because anal intercourse is practiced frequently among men who have sex with men, and occasionally among heterosexual couples, especially adolescents, there is a need for microbicides that are safe and effective for anal intercourse. However, specific product development strategies must be developed that account for the unique characteristics of a rectal microbicide. The differences in epithelium, flora, pH, volume, organic matter etc., are compelling. Effective vaginal microbicides will have to be clinically evaluated to determine their effectiveness for rectal use.

OVERARCHING ISSUES

The following summaries are derived from presentations made at the NIAID Topical Microbicide Workshop on Preclinical Evaluation, Atlanta, Georgia, May 1998. Presenters are acknowledged for their personal communication.

Normal Vaginal Ecosystem

The human vagina is a complex environment. Under normal circumstances a combination of physical, microbiological and chemical barriers acts synergistically to protect this mucosal surface from infection and injury. Disruption, destruction or functional loss of the mucous layer, vaginal pH, normal flora, anti-infective peptides (defensins), antibodies and/or epithelium are likely to increase the risk of infection. An understanding of the components of this ecosystem and their function, as well as the effect of microbicide use on them, is critical to the development of safe, effective topical microbicides.

The potential defenses of the vagina include the vaginal mucosa that acts as a physical barrier; the vaginal flora that produce antimicrobial compounds; the local environment which is acidic (pH 3.8–4.2); antimicrobial components of secretions, such as lysosymes, lactoferrins, zinc and peroxides; cell-mediated immunity; and humoral immunity.

Lactobacilli, Gram-positive rods found in the healthy vagina, interfere with pathogens by adhering competitively to receptors on the vaginal epithelium and by coaggregation with other bacteria, releasing antimicrobial compounds such as lactic acid, hydrogen peroxide, bacteriocins and biosurfactants. *Lactobacillus* species are found in 96% of normal women, versus 67% of women with bacterial vaginosis. Hydrogen peroxide-producing strains are found in 61% of women with normal vaginal flora, versus 5% of women with abnormal vaginal flora. Many bacterial species, including anaerobes, are present in the vagina, but are much more prominent in women with abnormal vaginal flora. Lactobacilli have been shown *in vitro* to inhibit bacteria associated with bacterial vaginosis, such as *Gardnerella*, *Mobiluncus* and *Bacteroides* spp. The hydrogen peroxide system can also inhibit HIV.

Other endogenous factors affecting the vaginal flora include hormones (estrogen), the growth substrate (glycogen), bacterial interactions (ecosystem) and the pH (acidic pH correlates with a healthy vagina). Exogenous factors that may affect the vaginal flora are douching, sexual intercourse, especially with a new partner, antibiotics and spermicides.

Loss of lactobacilli and the overgrowth of abnormal vaginal flora were found to be associated with gonorrhea, bacterial vaginosis, and HIV-1 infection. Women with abnormal vaginal flora were more likely to be seropositive for HIV (personal communication, Dr Jane Schweble, University of Alabama).

Epithelial Cells

Epithelial cells initiate the first response to pathogens and other proinflammatory stimuli. They produce antimicrobial peptides called defensins, which have a broad spectrum of antimicrobial activity and may help clear or reduce the pathogen load at the mucosal surfaces. Epithelial cells also rapidly produce an array of cytokines in response to infection and other stimuli. The cytokines are chemotactic to leukocytes and are involved in their activation and regulation.

The defensins are small (< 4 kDa) cationic, amphipathic peptides. They contain six cysteine residues forming three disulfide bridges, and are synthesized as propeptides which undergo sequential processing to the mature, biologically active peptide. Depending on the spacing of the cysteine residues and the connectivity of the disulfide bridges, defensins are classified into two families, the α- and β-defensins. Both are active against bacteria, fungi and enveloped viruses. They are membrane active, forming voltage-sensitive channels, and this probably requires dimer or tetramer formation. Some defensins are constitutively expressed, but others are inducible. Phagocytes and epithelial cells synthesize defensins. The female genital epithelium synthesizes α-defensin and the human kidney epithelium synthesizes a β-defensin.

The neutrophil defensins are the best characterized. In addition to their antimicrobial activity, a number of other functions have also been attributed to these molecules, including chemotaxis of mononuclear cells, wound healing, opsonization, suppression of endogenous corticosteroids, and cell-mediated cytoxicity. Whether defensins derived from epithelial cells have similar functions is unclear, and important questions are still to be answered.

Cytokines

Cytokines have been found in cervical vaginal lavages in normal ovulating women. Some have been found only during menses, e.g. IF, RANTES and MIP-1α. Interleukins 6, 8 and 1-β, and TGF-β are found throughout the cycle, but concentrations are highest at menses. Two cytokines, GM-CSF and EGF, peak during the late proliferative stage. Interestingly, interferon-γ production appears to be restricted to inflammatory conditions, including infections. The source of these cytokines is unknown; however cervical epithelial cells (HeLa cells) produce cytokines when infected with *Chlamydia trachomatis*. Spermicides containing N-9 will also induce a cascade of cytokines from epithelial cells *in vitro*, and when applied vaginally a similar profile of cytokine induction can be detected in cervical–vaginal lavages of women (personal communication, Dr Alison Quayle, Harvard Medical School).

The immune system of the reproductive tract

Studies indicate that: (a) the mucosal immune system is present throughout the human female reproductive tract; (b) IgA and IgG are present in cervical/vaginal secretions and are under hormonal control; (c) antigen-presenting cells in the cervix and vagina are able to present antigen, and the cervix and vagina are inductive sites for immune responses; (d) cytotoxic T-cell activity in the vagina and cervix is present throughout the menstrual cycle and following menopause; (e) the cervix and vagina contain the full spectrum of immune cells; and (f) mucosal immune responses in the reproductive tract are influenced by the stage of the menstrual cycle. Obviously, it is critically important that vaginal microbicides do not compromise the specific and non-specific defense mechanisms of the female reproductive tract. Similar concerns are relevant to use of rectal microbicides (personal communication, Dr Charles Wira, Dartmouth College).

Formulation

One of the critical steps in the development of safe, effective topical microbicides is combining the active ingredients into a vehicle, i.e. a gel or cream. This must deliver the active ingredients to the vagina so as to inactivate infectious agents in both ejaculate and cervical/vaginal secretions. Vaginal spermicides and medications have been successfully formulated and provide a framework within which to consider some of the characteristics of the vehicle and the formulated product. There is consensus on some of the characteristics of formulated products: it should maintain biological activity; it should be stable to heat and cold; and it should be non-toxic, both locally and systemically.

Other characteristics are not well understood and are the subject of investigation. Some of the unanswered aspects of structure/function include, but are not limited to, the need to coat the vaginal mucous membrane; the need to penetrate the cervical mucus; the ideal pH; the ideal time to persist in the vaginal vault. Clearly, the delivery system can enhance, detract or have no measurable impact on product performance. The science of vaginal formulations is a critical component of the effort to develop safe, effective topical microbicides.

The effectiveness of intravaginal microbicides depends upon interactions within the vaginal environment. The biology of such interaction has both chemical and physical components, including 'deployment' – the physical spreading and distribution of a formulation and its retention over time within the vagina; and 'delivery' – transfer of active ingredients and other components of a formulation to adjacent fluids and tissue. Both deployment and delivery can and should be characterized in terms of principles of chemistry and physics in order to help understand how formulation properties govern function. The impact upon vaginal ecology and overall physiology should also be considered. It is important to characterize how long the formulation needs to be in contact with the surface in order to prevent infection, the nature of the surface, and the forces that exist between the formulation, its active ingredients and the target organisms. Arguably, we do not wish to trap or contain pathogens (e.g. in a layer of formulation) if they are not killed or irreversibly neutralized.

The bioactivity of formulations should be assessed *in vitro* under conditions that simulate *in situ* conditions as much as possible. Deployment or distribution in the vagina is the result of several actions, including 'squeezing' – of a formulation between opposing epithelial surfaces; 'shearing' – between such surfaces and/or the penis; 'sliding' – due to gravity; 'seeping' – into vaginal rugae; and 'sticking' – the adhesion (or lack thereof) between a formulation and epithelial surfaces in the presence of mucus or other vaginal fluids. Each process can be mathematically modeled using basic physical and chemical principles, and tested to determine formulation properties.

Toxicity for Mucosal Surfaces of the Reproductive Tract

The safety considerations for topical microbicides include local inflammation and irritation of the mucous membranes of the vulva, vagina and cervix, as well as systemic toxicity. Gross and magnified (colsposcopy) examinations of the female lower reproductive tract and gross examination of the penis of the male partner are incorporated into phase I safety and subsequent safety studies. Although it is unclear what clinical significance can be attributed to epithelial erosion, full-thickness damage (ulceration) creates a window for viral exit and entry. Furthermore, recent studies suggest that focal cellular inflammation – the influx of neutrophils and mononuclear cells – may occur in areas of the mucous membrane which are grossly normal. Ulceration and the influx of inflammatory cells causes concern about the increased risk of HIV transmission and acquisition, similar to that seen in infectious processes.

Cats, dogs, rats, guinea pigs, rabbits and non-human primates have all been evaluated as potential models for vaginal irritation testing. Large differences between the species relative to the thickness and types of epithelium, as well as the vaginal architecture and physiology, make it difficult to readily identify the most appropriate animal model.

Developers of vaginal contraceptives have considered the rabbit the most practical model system owing to its availability, anatomy and size. The primary focus for most vaginal irritation tests is the histopathological evaluation. Scoring is based upon: the integrity of the epithelium; vascular congestion; leukocyte infiltration and other inflammatory responses; and edema. The scores are summed and a mean determined from three regions of the anterior vagina, and classified as minimal (1–4), mild (5–8), moderate (9–11) or marked (12–16), These scores are then interpreted as acceptable (score 0–8), borderline (9–11) or unacceptable (12–16). These interpretations are based primarily upon the past experience of surfactant effects upon the vaginal epithelium. The vaginal irritation rating scale may not be appropriate for non-detergent types of toxicity.

The vaginal wall of the rabbit is very thin and absorbs materials within the first 2–3 minutes, unlike the thicker human vagina, which can take up to 6 hours to absorb the same material. Irritation from repeated dosing can also change blood flow and alter the kinetics of absorption. The pH of the rabbit vagina is about neutral, whereas in women it is quite acidic (about 4). There are many examples of pH effects on the absorption of weakly acidic or weakly basic drugs from the intestinal tract and their elimination by the kidney. Interpretation of the absorption results for vaginal products should carefully consider whether pH may alter the ionization potential of the test substance and directly alter the absorption of the test substance in women.

In summary, the rabbit model for vaginal irritation testing has been the traditional test to perform. However, there are vast differences in the structure and anatomy of the rabbit: these limit the ability to directly extrapolate risk to the human vagina. The protective flora and the acidic pH of the human vagina are not present in the rabbit model. Agents that may have deleterious effects on these environmental protective mechanisms cannot be assessed in the rabbit model. For better or worse, a large amount of data has been generated over the past several years for vaginal products currently in the market using this model, and it is still considered an important comparative test for such products. Arguably, better models for screening, both *in vitro* and *in vivo*, should be developed to address the potential risks of vaginal irritation by new agents. There is a need for the validation of new models and a rating system not focused on detergent-induced vaginal toxicity (personal communication, Dr Donald Waller, University of Illinois at Chicago).

Surrogate Measures of Efficacy

The active ingredient in currently available spermicides, nonoxynol-9, is a potent antiseptic. In the test tube it inactivates viruses and kills bacteria and protozoa at very low concentrations. In tissue culture systems and animal models of SIV, HSV and chlamydial infection results may vary, depending upon the laboratory. How these *in vitro* and *in vivo* results will predict clinical trial results remains to be seen. Recent clinical trial results of the Vaginal Contraceptive Film (VCF) did not demonstrate the protection that was predicted by the laboratory studies. The development of surrogate models to predict clinical efficacy presents a formidable challenge when developing products to prevent infection, whether they are microbicides or vaccines. At a minimum, a product would have been demonstrated effective *in vitro*; if and how data from laboratory animal studies should be factored into decisions for efficacy studies remain subjects for discussion (and debate).

In vitro Model Systems

The advantages of *in vitro* microbicide testing using the active ingredient alone are that the inhibitory (killing) concentration can be determined and something may be learnt about the mechanism of action of the microbicide. There are also some disadvantages of *in vitro* models for microbes which colonize or grow inside eukaryotic cells. First, immortalized vaginal epithelial cell lines have not been available. Access to primary vaginal epithelial cells is necessary, or the vaginal epithelial cells must be transformed. However, recently a vaginal organ system has been developed using human vaginal tissue from postmenopausal women undergoing surgery for vaginal prolapse. The vaginal tissue is rolled up, sutured and inserted subcutaneously into the back of a nude mouse as a vaginal xenograph. An opening is maintained which permits the inoculation of sexually transmitted viruses or bacteria after inoculation with a microbicide. A second disadvantage is that there are some microbes for which vaginal epithelial cells are not the targets, but for which endocervical cells are, particularly ectopic endocervical epithelial cells. This presents two experimental problems: the cultured endocervical epithelial cells are sensitive to acid pH (obviously the ectopic cervical cells are not pH sensitive in the vagina), which makes it virtually impossible to test the microbicide for cytotoxicity against these cells at normal vaginal pH; and concentrations of microbicide which are antibacterial or antiviral can be cytotoxic to endocervical epithelial cells. Reduction of the microbicide concentration avoids host cell cytotoxicity, but the non-toxic dose may not prevent infection.

In vitro cells have different anatomic positions compared to cells *in situ*. These architectural and structural characteristics contribute to and are an integral part of function: the formation of a selective permeability barrier between the lumen and the cytosol to different ionic concentrations, and the regulation of the biological compartments dependent on vectorial function and secretion. The membrane proteins of non-polarized cells are the same as those of polarized cells, but are non-randomly distributed, which interferes with vectorial function. The consequence is that the data generated in such cultures may be misleading. For example, drugs tested for their antibacterial effect in polarized cells are five times more active than the same drugs tested in cells which are not polarized.

In order to overcome this, culture methods have been developed that allow endocervical epithelial cells to polarize and mimic *in situ* conditions. The polarized orientation of cytoplasmic organelles and cytoskeleton is important: actin is in the apical region in a vertical axis; microtubules are oriented in the apical-to-basal axis; the Golgi apparatus is supranuclear in the apical region; and cells are juxtaposed to a permeable substratum, the basement membrane, where tissue specificity is signaled and maintained.

Even though tissue culture systems are not ideal for reproducing the environment of the vaginal vault, new advances made in the last 5 years, such as the synthetic reproduction of cervical mucus and vaginal fluids, will be able to add a more realistic complexity to *in vitro* model systems for a more valid interpretation of preliminary microbicide efficacy (personal communication, Dr Priscilla Wyrick, University of North Carolina).

Small Animal Models for Microbicide Evaluations

The challenge is to determine whether potential microbicides can inactivate STD pathogens in the vaginal environment without toxicity to the vaginal epithelium and without interfering with normal defense mechanisms. What small animal models provide is the vaginal environment.

There are three requirements for using animals as microbicide efficacy models: the microbicide and the infectious dose should be vaginally inoculated with as little manipulation as possible; ideally, contact between the microbicide and the pathogen should occur in the vagina not in a test tube; and there should be microbiological and clinical endpoints.

Studies of herpes simplex virus in small animal models may be predictive of the activity of membrane-lysing microbicides *vis à vis* HIV as well as HSV. The mouse genital model of HSV-2 and the guinea-pig model of genital HSV-2 can be used sequentially. The advantages of the two-tiered strategy is that the mouse model is less expensive; vaginal inoculation does not disturb the vaginal vault (Depo-Provera is used to thin out the mucus on the epithelium to make the animals susceptible); viral replication is easily quantified; the animals develop clinical disease; the system lends itself to studying different formulations, doses, pHs, formulations and time courses; and the model has an easily discernible endpoint.

The murine model may be useful for screening, dose escalation studies, duration of effect studies, formulation studies, study of resistant viruses, and studies of the absorption of drugs.

In the guinea-pig model viral replication and shedding can be measured. The advantages of the guinea-pig model are that disease is easier to quantify; the vagina has a larger surface area; and latency and recurrences can be evaluated.

The effect of microbicides on chlamydial infections can also be studied in both the mouse and the guinea pig. In the murine model animals are also pretreated with Depo-Provera to increase infection rates with the mouse pneumonitis agent (*C. trachomatis* Moln), but human serovars can also be used. Both vaginal (lower tract) and uterine horn (upper tract) replication are used as endpoints. In the guinea-pig model the guinea-pig inclusion conjunctivitis agent (GPIC) is used for vaginal inoculation at 11 days post estrus, when they are most susceptible.

The feline model can be used for retroviral studies, especially with microbicides that do not work by interfering with host–gp 120 interaction.

Small animal models play an important role in screening potential compounds and evaluating new microbicides for the prevention of HIV/STDs, however the significance of the results vis a vis predicting clinical trial results is unknown (personal communication, Dr David Bernstein, Children's Hospital Medical Center).

The Primate Model

The primate model is analogous to humans *vis-à-vis* the anatomy, the physiology and the vaginal ecosystem, including microflora and pH. The primate is susceptible to human biovars of *C. trachomatis* and HIV/SIV without pretreatment to initiate, enhance and prolong infection.

Non-human primate models are well established for the study of genital transmission of simian immunodeficiency virus, using the female rhesus macaque. In this model cell-free SIV can be transmitted across intact vaginal mucosae. Once infection is established, both cell-free and cell-associated SIV can be isolated from vaginal secretions. SIV-infected cells are present in vaginal mucosae and submucosae, and epithelial cells of the cervix and uterus. Two N-9-containing products have been evaluated in this model: a Delfen Foam (12.5% solution of Non-9) and Gynol II gel (3% solution of Non-9). In the studies of the foam only half of the animals were protected, and in the animals exposed to the gel preparation before SIV challenge only one-third were protected. As expected, the animals in the control group all became SIV infected. This model uses 100 000 tissue culture infectious dose 50s ($TCID_{50}$) to challenge animals; this is a very high dose of virus and arguably does not reflect the ID_{50} in humans (Miller C.J. *et al.*, 1992).

Another primate model has been developed to study chlamydial infection. In the pigtail macaque model tissues are exposed to chlamydial elementary bodies and infection occurs minutes after exposure. Arguably, for a topical microbicide to be effective the product would have to coat mucosal tissues immediately after application, in order to prevent the actual uptake and internalization of the *Chlamydia*.

Non-9 products, benzalkonium chloride, and two concentrations of chlorhexidine gluconate (CHG), along with vehicle alone, have been evaluated in the model. In the N-9 studies four out of six animals were protected; in the benzalkonium chloride study two out of three animals were protected; with the 0.05% CHG three of four animals were protected, and in the 0.25% CHG study all five animals were protected. In each experiment all of the control animals (no product) became infected, and among those receiving placebo gels (boric acid vehicle or non-9 vehicle) all animals became infected.

In summary, the pigtailed macaque is an excellent model for the study of topical microbicides: it is analogous to humans in the anatomy and physiology of the reproductive tract, including vaginal microflora and pH; it is susceptible to chlamydial infection and SIV without hormonal or other pretreatments to increase susceptibility; it may also enable us to study other STDs in SIV-infected primates, and the converse (personal communication, Dr Dorothy Patton, University of Washington).

ISSUES IN CLINICAL RESEARCH

It is important to note that currently available spermicides have not been clinically evaluated for contraceptive efficacy. Without any history, the clinical efficacy trials of these products *vis-à-vis* the prevention of infection represent a new genre of clinical trials. As such, some of the fundamental methodological issues are being identified and approached *de novo*.

Methodological Issues

Acceptance and Adherence

If the microbicide is not used there is no chance that it will prevent infection. That said, measurement of use presents a formidable challenge: it is currently completely dependent upon self-reported data. Like the measurement of condom use, the accuracy an efficacy measurement is completely dependent on the validity of self-reported use. This is in marked contrast to the measurement of vaccine or drug efficacy, where clinical record of vaccination or laboratory measurement of drug blood levels coincide with adherence to the clinical protocol.

Epithelial Toxicity

The measurement of vaginal and cervical toxicity is also problematic. The architecture of the vagina makes thorough examination tedious and perhaps impossible. Furthermore, the discordance between gross epithelial lesions (erosions, ulcers, subepithelial hemorrhages etc.) and microscopic changes has been documented, making interpretation, as regards clinical significance, difficult. The discordance between observed lesions and symptoms is also perplexing. Careful studies have documented bidirectional discrepancies between the discomfort that women report and observed changes. Furthermore, other STDs and vaginal abnormalities can confound the interpretation of vaginal lesions. Microbiological outcomes present different logistical problems with respect to sample size and duration of study. If HIV infection is the primary outcome, screening and treatment of curable STDs must be conducted throughout the study; however, the anticipated HIV incidence rate may be decreased due to prevention and control of other STDs. Obviously, condom use must be encouraged throughout the study; however, HIV incidence rate will be altered depending upon condom use. Regardless of the effectiveness of the experimental microbicide, and regardless of adherence to clinical protocol, HIV incidence rates due to vaginal intercourse may be confounded by anal intercourse, intravenous drug use, genital herpes, dry sex practices and bacterial vaginosis. Finally, issues related to the male partner, including informed consent, risk perception and penile irritation, need to be addressed.

CLINICAL TRIALS OF SPERMICIDES FOR PREVENTION OF STD/HIV INFECTION

The results of a well designed, randomized, placebo-controlled clinical efficacy trial of an over-the-counter spermicide were published recently. The trial was designed to measure the efficacy of a spermicide to prevent HIV infection (gonorrhea and chlamydial infection were secondary endpoints). The trial, conducted in Cameroon, included a comprehensive HIV prevention program with condom-use education, condom distribution, and screening and treatment for curable STDs. In that context a contraceptive film (Vaginal Contraceptive

Film, VCF) containing approximately 70 mg of nonoxynol-9 had no impact on the rates of any of the measured infectious diseases. Based on the self-reported behavior of the 1200 participants, the women in both arms of the trial were able to persuade their partners to use the male condom 80% of the time. Women in both arms claimed to use the contraceptive film (experimental and placebo) 80% of the time. The biological outcomes in both arms were virtually identical, with one exception. The estimated HIV incidence rate was 6%; the point prevalence of gonorrhea and chlamydial infection was unaltered during the course of the study, and remained 30% and 22%, respectively. In addition to film and condom use, women were screened and treated for gonorrhea and chlamydial infection in both arms of the study. The only biological difference measured between women in the experimental and control arms was the presence of vaginal and vulvar ulcers. Approximately 25% of the women who used the experimental product had lesions, compared to those who used the placebo (Roddy *et al.*, 1998). Arguably the validity of all findings is completely dependent on the validity of the self-reported behavior. If film and condom use occurs less frequently among the population in general (i.e. in both arms), then the lack of efficacy may actually reflect lack of use-effectiveness. This explanation (overreported use of condoms and film) would explain the high incidence rate of HIV infection in both arms of the study. Furthermore, the bacterial STD outcomes are quite interesting. Male condom use should prevent HIV infection and gonorrhea (close to 100%); however, the prevention of chlamydial infection would be expected to be much less. These results call into question the validity of the self-reported condom use. Furthermore, it is remarkable that the incidence of vaginal and vulvar lesions occurred in the women who used the experimental film. Given the association of spermicides with such lesions, it is tempting to speculate that the rate of lesions is a surrogate for actual use of the experimental film. If film was used only 25% of the time, as opposed to 80% as reported, then the study was not adequately powered.

A randomized placebo-controlled trial of another contraceptive product (COL 1492, also Advantage) has recently been completed. Investigators measured the HIV incidence rate among women who used this bioadhesive gel containing approximately 50 mg of nonoxynol-9; results of the trial are expected to be published later this year.

Two additional clinical trials to measure the microbicidal efficacy of a gel containing 150 or 100 mg of nonoxynol-9 (Conceptrol) are in progress/planned. One trial will measure rates of gonorrhea or chlamydial infection; the other will measure HIV infection rates.

In the past several years a number of phase I and phase II safety studies have been conducted on new types of microbicides. Studies of a lactobacillus suppository (Lactovagin), Pro 2000, Acid Buffer gel, carrageenan and dextran sulfate have all demonstrated acceptable levels of safety for the products. Clinical efficacy trials are in progress or planned.

BEHAVIORAL RESEARCH

The effective use of any biomedical products, including topical microbicides, will depend on the development of behavioral interventions. The targeted population must perceive the need for the product, have the skills to use the product themselves, and have the skills to negotiate its use if it cannot be used clandestinely. There must also be a realistic appreciation of product performance, so that it is used correctly, consistently and realistically. Of particular concern is that the use of the product is not coincident with an increase in risk-related behavior that might 'swamp' the protective effect of the product.

In spite of the proven efficacy of correct and consistent use of the male condom to prevent HIV infection, condoms have not been used effectively in preventing HIV transmission. Arguably, the take-home message is that the best method in the world is worth nothing if it is not used. Correct and consistent use of a topical microbicide must be measurable in clinical trials. Issues related to acceptability must be addressed for both women and their partners. Important, albeit general, issues include informed consent for both the women and their partners, and loss to follow-up (this will introduce a bias into the study findings).

MICROBICIDES: REALISTIC PROSPECTS

As we move forward with clinical trials to measure the safety and efficacy of these products, two factors are particularly important. First, we must be certain that clinical trials are designed so that we are confident that the experimental product does not increase the risk of STDs or HIV infection. Second, we should not anticipate that we would develop products that are 100% effective for all infections. The product development and evaluation process is a long term one in which we make incremental gains. Over time, as we learn more about the basic scientific issues that need to be addressed, develop appropriate model systems to evaluate the products and address some of the methodological issues related to clinical trial design, we will identify products that are safe and effective. It is probably unrealistic to expect to develop a 'magic' gel that will protect against all infections and is safe for use dozens of times each day, but it is realistic to be able to develop a microbicide that will reduce the risk of one or more STDs, including HIV infection, and is safe for use once or twice a day. A microbicide is likely to be one of a repertoire of biomedical and behavioral tools available to *reduce* the risk of infection: it is unlikely that anyone alone would eliminate the *possibility* of infection, especially in those who engage in high-risk sexual behavior.

When one considers that we currently do not have the tools to make sex risk free, or even to guarantee an 'infection-free' pregnancy, the need for safe, effective microbicides is obvious. Delivering these products will not be cheap, nor will it be easy; but, given the necessary political, social and scientific commitments, it is a realistic possibility – certainly as realistic as developing safe and effective vaccines to prevent such infections.

REFERENCES

Elias C. and Heise L. (1993). The development of microbicides: a new method of HIV prevention for women. Program Division Working Papers [No. 6]. New York: *The Population Council*

Global Programme on AIDS, World Health Organization: Report on a meeting on the development of vaginal microbicides for the prevention of heterosexual transmission of HIV. Geneva: GPA/ WHO; 1993. [Document WHO/GPA/RID/CRD/94.1; 11–13 November 1993]

Hitchcock P.J. and Claypool L. (1994). HIV and other STD: Working Group Report. In: Barrier contraceptives. Edited by Måuck C., Cordero M., Gabelnik H., Spieler J.M., Rivera R. New York: Wiley-Liss, pp. 353–362

Holmes, K.K., Sparling, P.F., Mardh, P.-A. *et al.* (eds) (1999). Sexually transmitted diseases, 3rd edn. McGraw-Hill, New York

Miller C.J., Alexander, N.J., Gettie, Hendrickx, A.G. and Marx, P.A. (1992). The effect of contraceptives containing nonoxynol-9 on the genital transmission of similar immunodeficiency virus in rhesus macaques. *Fert. Ster.* **57**: 1126–1128

Roddy R.E., Zekeng L., Ryan K.A., Tamoufé U., Weir S.S. and Wong E.L. (1998). A controlled trial of nonoxynol-9 film to reduce male-to-female transmission of sexually transmitted diseases. *N Engl. J. Med.* **339**: 504–510

Stone A.B. and Hitchcock P.J. (1998). Vaginal microbicides for preventing the sexual transmission of HIV. *AIDS* **8** (suppl 1): S285–S293

Wasserheit J.N. (1992). Epidemiological synergy. Interrelationships between human immunodeficiency virus infection and other sexually transmitted diseases. *Sex. Transm. Dis.* **19**(2): 61–77

SPECIAL REPORT

Recommendations for the development of vaginal microbicides

The International Working Group on Vaginal Microbicides*

Vaginal microbicides are products for vaginal administration that can be used to prevent HIV infection and other sexually transmitted diseases (STD). We recognize two potential sources of vaginal microbicides: existing spermicides and new products (new products may or may not be spermicidal). This document is meant to serve as a general guide for development and evaluation of existing and new products. For new products preclinical studies will be required. Depending upon indication, in vitro activity against HIV, target STD, and sperm should also be assessed. Compatibility with barrier method materials should also be evaluated. The physical-chemical properties of the active agent and the clinical formulation should be assessed. Animals studies should be conducted to assess its safety and predict dosing; use of various models to assess local toxicity is indicated and microbicidal activity of the product may be evaluated if appropriate models are available. Carcinogenicity testing and segment III reproduction studies (perinatal and post-natal studies in rats) may be performed concurrently with Phase III clinical trials. All vaginal microbicides, including existing spermicides and new products, should be clinically evaluated for safety and efficacy. Safety studies are necessary because irritation of vaginal and cervical mucosae has been recently associated with spermicide use and those lesions might increase HIV transmission. Efficacy studies to assess prevention of HIV infection and/or STD, depending upon the product indication, should then be conducted with products that have been evaluated for safety and appear to be non-toxic to tissue. For spermicidal microbicides contraceptive efficacy studies will be needed.

AIDS 1996, **10:UNAIDS1-UNAIDS6**

Introduction

Heterosexual transmission of HIV is a serious public health concern, as it accounts for about three-quarters of all HIV-1 infections worldwide. Globally, there are currently three men infected for every two women, but by the year 2000 it is projected that number of new infections among women will be close to that among men [1]. While condoms, when used consistently and correctly, are effective in preventing the sexual spread of HIV, there is an urgent need for methods women can use for HIV prophylaxis, such as vaginal microbicides [2].

Vaginal microbicides are products for vaginal administration that can be used to

From the International Working Group on Vaginal Microbicides, c/o UNAIDS, Geneva, Switzerland. *See Appendix for contributors.

Note: The proposals in this document were endorsed by consultants to and members of the International Working Group on Vaginal Microbicides, but do not necessarily reflect the policies of their respective agencies and/or organizations. Requests for reprints to: J. Perrins, Policy Strategy and Research, UNAIDS, 20 Avenue Appia, CH-1211 Geneva 27, Switzerland; or P. Hitchcock, National Institute of Allergy and Infectious Diseases, National Institutes of Health, 9000 Rockville Pike, Solar 3A21, Bethesda, MD 20892, USA; or A. Stone, Medical Research Council, 20 Park Crescent, London W1N 4AL, UK.

ISSN 0269-9370

* Appendix 1 is reprinted, with permission, from The International Working Group on Vaginal Microbicides (1996) *AIDS* 10: UNAIDS1-UNAIDS6.

prevent HIV infection and/or other sexually transmitted diseases (STD). An ideal vaginal microbicide would be safe and effective, and also tasteless, colourless, odourless, nontoxyl, stable in most climates, and affordable. Currently available spermicides are being clinically evaluated for prevention of infection because their active ingredients, all surfactants such as nonoxynol-9, octoxynol-9, menfegol, benzalkonium. chloride or chlorhexidine, have been shown to have antiviral and antibacterial activity in vitro. Among the new classes of vaginal microbicides, some products are not spermicidal and may not interfere with procreation: for example, inhibitors of virus absorption, defensins and inhibitors of HIV reverse transcriptase [3–6]. The International Working Group on Vaginal Microbicides (IWGVM) was formed in November 1993 at a meeting held at the World Health Organization (WHO), Geneva, Switzerland. The goal of the IWGVM is to facilitate the development, production and distribution of safe acceptable, effective, and affordable vaginal microbicides to prevent HrV infection and other STDs. The working group members, experts in relevant scientific areas from governmental and non-governmental agencies, meet regularly to facilitate communication, consultation, and collaboration on scientific advances and technical problems in microbicide development, and to identify needs and opportunities to promote product development.

The present document outlines the guidance the IWGVM has developed on the evaluation of vaginal microbicides.

Preclinical considerations

This section presents steps to be considered in preclinical development of new vaginal microbicides for the prevention of HIV and STD. It is less relevant to the evaluation of existing spermicides as vaginal microbicides. For example, in the United States, the Food and Drug Administration (FDA) does not require pharmacology/toxicology studies for evaluation of nonoxynol-9 and octoxynol-9 as spermicides, and perhaps as vaginal microbicides [7]. In Europe and Japan regulatory authorities might rule similarly for approved spermicides. However, it might be wise to do some of the tests proposed in this section with spermicidal products to avoid toxicity problems in human studies and to improve prospects for developing a safe, effective microbicide. Indeed, several currently marketed spermicidal products have been associated with genital mucosal lesions [8–11].

The preclinical guidance provided here is a list of studies which may be valuable and performed prior to, or concurrently with, clinical studies. Since candidate products will have different active agents with unique mechanisms of action and formulations, preclinical approaches will have to be tailored for the specific product. The precise studies that will need to be performed will also depend on several factors, including (1) the indication for which the product will be tested and marketed, and (2) the regulatory requirements of the host/sponsoring country. Therefore, a product sponsor should contact regulatory agencies about specific requirements for approval of its product as soon as its potential has been delineated.

In vitro activity and safety

The tests described in Table I are designed to obtain information regarding the products' potential for prevention of HIV and STD. If the product indication is for the prevention of HIV infection, in vitro assays will include a variety of HIV strains (both laboratory-adapted and clinical

TABLE I Recommended tests for in vitro activity*.

Activity against HIV
- Laboratory-adapted HIV virus in T-cell lines
- Laboratory-adapted HIV virus in peripheral blood mononuclear cells
- Clinical HIV isolates (depending on the microbicide and the mechanism action, it may be appropriate to include drug-resistant isolates)
  Activity against cell-associated virus
  Antiviral activity in semen and, if possible, vaginal fluids or in an in vitro system that is physiologically appropriate

Activity against other sexually transmitted pathogens
- *Neisseria gonorrhoeae*
- *Chlamydia trachomatis*
- *Haemophilus ducreyi*
- *Trichomonas vaginalis*
- *Herpes simplex virus*

Activity against other vaginal organisms
- *Lactobacillus crispatus*
- *Candida albicans*

---

* Tests should be performed on active agent and clinical formulation; if the microbicide contains more than one active agent each must be tested separately.

isolates). Since the precise mechanism of mucosae transmission of HIV is unknown, tests for evaluating prevention of cell-associated HIV transmission are also recommended. It is desirable that an agent be evaluated for activity against HIV and other STD regardless of its intended HIV indication, since a clinical outcome of HIV prevention may be achieved by the prevention of other STD [12,13]. Similarly data on the effect of the agent on lactobacilli as a surrogate for normal vaginal flora is desirable.

As vaginal microbicides should be compatible with other methods, such as the male or female condom used to prevent HIV and STI transmission, it is suggested that both the active ingredient and the final formulation be evaluated for compatibility with latex, polyurethane or other physical barrier materials early on during preclinical development.

In vitro mutagenesis studies can be done in parallel with Phase I clinical studies, and should include gene mutation tests, tests for chromosomal abnormalities and primary DNA damage assays.

Physical–chemical considerations

Clinical formulations must be produced under good manufacturing practices. The properties of the active agent should be known. Well-characterized analytical methods to detect and measure the physical-chemical properties of the active agent are essential for product development. For example, quantitative analytical methods will be needed to determine drug diffusion and potential for systemic absorption of the active agent, and to meet the Standard Chemistry Requirements (Chemistry, Manufacturing, and Control section of the Investigational New Drug application to the FDA in the USA and similar requirements of regulatory bodies elsewhere).

As the combination of the active agent and a delivery vehicle may result in alterations in physical-chemical properties and microbicidal activity, it is recommended to assess the stability (real and accelerated time) of both clinical formulation and vehicle alone, to do formulation release studies (i.e., the kinetics of release of active agent from solid and semi-solid formulations), and to assess in vitro activity against the target pathogens, if possible, in comparison with active agent alone.

Animal studies

Due to the potential for increased HIV transmission in the presence of significant cervical/vaginal inflammation and ulceration, it is recommended that the active agent and the clinical formulation of the product be tested in a rabbit vaginal irritation model (standard 10-day application of active agent and/or preclinical formulation) early in the development process. A 4% nonoxynol-9-containing spermicidal gel should be used as a positive control. If significant irritation occurs, termination of product development may be warranted.

In addition, systemic absorption and the toxicity of the formulation to the rectal mucosa, pubic and other skin areas should be assessed in animal models. Oral toxicity studies, and gross necropsy or more specialized studies, if warranted (including penile irritation and absorption in male rabbits), may be performed concurrently with or after Phase I clinical studies, depending upon the topical microbicide. Systemic toxicity studies in one species with single dose administration, hypersensitivity and photosensitivity studies, and segment I reproductive toxicology (spermicidal activity) are required prior to Phase I, and segment II reproductive toxicology studies (standardized tests in two animal species) may be performed subsequent to Phase I

TABLE 2 Examples of animal models for testing the efficacy of clinical formulations against HIV and sexually transmitted diseases (STD).

HIV
- Simian immunodeficiency virus in macaques
- Feline immunodeficiency virus in cats

STD*
- Genital herpes in guinea pigs
- *Chlamydia trachomatis* in mouse or primate
- *Haemophilus ducreyi* in rabbits
- *Treponema pallidum* in rabbits
- Human papillomavirus in nude mice
- Rabbit papillomavirus model

* Newer models, such as the mouse herpes simplex virus genital infection models, are presently under development and may be considered.

studies in humans [14]. Toxicology studies in a rodent and non-rodent species (possibly concurrent with clinical trials) with duration equal to intended duration of clinical studies (up to 12 months) should be performed. Pharmacokinetics data (serum drug levels, maximum concentration, area under the curve, tissue distribution and metabolite profiles) should be presented and compared to human data. If Phase I trials establish that the drug is absorbed in humans and the mucosal route of delivery in animals cannot achieve much higher blood levels than those seen in Phase 1, a 1–3 month toxicity study during which the clinical formulation is given parenterally or orally may be required to identify all potential toxicities.

If possible, animal model data on the product's potential efficacy should be obtained: examples of two animal models to be considered are listed in Table 2. Carcinogenicity testing is necessary, but may be performed concurrently with Phase III clinical trials. Such testing should involve two animal species (rats and mice), and intravaginal administration for up to 2 years at maximum tolerated dose. Segment III reproduction studies (perinatal and postnatal studies in rats) can be performed concurrently with Phase III clinical trials [14].

Clinical considerations

Like all clinical trials, trials on vaginal microbicides should be conducted in accordance, with the current version of the Declaration of Helsinki and the Good Clinical Practice guidelines applicable in the country where the trial will take place [15]. In particular, the studies must be ethically sound and be reviewed in advance by an institutional (ethical) review board or other appropriate group, employ informed consent, assure confidentiality, and be monitored for the completeness and accuracy of study data. A standardized approach to collecting any clinical data, administering subject interviews and analysing data should be used. Laboratory tests should be conducted in laboratories with proper quality assurance procedures. A Data and Safety Monitoring Board (DSMB) composed of a group of physicians, epidemiologists, statisticians and ethicists should oversee the conduct of all long-term clinical trials. The DSMB should review the study at intervals specified in the protocol and advise the sponsors on modification or termination of the trial according to rules for termination of the study defined beforehand.

Use of currently available spermicides or new vaginal microbicides may cause changes and even lesions of the genital mucosa [8–101]. Not knowing what role, if any, these play in the transmission of HIV or STD, the IWGVM believes that information on the status of the genital mucosa must be collected both during safety and efficacy studies on vaginal microbicides. Because changes in genital mucosa may occur independently of symptoms, assessment of the genital mucosa should be by objective methods, such as vaginal speculum examination, while irritation/inflammation should be assessed in Phase I and II studies using colposcopy.

Community-based organizations should be involved as much as possible in the design and implementation of the studies, especially with regard to its feasibility and acceptability, including establishment and maintenance of the appropriate clinical and social infrastructure for the trial.

Design and implementation of Phase I, II and III clinical trials

Phase I trials

The main objectives of Phase I trials are usually: (1) to gather initial information on the incidence and extent of toxicity for a product that has never been used in humans; (2) to obtain data on the pharmacokinetics of the product; and (3) to aid in the selection of a dose for subsequent studies. Open-label studies, some probably with dose escalation, in a small number of healthy, non-pregnant female volunteers not at risk of pregnancy or STD, who would use the study drug for one to a few days (per dose-escalation level) should be adequate. If the main objective of the study is to gather safety information, colposcopy should be used to assess the status of the genital mucosa. Systemic toxicity of the product should be assessed by appropriate laboratory tests. Volunteers in early Phase I studies of a new product should abstain from vaginal intercourse, thereby avoiding its potential confounding effect on the incidence of genital lesions.

Phase II trials: further safety studies and pilot studies

After Phase I studies are completed, the product's safety should be confirmed in larger studies among healthy, non-pregnant female volunteers not at risk of pregnancy or STD. Ideally the study should be a randomized, double-blind comparison of the new product to placebo (i.e. the product formulation without its active ingredient) or to a reference product on one hand, and to an untreated group on the other. Use of a placebo group would enable assessment of the safety characteristics of the vehicle; this information is critical to future studies. A cross-over design, in which the product, placebo and non-use would be given sequentially, would also be acceptable. However, subject dropouts and the need to include a wash-out period for products with persistent effects limit the utility of this design and should be taken into consideration. Finally, the utility of open-label studies is limited because findings cannot be compared to the placebo. When participants in Phase II studies have sexual intercourse and lubricated condoms are used, the lubricant should contain no spermicidal products.

Study should be designed in collaboration of biostatisticians. The appropriate sample size, duration of the study, and virtually all other aspects of the study design will be affected by many factors, for example, product characteristics, outcome measures, and will require biostatistical expertise.

When the safety of the study drug has been documented in low-risk, healthy women, it is recommended that additional safety studies be conducted in other populations prior to or at the beginning of Phase III studies; safety should be monitored during the Phase III study as well. Examples of other populations of women include commercial sex workers, pregnant women, HIV-infected women, women using reproductive hormones, intrauterine device users, intravenous drugs or crack cocaine users, peri-menopausal or post-menopausal women, and breastfeeding women. Consideration should also be given to an assessment of the product's administration via the rectal route. Finally, it is suggested that all clinical studies should be regarded as opportunities to collect data about product acceptability and its ease of use.

Phase III studies

Phase III studies aim to assess the balance between efficacy and toxicity of a new product. For vaginal microbicides, Phase III studies should determine whether the product will prevent heterosexual transmission of HIV and/or other STD depending upon the product among women at high risk of acquiring those infections. The incidence of adverse events (i.e. potential toxicity) should be assessed to obtain information on the long-term safety of the product. Phase III studies should be randomized, double-blind, controlled trials. Again, the study should be designed with biostaticians. Phase III

studies will need to be sufficiently large to assess the risks and benefits of a product. Furthermore, inconsistent use of the product, and use of condoms or other interventions, such as treatment of incident STD, will complicate analyses. As male condoms are the only currently available effective method to prevent heterosexual transmission of HIV, condom use must be recommended and condoms made available to all study participants. In other words, the study design will test whether the microbicide will improve the protection afforded by condom use. If in the future a vaginal microbicide were shown to be effective in preventing HIV infection, placebo-controlled trials would become inappropriate and new microbicides should be compared to the vaginal microbicide with known efficacy. Phase III studies will also enable an assessment of product acceptability for women and men, compliance with product use, effects on the male partner, and pregnancy rates, if appropriate.

Study populations

Study participants for Phase I and initial Phase II studies should be volunteers at low risk of HIV infection and STD, ideally of reproductive age (i.e. 18–45 years of age). Participants in Phase III studies should be HIV-uninfected women at high risk of acquiring HIV infection and other STD through vaginal intercourse. Depending on the microbicide being evaluated and study type, women with systemic diseases, women who use other drugs that could modify the pharmacokinetics or the effects of the study product, or women who are pregnant or nursing (unless the product has been demonstrated to be safe for administration in these situations) may need to be excluded from the study. Women presenting with clinically apparent genital infections and/or lesions at enrolment should be treated prior to inclusion. For initial safety studies, the exclusion criteria may need to be more rigorous, and it may be necessary to exclude women with: (1) abnormal liver or renal function; (2) a history of genital problems; (3) colposcopic abnormalities; (4) other STD; and (5) continued use of other vaginal products such as douches, tampons, or spermicides (during the study period).

Excluding HIV-infected subjects from participation in Phase III studies may not be possible, mainly for confidentiality considerations. However, these participants will contribute safety information and may contribute to assessment of the product's efficacy against other pathogens. Because of the ease of intravenous or rectal HIV transmission, it is also preferable to exclude women who are intravenous drug users or who frequently and regularly practice anal intercourse. As it is not possible to ensure that the participants will not engage in these high-risk behaviours, it is recommended to quantify the incidence of high-risk behaviours in Phase III studies.

Study outcomes and their detection

It is recommended that histories of the incidence of vaginal discharge, genital discomfort (including vaginal pain and burning sensations in the vagina and/or the vulva), vaginal dryness, painful urination or intercourse, lower abdominal pain and genital infection be collected in all studies on vaginal microbicides. Reasons for discontinuation should be recorded and adverse events documented.

Toxicity to the genital mucosa should be assessed by objective methods. Colposcopy should be used in initial safety studies. In later studies, vaginal speculum examination may be sufficient, but it is probably wise to monitor local safety with colposcopy in a subset of participants, as

recommended for the development of new spermicides [4]. Colposcopic examination should follow the WHO manual for the standardization of colposcopy (available on request from UNAIDS). This manual describes the lesions as ulcers, abrasions, ecchymoses, and petechial haemorrhages, sub-epithelial haemorrhages plus oedema, erythema, oedema, or abnormal vaginal and cervical discharge. Signs of cervical/vaginal dysplasia, neoplasia or metaplasia are not described in the VMO manual but would be recognized by a skilled colposcopist and should also be recorded. In Phase I/II safety studies, the toxicity to the genital mucosa should be evaluated between menses, to avoid difficulties assessing the genital mucosa when menses are present.

In studies focusing on the assessment of safety of the product to the genital mucosa the frequency of clinical, gynaecological, and colposcopic examinations should be frequent, not less than once every 7–10 days. When genital lesions occur, more frequent examinations may be necessary. Systemic toxicity should be assessed using appropriate laboratory tests, at least at enrolment, and at the end of each observation period. Assessment of STD and other genital infections is recommended to exclude them as confounders in the analysis of the data. In Phase III studies, these evaluations should take place at study enrolment, and at regular intervals (possibly monthly) thereafter, for as long a follow-up period as possible. Currently the populations that would be appropriate for Phase III studies (those with high seroincidence rates) may be difficult to follow. It is anticipated that follow-up will not, in most cases, exceed 1 year; but every effort should be made to improve it by establishment of appropriate infrastructure. Evidence of local toxicity should be sought by pelvic examination (and colposcopy in a subset of participants if possible). Screening for incident HIV infection should be with state-of-the-art laboratory methods, such as third-generation HIV serologic tests, HIV culture, HIV antigen detection and polymerase chain reaction. The diagnosis of STD should be by using state-of-the-art laboratory techniques. Organisms to be considered to measure the incidence of STD include: (1) Trichomonas vaginalis; (2) Neisseria gonorrhoeae; (3) Chlamydia trachomatis; (4) Treponema pallidum; (5) Haemophilus ducreyi; and (6) herpes simplex virus. Compliance with both condom and microbicide use could be assessed with coital log charts and subject interviews. Incidence of anal intercourse and intravenous drug use during the study should be obtained and recorded. Pregnancies should be determined by testing. If the male partner can be evaluated, genital irritation and infections that are reported should be confirmed if possible.

Handling of participants who develop adverse reactions

Participants should discontinue use of study products immediately when a serious adverse reaction occurs [16].

In Phase I and Phase II studies, study products should be discontinued when genital ulceration occurs. When genital abrasion, petechial haemorrhage, ecchymosis, or subepithelial haemorrhage and swelling occurs, study products should be discontinued if after 24–72 h of continued study product use the condition worsens.

When in a Phase III study, a genital ulceration occurs, study products may continue but treatment for bacterial causes of genital ulceration (chancroid and/or syphilis) should be given immediately. If the ulcer worsens after 7 days, use of study products should be discontinued, at least until the lesion is cured. When other genital lesions occur, study products should be

discontinued if, after 7 days of continued study product use, the condition worsens. Product use may be resumed after the lesion resolves.

Women who discontinue therapy for any reason should be encouraged to continue follow-up to assess genital irritation, to detect potential toxicity that might arise after cessation of study drug use, and to assess their STD and HIV outcomes for intention-to-treat analyses.

References

1. Mertens T.E., Belsey E., Stonebuner R., *et al.*: Global estimates and epidemiology of HIV-11 infections and AIDS- further heterogeneity in spread and impact. *AIDS* 1995, 9 (suppl A):S259–S272.
2. Stein Z.A.: HIV prevention: the needs for methods women can use. *Am J Public Health* 1990, 80:460–462.
3. Elias C., Heise L.: *The Development of Microbicides: A New Method of HIV Prevention for women.* Program *Division* Working papers [No. 61. New York: The Population Council; 1993.
4. Global Programme on AIDS, World Health Organization: Report *on a* Meeting on the Development of Vaginal *Microbicides for* the Prevention *of* Heterosexual Transmission *of HIV*. Geneva: GPA/WHO; 1993. [Document WHO/GPA/RID/CRD/94.1; 11–13 November 1993).
5. Hitchcock P.J., Claypool L.: HIV and other STD: Working Group Report. In Barrier Contraceptives. Edited by Mauck C, Cordero M, Gabelnik H, Spieler JM, Rivera R. New York: Wiley-Liss; 1994:353–362.
6. Stone A.B., Hitchcock P.J.: Vaginal microbicides for preventing the sexual transmission of HIM AIDS 1994, 8 (suppl U5285–S293.
7. US Food and Drug Administration: Guidance for the Development *of* Vaginal Contraceptive Drugs (27 January 1995). [Clarifications: 8 *February 1995]*. Washington, DC: US Government; 1995.
8. Niruthisard S., Roddy R.E., Chutivongse S.: The effects of frequent -oxynol-9 use on the vaginal and cervical mucosa. *Sex Transm Dis* 1991, 18:176–179.
9. Roddy R.E., Cordero M., Cordero C., Fortney J.A.: A dosing study of nonoxynol-9 genital irritation. *Int J STD AIDS* 1993; 4:165–170.
10. Kreiss J., Ngugi E., Holmes K., *et al.*: Efficacy of nonoxynol-9 contraceptive sponge use in preventing heterosexual transmission of HIV in Nairobi prostitutes. *JAW* 1992, 268:477–482.

11. Goeman J., Ndoye 1, Sakho L.M., et al.: Frequent use of menfegol spermicidal vaginal foaming tablets associated with a high incidence of genital lesions. *J Infect Dis* 1995, 171:16–24.
12. Laga M., Manoka A., Kivuvu M., et al.: Nonulcerative sexually transmitted diseases as risk factors for HIV-1 transmission in women: results from a cohort study. *AIDS* 1993, 7:95–102.
13. Grosskurth H., Mosha F., Todd L. et al.: Impact of improved treatment of senially transmitted diseases on HIV infection in non-A Tanzanix randomised controlled trial. *Lancet* 199s, 346:530–536.
14. Federal Register Notice: International Conference on Harmonization – Guideline on Detection *of* Toxicity to Reproduction for *Medicinal* Products 122 September 1–994). Washington, DC: US Government; 1994.
15. World Medical Association Declaration of Helsinki. in International *Ethical* Guidelines for *Biomedical Research involving* Human *Subjects* Geneva: Council for International Organization of Medical Sciences; 1993:47–50.
16. World Health Organization: World Health Organization Guidelines on Good Clinical Practice. Geneva: WHO; iggs. [WHO Technical Report Series No. 850, Annex 31.

Appendix

Contributors to the International Working Group on Vaginal Microbicides

NJ. Alexander (National Institute of Child Health and Human Development, National Institutes of Health, Bethesda, Maryland, USA), V Alexander (Community Family Planning Council, New York, New York, USA), S. Allen (Advances in Health Technology, Washington, DC, USA), L. Dorffinger (Family Health International, Research Triangle Park, North Carolina, USA), WE Dommel (Office for Protection of Research Risks, National Institutes of Health, Bethesda, Maryland, USA), A. Duerr (Division of Reproductive Health, Centers for Disease Control and Prevention, Atlanta, Georgia, USA), C. Elias (the Population Council, Bangkok, Thailand), A Feigal (Division of Antiviral Drug Products, Food and Drug Administration, Rockville, Maryland, USA), E. Gollub (AIDS Coordinating Office, City of Philadelphia, Philadelphia, Pennsylvania, USA), H. Gabehlick (CONRAD Program, Arlington, Virginia, USA), M. Harris (Society for Women and AIDS in Africa, Freetown, Sierra Leone), P.J. Hitchcock (National Institute of Allergy and Infectious Diseases, National Institutes of Health, Bethesda, Maryland, USA), M. Karam (Global Programme on AIDS, World Health Organization, Genew, Switzerland), K. Kazempour (Division of Antiviral Drug Products, Food and Drug Administration, Rockville, Maryland, USA) T Lange (Global Programme on AIDS, World Health Organization, Geneva, Switzerland), R.- Roddy (Family Health International, Research Triangle Park, North Carolina, USA), P. Rowe (Special Programme for Research, Development and Training in Human Reproduction, World Health Organization, Geneva, Switzerland), Z. Rosenberg (National Institute of Allergy and Infectious Diseases, National Institutes of Health, Bethesda, Maryland, USA), J. Perrins (Policy, Strategy and Research, UNAIDS, Geneva, Switzerland), A. Stone (Medical Research Council, London, UK), R. Stratton (National Institute of Child Health and Human Development, National Institutes of Health, Bethesda, Maryland, USA).

APPENDIX 2

Topical microbicides:
NIAID considerations for preclinical product development

In 1993, the National Institute of Allergy and Infectious Diseases launched a research initiative aimed at the development of topical microbicides. These are products for intravaginal and or rectal use that protect against sexually transmitted infections, including human immunodeficiency virus (HIV) infection. The research initiative consists of three parallel tracks, including basic science, product development and clinical research.

This document presents an outline of the steps to be considered in preclinical development of topical microbicides for the prevention of STDs, including HIV infection. It is intended to be a comprehensive list of studies which could be performed with a potential topical microbicide prior to or concurrent with clinical studies in humans. The precise studies that may need to be performed will depend on several factors, including the indication for which the product will be marketed, and the regulatory requirements of the host/sponsoring country. In addition, as candidate products will have different active agents with unique mechanisms of action and formulations, relevant preclinical approaches will have to be tailored for the specific product. Again, this document is meant to serve as a general guideline for developing a relevant approach to product development. A product sponsor should contact the regulatory agency about specific requirements for licensure as soon as a product's potential has been delineated.

RECOMMENDATIONS FOR TESTS OF ACTIVE AGENTS AND CLINICAL FORMULATIONS

The tests described in this section are designed to obtain information regarding the spectrum of activity of the product in the prevention of HIV and STDs. If the product indication is for the prevention of HIV infection, in vitro assays will include a variety of HIV strains (both laboratory-adapted and clinical isolates). Because the precise mechanism of mucosal transmission of HIV is unknown, tests for evaluating the prevention of cell-associated HIV transmission are also needed. It is desirable that an agent be evaluated for activity against HIV and other STDs regardless of its intended indication, as a clinical outcome of HIV prevention will probably be affected by the prevention of other STDS. Similarly, data on the effect of the agent on lactobacilli as a surrogate for normal vaginal flora are desirable.

I Activity against HIV*

- (A) Laboratory-adapted virus in T-cell lines
- (B) Laboratory-adapted virus in peripheral blood mononuclear cells (PBMC)
- (C) Clinical HIV isolates (depending upon the microbicide and the mechanism of action, it may be appropriate to include drug-resistant isolates)
- (D) Activity against cell-associated virus; and
- (E) Antiviral activity in semen and, if possible, vaginal fluids or in an in vitro system that is physiologically appropriate.

* If the microbiocide contains more than one active agent, each must be tested separately.

II Activity against other STD pathogens

    (A)   *Neisseria gonorrhoeae*
    (B)   *Chlamydia trachomatis*
    (C)   *Haemophilus ducreyi*
    (D)   *Trichomonas vaginalis*
    (E)   Herpes simplex virus (HSV).

III Spermicidal activity (e.g. mucus penetration test; sperm motility)

IV Activity against *Lactobacillus crispatus* and *Candida albicans*

V Interactions with latex/polyurethane or other physical barrier materials as appropriate

VI Screening for vaginal irritation in a rabbit model (standard 10-day application of active agent and/or preclinical formulation).

Because of the potential for increased HIV transmission in the presence of significant cervical/vaginal inflammation and ulceration, it is recommended that the active agent be tested in a rabbit vaginal irritation model early in the development process. A 4% nonoxynol-9-containing spermicidal gel (e.g. Conceptrol) should be used as a positive control. If significant irritation occurs, termination of product development may be warranted.

VII Animal model efficacy testing of clinical formulation

This section describes animal model systems to be considered for assessing product efficacy via topical administration for prevention of HIV and STDs. This list is not intended to be comprehensive. For example, newer models which are presently under development, such as the mouse HSV skin or genital infection models, may be considered. Depending upon the indication, tests for animal model efficacy of the clinical formulation may include:

HIV	**STDs**
SIV in macaques	HSV in guinea pigs
FIV in cats	*C. trachomatis* in mouse or primate
	H. ducreyi in rabbits
	T. pallidum in rabbits
	Human papillomavirus (HPV) in nude mice
	Rabbit papillomavirus model

REQUIREMENTS FOR TESTS OF CLINICAL FORMULATIONS

(Clinical formulations must be produced under good manufacturing practices, GMP)
This section is based on US Food and Drug Administration (FDA) published documents (see references).

I Physical–chemical properties of active agent*

- (A) Solubility
- (B) Stability (pH rate profile; accelerated degradation)
- (C) Standard Chemistry Requirements, Chemistry, Manufacturing; and Control (CMC) section of Investigational New Drug (IND) application.

* Well-characterized analytical methods to detect and measure the physical–chemical properties of the active agent are essential for product development. For example, quantitative analytical methods will be needed to determine drug diffusion and potential for systemic absorption of the active agent.

II Define and assess characteristics of clinical formulation

The combination of the active agent and a delivery vehicle may result in alterations in physical–chemical properties and microbicidal activity. Thus, the following studies of the clinical formulation are recommended:

- (A) Stability, real and accelerated time of both clinical formulation and vehicle alone
- (B) Formulation release studies (i.e. the kinetics of release of active agent from solid and semisolid formulations)
- (C) *In vitro* activity against pathogen, if possible, in comparison with active agent alone.

III *In vitro* mutagenesis (can be done in parallel with phase I clinical studies)

- (A) Gene mutation tests
- (B) Tests for chromosomal abnormalities
- (C) Primary DNA damage assays.

IV Good laboratory practices (GLP) safety assessment of clinical formulation produced in compliance with current good manufacturing practices, GMP

- (A) Rabbit vaginal/cervical irritation model
- (B) Tests for systemic absorption
- (C) Studies of exposure to rectal mucosa, pubic and other skin areas in animal models
- (D) Oral toxicity studies*
- (E) Gross necropsy or more specialized studies if warranted, including penile irritation and absorption in male rabbit*
- (F) Systemic toxicity studies in one species with single-dose administration
- (G) Hypersensitivity and photosensitivity studies
- (H) Segment I reproductive toxicology as defined by the US FDA (spermicidal activity)

* May be done concurrently with or after phase I clinical studies, depending upon the topical microbicide.

V Segment II reproductive toxicology studies as defined by the FDA (standardized tests in two animal species performed subsequent to phase I studies in humans)

VI Toxicology studies in a rodent and non-rodent species (possibly concurrent with clinical trials)

(A) Duration equal to intended duration of clinical studies (up to 12 months)

(B) Pharmacokinetics study; serum drug levels, C_{max}, 'area under the curve', tissue distribution and metabolite profiles should be presented and compared to human data

(C) If phase I trials establish that the drug is absorbed in humans and the mucosal route of delivery in animals cannot achieve much higher blood levels than those seen in phase I, a 1–3-month toxicity study during which the clinical formulation is given parenterally or orally may be required to define all potential toxicities.

VII Carcinogenicity testing (concurrent with phase III clinical trials)

(A) Two animal species (rats and mice)

(B) Intravaginal administration for 2 years at maximum tolerated dose.

VIII Segment III reproduction studies as defined by the FDA (perinatal and postnatal studies in rats) can be done concurrently with phase III clinical trials.

Sources

1. US Food and Drug Administration Guidance for the Development of Vaginal Contraceptive Drugs (January 27, 1995; clarifications February 8, 1995)

2. U.S. Food and Drug Administration Points to Consider in the Preclinical Pharmacology/ Toxicology Development of Topical Microbicides Intended for Viral Sexually Transmitted Diseases (STD) Including HIV (DRAFT)

Federal Register Notice: International Conference on Harmonization – Guideline on detection of Toxicity to Reproduction for Medicinal Products (22 September 1994)

APPENDIX 3

Points to Consider in the Nonclinical Pharmacology/Toxicology Development of Topical Drugs Intended to Prevent the Transmission of Sexually Transmitted Diseases (STD) and/ or for the Development of Drugs Intended to Act as Vaginal Contraceptives

Division of Reproductive and Urologic Drug Products
Office of Drug Evaluation II

Division of Anti-infective Drug Products
and
Division of Antiviral Drug Products
Office of Drug Evaluation IV

1. This document is an informal communication under 21 CFR 10.90 (b) (9) that represents the best judgement of the Divisions of Reproductive and Urologic Drug Products, the Division of Antiinfective Drug Products and the Division of Antiviral Drug Products at this time. This document does not necessarily represent the formal position of the Center for Drug Evaluation and Research or the Food and Drug Administration, and it does not bind or otherwise obligate the Center or Agency to the views expressed.

Introduction

This document discusses several nonclinical pharmacology/toxicology issues that are inherent with the development of topical drugs that are intended to prevent sexually transmitted diseases (STD) or with the development of drugs proposed to act as vaginal contraceptives. It is not meant to replace existing guidance concerning the nonclinical pharmacology/toxicology safety data required to support the submission of Investigational New Drug (UsID) applications. Rather, it offers insight into some aspects of these classes of drugs that make them unique from systemically administered drugs, namely, the topical sites of their administration and action and the prophylactic nature of the indications. The publication of this document is a joint effort of three independent review divisions in the Center for Drug Evaluation and Research of the FDA, and represents their best harmonized view of the issues at this time. However, all the recommendations contained herein may be subject to modification and sponsors are encouraged to discuss the content of their nonclinical submissions with the review division appropriate to each individual application

Toxicology Studies

To support the safety of the individual phases of clinical development, studies in at least two animal species (one being a non-rodent) should be used to assess acute, subchronic and chronic toxicity at the proposed site of exposure. Potential sites of exposure to topical microbicides for the prevention of STIs include vaginal, cervical, penile, oral and rectal mucosal areas. For the purpose of this document, contraceptives are considered to expose vaginal and penile tissue. All toxicology studies should use at least three dose levels of the drug and an appropriate control, with the high dose showing frank toxicity (it is understood that there may be limits to the ability to achieve a toxic dose) and the low dose showing little or no toxicity. With the exception of the varying concentrations of the active product, the drug to be applied should be in its final formulation and the study duration should be equal to, or longer than, the proposed duration of treatment in the clinical trial. However, if in an early trial, the duration of treatment is for less than two weeks, a nonclinical study duration of two weeks should generally be used. All nonclinical studies carried out to assess safety of a drug should be performed according to Good Laboratory Practices as outlined in 21 CFR. 58. Sponsors should consult with the FDA reviewing pharmacologist concerning approaches and protocols regarding these issues.

Because the topical microbicides are intended for intermittent use during periods in which the individual will be at risk of contracting a STD and contraceptives are intended for intermittent use over long periods of reproductive competence, they are, for regulatory purposes, considered to be chronically administered drugs. Thus, in the latter stages of clinical development, six month studies in a rodent and one year toxicity studies in a non-rodent will be necessary to support phase $\frac{2}{3}$ human trials. Carcinogenicity studies in rats and mice should be performed during the late stages of the drug development program.

Pharmacokinetic/Toxicokinetic Studies

During the conduct of subchronic and chronic toxicity studies, concurrent pharmacokinetic/toxicokinetic analyses should be performed to evaluate the drug's ADME (absorption, disposition, metabolism and excretion) profile and to determine if the drug is systemically absorbed. Appropriate analytical methodology should be established as early as possible to provide for precise, consistent and reliable pharmacokinetic data. Major pharmacokinetic endpoints such as maximum plasma concentration, time to maximum plasma concentration, area under the concentration time curve, volume of distribution, bioavailability and clearance should be determined. To better assess margins of safety, the animal pharmacokinetic data should be compared with data in humans. For instance, a drug might be absorbed in humans and the vaginal route of delivery in animals cannot achieve much higher systemic drug levels than those seen in humans. In that case, a one to three-month toxicology study in one species, with parenteral or oral administration of the drug, may be requested to produce sufficiently high blood levels to identify all potential toxicities. If the drug is absorbed, histopathology should be carried out on the full spectrum of organs and tissues.

Irritation

The likelihood of drug-induced irritation (adverse reactions such as inflammatory responses) to drug exposed tissues should be evaluated in animals at both the macroscopic and microscopic levels. Vaginal irritation tests should be carried out in rabbits with daily applications for ten days. If rabbits are adequately tested by the vaginal route for adverse effects in toxicology studies, no separate vaginal irritation study is required.

If the drug is found to be irritating to mucosal tissues and the sponsor decides to pursue further development, additional pharmacokinetics should be carried out to compare the extent of penetration of the drug across the inflamed tissue with that occurring in intact tissue. These data will help to determine whether the absorption kinetics of the drug is facilitated when the mucosa is compromised.

Hypersensitivity and Photosensitivity

If appropriate, the potential to produce hypersensitivity and photosensitivity of the drug should be evaluated. Examples of animal models and protocols for tests related to these toxicities can be found in references such as Marzulli and Maibach's 'Dermatotoxicology' or Burleson, Dean and Munson's 'Methods in Immunotoxicology.'

Genotoxicity

A standard battery of tests should be carried out to evaluate all new molecular entities for the potential to induce genotoxic effects. A drug should be evaluated for potential genetic toxicity prior to the submission of the IND. The standard battery consists of the three following tests:

(1)   A gene mutation test in bacteria
(2)   An in vitro mouse lymphoma tk assay *or* a mammalian cell *in vitro* cytogenetic test for chromosomal aberrations
(3)   An in vivo test for chromosomal damage to rodent bone marrow cells *or* a mouse n-ticronucleus test.

If the three tests chosen indicate that the drug is devoid of genetic toxicity, no additional studies need to be carried out. If one or more of the battery is positive, the sponsor will be expected to carry out additional genotoxicity tests in consultation with the appropriate review division. For detailed information regarding genotoxicity, sponsors should refer to the document entitled 'Guidance on Specific Aspects of Regulatory Genotoxicity Tests for Pharmaceuticals.' Copies of the Guideline are available from the CDER Consumer Affairs Branch, BFD-210, Center for Drug Evaluation and Research, 5600 Fishers Lane, Rockville, MD 20857.

Reproductive Toxicology

Reproductive toxicology studies should be carried out to explore the possible effects of the drug on fertility and reproductive performance. Additional studies should be performed to examine whether the drug is teratogenic or has an affect on perinatal/postnatal development. Studies designed to assess the teratogenic potential should be carried out in two species, usually in rats and rabbits. It is expected that reproductive toxicology studies will be completed prior to phase $\frac{2}{3}$ trials. For detailed information regarding reproductive toxicology, sponsors should refer to the document entitled, 'Guideline for Industry, Detection of Toxicity to Reproduction for Medicinal Products.' Copies of the Guideline are available from the CDER Consumer Affairs Branch, FIFD-210, Center for Drug Evaluation and Research, 5600 Fishers Lane, Rockville, MD 20857.

Carcinogenicity

Carcinogenicity information is usually obtained during the final phase of the clinical development program. For a vaginal contraceptive, carcinogenicity tests in rats and mice should be submitted with the NDA. For a microbicide, carcinogenicity studies are also required, but the expected date of completion of the studies in relation to the filing of the NDA may vary. The drug should be administered for 2 years and the doses used should be chosen according to the principles outlined in the document entitled, 'Guideline for Industry, Dose Selection for Carcinogenicity Studies of Pharmaceuticals.' Copies of the

Guideline are available from the CDER Consumer Affairs Branch, HFD-210, Center for Drug Evaluation and Research, 5600 Fishers Lane, Rockville, MD 20857. Sponsors should consult with the appropriate reviewing division for the current position and strategic approaches to these issues. Currently, it is recommended that sponsors request that the Executive Committee of the Carcinogenicity Assessment Committee of the Center for Drug Evaluation and Research review the protocols for the carcinogenicity studies prior to their initiation and concur on their content.

Already Marketed Drugs

On 2/3/95, a notice of proposed rulemaking from the Food and Drug Administration entitled 'Vaginal Contraceptive Drug Products for Over-the-Counter Human Use' appeared in the *Federal Register* (Vol. 60, No. 23, pp, 6892–6903). The proposed rule stated that manufacturers of over-the-counter (OTC) vaginal contraceptive drug products would be required to obtain approved applications for marketing of their products. The agency took this action because the effectiveness of these products is dependent upon the final formulation. Therefore, each product must be tested in appropriate clinical trials under actual conditions of use. The proposed rulemaking does not affect the current marketing status of OTC vaginal contraceptives. However, on the effective date of a final regulation, an OTC vaginal contraceptive drug product that is not the subject of an approved application would be regarded as a new drug and be subjected to regulatory action regardless of its status prior to the publishing of the final regulation.

The agency has determined that nonoxynol-9 and octoxynol-9 would be appropriate ingredients for an approved application and has concluded that although nonoxynol-9 and octoxynol-9 kill sperm in vitro and in vivo, the spermicidal activity and resulting effectiveness of these contraceptive active ingredients cannot be considered separately from a product's vehicle. These active ingredients lose some of their effectiveness in humans when the spermicide in final formulation is diluted by various amounts of genital secretions during coitus. Thus, clinical studies are necessary to establish effectiveness of the spermicide's final formulation when used in humans. *However, applications for currently marketed OTC products containing these ingredients require no additional nonclinical data,* but, instead, may refer to (1) the recommendations of the Advisory Review Panel on OTC Contraceptives and Other Vaginal Drug Products that appeared in an advanced notice of proposed rulemaking in the Federal Register (12/12/1980, Vol. 45, No.24 1, pp 82014–82049), entitled 'Vaginal Contraceptive Drug Products for Over-the-Counter Human Use; Establishment of a Monograph; Proposed Rulemaking,' and (2) the history of use of approved drug products containing nonoxynol-9.

The quest to identify safe and effective products for the prevention of HIV and other STIs is a high priority public health concern. Sponsors are strongly encouraged to evaluate OTC contraceptive products towards this use. However, it should be recognized that studies designed to assess the safety and effectiveness of a formulated drug to prevent pregnancy and to prevent the transmission of an STD may be different. Manufacturers wishing to develop a formulation towards one or the other are encouraged to enter into a dialog with the appropriate reviewing divisions concerning the content of their applications.

References

Dermatotoxicolgie, Eds. Marzulli, F.N. and H.I. Maibach. New York: Hemisphere Publishing, 1991.

Methods in Immunotoxicology. Eds. Burleson, G.R., J.H. Dean and A.E. Munson. New York: Wiley-Liss Inc., 1995.

III

PATHOGENS AND VACCINES

8

GENITAL AND PERINATAL HERPES SIMPLEX VIRUS INFECTIONS: PROPHYLACTIC VACCINES

LAWRENCE R. STANBERRY

Division of Infectious Diseases,
University of Cincinnati College of Medicine,
Children's Hospital Medical Center,
Cincinnati, Ohio, USA

DISEASE MANIFESTATIONS OF GENITAL AND PERINATAL HERPES

Genital herpes can be caused by either herpes simplex virus type 1 (HSV-1) or type 2 (HSV-2). The pathogenesis of HSV genital infection is complex and involves primary, latent and recurrent infections (Stanberry, 1996). As summarized in Figure 8.1, the primary infection begins with the sexual transmission of HSV to anogenital sites. The virus replicates at these mucocutaneous sites but also spreads via sensory nerve fibers beyond the portal of entry to sacral dorsal root ganglia. Once in the ganglia, HSV actively replicates in some neurons while establishing a lifelong latent infection in other neurons. After recovery from the primary infection, latent infection can reactivate and spread back along sensory nerve fibers to mucocutaneous sites to cause recurrent anogenital HSV infections (Figure 8.2). Genital HSV infections, both primary and recurrent, can produce recognizable signs and symptoms or can be asymptomatic. During the infection, whether symptomatic or not, virus is present in the genital tract and thus there is the potential for the spread of virus to susceptible sexual partners or, in the case of the pregnant woman, to her fetus or newborn infant.

The incubation period for primary genital herpes ranges from 2 to 20 days, with an average of 6 days (Wald and Corey, 1996). For symptomatic primary infection the clinical signs and symptoms may range from very mild to extremely severe. Individuals may experience itching, tingling, burning and pain in the anogenital area. Some develop classic vesiculoulcerative lesions. On dry skin the lesions may begin as small clear vesicles with an erythematous base. These may persist for several days. When the thin skin covering the vesicle is lost the lesions become shallow yellow-gray ulcers, which eventually crust over; healing occurs with loss of the crust. On chronically moist tissues vesicles are

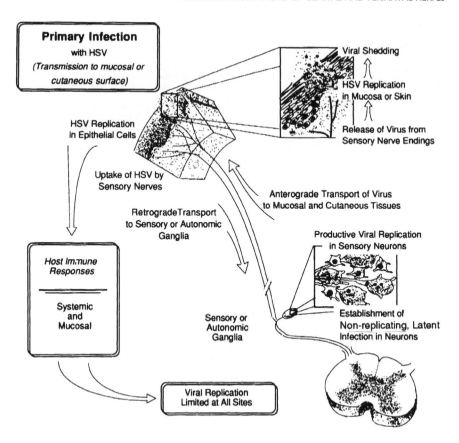

FIGURE 8.1 The pathogenesis of primary genital HSV infection (from Stanberry LR, Jorgensen DM, Nahmias AJ. Herpes simplex viruses 1 and 2. In: Evans AS, Kaslow RA (eds) Viral infection of humans: epidemiology and control, 4th ed. New York, Plenum Press, pp. 419–54, 1997).

very short-lived and rapidly progress to the ulcer stage. Crusts rarely develop: rather, healing occurs as the ulcer re-epithelializes. It takes 7–10 days for progression from vesicle to complete healing, but during the primary infection individuals typically develop new crops of vesicles for the first 1–2 weeks of the illness. Consequently, the average duration of lesions, from first vesicle to complete healing, is 16 days for males and 20 days for females in subjects experiencing classic genital herpes (Wald and Corey, 1996; Brown *et al.*, 1979). Other local symptoms may include urethritis, dysuria and lymphadenopathy. It should be noted that there is great variability in the course of primary genital herpes. In some the infection may be so mild as to not be recognized as genital herpes, whereas in others the classic lesions may last for more than 35 days (Wald and Corey, 1996; Benedetti *et al.*, 1994).

During classic primary infection approximately 40% of males and 70% of females also experience fever, headache and myalgias. This flu-like illness typically lasts 2–7 days. About 10% of males and 40% of females will also

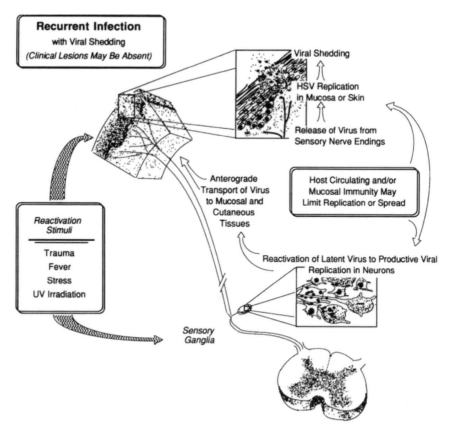

FIGURE 8.2 The pathogenesis of recurrent genital HSV infection (from Stanberry LR, Jorgensen DM, Nahmias AJ. Herpes simplex viruses 1 and 2. In Evans AS, Kaslow RA (eds) Viral infection of humans: epidemiology and control, 4th ed. New York, Plenum Press, pp. 419–54, 1997).

develop aseptic meningitis. Other, less common, complications include HSV oropharyngeal infection, HSV pelvic inflammatory disease, hyper- or hypo-ethesia of the anogenital area, urinary retention, temporary impotence, transverse myelitis, pneumonitis, hepatitis, arthritis, cutaneous and/or systemic dissemination, and vaginal candidiasis (Corey and Wald, 1999).

Recurrent genital HSV infections are generally less severe than the classic primary infection. Approximately 90% of subjects with recognized recurrent genital herpes report that they may experience prodromal symptoms 1–2 days before the development of recurrent lesions. Classically, recurrent lesions begin as a discrete erythematous area that progresses over several hours to form one or a few vesicles that subsequently evolve to ulcer and crust stages. Recurrent lesions tend to be clustered together and involve only a very limited area of the anogenital region. All lesions can arise within a few hours, or new crops of vesicles can develop over several days. An episode of classic recurrent genital herpes lasts about 10 days, although the range varies from 4 to 29 days (Corey and Wald, 1999). As with the primary infection, recurrent infections may be so mild or non-classic as to be unrecognized as genital herpes. Recurrent genital herpes can be mistaken for a variety of cutaneous and urogenital disorders.

The number of symptomatic recurrences experienced by an individual varies greatly from person to person. Three factors – gender, virus type and severity of the primary infection – somewhat predict the frequency of recurrent infections (Benedetti *et al.*, 1994; Reeves *et al.* 1981a). Males tend to have recognizable recurrences more frequently than do females. Individuals who have primary genital herpes caused by HSV-1 are less likely to experience recurrent infections and have fewer recurrences than do those with genital HSV-2 infection. People who have very severe primary infection lasting longer than 35 days also tend to have more recurrences (Benedetti *et al.*, 1994).

Recurrent HSV genital infections may also be asymptomatic or unrecognized (Rattray *et al.*, 1978; Ekwo *et al.*, 1979; Ferrer *et al.*, 1983). Recent studies have found that asymptomatic HSV shedding from anogenital sites is a frequent occurrence (Wald *et al.*, 1995, 1996, 1997, 1998; Sacks *et al.*, 1997; Sacks and Shafran, 1998). In one study, 65% of women with a history of recurrent genital herpes had HSV-2 detected in genital tract swab samples at times when no lesions were noted, and 11% of the subjects experienced asymptomatic shedding on more than 5% of days (Wald *et al.*, 1995). Research using sensitive polymerase chain reaction (PCR)-based assays that detect HSV DNA found that asymptomatic HSV shedding from the genital tract occurred on about 20% of days, even in HSV-2-seropositive subjects who lacked a history of genital herpes (Wald *et al.*, 1997; Corey, 1998). The risk of asymptomatic shedding was greater in subjects infected with HSV-2 than HSV-1, and in those experiencing frequent symptomatic recurrences. Also, subjects with newly acquired infections were more likely to shed HSV than those who had been infected for more than 1 year. The high frequency with which HSV-2 is shed from the genital tract supports the opinion that asymptomatic or unrecognized shedding is responsible for most perinatal and sexual transmission of HSV-2 (Corey, 1998; Mertz *et al.*, 1985). This observation has important implications for the development of vaccines to control genital herpes.

Besides the physical morbidity caused by genital herpes, genital HSV infection is also an important cofactor in HIV acquisition and transmission (Cameron *et al.*, 1989; Hook *et al.*, 1992; Gwanzura *et al.*, 1998), is the most common cause of neonatal herpes (Jacobs, 1998; Kohl, 1997; Arvin, 1996), and in some, can cause psychological morbidity that is more debilitating than the physical morbidity (Carney *et al.*, 1993, 1994; Longo and Koehn, 1993; Mindel, 1996). Ideally, an effective vaccine would prevent or reduce the complications that can result from genital HSV infection.

Perinatal herpes infection – infection of the fetus or neonate (i.e. an infant usually less than 1 month of age) – is one of the most dreaded complications of genital herpes. Intrauterine infection can occur, but most cases of perinatal herpes result from intrapartum transmission of the virus during labor and delivery (Arvin, 1996; Whitley and Arvin, 1995). Genital infection in the mother is often asymptomatic (Whitley *et al.*, 1980; Brown *et al.*, 1991, 1997). Infection in the neonate is most likely initiated at mucosal surfaces (e.g. conjunctiva, oral and nasopharyngeal sites) or cutaneous sites where the integrity of the skin has been damaged (e.g. scalp electrode site) (Bravo *et al.*, 1994). Virus replicates at the mucocutaneous portals of entry but can also spread to visceral organs through hematogenous dissemination, to the peripheral and central nervous systems via hematogenous and/or neural spread, and to the lungs through hematogenous spread and possibly by aspiration of HSV-contaminated secretions (Stanberry, 1996). Based on the extent of spread, infants with perinatal herpes may have any combination of mucocutaneous manifestations, meningoencephalitis and disseminated disease (Prober, 1996). Without intervention the disease typically progresses to a fatal outcome (Whitley *et al.*, 1980). Antiviral therapy with aciclovir or vidarabine can improve the outcome, with the best prognosis in those with limited mucocutaneous disease and the poorest prognosis in those with disseminated infection (Whitley *et al.*, 1991, 1988). Even with antiviral therapy, survivors of encephalitis or disseminated disease often manifest neurological sequelae. Because the early manifestations of perinatal herpes can be very non-specific and easily attributed to other more common disorders, there can be a delay in establishing the diagnosis. Such delays may contribute to the poor prognosis. Even with antiviral therapy there is a clear need for more effective strategies to control perinatal herpes. Vaccines to prevent or limit maternal genital herpes may be such a strategy.

EPIDEMIOLOGY OF GENITAL AND PERINATAL HERPES

Over the past two decades the prevalence of genital herpes has increased worldwide (Stanberry *et al.*, 1997). In the United States the proportion of HSV-2 seropositive adults increased from 16.4% for the period 1976–1980 to 21.9% for 1988–1994 (Johnson *et al.*, 1989; Fleming *et al.*, 1997). Particularly troubling is the fact that this 30% increase occurred during a time of increased awareness regarding the importance of safe sexual practices, and with the availability of an effective antiviral treatment, aciclovir. Similar trends have been reported in the United Kingdom and Sweden (Cowan *et al.*, 1994; Forsgren *et al.*, 1994). Because HSV-2 is transmitted almost exclusively sexually, detection of HSV-2 antibodies beyond infancy is a biomarker for genital HSV-2 infection. Using this marker, it is estimated that approximately 45 million adults in the United States have genital herpes caused by HSV-2 (Fleming *et al.*, 1997). Estimates of the prevalence of genital herpes based on detection of HSV-2 antibodies underrepresents the magnitude of the problem because it excludes the cases caused by HSV-1. In the United States, up to 30% of the cases of first-episode genital herpes are caused by HSV-1 (Corey and Wald, 1999; Stanberry *et al.*, 1999c), and in the United Kingdom up to half of the cases of primary genital herpes result from HSV-1 infection (Government Statistical Service, 1995; Rodgers and O'Mahony, 1995; Ross *et al.*, 1993; Tayal and Pattman, 1994). Risk factors for HSV-2 seropositivity vary with the population and country studied, but may include female gender, being a member of selected racial or ethnic groups, limited education, poverty, drug use, and having a greater number of lifetime sexual partners (Stanberry *et al.*, 1997; Fleming *et al.*, 1997; Stanberry and Rosenthal, 1999). In the United States there has been an increase in the number of teenagers with HSV-2 infection (Fleming *et al.*, 1997). Worldwide, HSV-2 seroprevalence in adolescents 20 years of age or under ranges from approximately 25–30% in central African villagers to 1–5% in Europeans, to 5–14% in US teenagers (Stanberry and Rosenthal, 1999). These high rates suggest that, to be maximally effective, vaccines to control genital HSV infections may need to be used in the pre-teen or early teen years, before the subjects become sexually active. Delaying vaccination until adulthood may have less impact on the spread of disease, as approximately 90% of HSV-2 infected individuals are unaware they are infected; nevertheless, they are capable of transmitting the virus to susceptible partners (Corey, 1998; Mertz et al., 1985; Fleming et al., 1997; Stanberry and Rosenthal, 1999).

Perinatal HSV infection is generally a complication of maternal genital herpes, although infection of the neonate can also result from exposure to virus transmitted from a source other than the mother. Recent studies estimate the incidence of perinatal herpes in the United States to be about 11–18 cases per 100 000 live births, with approximately 75–80% being caused by HSV-2 (Schomogyi et al., 1998; Brady et al., submitted). The incidence is lower in Europe and Japan, in the range of 1–7 cases per 100 000 live births, with HSV-1 accounting for a greater percentage of cases (Kroon and Whitley, 1995; Tookey and Peckham, 1996). The reason for the difference in the incidence of disease in the United States compared to other developed countries is unclear, but is not explained simply by the difference in the prevalence of genital herpes in the different countries.

Perinatal HSV infection may result from either primary or recurrent maternal genital herpes, with transmission to the newborn generally occurring during birth. In most cases there is no history of genital HSV infection in either the mother or the father. The risk of transmission from mother to newborn is greatest when primary genital herpes occurs at or near term (up to 50%), but the risk is substantially lower for women who are experiencing recurrent infections, whether symptomatic or unrecognized (less than 4%) (Nahmias et al., 1971; Yeager et al., 1980). It is thought that exposure to higher viral loads increases the risk of perinatal infection, whereas the presence of passively acquired maternally derived HSV-specific antibody decreases the risk (Stanberry, 1996; Nahmias et al., 1971; Yeager et al., 1980; Stone et al., 1989; Sweet and Gibbs, 1990). This model has important implications for how a vaccine against genital herpes might affect perinatal HSV infection.

EXPECTED IMPACT OF AN EFFECTIVE VACCINE

It is unlikely that vaccines intended to control HSV infections can completely protect the genital tract against infection (i.e. prevent virus replication in genital tract tissues); however, the amount of virus needed to cause infection (i.e. the threshold of infection) will probably be greater in vaccinated individuals, thus early making them more resistant to infection (Stanberry et al., 1999a). Animal studies have shown that vaccines do not prevent infection but can significantly affect the natural history of experimental genital herpes (Stanberry et al., 1987; Meignier et al., 1988; Spector et al., 1998; Da Costa et al., 1997; Boursnell et al., 1997; Bourne et al., 1996a,b; Bernstein et al., 1999; Kuklin et al., 1998). In experimental studies HSV vaccines have substantially reduced virus replication in the genital tract and prevented or significantly ameliorated the symptomatic disease that can result from primary infection. It also appears that vaccine-induced immunity can at least partially protect against latent infection, thereby reducing or preventing the subsequent development of recurrent infections (Bourne *et al.*, 1996a). If the results seen in animals can be extended to humans, then HSV vaccines will probably provide substantial clinical benefit. Although some vaccinated individuals might develop symptoms if infected, the severity of their disease would probably be substantially less than in infected individuals who had not been vaccinated. Vaccinated individuals who became infected would probably be partially protected against latent infection, and would therefore be expected to have fewer and less severe recurrences, to shed less virus, and consequently to be less contagious than non-vaccinated people who become infected. They would nevertheless be periodically contagious, and could transmit virus to an uninfected partner. Universal vaccination would be a strategy to minimize the consequence of asymptomatic transmission, as vaccinated individuals who become infected would probably not experience symptomatic disease. Furthermore, if vaccination altered the natural history of genital herpes so that those who became infected were less contagious, and also induced immunity that rendered those who were uninfected less susceptible, the net effect of universal vaccination might be a reduction in the spread of genital HSV infection.

The incidence of perinatal HSV infections resulting from primary genital herpes during pregnancy might also be reduced by widespread use of a genital HSV vaccine. Because of the natural history of perinatal herpes, vaccination of the newborn would probably have little benefit; however, vaccination of uninfected women prior to or possibly during pregnancy might interrupt or at least minimize the risk of vertical transmission from mother to baby. If vaccination prevented maternal genital HSV-1 and HSV-2 infection, over time it would be expected to reduce the incidence of perinatal herpes by as much as 90%. Vaccination of women would probably provide some protection to the fetus or newborn even if the vaccine did not completely prevent maternal HSV genital infection. This protection would result from protective, maternally derived, transplacentally passaged antibody and/or a reduction in the amount of virus present in the genital tract around the time of delivery (Brown *et al.*, 1991; Maccato, 1993). Depending on the effectiveness of the passively acquired antibody, maternal vaccination might also afford the infant some protection against infection acquired in the postpartum period from non-maternal sources.

MICROBIOLOGY

Herpes simplex viruses types 1 and 2 are members of the Herpesviridae family of viruses. The HSV virion has a dense core surrounded by three layers: an icosahedral protein capsid, an amorphous protein layer called the tegument, and an outer envelope studded with glycoproteins. The core contains a linear double-stranded DNA genome of about 150 kilobase pairs which encodes at least 81 genes (Whitley and Roizman, 1997). Approximately 50% of the HSV-1 and HSV-2 DNA sequences are homologous, and the two viruses are closely related both genetically and biologically (Stanberry *et al.*, 1997). Although more than half of the HSV genes are not required for virus replication in cell culture (Whitley and Roizman, 1997), these 'non-essential' genes play important roles in the pathogenesis of infection in vivo, including facilitating immune evasion (Roizman and Sears, 1996; Dubin *et al.*, 1992; Ward and Roizman, 1998). A wide variety of structural and regulatory viral proteins are targets for the host's adaptive immune system (Mester *et al.*, 1996; Simmons *et al.*, 1992; Schmidt and Rouse, 1992; Norrild, 1985). Infected individuals develop serum antibodies to viral envelope proteins, including glycoproteins gB, gC, gD and gG (Zweerink and Corey, 1982; Eberle and Mou, 1983; Eberle *et al.*, 1984, 1985; Ashley *et al.*, 1988). Antibodies to these glycoproteins are also found in the cervicovaginal secretions of HSV-2 infected women (Ashley *et al.*, 1994). Serum antibodies are also produced to capsid proteins, including the major capsid protein VP5 and the capsid assembly protein ICP35, as well as to a variety of regulatory proteins, including ICP0, ICP4, ICP6 and VP16 (Zweerink and Corey, 1982; Eberle and Mou, 1983; Eberle *et al.*, 1984, 1985; Ashley *et al.*, 1988). Less is known regarding the antigen specificity of the T-lymphocyte response to HSV. Viral glycoproteins, including gB, gC and gD, and several regulatory proteins, including ICP4, ICP6 and ICP27, have been shown to induce cytotoxic T-lymphocyte (CTL) responses in mice (Glorioso *et al.*, 1985; Martin *et al.*, 1988; Johnson *et al.*, 1990; Banks *et al.*, 1993; Salvucci *et al.*, 1995; Cose *et al.*, 1997; Blaney *et al.*, 1998). In humans, gD-specific CD4+ and CD8+ T-cell clones have been isolated, as well as CD4+ T-cell clones with specificity for gB, gC and VP16 (Tigges *et al.*, 1992; Koelle *et al.*, 1994, 1988a,b; Mikloska and Cunningham, 1998; Doherty *et al.*, 1998). It is likely that induction of virus-specific cell-mediated immune responses will be very important in the development of effective HSV vaccines. More research is needed to define the antigen specificity of the human T-cell response to HSV infection in order to identify the viral gene products that are broadly immunogenic and thus potentially useful as vaccine antigens.

IMMUNOLOGY/IMMUNOBIOLOGY OF INFECTION

Based on the pathogenesis of genital herpes (Stanberry, 1996), vaccine-induced host immune responses might act at one or more key steps to prevent or limit genital HSV infection (Table 8.1). Ideally, immune responses elicited by vaccination would eliminate transmitted virus at the skin or mucous membrane

TABLE 8.1 Steps in the pathogenesis of genital herpes where vaccine-induced immune responses might act

At time of transmission neutralize virus at the skin or mucous membrane before it can replicate

Block entry of virus into the endings of the sensory nerves that innervate the genital tract

Act at the level of the sensory ganglia to prevent or reduce virus replication in neurons and/or glial cells

Neutralize virus exiting nerve fibers in the dermis

before it could replicate or enter sensory nerve endings. Such 'sterilizing' immunity would prevent both acute disease and the establishment of the latent infection. It is uncertain whether complete protection of mucosal surfaces can be achieved, but in order to obtain maximum protection against initial virus replication it is likely that a vaccine will need to induce genital tract mucosal immune responses.

Another way that vaccine-induced immune responses might affect genital herpes is by preventing virus from entering the endings of the sensory nerves that innervate the genital tract. Because HSV is likely to be extracellular at this point, it presents an opportunity for the immune system to eliminate the virus. If virus does not enter the nerve fiber, the acute disease will be reduced because most of the lesions develop from virus completing the neural arc, i.e. returning from the acutely infected dorsal root ganglia (Stanberry, 1996). Further, if vaccine-induced immunity interfered with virus entry into sensory nerves, then the latent infection would be prevented or reduced. For a vaccine to interfere with virus entry into sensory nerves it would probably need to induce mucosal immune responses.

Vaccine-induced immunity could also work at the level of the sensory ganglia by reducing the virus replication that occurs in neurons and/or glial cells. This would be expected to reduce the amount of virus that returns to the skin to form genital lesions, and thus prevent or reduce the severity of acute disease. An effect on the acute ganglionic infection might also reduce the magnitude of the latent infection, which in turn could result in fewer and/or less severe recurrent infections.

Virus exiting nerve fibers represents another point where vaccine-induced immunity might alter the pathogenesis of genital herpes. At least some of the lesions develop as a consequence of virus that descends from the acutely infected sensory ganglia to the dermis, where it is released and infects adjacent epithelial cells, with subsequent lesion formation. Host responses could intervene and prevent or reduce the number of lesions that develop by eliminating virus before infection of the skin, or by reducing the magnitude of replication after exit from the nerve fiber.

Studies in laboratory animals have examined whether it is possible to induce immunity that protects the genital tract against infection. Animals inoculated intravaginally with HSV develop evidence of genital herpes, and virus can be demonstrated in genital secretions and in sensory ganglia. With recovery from the initial infection animals develop both local and systemic HSV-specific immune responses (Kuklin et al., 1998; Bernstein et al., 1986, 1991; McDermott et al., 1989; Milligan and Bernstein, 1995a,b; Parr and Parr, 1997; Parr et al., 1998). In animals rechallenged with a different HSV strain, immunity induced by the initial infection does not prevent reinfection of the genital mucosa but does significantly reduce the magnitude and duration of virus replication occurring in the cervicovaginal tissue after reinoculation, prevents clinical disease associated with infection, and protects the ganglia from reinfection (Stanberry et al., 1986; Parr et al., 1994; Milligan et al., 1998). It has been observed that reinfection of the human genital tract with a second strain of virus occurs very rarely (Schmidt et al., 1994; Sucato et al., 1998), suggesting that immunity resulting from the initial episode of genital herpes may also protect humans against genital reinfection. This observation has important implications for vaccine development, as prior non-genital HSV infection appears to afford little or no protection against genital HSV infection (Reeves *et al.*, 1981b; SCRIP, 1997), suggesting that local (genital) immune responses may be critical for protection. The importance of pre-existing systemic virus-specific immune responses in protecting humans (other than neonates) against HSV infection is uncertain. However, the observation that a patient's own virus can establish infection and cause disease at a secondary site, despite the presence of circulating antibody and virus-specific cell-mediated immunity (LaRossa and Hamilton, 1971; Blank and Haines, 1973), further argues that local anti-HSV immune responses may be key to developing protection against mucosal infection.

Studies in mice have shown that T lymphocytes play a critical role in protecting the genital tract and sensory ganglia against HSV infection (Kuklin *et al.*, 1998; Milligan *et al.*, 1998; Gallichan and Rosenthal, 1998). The protection against virus replication in the genital tract induced by vaccination or prior genital HSV infection was lost when animals were depleted of T lymphocytes, despite the presence of high levels of antibody in serum and cervicovaginal secretions. Depletion of either CD4+ or CD8+ T cells reduced protection of the genital mucosa, but depletion of CD4+ cells had the greatest negative effect (Milligan *et al.*, 1998). Protection was also lost when interferon-γ was neutralized. In immune mice depleted of T lymphocytes HSV-2 could be recovered from sensory ganglia following rechallenge, but no infectious virus was found in control immune mice. In contrast to protection of the vaginal mucosa, however, either CD4+ or CD8+ cells were able to protect the ganglia (Milligan *et al.*, 1998). These results suggest that CD4+ lymphocytes and interferon-γ may be more important in protection against genital herpes than is antibody. Supporting the importance of local responses is the observation that in mice protection was greater with mucosal vaccination than with systemic immunization (Gallichan and Rosenthal, 1998).

OTHER STRATEGIES FOR CONTROL OF GENITAL AND PERINATAL HSV INFECTIONS
(Table 8.2)

Effective treatment for genital herpes has been available since the mid-1980s (Whitley and Gnann, 1992). Three licensed antiviral drugs, aciclovir, valaci-clovir and famciclovir, have been proved useful in the treatment of primary and

TABLE 8.2 Strategies for limiting the spread of genital HSV infection

Abstinence

Antiviral drug therapy

Safe sex practices, including limiting number of partners and use of condoms

Serological screening to identify those with unrecognized genital herpes

Topical microbicides

HSV vaccines

recurrent genital herpes and in the suppression of recurrent genital herpes (Barton, 1998; Wald, 1999; Wald and Corey, 1999; Stanberry *et al.*, 1999b). Chronic suppressive antiviral therapy has also been shown to reduce asymptomatic shedding of virus from the genital tract (Wald *et al.*, 1996, 1997, 1998; Sacks *et al.*, 1997, 1998). Whether suppression of shedding is sufficient to reduce the risk of HSV transmission to a susceptible sexual partner is unknown. What is clear, however, is that the prevalence of genital herpes in the United States has continued to increase despite the availability of effective antiviral therapy (Johnson *et al.*, 1989).

The practice of safer sexual behaviors may reduce the risk of acquiring genital herpes. In most populations, limiting the number of lifetime sexual partners reduces the likelihood of acquiring genital herpes (Fleming *et al.*, 1997; Stanberry and Rosenthal, 1999; Breinig *et al.*, 1990; Becker *et al.*, 1996; Obasi *et al.*, 1999; Rosenthal *et al.*, 1997). The exception is in populations where the prevalence of genital herpes in the partner pool is very high, under which circumstances even limiting oneself to a single partner carries a substantial risk (Lewis *et al.*, 1999). Condom use has been advocated as a strategy to control the spread of genital herpes (Centers for Disease Control, 1998). Although condoms have been shown to be impervious to HSV in the laboratory (Conant *et al.*, 1984), data supporting their effectiveness in preventing genital herpes transmission are lacking (Oberle *et al.*, 1989; Mertz *et al.*, 1992; Huerta *et al.*, 1996).

Serological screening to identify those with unrecognized genital HSV-2 infection could theoretically affect the spread of genital herpes (Schomogyi *et al.*, 1998; Handsfield *et al.*, 1999). This approach would require inexpensive and accurate type-specific serological tests, which are currently unavailable although under development (Ashley *et al.*, 1988; Ashley, 1998; Munday *et al.*, 1998; Schmidt *et al.*, 1999; Ashley and Wald, 1999). To be effective, a serological screening program for genital herpes would need to be widespread, coupled with counseling and extensive use of suppressive antiviral therapy. Such a program would probably be very expensive.

Topical microbicides are products for intravaginal use that are intended to protect against sexually transmitted infections (Elias and Coggins, 1996; Rosenthal *et al.*, 1998). There are none currently on the market, but several in development. A variety of formulations are being explored, including gels, foams, suppositories and films. These products are designed to provide a chemical barrier between the pathogen and the cells of the female genital tract that are susceptible to infection. There are several mechanisms by which the microbicide might act, including disruption of the organism's cell membrane or envelope, blocking of the receptor–ligand interactions essential for infectivity, inhibition of intra- or extracellular replication, alteration of the vaginal environment so as to reduce susceptibility, or enhancement of local immunity (Rosenthal *et al.*, 1998). Several candidate microbicides have been shown to have activity in vivo in mouse and guinea-pig models of genital herpes (Zeitlin *et al.*, 1997; Maguire *et al.*, 1998; Bourne *et al.*, 1999, 1999a; Bourne KZ *et al.*, 1999). In these experimental studies the products were used to protect female animals against intravaginal HSV-2 challenge. It is likely that some of these products would also be useful in preventing virus transmission from an infected female who may be shedding virus asymptomatically to her susceptible male partner. Studies on the effect of microbicides on asymptomatic shedding are needed. If effective, the widespread use of microbicides might affect the current genital herpes epidemic.

Strategies to prevent perinatal HSV infection have focused on the management of pregnant women with a history of genital herpes. As up to two-thirds of cases of perinatal infection occur in a setting where genital herpes is unrecognized, these approaches will have a limited impact on the problem. The most widely used strategy has been to deliver the infant by cesarean section if the woman has evidence of active genital herpes at labor (American College of Obstetricians and Gynecologists, 1988). This approach is not entirely effective, as 20–30% of infants who develop neonatal herpes have been delivered by cesarean section (Stone *et al.*, 1989; Tookey and Peckham, 1996). Furthermore, there is considerable morbidity and mortality associated with cesarean delivery (Stanberry *et al.*, 1999a). For every seven herpes-related neonatal deaths or severe disabilities prevented by cesarean delivery for recurrent genital herpes, four maternal deaths are estimated to occur (Randolph *et al.*, 1993). In addition, this is an expensive strategy costing an estimated $2.5 million to prevent each neonatal death resulting from recurrent genital herpes (Randolph *et al.*, 1993).

Another approach, which can be used alone or in conjunction with cesarean delivery, is the use of aciclovir suppression in the pregnant woman with genital herpes in order to prevent recurrent genital HSV at delivery. Three small studies have shown that this approach reduces recurrent infections around the time of delivery (Stray-Pedersen, 1990; Scott *et al.*, 1996; Brocklehurst *et al.*, 1998); however, its efficacy in preventing vertical transmission is unknown, and there is at least one documented case of transmission to the neonate occurring while the mother was on suppressive aciclovir (Haddad *et al.*, 1993). At present aciclovir is not approved for use during pregnancy, and before being used widely in pregnant women its effect on the fetus/neonate should be established.

HSV VACCINE DEVELOPMENT: AN OVERVIEW

There have been attempts to develop HSV vaccines for more than 60 years. Six different types of vaccines have been investigated (Table 8.3): inactivated whole virus vaccines; virion component or subunit vaccines; genetically

TABLE 8.3 General categories of HSV vaccines

Inactivated whole virus vaccines

Virion component or subunit vaccines

Genetically attenuated live virus vaccines

Vaccines consisting of replication-limited or replication-impaired HSV mutants

Replicating vectors that express one or more HSV antigens

DNA vaccines, plasmids that express one or more HSV genes

attenuated live virus vaccines; vaccines consisting of replication-limited or replication-impaired virus mutants; replicating non-pathogenic vectors that express one or more HSV antigens; and DNA vaccines, i.e. plasmids expressing one or more HSV genes. The earliest efforts to develop anti-herpes vaccines consisted of preparing crude vaccine formulations from HSV-infected animal tissues and testing them directly in humans for their ability to control recurrent herpetic disease. The clinical studies were open, uncontrolled trials that used reporting of subjective improvement as the primary outcome measure. Most early trials did not even assess the immunogenicity of the experimental vaccine.

The current requirements for vaccine development are substantially more rigorous. Vaccines must be shown to be free of adventitious agents or undesirable chemical residues. Manufacturing conditions must be developed that permit production of a consistent product with regard to safety and immunogenicity. For the past two decades candidate vaccines have been studied first in animals before proceeding to human clinical trials. The preclinical evaluation of most experimental vaccines has included their assessment in mouse and/or guinea-pig models of genital herpes (Bourne et al., 1996b; Kuklin et al., 1998; Stanberry, 1991a). These models are used to demonstrate the immunogenicity of the vaccine and establish its ability to protect female animals from intravaginal challenge with HSV. The initial clinical evaluations (phase I and II trials) of candidate vaccines are intended to establish their safety and immunogenicity in healthy subjects. Subsequent (phase III) studies are conducted to assess the effectiveness of the vaccine. For prophylactic vaccines these trials are complex, typically require multiple study sites, and are very expensive. The efficacy trials generally use volunteers who are thought to be at increased risk for acquiring genital herpes. These may include the consorts of patients known to have genital herpes, subjects who have recently acquired another sexually transmitted disease, or those with multiple sexual partners. Ethical design requires that the volunteers be counseled regarding safe sex practices that will hopeful prevent or reduce their risk of acquiring genital herpes. The impact of safer sexual practices on the attack rate must be considered when estimating the sample size needed for the clinical trial. Confounding variables that affect the likelihood of acquiring genital herpes are also considered when designing the efficacy trial. Such variables include gender and pre-existing immunity to HSV-1. In order to prove effectiveness there must be a comparator group, therefore phase III trials are placebo controlled and, in order to minimize bias, the studies are double blinded. Efficacy trials measure the ability of the vaccine to protect against infection and disease. For some vaccines infection can be detected by serological methods (serocoversion to HSV antigens not contained in the vaccine). With regard to disease, i.e. symptomatic genital herpes, the study cannot rely simply on self-reported illness but requires investigator and/or laboratory confirmation that the signs and symptoms are due to genital HSV infection. Because genital HSV infection has myriad clinical manifestations, volunteers in phase III trials must be evaluated by the investigator for any genitourinary sign or symptom. Although phase III trials to assess the efficacy of prophylactic HSV vaccines are complex they are feasible, and over the past 15 years five such trials have been conducted or are ongoing.

HISTORIC APPROACHES TO THE DEVELOPMENT OF HERPES SIMPLEX VIRUS VACCINES

HSV vaccine development has a long and somewhat chequered history (Meignier, 1985; Hall and Katrak, 1986; Dix, 1987; Stanberry, 1991b; Burke, 1992; Whitley, 1997). For the first four decades all the candidate vaccines were inactivated whole virion preparations. The earliest vaccines were developed in the 1930s and consisted of crude homogenate preparations of HSV-infected animal tissues inactivated by chemical treatment (Bernstein and Stanberry, 1999; Brain, 1936; Frank, 1938). In the 1950s, vaccines were prepared from virus grown in embryonated eggs and inactivated by physical means (Jawetz *et al.*, 1955). By the 1960s vaccines were being prepared from virus grown in cultured cells and inactivated by a variety of chemical and physical methods (Kern and Schiff, 1964; Henocq, 1967). These early vaccines were intended for both prophylactic and therapeutic use, although the majority of the clinical trials examined their effectiveness in treating patients with recurrent herpetic infections. Unfortunately, most of the early clinical trials had significant design flaws, such as failure to include necessary control groups (Stanberry, 1991b). None of the early vaccines were ever proved efficacious in well controlled trials. It was not until the 1970s and 1980s that greater emphasis was placed on developing vaccines to prevent genital herpes. The remainder of this chapter deals with modern aspects of prophylactic HSV vaccine development. Readers interested in therapeutic HSV vaccines should refer to Chapter 9.

MODERN VACCINE STRATEGIES

The HSV vaccines currently in development can be broadly divided into three categories: virion constituent vaccines, replication-competent vaccines, and nucleic acid-based vaccines (Table 8.4). Virion component or subunit vaccines contain viral proteins or peptides, either derived from inactivated whole virion preparations (inactivated vaccines) or produced by recombinant DNA or synthetic methods (synthetic subunit vaccines). By themselves, virion components are not broadly immunogenic. The exogenous antigen can engender B lymphocytes to produce antibody and, to a lesser extent, can be processed by antigen-presenting cells through an endosomal pathway to generate antigen-specific T-lymphocyte responses, primarily MHC class II restricted CD4+ T cells (Lanzavecchia, 1996; Gordon and Ramsay, 1997). By combining the virion components with adjuvants the immune responses can be both broadened and enhanced, but generally with a concomitant increase in reactogenicity (Hughes and Babiuk, 1994; Bussiere *et al.*, 1995). Replication-competent vaccines are capable of at least limited replication in the host. They induce B-cell responses and, because they replicate intracellularly, their antigens are processed by antigen-presenting cells via a cytosolic pathway to produce T-lymphocyte responses, including MHC class I restricted CD8+ cytotoxic T cells (Schmidt and Rouse, 1992; Gordon and Ramsay, 1997; Robinson *et al.*, 1994). Three types of replication-competent vaccines are in development: genetically attenuated replication-competent HSV vaccines (Meignier *et al.*, 1988; Spector *et al.*, 1998); replication-limited or single-cycle virus vaccines (Boursnell *et al.*, 1997; Farrell *et al.*, 1994; Morrison and Knipe, 1996, 1997; Da Costa *et al.*, 1997); and

▬ **TABLE 8.4 Modern HSV vaccines**

Virion component vaccines

Inactivated virion-derived vaccines
  Lupidon vaccines
  Skinner vaccine
  Merck HSV-2 GS vaccine

Recombinant or synthetic subunit vaccines
  Chiron HSV-2 gB/gD vaccine
  SmithKline HSV-2 gD vaccine
  Tizard/Chan HSV-2 gB p59

Replication-competent vaccines

Genetically attenuated live virus vaccines
  Pasteur-Mérieux R7020
  Aviron RAV 9395

Replication-limited or replication-impaired virus vaccines
  Cantab/Glaxo-Wellcome TA-HSV vaccine
  AVANT vaccine

Vectored vaccines

Viral vectors
  Vaccinia virus
  Adeno-associated virus
  Adenovirus
  Canarypox

Bacterial vectors
  Salmonella

Nucleic acid-based vaccines

DNA vaccines
  Vical HSV-2 gD vaccine
  Merck HSV-2 gD vaccine
  Apollon HSV-2 gD vaccine

RNA vaccines
  Chiron Sindbis virus-based vector system expressing HSV gB

---

vectored vaccines that express one or more HSV antigens (Gallichan *et al.*, 1993; Fleck *et al.*, 1994; Heineman *et al.*, 1995; Karem *et al.*, 1997). Nucleic acid-based vaccines (i.e. naked DNA or genetic vaccines) are a recent development (McClements *et al.*, 1996; Bourne *et al.*, 1996a,b; Kriesel *et al.*, 1996; Rouse *et al.*, 1998; Pachuk *et al.*, 1998; Bernstein *et al.*, 1999). They consist of a plasmid engineered to express one or two genes from the pathogenic organism targeted for vaccine control. Injection of the plasmid DNA results in the in vivo expression of the encoded protein in animals, with subsequent development of T-lymphocyte and, to a lesser extent, B-lymphocyte responses (Liu *et al.*, 1998).

INACTIVATED VIRION-DERIVED VACCINES

Over the past three decades several inactivated HSV vaccines have been developed, consisting of virus grown in cell culture and inactivated by either physical or chemical means. In most cases the inactivated virus is fractionated to remove viral DNA. This is considered an important safety step, as HSV DNA has been shown to transform cells in culture (Knipe, 1982). In some cases the inactivated virion products are fractionated further to obtain mainly the envelope glycoproteins, a highly immunogenic component of the virion. Examples of inactivated HSV vaccines include the Lupidon vaccines (Germany) (De Maria *et al.*, 1995); the Dundarov vaccine (Bulgaria) (Kavaklova *et al.*, 1986; Dundarov and Andonov, 1994); the Skinner vaccine (United Kingdom) (Skinner *et al.*, 1987, 1992); the Cappel vaccine (Belgium) (Cappel *et al.*, 1982a,b, 1985); the Kutinova vaccine (Czech Republic) (Kutinova *et al.*, 1982); and the Ivanovsky vaccine (Russia). Examples of virion-derived glycoprotein vaccines include the HSV-2 GS vaccine developed by Merck (United States) (Mertz *et al.*, 1990) and the Lederle vaccine (United States) (Stanberry, 1991b).

Inactivated vaccines have been shown to protect animals from developing symptomatic disease following intravaginal HSV challenge (Skinner *et al.*, 1980; Phillpotts *et al.*, 1988; Erturk *et al.*, 1991). Unfortunately, there have been problems with the clinical trials conducted to assess the effectiveness of the inactivated vaccines in preventing genital herpes. One example, the inactivated HSV-1 Skinner vaccine, was tested in an open trial (Skinner *et al.*, 1987, 1992). All the volunteers (consorts of patients with genital herpes) received the vaccine, and without a control group for comparison purposes it is not possible to determine whether the vaccine was effective or not. Another problem developed in the trial to assess the Merck HSV-2 GS vaccine (Mertz *et al.*, 1990). Although the trial utilized a double-blind placebo-controlled design, the lot of vaccine used in the efficacy trial was less immunogenic than that used in earlier safety and immunogenicity trials. The poorly immunogenic vaccine failed to protect against genital herpes and Merck abandoned its further development. Unfortunately, there has been no well designed clinical trial of a highly immunogenic inactivated HSV vaccine, therefore the potential of inactivated vaccines in controlling genital herpes is unknown.

SYNTHETIC VIRION COMPONENT (SUBUNIT) VACCINES

Virion component vaccines are viral proteins or peptides, usually from the virion envelope. Unlike inactivated vaccines they are not derived from HSV-infected cells, but rather produced by recombinant DNA methods or prepared by chemical synthesis (Burke, 1992; Watson and Enquist, 1985). They are generally regarded as safer than inactivated vaccines because there is no possibility of the vaccine containing contaminating HSV DNA or infectious virus that was not adequately inactivated. Subunit vaccines contain fewer antigens than inactivated whole virion or envelope vaccines, but whether more antigens are better is unknown. Like inactivated vaccines subunit vaccines are weakly immunogenic, but the immune responses to the antigens can be enhanced and broadened by including an adjuvant in the formulation (Hughes and Babiuk, 1994; Stanberry *et al.*, 1989).

HSV subunit vaccine development has focused on two envelope glycoproteins, gB and gD (Burke, 1992). Subunit vaccines containing gD or a combination of gB and gD have been shown to be immunogenic in animals and humans (Burke, 1992; Stanberry, 1991c; Langenberg *et al.*, 1995; Leroux-Roels *et al.*, 1994). As illustrated in Table 8.5, when administered prophylactically to animals glycoprotein vaccines protected against the development of clinical disease but did not prevent the animal from becoming asymptomatically infected (Stanberry *et al.*, 1987; Bernstein *et al.*, 1987, 1990). These results suggest that subunit vaccines may protect against disease without preventing infection (Stanberry *et al.*, 1999a).

Two major vaccine developers, the Chiron Corporation and SmithKline Beecham Biologicals, have conducted phase III clinical trials to assess the efficacy of glycoprotein subunit vaccines in preventing genital herpes. The Chiron vaccine contained HSV-2 gD and gB with MF59, an immunopotentiating emulsion, and was shown to be immunogenic in humans (Langenberg *et al.*, 1995). Chiron conducted two concurrent phase III clinical trials, one in the consorts of patients known to have genital herpes, and the second in subjects with a history of another sexually transmitted disease. Although the details of the studies have not yet been published, the vaccine was reported to be ineffective in preventing genital HSV-2 infection (SCRIP, 1997). When interpreting the results it will be important to determine whether the vaccine failed to prevent disease as assessed by clinical evaluation, or failed to prevent

TABLE 8.5 Effect of HSV glycoprotein vaccines on infection and disease in a guinea-pig model of genital herpes[1]

Vaccine or control	Concentration (μg)	Incidence of clinical disease[2]	Severity of clinical disease[3]	Incidence of infection[4]	Magnitude of infection[5]
Unimmunized control	0	16/16	17.9 ± 1.6	16/16	29.1 ± 0.8
Cell extract control plus CFA[7]	32	10/11	23.6 ± 12.7	11/11	30.3 ± 2.4
HSV-1 gB plus CFA	20	1/7	0.8 ± 0.8[6]	7/7	13.2 ± 2.5[6]
HSV-1 gD plus CFA	100	3/7	1.6 ± 1.1[6]	7/7	21.8 ± 2.5[6]
HSV-2 glycoproteins plus CFA	50	0/7	0	7/7	3.4 ± 0.9[6]

[1] Adapted from Stanberry *et al.*, 1987.
[2] Number of animals exhibiting herpetic skin lesions over number of animals intravaginally inoculated with HSV-2.
[3] Value is the mean area under the lesion score–day curve ± standard error.
[4] Infection defined as recovery of HSV from cervicovaginal swab samples on at least two occasions 24 hours apart. Incidence is the number of infected animals over the number of inoculated animals.
[5] Value is the mean area under the virus titer–day curve ± standard error.
[6] Significantly different from unimmunized control, $P < 0.01$.
[7] CFA = complete Freund's adjuvant

infection as measured by seroconversion to HSV-2 antigens not present in the vaccine. The results of animals studies predict that the vaccine probably cannot prevent infection, but might prevent or ameliorate disease. The SmithKline Beecham vaccine contains HSV-2 gD as the antigen, combined with two adjuvants, alum and 3-De-O-acylated monophosphoryl lipid A, and induces humoral and cellular immune responses superior to those produced by gD/alum alone (Leroux-Roels *et al.*, 1994). The SmithKline Beecham vaccine is being evaluated in phase III trials involving HSV-1/HSV-2 seronegative consorts of subjects with documented genital herpes, and subjects who perceive themselves at risk for acquiring genital herpes. The results of the clinical trials are expected by early 2000. HSV peptides have also been shown to be immunogenic and at least partially protective in studies using experimental animal models (Stanberry, 1991c; Tizard and Chan, 1997). It is possible that a cocktail of peptides might induce broad B- and T-cell responses that could afford humans protection against genital HSV infection. However, at present there is no indication that this approach is advancing towards clinical trials.

GENETICALLY ATTENUATED LIVE VIRUS VACCINES

Live attenuated viral vaccines have been very effective in controlling common childhood diseases. The traditional approach to producing an attenuated virus has been to repeatedly passage it in cell culture until it becomes less virulent but still retains the ability to induce protective host immunity. This strategy has not been successful for HSV because cell culture passage does not result in stable attenuation, and therefore the attenuated virus has the potential to unexpectedly revert to a fully virulent pathogen capable of causing disease. In addition to the problem of genotypic stability there are other safety concerns, including the potential of vaccine virus to establish latency and to reactivate, to recombine with wild-type virus to yield new virulent mutants, and issues regarding the potential of some HSV gene products to cause cancer. Some of the problems of live HSV vaccines are being overcome through genetic engineering. Using recombinant DNA technology it is possible to delete viral genes that are important in disease pathogenesis while retaining those that are essential for replication in vivo. It is also possible to add genes that might enhance the immunogenicity of the vaccine. Engineered vaccine viruses are genetically stable and therefore overcome the problem associated with cell culture attenuation. The first candidate genetically engineered vaccines were safe and effective in animals (Meignier *et al.*, 1988), but a study by Pasteur-Mérieux found that the lead candidate (R7020) was poorly immunogenic in humans, suggesting that the engineered virus was overly attenuated (Whitley, 1997). The development of genetically engineered HSV vaccines was discontinued by Pasteur-Mérieux, but this strategy is being pursued by an American company, Aviron Inc. One candidate vaccine developed by Aviron was shown to be safe and effective in animals (Spector *et al.*, 1998). Further studies are anticipated.

REPLICATION-LIMITED OR REPLICATION-IMPAIRED VIRUS VACCINES

Through genetic engineering it has been possible to create viral mutants that are restricted in their ability to replicate and therefore to cause disease. This approach has been extensively exploited in developing viral vectors for gene therapy, and has recently been used to develop a new class of HSV vaccines. The approach involves the deletion of one or two genes that are essential for virus replication. The defective mutant can only replicate in genetically engineered cells that express the missing gene product(s). The vaccine virus is produced by growing the mutant in the complementing cells, where it acquires the missing gene product(s) but not the missing gene(s). The virus that is grown in the complementing cells is fully infectious but, because it lacks the genetic information necessary to produce the missing gene product(s), it cannot produce progeny virus in normal human cells. Consequently, when injected into a host the mutant virus can undergo only a single replication cycle, and because no new virus is produced the infection is terminated. Studies of several replication-limited HSV mutants have shown that they do not cause disease in animals but are immunogenic and can protect against experimental genital HSV infection (Boursnell *et al.*, 1997; Farrell *et al.*, 1994; Morrison and Knipe, 1996, 1997; Da Costa *et al.*, 1997). Phase I clinical studies of an HSV-2 vaccine developed by the British company Cantab Pharmaceuticals plc have shown that their TA-HSV vaccine is safe, well tolerated and immunogenic in healthy volunteers (Hill *et al.*, 1997; Hickling *et al.*, 1998; Roberts *et al.*, 1998). The TA-HSV vaccine has been licensed to Glaxo-Wellcome, where further development is planned. Another replication-limited vaccine is being developed by an American company, AVANT Inc.

VECTORED VACCINES

Non-pathogenic viral and bacterial organisms have been used as vectors for the delivery of immunogenic foreign genes. Using this approach, one or two HSV genes are inserted into the vector. When the vaccine is injected into a host the vector replicates and expresses the HSV gene product(s), inducing both humoral and cellular HSV-specific immunity. These vectored vaccines have some of the advantages of live attenuated vaccines while avoiding the problems associated with HSV virulence, latency, reactivation and oncogenicity. Experimental HSV vaccines have been constructed using a variety of vectors, including adenovirus, adeno-associated virus, canarypox, vaccinia, varicella-zoster virus and *Salmonella typhimurium* (Gallichan *et al.*, 1993; Fleck *et al.*, 1994; Heineman *et al.*, 1995; Karem *et al.*, 1997; Manning *et al.*, 1997; Manickan *et al.*, 1995b; Gallichan and Rosenthal, 1995). Some vectored vaccines have been shown to protect animals against HSV disease (Gallichan *et al.*, 1993; Fleck *et al.*, 1994; Heineman *et al.*, 1995; Karem *et al.*, 1997; Manning *et al.*, 1997; Manickan *et al.*, 1995b). One interesting vectored vaccine has made use of the licensed, live attenuated Oka varicella-zoster virus vaccine to express the HSV-2 glycoprotein D gene. This vaccine was shown to be immunogenic and partially protective in a guinea-pig model of genital HSV-2 disease (Heineman *et al.*, 1995). Several laboratories are continuing work on the preclinical development of vectored vaccines, but at present there are no vectored HSV vaccines being evaluated in clinical trials.

DNA VACCINES

Nucleic acid-based vaccines, also known as naked DNA vaccines, are a new development. Studies in animals have shown that injection of plasmid DNA constructs containing viral or bacterial genes results in in-vivo expression of the encoded protein and the development of both cell-mediated and humoral immune responses to the expressed protein (Rouse *et al.*, 1998; Liu *et al.*, 1998). HSV DNA vaccine development has focused on the construction of expression plasmids containing HSV genes encoding envelope glycoproteins gB, gC, gD and gE. These vaccines have been reported to induce both humoral and cellular immune responses in animals, and to protect mice and guinea pigs from disease resulting from cutaneous or intravaginal HSV challenge (McClements *et al.*, 1996; Bourne *et al.*, 1996a,b; Kriesel *et al.*, 1996; Pachuk *et al.*, 1998; Bernstein *et al.*, 1999; Manickan *et al.*, 1995a; McClements *et al.*, 1997; Kuklin *et al.*, 1997; Nass *et al.*, 1998). The protection afforded by HSV DNA vaccines is thought to be mediated by HSV-specific CD4+ T lymphocytes (Manickan *et al.*, 1995a). Like subunit vaccines, nucleic acid-based vaccines can only express one or two viral antigens; however, they are capable of inducing cell-mediated responses without requiring the potent adjuvants necessary with subunit vaccines. It is possible, however, to enhance the immunogenicity of DNA vaccines by including an adjuvant. A recent study showed that coadministration of a plasmid encoding interleukin-12 (IL-12) enhanced cell-mediated responses to a DNA vaccine containing HSV-2 gD and afforded mice greater protection against lethal HSV challenge (Sin *et al.*, 1999). Although DNA vaccines do not replicate in the host, there are other concerns, such as possible integration into the host genome, that require that the safety of these promising new vaccines be carefully evaluated. One company, Apollon Inc., has initiated clinical trials with an HSV-2 gD DNA vaccine, and two other companies, Vical Inc. and Merck, are proceeding with the preclinical development of DNA vaccines. In addition, Chiron Technologies has developed a strategy that combines the DNA and vectored vaccine approaches. They have engineered a Sindbis virus-based vector system that permits high-level expression of a foreign gene via self-amplifying cytoplasmic RNA amplification. They have shown that a Sindbis virus-based vaccine expressing HSV gB induces virus-specific antibody and cytotoxic T-lymphocyte responses and protects mice against lethal challenge (Hariharan *et al.*, 1998). Although nucleic acid-based vaccines appear promising, they have not yet been shown to be highly immunogenic in humans.

FUTURE PROSPECTS

Despite the availability of effective antiviral drug therapy, genital and perinatal HSV infections remain a significant health problem. As summarized in this chapter, the development of safe and effective HSV vaccines is the only practical strategy for controlling disease and possibly limiting the spread of infection. Although previous efforts to develop HSV vaccines were unsuccessful, there are reasons to be optimistic concerning the future. A better understanding of the pathogenesis and immunobiology of genital herpes, a growing body of information regarding the human immune system, including mucosal immunity, the discovery of new adjuvants, technological developments that permit the design of novel vaccines, and growing appreciation of the importance of clinical trial design all contribute to likelihood that we will be successful in developing effective vaccines for the control of genital and perinatal herpes. Currently three types of HSV vaccines are being evaluated in clinical trials. A promising subunit gD vaccine is being tested in large phase III clinical trials, a novel replication-impaired HSV vaccine is entering phase II trials, and a DNA-based vaccine is being tested in phase I studies. In preclinical development there are vectored vaccines, genetically attenuated live viral vaccines, and other genetically engineered products. With the prospect of a vaccine that can prevent or ameliorate genital herpes being available within the next few years, it is important that we begin discussions regarding how these vaccines may best be used and whether they might be of value in the control of non-genital HSV infections, such as herpes gingivostomatitis or herpes keratitis. Informed decisions must be made regarding how widely and at what age the vaccines are used, i.e. universal vaccination or vaccination of selected groups; vaccination of infants or vaccination of adolescents or adults. Targeted vaccination programs will benefit the recipients but may have do little to affect the spread of genital herpes. Alternatively, universal vaccination in infancy with a booster in adolescence might protect the recipient against symptomatic oral, ocular and genital herpes, and also reduce the risk of spread to unimmunized people. Although the latter strategy is the most expensive, the prevention of all HSV disease would have significant medical and economic benefits.

REFERENCES

American College of Obstetrician and Gynecologists (1988). Perinatal herpes simplex virus infections. ACOG Technical Bulletin 122, ACOG, Washington DC

Arvin, A.M. (1996). The epidemiology of perinatal herpes simplex virus infections. In: Stanberry L.R. (ed) Genital and neonatal herpes. London, John Wiley and Sons, pp. 179–192

Ashley, R.L. (1998). Genital herpes. Type-specific antibodies for diagnosis and management. *Dermatol. Clin.* **16**: 789–93

Ashley, R.L. and Wald, A. (1999). Genital herpes: review of the epidemic and potential use of type-specific serology. *Clin. Microbiol. Rev.* **12**: 1–8

Ashley, R.L., Corey, L., Doles, J. *et al.* (1994). Protein-specific cervical antibody responses to primary genital herpes simplex virus type 2 infections. *J. Infect. Dis.* **170**: 20–26

Ashley, R., Militoni, J., Lee, F., Nahmias, A. and Corey, L. (1988). Comparison of western blot (immunoblot) and glycoprotein G-specific immunodot enzyme assay for detecting antibodies to herpes simplex virus types 1 and 2 in human sera. *J. Clin. Microbiol.* **26**: 662–667

Banks, T.A., Nair, S. and Rouse, B.T. (1993). Recognition by and in vitro induction of cytotoxic T lymphocytes against predicted epitopes of the immediate-early protein ICP27 of herpes simplex virus. *J. Virol.* **67**: 613–616

Barton, S.E. (1998). Herpes management and prophylaxis. *Dermatol. Clin.* **16**: 799–803

Becker, T.M., Lee, F., Daling, J.R. and Nahmias, A.J. (1996). Seroprevalence of and risk factors for antibodies to herpes simplex viruses, hepatitis B, and hepatitis C among Southwestern Hispanic and non-Hispanic White women. *Sex. Transm. Dis.* **23**: 138–144

Benedetti, J., Corey, L. and Ashley, R. (1994). Recurrence rates in genital herpes after symptomatic first-episode infection. *Ann. Intern. Med.* **121**: 847–854

Bernstein, D.I. and Stanberry, L.R. (1999). Herpes simplex virus vaccines – new developments. *Vaccine* **17**: 1681–1689

Bernstein, D.I., Ashley, R.L., Stanberry, L.R. and Myers M.G. (1990). Detection of asymptomatic initial HSV infection in animals previously immunized with subunit HSV glycoprotein vaccines. *J. Clin. Microbiol.* **28**: 11–15

Bernstein, D.I., Harrison, C.J., Jenski, L., Myers, M.G. and Stanberry, L.R. (1991). Cell mediated immunological responses and recurrent genital herpes in the guinea pig: effects of glycoprotein immunotherapy. *J. Immunol.* **146**: 3571–3577

Bernstein, D.I., Stanberry, L.R., Harrison, C.J., Kappes, J. and Myers, M.G. (1986). Effects of oral acyclovir treatment of initial genital HSV-2 infection upon antibody response, recurrence patterns and subsequent HSV-2 reinfection in guinea pigs. *J. Gen. Virol.* **67**: 1601–1612

Bernstein, D.I., Stanberry, L.R., Harrison, C.J., Kappes, J.C. and Myers, M.G. (1987). Antibody response to herpes simplex virus glycoprotein D: effects of acyclovir and relationship to recurrence. *J. Infect. Dis.* **156**: 423–429

Bernstein, D.I., Tepe, E.R., Mester, J.C., Arnold, R.L., Stanberry, L.R. and Higgins, T. (1999). Effects of DNA immunization formulated with bupivacaine in murine and guinea pig models of genital herpes simplex virus infection. *Vaccine* **17**: 1964–1969

Blaney, J.E. Jr, Nobusawa, E., Brehm, M.A. *et al.* (1998). Immunization with a single major histocompatibility complex class I-restricted cytotoxic T-lymphocyte recognition epitope of herpes simplex virus type 2 confers protective immunity. *J Virol* **72**: 9567–9574

Blank, H. and Haines, H.G. (1973). Experimental human reinfection with herpes simplex virus. *J Invest Dermatol* **61**: 223–5

Bourne, K.Z., Bourne, N., Reising, S.F. and Stanberry, L.R. (1999). Plant products as topical microbicide candidates: assessment of in vitro and in vivo activity against herpes simplex virus type 2. *Antiviral Res.* **42**: 219–226

Bourne, N., Ireland, J., Stanberry, L.R. and Bernstein, D.I. (1999). Effect of undecylenic acid as a topical microbicide against genital herpes infection in mice and guinea pigs. *Antiviral Res* **40**: 139–144

Bourne, N., Stanberry, L.R., Bernstein, D.I. and Lew, D. (1996a). DNA immunization against experimental genital herpes simplex virus infection. *J Infect Dis* **173**: 800–807

Bourne, N., Milligan, G.N., Schleiss, M.R., Bernstein, D.I. and Stanberry, L.R. (1996b). DNA immunization confers protective immunity on mice challenged intravaginally with herpes simplex virus type 2. *Vaccine* **14**: 1230–1234

Bourne, N., Bernstein, D.I., Ireland, J., Sonderfan, A.J., Profy, A.T. and Stanberry, L.R. (1999). The topical microbicide Pro 2000 protects against genital herpes infection in a mouse model. *J Infect Dis.* **180**: 203–205

Boursnell, M.E., Entwisle, C., Blakeley, D. *et al.* (1997). A genetically inactivated herpes simplex virus type 2 [HSV-2] vaccine provides effective protection against primary and recurrent HSV-2 disease. *J Infect Dis* **175**: 16–25

Brady, R.C., Bernstein, D.I., Conway, A.M., Connelly, B.L. and Stanberry, L.R. (Submitted). Neonatal herpes simplex virus infections in Hamilton County Ohio

Brain, R.T. (1936). Biological therapy in virus diseases. *Br J Dermatol Syphilis* **48**: 21–26

Bravo, F.J., Myers, M.G. and Stanberry, L.R. (1994). Neonatal herpes simplex virus infection: pathogenesis and treatment in the guinea pig. *J Infect Dis* **169**: 947–955

Breinig, M.K., Kingsley, L.A., Armstrong, J.A., Freeman, D.J. and Ho, M. (1990). Epidemiology of genital herpes in Pittsburgh: serologic, sexual, and racial correlates of apparent and inapparent herpes simplex infections. *J Infect Dis* **162**: 299–305

Brocklehurst, P., Kinghorn, G., Carney, O. *et al.* (1998). A randomised placebo controlled trial of suppressive acyclovir in late pregnancy in women with recurrent genital herpes infection. *Br J Obstet Gynaecol* **105**: 275–280

Brown, Z.A., Kern, E.R., Spruance, S.L. and Overall, J.C. Jr. (1979). Clinical and virological course of herpes simplex genitalis. *West J Med* **130**: 414–421

Brown, Z.A., Benedetti, J., Ashley, R. *et al.* (1991). Neonatal herpes simplex virus infections in relation to asymptomatic maternal infection at the time of labor. *N Engl J Med* **324**: 1247–1252

Brown, Z.A., Selke, S., Zeh, J. *et al.* (1997). The acquisition of herpes simplex during pregnancy. *N Engl J Med* **337**: 509–515

Burke, R.L. (1992). Contemporary approaches to vaccination against herpes simplex virus. *Curr Topics Microbiol Immunol* **179**: 137–158

Bussiere, J.L., McCormick, G.C. and Green, J.D. (1995). Preclinical safety assessment: considerations in vaccine development. In: Powell M.F., Newman M.J. (eds) Vaccine design: the subunit and adjuvant approach. New York, Plenum Press, pp. 61–80

Cameron, D.W., Simonsen, J.N., D'Costa, L.J. *et al.* (1989). Female to male transmission of human immunodeficiency virus type 1: risk factors for seroconversion in men. *Lancet* **2**: 403–407

Cappel, R., Sprecher, S. and de Cuyper, F. (1992a). Immune responses to DNA free herpes simplex proteins in man. *Dev Biol Standards* **52**: 345–350

Cappel, R., Sprecher, S., de Cuyper, F. and De Braekeleer, J. (1985). Clinical efficacy of a herpes simplex subunit vaccine. *J Med Virol* **16**: 137–145

Cappel, R., Sprecher, S., Rickaert, F. and de Cuyper, F. (1982b). Immune response to a DNA free herpes simplex vaccine in man. *Arch Virol* **73**: 61–67

Carney, O., Ross, E., Ikkos, G. and Mindel, A. (1993). The effect of suppressive oral acyclovir on the psychological morbidity associated with recurrent genital herpes. *Genitourinary Med* **69**: 457–459

Carney, O., Ross, E., Bunker, C., Ikkos, G. and Mindel, A. (1994). A prospective study of the psychological impact on patients with a first episode of genital herpes. *Genitourinary Med* **70**: 40–45

Centers for Disease Control and Prevention. (1998). 1998 Guidelines for treatment of sexually transmitted diseases. *MMWR* **47**: 22

Conant, M.A., Spicer, D.W. and Smith, C.D. (1984). Herpes simplex virus transmission: condom studies. *Sex Transm Dis* **11**: 94–95

Corey, L. (1998). Raising the consciousness for identifying and controlling viral STDs: fears and frustrations. *Sex Transm Dis* **25**: 58–69

Corey, L. and Wald, A. (1999). Genital Herpes. In: Holmes K.K., Sparling P.F., Mardh P.-A., Lemon S.M., Stamm W.E., Piot P., Wasserheit J.W. (eds). Sexually transmitted diseases, 3rd edn. New York, McGraw-Hill, pp. 285–312

Cose, S.C., Jones, C.M., Wallace, M.E., Heath, W.R. and Carbone, F.R. (1997). Antigen-specific CD8+ T cell subset distribution in lymph nodes draining the site of herpes simplex virus infection. *Eur J Immunol* **27**: 2310–2316

Cowan, F.M., Johnson, A.M., Ashley, R., Corey, L. and Mindel, A. (1994). Antibody to herpes simplex virus type 2 as serological marker of sexual lifestyle in populations. *Br. Med. J.* **309**: 1325–1329

Da Costa, X.J., Bourne, N., Stanberry, L.R. and Knipe, D.M. (1997). Construction and characterization of a replication-defective herpes simplex virus 2 ICP8 mutant strain and its use in immunization studies in a guinea pig model of genital disease. *Virology* **232**: 1–12

De Maria, A., Tundo, P., Romano, A. and Grima, P. (1995). Anti-HSV-1 herpes vaccination by LUPIDON H: preliminary results. *Adv Exp Med Biol* **371b**: 1599–1600

Dix, R.D. (1987). Prospects for a vaccine against herpes simplex virus types 1 and 2. *Prog Med Virol* **34**: 89–128

Doherty, D.G., Penzotti, J.E., Koelle, D.M. *et al.* (1998). Structural basis of specificity and degeneracy of T cell recognition: pluriallelic restriction of T cell responses to a peptide antigen involves both specific and promiscuous interactions between the T cell receptor, peptide, and HLA-DR. *J Immunol* **161**: 3527–3535

Dubin, G., Fishman, N.O., Eisenberg, R.J. and Cohen, G.H. (1992). The role of herpes simplex virus glycoproteins in immune evasion. *Curr Topics Microbiol Immunol* **179**: 111–120

Dundarov, S. and Andonov, P. (1994). Seventeen years of application of herpes vaccines in Bulgaria. *Acta Virol* **38**: 205–208

Eberle, R. and Mou, S.W. (1983). Relative titers of antibodies to individual polypeptide antigens of herpes simplex virus type 1 in human sera. *J Infect Dis* **148**: 436–444

Eberle, R., Mou, S.W. and Zaia, J.A. (1984). Polypeptide specificity of the early antibody response following primary and recurrent genital herpes simplex virus type 2 infections. *J Gen Virol* **65**: 1839–1843

Eberle, R., Mou, S.W. and Zaia, J.A. (1985). The immune response to herpes simplex virus: comparison of the specificity and relative titers of serum antibodies directed against viral polypeptides following primary herpes simplex virus type 1 infections. *J Med Virol* **16**: 147–162

Ekwo, E., Wong, Y.W. and Myers, M.G. (1979). Asymptomatic cervicovaginal shedding of herpes simplex virus. *Am J Obstet Gynecol* **134**: 102–103

Elias, C.J. and Coggins, C. (1996). Female-controlled methods to prevent sexual transmission of HIV. *AIDS* **10** (Suppl 3): S43–51

Erturk, M., Phillpotts, R.J., Welch, M.J. and Jennings, R. (1991). Efficacy of HSV-1 ISCOM vaccine in the guinea-pig model of HSV-2 infection. *Vaccine* **9**: 728–734

Farrell, H.E., McLean, C.S., Harley, C., Efstathiou, S., Inglis, S. and Minson, A.C. (1994). Vaccine potential of a herpes simplex virus type 1 mutant with an essential glycoprotein deleted. *J Virol* 1994 **68**: 927–932

Ferrer, R.M., Kraiselburd, E.N. and Kouri, Y.H. (1983). Inapparent genital herpes simplex virus infection in women attending a venereal disease clinic. *Sex. Transm. Dis.* **10**: 91–92

Fleck, M., Podlech, J., Weise, K. and Falke, D. (1994). A vaccinia virus-herpes simplex virus (HSV) glycoprotein B1 recombinant or an HSV vaccine overcome HSV type 2 induced humoral immunosuppression and protect against viral challenge. *Med. Microbiol. Immunol.* **183**: 87–94

Fleming, D.T., McQuillan, G.M., Johnson, R.E. *et al.* (1997). Herpes simplex virus type 2 in the United States, 1976 to 1994. *N. Engl. J. Med.* **337**: 1105–1111

Forsgren, M.F., Skoog, E., Jeansson, S., Olofsson, S. and Giesecke, J. (1994). Prevalence of antibodies to herpes simplex virus in pregnant women in Stockholm in 1969, 1983 and 1989: implications for STD epidemiology. *Int. J. STD AIDS* **5**: 113–116

Frank, S.B. (1938). Formolized herpes virus therapy and neutralizing substance in herpes simplex. *J. Invest. Dermatol.* **1**: 267–282

Gallichan, W.S. and Rosenthal, K.L. (1995). Specific secretory immune responses in the female genital tract following intranasal immunization with a recombinant adenovirus expressing glycoprotein B of herpes simplex virus. *Vaccine* **13**: 1589–1595

Gallichan, W.S. and Rosenthal, K.L. (1998). Long-term immunity and protection against herpes simplex virus type 2 in the murine female genital tract after mucosal but not systemic immunization. *J. Infect. Dis.* **177**: 1155–1161

Gallichan, W.S., Johnson, D.C., Graham, F.L. and Rosenthal, K.L. (1993). Mucosal immunity and protection after intranasal immunization with recombinant adenovirus expressing herpes simplex virus glycoprotein B. *J. Infect. Dis.* **168**: 622–629

Glorioso, J.C., Kees, U., Kumel, G., Kirchner, H. and Krammer, P.H. (1985). Identification of herpes simplex virus type 1 (HSV-1) glycoprotein gC as the immunodominant antigen for HSV-1-specific memory cytotoxic T lymphocytes. *J. Immunol.* **135**: 575–582

Gordon, A. and Ramsay, A. (1997). Vaccines, Vaccination and the Immune Response. Philadelphia, Lippincott-Raven

Government Statistical Service. (1995). Sexually transmitted diseases England, 1994, p. 16

Gwanzura, L., McFarland, W., Alexander, D., Burke, R.L. and Katzenstein, D. (1998). Association between human immunodeficiency virus and herpes simplex virus type 2 seropositivity among male factory workers in Zimbabwe. *J. Infect. Dis.* **177**: 481–484

Haddad, J., Langer, B., Astruc, D., Messer, J. and Lokiec, F. (1993). Oral acyclovir and recurrent genital herpes during late pregnancy. *Obstet. Gynecol.* **82**: 102–104

Hall, M.J. and Katrak, K. (1986). The quest for a herpes simplex virus vaccine: background and recent developments. *Vaccine* **4**: 138–150

Handsfield, H.H., Stone, K.M. and Wasserheit, J.N. (1999). Prevention agenda for genital herpes. *Sex. Transm. Dis.* **26**: 228–231

Hariharan, M.J., Driver, D.A., Townsend, K. *et al.* (1998). DNA immunization against herpes simplex virus: enhanced efficacy using a Sindbis virus-based vector. *J. Virol.* **72**: 950–958

Heineman, T.C., Connelly, B.L., Bourne, N., Stanberry, L.R. and Cohen, J.I. (1995). Immunization with recombinant varicella-zoster virus expressing herpes simplex virus type 2 glycoprotein D reduces the severity of genital herpes in guinea pigs. *J. Virol.* **69**: 8109–8113

Henocq, E. (1967). Vaccin anti-herpétique. *Rev. Méd.* (Paris) **8**: 695–708

Hickling, J.K., Chrisholm, S.E., Duncan, *et al.* (1998). Immunogenicity of a disabled infectious single cycle HSV-2 vaccine in phase I clinical trials in HSV-2 seropositive and seronegative volunteers. Abstracts of the 8th International Congress on Infectious Diseases. Boston, Massachusetts, Abstract 22.008, p. 89, May 1998

Hill, S., Zhang, X., Boursnell, M.E., Shields, J.G., Ricciardi-Castagnoli, P. and Hickling, J.K. (1997). Generation of a primary immune response to a genetically inactivated (DISC) herpes simplex virus and wild type virus. *Biochem. Soc. Trans.* **25**: 200S

Hook, E.W. III, Cannon, R.O., Nahmias, A.J. *et al.* (1992). Herpes simplex virus infection as a risk factor for human immunodeficiency virus infection in heterosexuals. *J. Infect. Dis.* **165**: 251–255

Huerta, K., Berkelhamer, S., Klein, J., Ammerman, S., Chang, J. and Prober, C.G. (1996). Epidemiology of herpes simplex virus type 2 infections in a high-risk adolescent population. *J. Adolesc. Health* **18**: 384–386

Hughes, H.P.A. and Babiuk, L.A. (1994). Potentiation of vaccines through effective adjuvant formulations and manipulation of the immune response. In: Kurstak E. (ed) Modern vaccinology. New York, Plenum Medical Book Co, pp. 87–120

Jacobs, R.F. (1998). Neonatal herpes simplex virus infections. *Semin. Perinatol.* **22**: 64–71

Jawetz, E., Allende, M.F. and Coleman, V.R. (1955). Studies on herpes simplex virus. VI. Observations on patients with recurrent herpetic lesions injected with herpes viruses or their antigens. *Am. J. Med. Sci.* **229**: 477–485

Johnson, R.E., Nahmias, A.J., Magder, L.S., Lee, F.K., Brooks, C.A. and Snowden, C.B. (1989). A seroepidemiologic survey for the prevalence of herpes simplex virus type 2 infection in the United States. *N. Engl. J. Med.* **321**: 7–12

Johnson, R.M., Lancki, D.W., Fitch, F.W. and Spear, P.G. (1990). Herpes simplex virus glycoprotein D is recognized as antigen by CD4+ and CD8+ T lymphocytes from infected mice. *J. Immunol.* **145**: 702–710

Karem, K.L., Bowen, J., Kuklin, N. and Rouse, B.T. (1997). Protective immunity against herpes simplex virus (HSV) type 1 following oral administration of recombinant *Salmonella typhimurium* vaccine strains expressing HSV antigens. *J Gen Virol* **78**: 427–434

Kavaklova, L., Dundarov, S., Andonov, P., Bakalov, B., Dundarova, D. and Brodvarova I. (1986). Preparation and efficacy of antiherpes type 1 and 2 subunit vaccines. *Acta Virol* **30**: 402–410

Kern, A.B. and Schiff, B.L. (1964). Vaccine therapy in recurrent herpes simplex. *Arch Dermatol* **89**: 844–845

Knipe, D.M. (1982). Cell growth transformation by herpes simplex virus. *Prog Med Virol* **28**: 114–144

Koelle, D.M., Frank, J.M., Johnson, M.L. and Kwok, W.W. (1998a). Recognition of herpes simplex virus type 2 tegument proteins by CD4 T cells infiltrating human genital herpes lesions. *J Virol* **72**: 7476–7483

Koelle, D.M., Posavad, C.M., Barnum, G.R., Johnson, M.L., Frank, J.M. and Corey, L. (1998b). Clearance of HSV-2 from recurrent genital lesions correlates with infiltration of HSV-specific cytotoxic T lymphocytes. *J Clin Invest* **101**: 1500–1508

Koelle, D.M., Corey, L., Burke, R.L. *et al.* (1994). Antigenic specificities of human CD4+ T-cell clones recovered from recurrent genital herpes simplex virus type 2 lesions. *J Virol* **68**: 2803–2810

Kohl, S. (1997). Neonatal herpes simplex virus infection. *Clin Perinatol* **24**: 129–150

Kriesel, J.D., Spruance, S.L., Daynes, R.A. and Araneo, B.A. (1996). Nucleic acid vaccine encoding gD2 protects mice from herpes simplex virus type 2 disease. *J Infect Dis* **173**: 536–541

Kroon, S. and Whitley, R. eds. (1995). Management strategies in herpes: can we improve management of perinatal herpes simplex virus infections? PPS Europe Ltd, Worthing

Kuklin, N.A., Daheshia, M., Chun, S. and Rouse, B.T. (1998). Role of mucosal immunity in herpes simplex virus infection. *J Immunol* **160**: 5998–6003

Kuklin, N., Daheshia, M., Karem, K., Manickan, E. and Rouse, B.T. (1997). Induction of mucosal immunity against herpes simplex virus by plasmid DNA immunization. *J Virol* **71**: 3138–3145

Kutinova, L., Slichtova, V. and Vonka, V. (1982). Subviral herpes simplex vaccine. *Dev Biol Standards* **52**: 313–319

Langenberg, A.G., Burke, R.L., Adair, S.F. *et al.* (1995). A recombinant glycoprotein vaccine for herpes simplex virus type 2: safety and immunogenicity. *Ann Intern Med* **122**: 889–898

Lanzavecchia, A. (1996). Mechanisms of antigen uptake for presentation. *Curr Opin Immunol* **8**: 348–354

LaRossa, D. and Hamilton, R. (1971). Herpes simplex virus infections of the digits. *Arch Surg* **102**: 600–603

Leroux-Roels, G., Moreau, E. and Desombre, I. (1994). Persistence of humoral and cellular immune response and booster effect following vaccination with herpes simplex (gD2) candidate vaccine with MPL. In: Abstracts of the 34th Interscience Conference on Antimicrobial Agents and Chemotherapy, Orlando, FL, p. 205 (abstract H57)

Lewis, L.M., Bernstein, D.I., Rosenthal, S.L. and Stanberry, L.R. (1999). Seroprevalence of herpes simplex virus type 2 in African-American college women. *J Natl Med Assoc.* **91**: 210–212

Liu, M.A., Fu, T.-M., Donnelly, J.J., Caulfield, M.J. and Ulmer, J.B. (1998). DNA vaccines: mechanism for generation of immune responses. In: Gupta S., Sher A., Ahmed R. (eds) Mechanism of lymphocyte activation and immune regulation VII. New York, Plenum Press, pp. 187–192

Longo, D. and Koehn, K. (1993). Psychosocial factors and recurrent genital herpes: a review of prediction and psychiatric treatment studies. *Int J Psychiatry Med* **23**: 99–117

Maccato, M. (1993). Herpes in pregnancy. *Clin Obstet Gynecol* **36**: 869–77

Maguire, R.A., Zacharopoulos, V.R. and Phillips, D.M. (1998). Carrageenan-based nonoxynol-9 spermicides for prevention of sexually transmitted infections. *Sex Transm Dis* **25**: 494–500

Manickan, E., Rouse, R.J., Yu, Z., Wire, W.S. and Rouse, B.T. (1995a). Genetic immunization against herpes simplex virus is mediated by CD4+ T lymphocytes. *J Immunol* **155**: 259–265

Manickan, E., Francotte, M., Kuklin, D. *et al.* (1995b). Vaccination with recombinant vaccinia viruses expressing ICP27 induces protective immunity against herpes simplex virus through CD4+ Th1+ T cells. *J Virol* **69**: 4711–4716

Manning, W.C., Paliard, X., Zhou, S. *et al.* (1997). Genetic immunization with adeno-associated virus vectors expressing herpes simplex virus type 2 glycoprotein B and D. *J Virol* **71**: 7960–7962

Martin, S., Courtney, R.L., Fowler, G. and Rouse, B.T. (1988). Herpes simplex virus type 1-specific cytotoxic T lymphocytes recognize non-structural proteins. *J Virol* **62**: 2265–2273

McClements, W.L., Armstrong, M.E., Keys, R.D. and Liu, M.A. (1996). Immunization with DNA vaccines encoding glycoprotein D or glycoprotein B, alone or in combination, induces protective immunity in animal models of herpes simplex virus-2 disease. *Proc Natl Acad Sci USA* **93**: 11414–11420

McClements, W.L., Armstrong, M.E., Keys, R.D. and Liu, M.A. (1997). The prophylactic effect of immunization with DNA encoding herpes simplex virus glycoproteins on HSV-induced disease in guinea pigs. *Vaccine* **15**: 857–860

McDermott, M.R., Goldsmith, C.H., Rosenthal, K.S. and Brais, L.J. (1989). T lymphocytes in the genital lymph nodes protect mice from intravaginal infection with herpes simplex virus type 2. *J. Infect. Dis.* **159**: 460–466

Meignier, B. (1995). Vaccination against herpes simplex virus infections. In: Roizman B., Lopez C. (eds) The herpesviruses Vol 4. New York, Plenum, pp. 256–312

Meignier, B., Longnecker, R. and Roizman, R. (1988). In vivo behavior of genetically engineered herpes simplex viruses R7017 and R7020: construction and evaluation in rodents. *J. Infect. Dis.* **158**: 602–614

Mertz, G.J., Ashley, R., Burke, R.L. *et al.* (1990). Double-blind, placebo-controlled trial of a herpes simplex virus type 2 glycoprotein vaccine in persons at high risk for genital herpes infection. *J. Infect. Dis.* **161**: 653–660

Mertz, G.J., Benedetti, J., Ashley, R., Selke, S.A. and Corey, L. (1992). Risk factors for the sexual transmission of genital herpes. *Ann. Intern. Med.* **116**: 197–202

Mertz, G.J., Schmidt, O., Jourden, J.L. *et al.* (1985). Frequency of acquisition of first-episode genital infection with herpes simplex virus from symptomatic and asymptomatic source contacts, *Sex Transm. Dis.* **12**: 33–39

Mester, J.C., Milligan, G.N. and Bernstein, D.I. (1996). The immunobiology of herpes simplex virus. In: Stanberry L.R. (ed) Genital and neonatal herpes. Chichester, John Wiley and Sons, pp. 49–92

Mikloska, Z. and Cunningham, A.L. (1998). Herpes simplex virus type 1 glycoproteins gB, gC, and gD are major targets for CD4 lymphocyte cytotoxicity in HLA-DR expressing human epidermal keratinocytes. *J. Gen. Virol.* **79**: 353–361

Milligan, G.N. and Bernstein, D.I. (1995a). Analysis of herpes simplex virus-specific T cells in the murine female genital tract following genital infection with herpes simplex virus type 2. *Virology* **212**: 481–489

Milligan, G.N. and Bernstein, D.I. (1995b). Generation of humoral immune responses against herpes simplex virus type 2 in the murine female genital tract. *Virology* **206**: 234–241

Milligan, G.N., Bernstein, D.I. and Bourne N. (1998). T lymphocytes are required for protection of the vaginal mucosae and sensory ganglia of immune mice against reinfection with herpes simplex virus type 2. *J Immunol* **160**: 6093–6100

Mindel, A. (1996). Psychological and psychosexual implications of herpes simplex virus infections. *Scand. J. Infect. Dis. Suppl.* **100**: 27–32

Morrison, L.A. and Knipe, D.M. (1996). Mechanisms of immunization with a replication-defective mutant of herpes simplex virus 1. *Virology* **220**: 402–413

Morrison, L.A. and Knipe, D.M. (1997). Contributions of antibody and T cell subsets to protection elicited by immunization with a replication-defective mutant of herpes simplex virus type 1. *Virology* **239**: 315–326

Munday, P.E., Vuddamalay J., Slomka M.J. and Brown D.W. (1998). Role of type specific herpes simplex virus serology in the diagnosis and management of genital herpes. *Sex Transm Infect* **74**: 175–178

Nahmias, A.J., Josey, W.E., Naib, Z.M., Freeman, M.G., Fernandez, R.J. and Wheeler, J.H. (1971). Perinatal risk associated with maternal genital herpes simplex virus infection. *Am J Obstet Gynecol* **110**: 825–837

Nass, P.H., Elkins, K.L. and Weir, J.P. (1998). Antibody response and protective capacity of plasmid vaccines expressing three different herpes simplex virus glycoproteins. *J Infect Dis* **178**: 611–617

Norrild, B. (1985). Humoral response to herpes simplex virus infections. In: Roizman B, and Lopez C. (eds) The herpesviruses, vol 4. New York, Plenum Press, pp. 69–86

Obasi, A., Mosha, F., Quigley, M. *et al.* (1999). Antibody to herpes simplex virus type 2 as a marker of sexual risk behavior in rural Tanzania. *J Infect Dis* **179**: 16–24

Oberle, M.W., Rosero-Bixby, L., Lee, F.K., Sanzchez-Braverman, M., Nahmias, A.J. and Guinan, M.E. (1989). Herpes simplex virus type 2 antibodies: high prevalence in monogamous women in Costa Rica. *Am J Trop Med Hyg* **41**: 224–229

Pachuk, C.J., Arnold, R., Herold, K., Ciccarelli, R.B. and Higgins, T.J. (1998). Humoral and cellular immune responses to herpes simplex virus -2 glycoprotein D generated by facilitated DNA immunization of mice. *Curr Topics Microbiol Immunol* **226**: 79–89

Parr, E.L., Bozzola, J.J. and Parr, M.B. (1998). Immunity to vaginal infection by herpes simplex virus type 2 in adult mice: characterization of the immunoglobulins in vaginal mucus. *J Reprod Immunol* **38**: 15–30

Parr, M.B. and Parr, E.L. (1997). Protective immunity against HSV-2 in the mouse vagina. *J Reprod Immunol* **36**: 77–92

Parr, M.B., Kepple, L., McDermott, M.R., Drew, M.D., Bozzola, J.J. and Parr, E.L. (1994). A mouse model for studies of mucosal immunity to vaginal infection by herpes simplex virus type 2. *Lab Invest* **70**: 369–380

Phillpotts, R.J., Welch, M.J., Ridgeway, P.H., Walkland, A.C. and Melling, J. (1998). A test for the relative potency of herpes simplex virus vaccines based upon the female guinea-pig model of HSV 2 genital infection. *J Biol Standards* **16**: 109–118

Prober, C.G. (1996). Clinical features and diagnostic evaluation of perinatal herpes simplex virus infections. In: Stanberry L.R. (ed) Genital and neonatal herpes. Chichester, John Wiley and Sons, Ltd., pp. 193–208

Randolph, A.G., Washington, A.E. and Prober, C.G. (1993). Caesarean delivery for women presenting with genital herpes lesions. Efficacy, risks, and costs. *JAMA* **270**: 77–82

Rattray, M.C., Corey, L. and Reeves, W.C. (1978). Recurrent genital herpes among women: symptomatic versus asymptomatic shedding. *Br J Vener Dis* **54**: 262–265

Reeves, W.C., Corey, L. and Adams, H.G. (1981a). Risk of recurrence after first episodes of genital herpes: relation of HSV type and and antibody response. *N Engl J Med* **305**: 315–319

Reeves, W.C., Corey, L., Adams, H.G., Vontver, L.A. and Holmes, K.K. (1981b). Risk of recurrence after first episodes of genital herpes. Relation to HSV type and antibody response. *N Engl J Med* **305**: 315–9

Roberts, J.S.C., Uttridge, J.A., Hickling, J.K. *et al.* (May, 1998). Safety tolerability and viral containment of a disabled infectious single cycle HSV-2 vaccine evaluated in phase 1 clinical trials in HSV-2 seropositive and seronegative volunteers. Abstracts of the 8th International Congress on Infectious Diseases. Boston, Massachusetts, Abstract 22.009, p.89

Robinson, A.J., Nicholson, B.H. and Lyttle, D.J. (1994). Engineered viruses and vaccines. In: Nicholson B.H. (ed) *Synthetic vaccines*. Blackwell Scientific Publishers, Oxford, pp. 331–375

Rodgers, C.A., O'Mahony, C. (1995). High prevalence of herpes simplex virus type 1 in female anogenital herpes simplex. *Int J STD AIDS* **6**: 144–146

Roizman, B. and Sears A.E. (1996). Herpes simplex viruses and their replication. In: Fields B.N., Knipe D.M., Howley P.M. (eds) *Fields virology*, 3rd edn. Lippincott-Raven, Philadelphia. pp. 2231–2295

Rosenthal, S.L., Cohen, S.S. and Stanberry, L.R. (1998). Topical microbicides. current status and research considerations for adolescent girls. *Sex Transm Dis* **25**: 368–377

Rosenthal, S.L., Stanberry, L.R., Biro, F.M. *et al.* (1997). Seroprevalence of herpes simplex virus type 1 and type 2 and cytomegalovirus in adolescents. *Clin Infect Dis* **24**: 135–139

Ross, J.D.C., Smith, I.W. and Elton, R.A. (1993). The epidemiology of herpes simplex types 1 and 2 infection of the genital tract in Edinburgh 1978–1991, *Genitourinary Med* **69**: 381–383

Rouse, B.T., Nair, S., Rouse, R.J. *et al.* (1998). DNA vaccines and immunity to herpes simplex virus. *Curr Topics Microbiol Immunol* **226**: 69–78

Sacks, S.L. and Shafran, S.D. (1998). BID famciclovir suppression of asymptomatic genital herpes simplex virus shedding in men. In: Programs and abstracts of the 38th Interscience Conference on Antimicrobial Agents and Chemotherapy, San Diego, California

Sacks, S.L., Hughes, A., Rennie, B. and Boon, R. (1997). Famciclovir for suppression of asymptomatic and symptomatic recurrent genital herpes shedding: a randomized, double-blind, double dummy, parallel group, placebo-controlled trial. In: Programs and abstracts of the 37th Interscience Conference on Antimicrobial Agents and Chemotherapy, Toronto, Canada

Salvucci, L.A., Bonneau, R.H. and Tevethia, S.S. (1995). Polymorphism within the herpes simplex virus (HSV) ribonucleotide reductase large subunit (ICP6) confers type specificity for the recognition by HSV type 1-specific cytotoxic T lymphocytes. *J Virol* **69**: 1122–1131

Schmidt, D.S. and Rouse, B.T. (1992). The role of T cell immunity in the control of herpes simplex virus. *Curr. Topics Microbiol Immunol* **179**: 57–74

Schmidt, D.S., Brown, D.R., Nisenbaum, R. *et al.* (1999). Limits in reliability of glycoprotein G-based type-specific serologic assays for herpes simplex virus types 1 and 2. *J Clin Microbiol* **37**: 376–379

Schmidt, O.W., Fife, K.H. and Corey, L. (1998). Reinfection is an uncommon occurrence in patients with symptomatic recurrent genital herpes. *J Infect Dis* **149**: 645–6

Schomogyi, M., Wald, A. and Corey, L. (1998). Herpes simplex virus-2 infection: an emerging disease? *Infect Dis Clin* **12**: 47–61

Scott, L.L., Sanchez, P.J., Jackson, G.L., Zeray, F. and Wendel, G.D. (1996). Acyclovir suppression to prevent cesarean delivery after first episode genital herpes. *Obstet Gynecol* **87**: 69–73

SCRIP, No. 2279 (October 28, 1997) PJB Publications, UK, p. 22

Simmons, A., Tscharke, D. and Speck, P. (1992). The role of immune mechanisms in control of herpes simplex virus infection of the peripheral nervous system. *Curr Topics Microbiol Immunol* **179**: 32–56

Sin, J.I., Kim, J.J., Arnold, R.L. *et al.* (1999). IL-12 gene as a DNA vaccine adjuvant in a herpes mouse model: IL-12 enhances Th1-type CD4+ T cell-mediated protective immunity against herpes simplex virus-2 challenge. *J Immunol* **162**: 2912–2921

Skinner, G.R., Buchan, A., Hartley, C.E., Turner, S.P. and Williams, D.R. (1980). The preparation, efficacy and safety of 'antigenoid' vaccine NFU1 (S-L+) MRC toward prevention of herpes simplex virus infections in human subjects. *Med Microbiol Immunol* (Berlin) **169**: 39–5

Skinner, G.R., Fink, C.G., Cowan, M. *et al.* (1987). Follow-up report on 50 subjects vaccinated against herpes genitalis with Skinner vaccine. *Med Microbiol Immunol* (Berlin) **176**: 161–168

Skinner, G.R., Fink, C., Melling, J. *et al.* (1992). Report of twelve years experience in open study of Skinner herpes simplex vaccine towards prevention of herpes genitalis. *Med Microbiol Immunol* (Berlin) 180: 305–320

Spector, F.C., Kern, E.R., Palmer, J. *et al.* (1998). Evaluation of a live attenuated recombinant virus RAV 9395 as a herpes simplex virus type 2 vaccine in guinea pigs. *J Infect Dis* 177: 1143–1154, 1998

Stanberry, L.R. (1991a). Herpes simplex virus vaccine evaluation in animals: the guinea pig model. *Rev Infect Dis* 13 (Suppl 11): S920–923

Stanberry, L.R. (1991b). Herpes simplex virus vaccines. *Semin Pediatr Infect Dis* 2: 178–185

Stanberry, L.R. (1991c). Subunit viral vaccines: therapeutic and prophylactic uses. In: Aurelian L. (ed) Herpesviruses, the immune system and Aids. Boston, Massachusetts, Kluwer Academic Publishers, pp. 309–341

Stanberry, L.R. (1996). The pathogenesis of herpes simplex virus infections. In: Stanberry L.R. (ed) Genital and neonatal herpes. Chichester, John Wiley and Sons, pp. 31–48

Stanberry, L.R. and Rosenthal, S.L. (1999). Epidemiology of herpes simplex virus infections in adolescents. *Herpes*. 6: 12–15

Stanberry, L.R., Jorgensen, D.M. and Nahmias, A.J. (1997). Herpes simplex viruses 1 and 2. In: Evans A.S. and Kaslow R.A. (eds) Viral infection of humans: epidemiology and control, 4th edn. New York, Plenum Press, pp. 419–54

Stanberry, L.R., Bernstein, D.I., Kit, S. and Myers, M.G. (1986). Genital reinfection after recovery from initial herpes simplex virus type 2 genital infection in guinea pigs. *J Infect Dis* 153: 1055–61

Stanberry, L.R., Myers, M.G., Stephanopoulos, D.I. and Burke, R.L. (1989). Preinfection prophylaxis with herpes simplex virus glycoprotein immunogens: factors influencing efficacy. *J Med Virol* 70: 3177–3185

Stanberry, L.R., Bernstein, D.I., Burke, R.L., Pachl, C. and Myers, M.G. (1987). Recombinant herpes simplex virus glycoproteins: protection against initial and recurrent genital herpes. *J Infect Dis* 155: 914–920

Stanberry, L., Handsfield, H., Sacks, S., Peters, B. and Chinn, C. (1999c). The health care costs of genital herpes in the United States – a pharmacoeconomic model. *Antiviral Res* 41: A58

Stanberry, L., Cunningham, A., Mindel, A. *et al.* (in press 1999a). Prospects for control of herpes simplex virus disease through immunization. *Clin Infect Dis*

Stanberry, L., Cunningham, A., Mertz, G. *et al.* (1999b). New developments in the epidemiology, natural history and management of genital herpes. *Antiviral Res*. 42: 1–14

Stone, K.M., Brooks, C.A., Guinan, M.E. and Alexander, E.R. (1989). National surveillance for neonatal herpes simplex virus infections. *Sex Transm Dis* 16: 152–156

Stray-Pedersen, B. (1990). Acyclovir in late pregnancy to prevent neonatal herpes simplex. *Lancet* 336: 756

Sucato, G., Wald, A., Wakabayashi, E., Vieira, J. and Corey, L. (1998). Evidence of latency and reactivation of both herpes simplex virus (HSV)-1 and HSV-2 in the genital region. *J Infect Dis* 177: 1069–1072

Sweet, R.L., Gibbs, R.S. (1990). Infectious diseases of the female genital tract, 2nd edn. Baltimore: Williams and Wilkins

Tayal, S.C., Pattman, R.S. (1994). High prevalence of herpes simplex virus type 1 in female anogenital herpes simplex in Newcastle upon Tyne 1983–92. *Int J STD AIDS* 5: 359–361

Tigges, M.A., Koelle, D., Hartog, K., Sekulovich, R.E., Corey, L. and Burke, R.L. (1992). Human CD8+ herpes simplex virus specific cytotoxic T-lymphocyte clones recognize diverse virion protein antigens. *J Virol* 66: 1622–1634

Tizard, M.L. and Chan, W.L. (1997). Differential T cell response induced by certain recombinant oligopeptides of herpes simplex virus glycoprotein B in mice. *J Gen Virol* 78: 1625–1632

Tookey, P. and Peckham, C.S. (1996). Neonatal herpes simplex virus infection in the British Isles. *Paediatr Perinat Epidemiol* 10: 432–442

Wald, A. (1999). New therapies and prevention strategies for genital herpes. *Clin Infect Dis* 289 (Suppl 1): S4–13

Wald, A. and Corey, L. (1996). The clinical features an diagnostic evaluation of genital herpes. In: Stanberry L.R. (ed) Genital and neonatal herpes. Chichester, John Wiley and Sons, pp. 109–138

Wald, A. and Corey, L. (1999). Antiviral therapies for long-term suppression of genital herpes. *JAMA* **281**: 1169–70

Wald, A., Zeh, J., Selke, S., Ashley, R.L. and Corey, L. (1995). Virologic characteristics of subclinical and symptomatic genital herpes infections, *N Engl J Med* **333**: 770–775

Wald, A., Zeh, J., Barnum, G., Davis, L.G. and Corey, L. (1996). Suppression of subclinical shedding of herpes simplex virus type 2 with acyclovir. *Ann Intern Med* **124**, 8–15

Wald, A., Corey, L., Cone, R., Hobson, A., Davis, G. and Zeh, J. (1997). Frequent genital herpes simplex virus 2 shedding in immunocompetent women. Effect of acyclovir treatment. *J Clin Invest* **99**: 1092–1097

Wald, A., Warren, T., Hu, H. *et al.* (1998). Suppression of subclinical shedding of herpes simplex virus type 2 in the genital tract with valacyclovir. In: Programs and abstracts of the 38th Interscience Conference on Antimicrobial Agents and Chemotherapy, San Diego, California

Ward, P.L. and Roizman, B. (1998). Evasion and obstruction: the central strategy of the interaction of human herpesviruses with host defenses. In: Medvecky P.G., Friedman H., Bendinelli M. (eds) Herpesviruses and immunity. New York, Plenum Press, pp. 1–32

Watson, R.J. and Enquist, L.W. (1985). Genetically engineered herpes simplex virus vaccines. *Prog Med Virol* **31**: 84–108

Whitley, R.J. (1997). Herpes simplex vaccines. In: Levine M.M., Woodrow G.C., Kaper J.B., Cobon G.S. (eds). New generation vaccines, 2nd edn. New York, Marcel Dekker, pp. 727–748

Whitley, R.J. and Arvin, A.M. (1995). Herpes simplex virus infections. In: Remington J.S., Klein J.O. (eds) Infectious diseases of the fetus and newborn infant. Philadelphia, W.B. Saunders, pp. 354–76

Whitley, R.J. and Gnann, J.W. Jr. (1992). Acyclovir: a decade later. *N Engl J Med* **327**: 782–789

Whitley, R.J. and Roizman, B. (1997). Herpes simplex viruses. In: Richman D.D., Whitley R.J., Hayden F.G. (eds) Clinical virology. New York, Churchill Livingstone, pp. 375–410

Whitley, R.J., Nahmias, A.J., Visintine, A.M., Fleming, C.L. and Alford, C.A. (1980). The natural history of herpes simplex virus infection of mother and newborn. *Pediatrics* **66**: 489–94

Whitley, R., Arvin, A., Prober, C. *et al.* (1991). A controlled trial comparing vidarabine with acyclovir in neonatal herpes simplex virus infection. *N. Engl J Med* **324**: 444–449

Whitley, R.J., Corey, L., Arvin, A. *et al.* (1988). Changing presentation of herpes simplex virus infection in neonates. *J Infect Dis* **158**: 109–116

Yeager, A.S., Arvin, A.M., Urbani, L.J. and Kemp, III, J.A. (1980). The relationship of antibody to outcome in neonatal herpes simplex virus infections. *Infect Immun* **29**: 532–538

Zeitlin, L., Whaley, K.J., Hegarty, T.A., Moench, T.R. and Cone, R.A. (1997). Tests of vaginal microbicides in the mouse genital herpes model. *Contraception* **56**: 329–335

Zweerink, H.J. and Corey, L. (1982). Virus-specific antibodies in sera from patients with genital herpes simplex virus infection. *Infect Immun* **37**: 413–421

9

THERAPEUTIC VACCINES FOR CONTROL OF HERPES SIMPLEX VIRUS CHRONIC INFECTIONS

PIERRE VANDEPAPELIÈRE

SmithKline Beecham Biologicals, Rixensart, Belgium

EPIDEMIOLOGICAL AND CLINICAL ASPECTS

Introduction

Genital herpes disease represents an important problem both in the US and worldwide (ASHA, 1998; Nahmias *et al.*, 1990). There are two types of herpes simplex viruses (HSV): HSV1 and HSV2. Both can cause a primary disease, followed at variable intervals by recurrent episodes in the same anatomical area. For both viruses the disease following primary infection is more severe than the recurrent disease, although the frequency of recurrent HSV1 genital infection is much less than with HSV2 (Anker Peterslund, 1991; Kinghorn, 1993; Reeves *et al.*, 1981). However, in all cases the infection can go unrecognized for many years and will be detected by serological testing or swabbing and culture (Koutsky *et al.*, 1990, 1992).

Apart from causing recurrent lesions that can be painful, recurrent disease also has important psychosociosexual consequences (Mindel, 1993). Viral shedding is maximum during clinical episodes; however, unrecognized or asymptomatic recurrences also cause viral shedding, although less important. It is therefore likely that the majority of transmission occurs outside the clinical episodes. It is not known, however, whether a minimum viral load is needed for transmission, or, in other words, if transmission is attenuated by lower levels of virus in mucosal secretions. An important complication of genital herpes infection is neonatal herpes disease (Stanberry, 1993), which is rare – 1/2500 to 1/10000 live births – but with a very high mortality of 65% (Arvin, 1991; Whitley, 1993c). Genital HSV infection also increases the risk of acquisition of HIV infection, like all ulcerative sexual diseases (Hook *et al.*, 1992), but also through a reciprocal enhancement of viral replication (Heng *et al.*, 1994).

Recurrent genital herpes can be managed with antiviral drugs such as aciclovir, famciclovir and valaciclovir. Episodic treatment, i.e. drug intake during the clinical episodes, will only partially reduce the duration of the episodes. To affect the frequency of the recurrences continuous suppressive therapy is needed, but this imposes the necessity to take the drugs daily for months or years. No treatment exists that can eradicate the latent infection and thereby cure the disease.

Pathogenesis of the Disease

Herpes simplex viruses are human DNA viruses of the α Herpes viridae family. Both HSV1 and HSV2 cause primary disease upon primary infection of mucocutaneous surfaces, and establish a clinically silent latent infection in the neuronal cells of the ganglia innervating the mucocutaneous zone of infection. This latent infection can reactivate spontaneously or following various stimuli, and, through anterograde axonal transfer of virus to cutaneous sites, causes recurrent infections that can be either clinically apparent or result in asymptomatic virus shedding. The mechanisms of establishment, maintenance and reactivation of latent virus are not fully understood, but it is accepted that both viral and host factors are important (Figure 9.1).

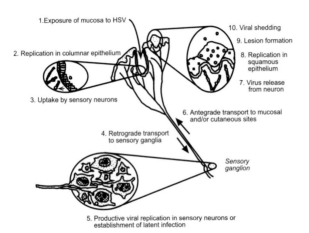

FIGURE 9.1 Schematic representation of the pathogenesis of primary, latent and recurrent genital herpes (Reproduced, with permisssion, from *J. Eur. Acad. Dermatol. Venereol.* 7: 120–128, 1996).

The two main arms of the immune system appear to play a role in preventing and/or controlling HSV infection. The humoral response mediated by antibodies primarily directed against various envelope glycoproteins is important for the prevention of infection of mucocutaneous cells. Glycoproteins gB, gD and gH are major targets of neutralizing antibodies, and HSV1 and HSV2 can induce cross-neutralizing antibodies (Eing *et al.*, 1989; Spear, 1984). The cellular immune response against HSV proteins, both CD4+ and CD8+ T-cell responses, plays a major role in the control of disease (Scott Schmid, 1991; Kohl, 1991; Manickan *et al.*, 1995; Cunningham and Merigan, 1983; Sin *et al.*, 1999). It is important to note that both latent and recurrent infections occur in patients who have made both humoral and cellular responses to a wide array of HSV proteins. This has raised doubts as to whether a therapeutic vaccine for HSV could be effective. Animal studies, however, indicate that the cellular immune response is key to inducing a therapeutic effect capable of modifying recurrent disease (Bernstein *et al.*, 1991). Indeed, only vaccine formulations inducing a strong Th1 response have had a beneficial effect on the course of the disease. For example, in guinea pigs and mice, glycoproteins need to be combined to Th1-inducing adjuvants such as muramyl tripeptide (MTP-PE), monophosphoryl lipid A (MPL) or IL-2 to have a good therapeutic effect (Stanberry, 1989; Ho *et al.*, 1989; Leroux-Roels *et al.*, 1993; Burke *et al.*, 1994). When the adjuvant has a predominant Th2 effect, such as MF-59 or alum, the therapeutic efficacy is decreased or absent (Burke *et al.*, 1994).

RATIONALE FOR HSV THERAPEUTIC VACCINE

Among chronic viral infections, therapeutic vaccines for herpes simplex virus infections have a long history. Indeed, therapeutic vaccines have been tried since the 1920s and a wide variety of vaccine types have been investigated, unfortunately with no or only very limited success. Several reviews have been written in the recent years summarizing animal data (Stanberry, 1990; Dix, 1987) and clinical evaluation in humans (McKenzie and Straus, 1996a,b; Burke, 1993; Stanberry, 1995, 1996; Jennings *et al.*, 1998).

The main obstacle to any immunotherapeutic approach is the establishment of HSV latency in the nucleus of the sensory neuronal cells. This location protects the virus from any kind of immune attack. Indeed, neuronal cells do not display MHC class I nor MHC class II antigens, which are needed for stimulation of CD4+ or CD8+ T cells. Outside periods of reactivation, only the genetic information of the latent virus is present in the neuronal cells, where no virion or capsid can be detected. In addition, destruction of the infected neuronal cells by the immune attack is not a desirable goal, as these cells will not be replaced and such destruction might therefore lead to neurosensorial defects.

Therefore the immune response could probably only reach the virus after reactivation, either during the axonal migration towards the sensorial endings, or at the mucocutaneous site before clinical lesions develop. As this process only lasts for a few hours to a few days, the destruction of the virus by an adequate immune response during this short process is a major challenge. Knowing that the majority of the reactivations go clinically unrecognized or are totally asymptomatic, and that viral shedding can occur in the absence of any clinical sign of disease, the potential objectives for immunotherapy can be subdivided as follows:

> suppression or reduction in the severity and/or frequency of symptomatic recurrent episodes;
>
> reduction of virus transmission (duration and/or quantity) during the clinical episodes;
>
> reduction of asymptomatic shedding to reduce viral transmission outside the clinical recurrences (frequency, duration, quantity). Demonstrating a reduction in transmission is a difficult objective since it is not proven that decreasing viral shedding will be correlated with a decrease in transmission;
>
> reduction/suppression of viral reactivation: this objective is probably not achievable with current immunological tools.

There are some other arguments outlining the difficulty of therapeutic HSV vaccines. Spontaneous, complete clearance of latent infection has not been demonstrated. Even after many years of successful complete control of the disease recurrences can occur, either spontaneously or following immunosuppression, or after triggering by stimulating factors such as stress or fever. Moreover, the absence of clinical manifestations could be due to mechanisms other than immune control, as the level of circulating humoral and cellular immunity to HSV is usually much higher in the presence of active clinical disease than during asymptomatic periods. Another problem relates to the placebo effect: up to 70% reduction in the subjective reporting of the severity and frequency of the recurrences has been induced by placebos (Kern and Schiff, 1964). Indeed, even differences in the severity of local or general reactions induced by the active or control compounds could in theory affect the outcome of the study. Finally, with the exception of a few studies that need to be confirmed, all well designed, controlled studies with any type of vaccines have failed to affect significantly the clinical manifestations of the disease.

Despite these problems, there are some facts supporting the concept. There have been numerous reports of positive outcome from limited, uncontrolled studies from the 1930s to the 1990s. Even if no definitive conclusion can be drawn from these studies because of the important placebo effect, they are encouraging.

Two inactivated vaccines have shown some efficacy in controlled studies: one whole-cell inactivated vaccine, which is marketed in some European countries under the name of Lupidon (Weitgasser, 1977); and a purified envelope vaccine (Skinner vaccine) which showed limited efficacy (Benson et al., 1995). More recently, a recombinant subunit vaccine showed some efficacy (gD2-alum (Straus et al., 1994), but these results could not be confirmed with another adjuvant formulation (gB2-gD2-MF59) (Straus et al., 1997).

Strong evidence supporting the concept has come from studies in animal models, especially in guinea pigs, where a reduction of up to 94% in frequency and severity of recurrences could be induced with subunit glycoprotein vaccines (Stanberry et al., 1988, 1989; Myers et al., 1988). This efficacy was correlated mainly with the type of adjuvant used (Burke et al., 1994; Stanberry et al. 1989). Efficacy was also demonstrated in other animal models, including the rabbit corneal model (Nesburn et al., 1994, 1998) and the mouse zosteriform infection model (Rouse et al., 1985). Some investigators have found a correlation between the level of circulating α-interferon (Cunningham and Merigan, 1983; Green et al., 1985) or of interleukin 2 (Rouse et al., 1985; Ho et al., 1992; Bernstein et al., 1991) and the time to the next herpetic recurrence in humans. This and results in guinea pig studies indicate the need for an adequate cellular response for efficacy. Furthermore, the worsening of disease following immunosuppression also supports the role of cellular immunity in the control of recurrent HSV infections.

In conclusion, the development of therapeutic vaccines for the control of chronic HSV disease is confronted by major obstacles, the most important being the establishment of virus latency in neuronal cells (Lamontagne, 1991). Numerous attempts for more than 60 years have been minimally successful at best. The future may lie in more novel concepts based on molecular immunobiological and virological methods (Glorioso *et al.*, 1995; Isfort *et al.*, 1994).

ANIMAL MODELS

The most widely used and probably the most useful animal model for genital herpes is the guinea pig model. This has been extensively developed and characterized by Stanberry and colleagues (Stanberry *et al.*, 1982, 1985). It shares several features with human disease, such as an initial infection characterized by severe genital lesions and cutaneous, neurologic and urologic manifestations, and spontaneous recurrent episodes. Guinea pigs have been extensively used for the evaluation of antiviral drugs and both prophylactic and therapeutic HSV vaccines. It is also a good model to study the pathogenesis of the disease.

Mice and rabbits are used to evaluate the HSV ocular disease. Protection against ocular infection has been demonstrated by both periocular immunization (Nesburn *et al.*, 1994, 1998) and mucosal administration (Richards *et al.*, 1998).

THERAPEUTIC VACCINES

As mentioned earlier, a large number of different therapeutic vaccines have been evaluated in patients with recurrent HSV disease, starting in the 1920s. Most of them used active specific immunotherapeutic approaches. Recent reviews exist and describe in detail all these studies, up to 1996 (McKenzie and Straus, 1996a,b; Stanberry, 1990; Burke, 1993). Instead of repeating these reviews, all these results have been summarized in Tables 9.1 to 9.8 and will be commented as in the text.

Active Specific Immunotherapy

Whole-Cell Vaccines

Live Vaccines (Table 9.1)

For more than 40 years, from the 1920s until the 1960s, attempts were made to vaccinate recurrent herpes patients using non- or partially attenuated virus from the patients' own lesions, from heterologous lesions or from infected animal tissue. In the latter case, a typical example is a vaccine preparation from the brain of rabbits that died from herpetic encephalitis. Expecting local reactions similar to those caused by the smallpox vaccine, vesicles developing at the site of injection were seen as a positive 'take' of the vaccine, but they were most likely new herpetic lesions. Subjective improvements were repeatedly reported in these limited, non-controlled trials. This avenue has been abandoned because of the poor efficacy of this approach, the development of new lesions at the site of injection, and because we have developed a better understanding of the virus pathobiology.

■ **TABLE 9.1 Active specific immunotherapy: whole cell vaccines**

Category	Vaccine description	Clinical outcome	References
Whole-cell live vaccine	Virulent HSV from vesicular fluid from infected rabbit heterologous lesions autologous lesions	*Outcome* Development of lesions at the site of injection the site of pre-existing lesions Subjective improvement of the disease	Freund, 1928 Frank, 1938 Lazar, 1956 Blank and Haines, 1973
Whole-cell attenuated vaccine	Attenuated HSV from infected animal tissue: brain of rabbit dead from herpetic encephalitis infected guinea pig pad Attenuated by heat, phenol or formalin	*Non-controlled studies* Subjective improvement after variable number of injections	Biberstein and Jessner, 1935 Brain, 1936 Frank, 1938

Inactivated Whole-Cell Vaccines (Table 9.2)

Whole-cell vaccines preparations inactivated by chemical treatment or by heat have been explored since the 1960s. Two whole-cell vaccines have been evaluated in double-blind placebo-controlled studies. The first, prepared by Ely Lilly, provided striking results, with a 70% reduction in recurrent episodes in the vaccine group versus 76% in the placebo group (Kern and Schiff, 1964). This clearly demonstrated the importance of the placebo effect in this disease and the need for well controlled, blinded studies to evaluate any therapy of recurrent herpes disease. Lupidon H (for HSV1 virus) and Lupidon G (for HSV2 virus) are two vaccine preparations produced in Germany and marketed for more than 30 years in some European countries, and numerous open studies described positive results. Weitgasser conducted a double-blind placebo-controlled study in 1977 and reported an improvement in 82% of the patients in the vaccine group and in 30% of the subjects in the control group (Weitgasser, 1977). The heterogenicity of the study population, as well as weaknesses in the methodology of disease assessment, raise some questions as to the validity of the results.

In Bulgaria, Dundarov also reported positive results using a formalin-inactivated whole-cell vaccine (Dundarov et al., 1982), but unfortunately the results were generated in an open, non-controlled study.

Inactivated Purified Envelope Vaccines (Table 9.3)

Following the concern that HSV2 could be an oncogenic virus capable of causing cancer (Melnick et al., 1974), it has been a rule to avoid any trace of an HSV DNA in recently developed vaccines. Several purified envelope vaccines were developed, composed of envelope glycoproteins. These are major candidate antigens, as they are important targets for neutralizing antibodies (Norrild, 1980), they protect mice against lethal challenge (Laskey et al., 1984; Roberts

TABLE 9.2 Active specific immunotherapy: inactivated whole-cell virus vaccine

Vaccine company	Vaccine description	Population (n)	Design end points (EP)	Results	References
Eli Lilly	Formalin-inactivated vaccine Rabbit kidney cells	52 patients with recurrent HSV disease	Vaccine or placebo: at 2-week intervals, increasing doses (0.1–1 ml) at least 10 injections (SC) EP: no or fewer recurrences	Improvement in: Vaccine: 70% Placebo: 76%	Kern and Schiff, 1964 Hull and Peck, 1966
Lupidon®	Heat-killed virus H: HSV-1 G: HSV-2	*Open study* 1059 subjects	Improvement in severity or no. of recurrences	81% reported improvement	Altomare *et al.*, 1986 Schmersamz and Rudiger, 1975 Kitagawa, 1973
		Double-blind-randomized study 94 patients with HSV eruptions on various sites (genital, facial, both, gluteal, others)	Vaccines or placebo, given sequentially during 6 months: weekly for 3 months biweekly for 2 months monthly EP: prolongation of recurrence intervals decrease severity during the 6 months	Improvement in: Vaccine: 28/34 (82%) Placebo: 18/60 (30%)	Weitgasser, 1977
Dundarov vaccine	Formalin-inactivated vaccine Rabbit kidney culture 5 strains of HSV-1 and/or HSV-2	2350 patients with any HSV disease	Open, non-controlled trial Intracutaneous injection	65–80% reported improvements	Dundarov *et al.*, 1982

TABLE 9.3 Active specific immunotherapy: inactivated envelope vaccines

Type of Vaccine	Vaccine Description	Population (n)	Design end points	Results	Refs
Kutinova vaccine	HSV-1 lectin purified glycoproteins on Al(OH)3 Formalin inactivated	42 patients with at least 4 recurrences/year (62% had one recurrence/month) 19 facial HSV 23 genital HSV	Double-blind placebo-controlled baseline: at least 6 months prior entry 1 ml injected 3 times at 3 week intervals (im) 4th dose 6 months after first injection, only if no worsening of disease Evaluation at month 6 and 14	Weak antibody response Around 50% improvement, in both vaccine and placebo groups	Kutinova et al., 1988
Ac NFU₁ (S) MRC5 (Skinner vaccine)	HSV-1 mixed glycoproteins Formalin inactivated Detergent extraction	Patients with recurrent genital herpes Woodman: + untreated group Double-blind, randomized, controlled study:	Injection at 0, 1, 2 months subcutaneously Injection at 0, 1, 2 months. Evaluation at 1, 3 and 6 months post 3rd vaccine dose	'Improvement' in frequency, severity, extent of lesions. Significant reduction of mean n of lesions per recurrence mean n of reactions associated with recurrences frequency of recurrences at 3 and 6 months in women only Significant increase of proportion of subjects with time to 1st recurrence > 60 days. Time to first recurrences > 60 days Correlation between cellular immune response and modulation of disease pattern No decrease in overall frequency	Skinner et al., 1982, 1989 Woodman et al., 1983 Benson et al., 1995
Cappel	HSV-2 glycoproteins	Patients with genital HSV n = 44	Open, non-controlled study	Subjective improvements	Cappel et al., 1985
Frenkel (Lederle)	gD1 purified by immunoaffinity chromatography + Alum	HSV seropositive patients	Phase I: single dose 10 μg	Poorly immunogenic	Frenkel et al., 1990

et al., 1985), and they effectively decrease the frequency and severity of recurrent episodes in the guinea pig model (Stanberry *et al.*, 1987, 1988).

Three of these inactivated vaccines have been tested for the treatment of recurrent genital herpes. The Skinner vaccine is a mixture of HSV1 glycoproteins which has been repeatedly tested in open, non-controlled trials (Skinner *et al.*, 1989). Benson *et al.*, in 1995, reported the results of a placebo-controlled trial with a positive effect, mainly in the early period following vaccination. In the whole study population, there was no statistically significant reduction in the frequency of the recurrences. However, in women 1 month and 4 months after the last vaccinations, the reduction was significant. In addition, there was a significant reduction in the mean number of lesions per recurrence and in the mean number of reactions associated with recurrences. The proportion of vaccinated subjects with a time to first recurrence of larger than 60 days was significantly greater than in placebo recipients. In Belgium, Cappel reported in 1995 a subjective improvement in patients treated openly with an HSV2 glycoprotein preparation. Another purified HSV1 glycoprotein vaccine preparation was evaluated in Czechoslovakia in patients with genital herpes, but no difference could be detected (Kutinova *et al.*, 1988).

Recombinant Subunit Vaccines (Table 9.4 and 9.5)

In the 1980s, with the availability of genetic engineering for the production of antigens, several companies developed recombinant subunit vaccines (Laskey, 1990). These were based on glycoprotein D (gD), with or without glycoprotein B (gB), together with specific adjuvants. Indeed, the need for immunostimulants was clearly demonstrated in the guinea pig model (Burke *et al.*, 1994; Sanchez-Pescador *et al.*, 1988).

Chiron Biocine developed HSV2 subunit vaccines produced in Chinese hamster ovary (CHO) cells. These vaccines contained glycoprotein D, with or without glycoprotein B, and various types of adjuvants (Straus *et al.*, 1993). In an initial clinical study promising results were obtained with a vaccine containing 100 μg of gD with alum. The vaccine produced a significant decrease in the frequency of recurrences in patients with genital herpes (Straus *et al.*, 1994). The effect was moderate and transient, occurring mainly in the first months after vaccination. A second vaccine, composed of gD and gB, using MF59 (an oil-in-water emulsion) as adjuvant, was ineffective in humans (Straus *et al.*, 1997) and guinea pigs (Burke *et al.*, 1994).

The failure of this second vaccine may be due to the selection of MF59 as adjuvant. MF59 induces a Th2-type response (Singh, 1998) and, as such, is a very good antibody producer; however, neutralizing antibody titer was not correlated with clinical efficacy, neither in humans nor in the guinea pig model. On the contrary, infected subjects with severe disease and frequent recurrences have the highest antibody titers, and early natural history studies suggested a correlation with a higher Th1-type CMI immunity and a less severe disease (Cunningham and Merigan, 1983).

The positive results using 100 μg of gD with alum is interesting as again, this effect is limited to the early period after vaccination. This could mean that a therapeutic vaccine would need to be administered at regular intervals to be efficacious. The concept evolved from prophylactic vaccination – that a short

TABLE 9.4 Active specific immunotherapy: recombinant subunit vaccines

Company	Vaccine	Outcome	References
Chiron-Biocine	gD 100 µg + alum gBgD + MTP + PE + MF-59 gBgD + MF-59 expressed in CHO cells	*gD-alum:* *gDgB MTP-PE-MF-59* Guinea pigs: up to 80% reduction in frequency of disease Humans: toxic *gDgB MF-59* Guinea pig: no therapeutic efficacy Humans: see Table 9.5 for double-blind trials results	Stanberry *et al.*, 1988 Stanberry *et al.*, 1989 Burke, 1991 Straus *et al.*, 1993 Burke *et al.*, 1994
Smithkline Beecham Biologicals	Truncated gD2 (20 µg) produced on CHO + alum + monophosphoryl lipid A (MPL)	*Safety–immunogenicity in HSV seronegative and seropositive volunteers:* higher Ab, Nab, IL2, γIFN secretion with MPL than without	Leroux-Roels *et al.*, 1993 Leroux-Roels *et al.*, 1994
Genentech	Truncated gD1 produced on CHO cells	*Guinea pigs:* gD1 + CFA: complete protection gD1 + Alum: partial protection No therapeutic efficacy	Berman *et al.*, 1984 Berman *et al.*, 1988 Laskey, 1990
Lederle Praxis Biologicals	HSV-2 glycoprotein expressed on baculovirus + Al (PO4) gD related peptides	*In mice infected with HSV:* shift Th1 towards Th2 profile *Protection against footpad challenge*	York *et al.*, 1995 Eisenberg *et al.*, 1985
Takeda	Recombinant + tgD1-IL-2	*Guinea pigs:* 65% reduction in number of days with lesions	Nakao *et al.*, 1993

TABLE 9.5 Recombinant subunit vaccines/2: double-blind, placebo-controlled studies

Company	Vaccine description	Population (*n*)	Design/endpoints	Results	References
Chiron–Biocine	gD 100 µg on aluminium hydroxyde or Alum alone im injection (deltoid)	98 patients with genital HSV (4–14 recurrences/year) (49 in each group)	Injection at 0–2 months Follow-up for 12 months post 1st injection EP: Time to first recurrence Median *n* of recurrences *n* recurrences/month	Vaccine *vs* placebo: Significant (24%) decrease in cumulative *n* of recurrences Greater difference during first 4 months (34%) Higher rate of prodrome only recurrences	Straus *et al.*, 1994
Chiron–Biocine	gD 10 µg + gB 10 µg with MF-59 or MF-59 alone IM injection	202 patients with genital HSV (4–14 recurrences/year)	Injection at 0–2–12 months Follow-up for 18 months Primary EP: *n* of recurrences during the first 8 months	No significant decrease of recurrences frequency Significant improvement of severity/duration of first recurrences post vaccination	Straus *et al.*, 1997

administration course is sufficient for long-lasting efficacy – may be incorrect for HSV therapeutic vaccines.

SmithKline Beecham Biologicals is also developing a herpes vaccine based on a recombinant glycoprotein D for HSV2 produced on CHO cells, mainly for prophylactic use. A specific adjuvant, 3D-monophosphoryl lipid A with alum, is added to the antigen, with the objective of inducing a Th1-type response. Immunogenicity data in HSV-seropositive and HSV-seronegative subjects demonstrated that the formulation with MPL induced higher titers of neutralizing antibodies, but most importantly a stronger Th1 cellular response, as evidenced by high levels of IL-2 and γ-IFN, than the glycoproteins with alum alone (Leroux-Roels *et al.*, 1993, 1994).

Three other recombinant subunit vaccines have been tested in animal models. The results, summarized in Table 9.4, support the need for enhancement of specific Th1 responses and indicate that alum alone is not sufficient for therapeutic efficacy. Based on this concept, a fusion protein composed of gD and interleukin 2 was developed and gave positive results in guinea pigs (Nakao *et al.*, 1993).

Genetically Engineered HSV Mutant Vaccines

More recently, several groups have worked on a different approach using genetically engineered replication-impaired mutants of HSV. In principle such vaccines have the advantages of endogenous expression of viral proteins and therefore of a broad, multiclonal immune stimulation, and the safety of inactivated virus vaccines. The sequences potentially linked with oncogenicity can be deleted, in addition to the sequences needed for replication.

Two general forms of replication-impaired mutants are being developed: replication-defective mutants and single-cycle mutants. Replication-defective mutants are defective for a non-structural viral protein, such as infected cell protein 8 or 27 (ICP 8 or 27), and when a normal cell is infected by such a mutant incomplete viral replication occurs. This allows expression of many viral proteins, but no complete viral particle is produced. Single-cycle mutants are defective for a structural protein, such as glycoproteins H (gH), and when these viruses infect a normal cell, a single full replication cycle occurs, producing viral particles lacking gH that are non-infectious.

Several of these vaccines have demonstrated good efficacy in preventing infection, disease or latency in various animal models (Tables 9.6 and 9.7). Only one of them (DISC vaccine) has been tested therapeutically in the guinea pig model, but it had a limited effect on the rate of recurrences (McLean *et al.*, 1994).

This approach is probably more appealing as a prophylactic compound than for therapeutic vaccination. One interesting and surprising feature is the fact that a virus that can only infect a limited number of cells with no or only one replication cycle is in animals as immunogenic as a fully replicative virus (Knipe *et al.*, 1995). Understanding this factor as well as others would probably be very important for the development of HSV vaccines (Rollison, 1998).

Another vaccine mutant, R7020, which was a replication-competent intertypic mutant, has been developed by Pasteur Mérieux. This vaccine was evaluated in healthy HSV-seronegative and HSV-seropositive volunteers for

TABLE 9.6 Active specific immunotherapy: genetically engineered HSV mutants: replication-defective mutants

	Vaccine Description	Outcome	References
DISC Vaccine Replication-defective HSV-1 mutant	Deletion of non-structural protein ICP 4-8-27	Protection against encephalitis, keratitis, latent infection in the mouse corneal model	Nguyen *et al.*, 1992 Morisson and Knipe, 1994 Farrell *et al.*, 1994
		No therapeutic set-up	
Replication-defective HSV-2 mutant	5 Black Z virus: recombination of a mutant HSV-1 ICP8 gene into a HSV-2 wild-type strain	Reduction in genital lesions and virus shedding in guinea pigs in prophylactic set-up	Knipe *et al.*, 1995

TABLE 9.7 Active specific immunotherapy: genetically engineered HSV mutants – single-cycle mutants

Vaccines	Vaccine description	Clinical outcome	References
R7017–R7020	Deletion of sequences in HSV-1 to reduce virulence and neutrotropin Addition of gG and gD from HSV-2	*Animal:* Protection from challenge Reduced establishment of latency *Human:* HSV seronegative: safe & immunogenic HSV seropositive: local reactions, lymphadenopathy, fever, systemic reactions No increase in humoral response	Meignier *et al.*, 1987, 1988, 1990 Cadoz *et al.*, 1992
ICP34.5 deleted HSV	Deletion of neurovirulence gene ICP34.5		Whitley, 1993b
gH deleted mutant (SC16AgH) DISC Vaccine	HSV-1 mutant lacking glycoprotein H, necessary for virion entry into cells	*Guinea pigs:* Prevention of establishment of infection Therapeutically: limited effect on rate of recurrences	Forrester *et al.*, 1992 McLean *et al.*, 1994
RAV9395	Deletion of UL55, UL56 and γ1 34.5 genes Functions as a live attenuated HSV-2 vaccine	Decreased reactivation and establishment of latency in guinea pigs	Spector *et al.*, 1998

both safety and immunogenicity (Cadoz *et al.*, 1992). In HSV-seronegative subjects the vaccine was safe and immunogenic, but in HSV-seropositive subjects it induced systemic reactions that prevented continuation of the trial, and furthermore it did not increase the pre-existing antibody levels.

Viral Vector Vaccines

Several vaccines have been developed using viral vectors such as adenovirus or vaccinia (Table 9.8). They are usually immunogenic and protective against challenge in animal models. None of them has reached human testing, mainly for safety reasons, nor have they been evaluated therapeutically in animals.

DNA Vaccines

As opposed to other diseases, the development of HSV DNA vaccines has been surprisingly limited (Table 9.8). Plasmids encoding for gC, gD and gE from HSV1 have been evaluated in mice, but they were less effective than vaccinia recombinant vaccines (Nass *et al.*, 1998).

Active Non-Specific Immunotherapy

In the 1930s and up until 1960, several studies were conducted to evaluate the potential therapeutic effect of the smallpox vaccine (Wise and Sulzberger, 1934;

TABLE 9.8 Active specific immunotherapy: other vaccines

Approach	Description of vaccine	Outcome	References
Viral Vectors	Vaccinia: VgD52 (HSV-1)	*Mice:* Prevention of latency Protection against lethal challenge No therapeutic study	Cremer *et al.*, 1985 Rooney *et al.*, 1988 Rooney *et al.*, 1989
	VP176(gD1) VP221 (Gd2)	*Guinea pigs:* Protection against challenge, correlated with lymphoproliferation and with NK cytolysis enhancing cytokines	Wachsman *et al.*, 1997 Wachsman *et al.*, 1989
	Adenovirus: gB1 (AdgB8)	**Mucosal immunization:** *Mice:* Induction of CTLs Protection against challenge No therapeutic set-up	Gallichan, 1993 Gallichan and Rosenthal, 1998
DNA vaccine	Plasmids encoding for gC, gD, gE from HSV-1 (50µg)	*Lethal challenge in mice:* gD: 50 µg is protective gC-gE: 250 µg needed for protective antibody response IgG2 as isotype Weaker protection than vaccinia recombinant No therapeutic trial	Nass *et al.*, 1998

Foster and Abshier 1937). This was initially based on the belief that both vaccinia and herpes viruses were closely related, and that crossprotection could be established. Better knowledge of both viruses and the absence of efficacy have terminated this approach (Lyon, 1961; Schiff and Kern, 1954; Kern and Schiff, 1959). The tuberculosis vaccine (Bacillus Calmette–Guérin: BCG) was also evaluated as an immunostimulant (Douglas *et al.*, 1985). It was inefficient to improve the HSV disease and, when a second injection was given, persistent granulomas developed in some subjects.

In conclusion, non-specific immune stimulation has been totally ineffective in improving HSV disease.

Passive Specific Immunotherapy

Polyclonal and Monoclonal Antibodies

Several animals studies have shown prophylactic or therapeutic benefit of passive immunization with polyclonal or monoclonal antibodies. In 1940, Evans administered hyperimmune rabbit serum to mice which protected them against lethal HSV challenges (Evans *et al.*, 1946). Most studies in mice demonstrated that antibodies can be protective only if administered prior to and up to 48 hours after virus challenge (Cheever and Daikos, 1950; Luyet *et al.*, 1975). This suggests that antibodies are effective in mice before the establishment of latency, i.e. in the epithelium or even in the peripheral nerves (Simmons and Nasch, 1985). However, in the mouse zosteriform model antibodies are still protective when given more than 60 hours after challenge, i.e. when latency is established. This could suggest that antibodies may also act within infected ganglia (Oakes and Launch, 1984). In herpetic ocular infection the positive results were obtained in the corneal infectious mice model (Metcalf *et al.*, 1988), but no clinical study has been done in humans.

This approach has limited application for human use. The administration should happen rapidly after exposure, which is rarely a clearly identified event.

Transfer of Cellular Immunity

In various animal models the transfer of mononuclear cells, especially activated macrophages, increased survival after lethal challenge (Hirsch *et al.*, 1970). Most recent studies showed that this protection could be correlated with a CD4+ cellular stimulation (Kohl, 1990). This is in agreement with the observation of Cunningham (1983) that the time to next recurrence was correlated with the production of α-interferon. Again, experience with this approach remains limited to animal experimentation.

Passive Non-Specific Immunotherapy

Systemic interferon therapy was evaluated clinically in various situations, such as before and after trigeminal nerve root decompression (Pazin *et al.*, 1979), in renal transplant patients (Cheeseman *et al.*, 1979), or as suppressive therapy in patients with recurrent herpes (Kuhls *et al.*, 1986; Mendelson *et al.*, 1986). Interferon therapy had sometimes a moderate effect in shortening the duration of recurrences, but it did not decrease their number. Given the toxicity and cost of interferon compared to antivirals such as aciclovir, interferon therapy is unlikely to play a major therapeutic role in HSV disease (Levin, 1989). Interleukin 2 is another cytokine that could be beneficial in HSV disease, and in guinea pigs it reduces the frequency and severity of recurrences (Rouse et al., 1985; Ho et al., 1992; Weinberg et al., 1986). However, it has the same limitations of safety and costs as interferon. The current approaches are rather to use it as immunostimulant to active immunotherapy, such as the recombinant fusion protein of gD + IL-2 (Nakao et al., 1993) or co-expression of IL-2 and HSV proteins in recombinant vaccinia vectors (Allen et al. 1990). Other cytokines are being evaluated in animals, such as tumor necrosis factor-α (TNF-α), macrophage colony-stimulating factor (M-CSF) or R837, an immune modulator from 3M Riker (Harrison et al., 1991).

Imiquimod (Aldara – 3M Pharmaceuticals), a potent cytokine inducer, has been evaluated in guinea pigs in coadministration with purified glycoproteins (Bernstein et al., 1995), and this combination decreased the severity of the disease significantly more than the glycoprotein, either alone or with the complete Freund adjuvant (CFA).

CONCLUSIONS

The track record of the therapeutic vaccines for treatment of herpes simplex virus disease is very long and is rich in attempts using most possible vaccinal approaches. Unfortunately, until now, these efforts have not been or are only weakly successful. The main challenge faced by any therapeutic approach is the establishment of virus latency in the nucleus of neuronal cells, where it is unlikely that any type of immune response would have a definite curative effect. However, the feasibility of an HSV therapeutic vaccine is strongly supported by some clinical results and by a large body of data generated in animal models. These results suggest that an adequate immune stimulation could positively affect the clinical pattern of recurrent HSV disease. To reach this target, much remains to be understood on the biology of the virus and on the interactions between the virus and the immune system. Valuable information can be drawn from the studies conducted so far. It is almost certain that the immune stimulation must predominantly be of the Th1 type. This is sustained by animal experiments and by the failure of a vaccine using MF59, an adjuvant that induces a strong Th2-type response (Straus et al., 1997). From the antigen point of view, all studies that generated positive results used vaccines containing more than one antigen, from only two at high dose (gB + gD, by Straus et al., 1994) to most surface glycoproteins in purified envelope vaccines. Current approaches using genetically modified virus vaccines also induce immune responses to most viral epitopes. Therefore, a vaccine based on a single antigen will probably have less therapeutic potential. The design of administration is probably also important. Indeed, HSV infection is not fully curable because of the latency, and its chronic control will require chronic immune stimulation. Therefore, vaccine efficacy should perhaps be evaluated during chronic administration and not after a limited prophylactic type of administration. One study (Benson et al., 1995) has identified a gender effect, which is also clear in the natural history of infectivity and of the natural clinical course. Studies should take this factor into account as well as the major placebo effect. Finally, the clinical endpoints should also be reconsidered on a broader scale and not be limited to classic clinical endpoints developed for antiviral drugs. In conclusion, despite a long history, much remains to be learned and done for the development of HSV therapeutic vaccines.

ACKNOWLEDGEMENTS

My very personal thanks to my friend Larry Stanberry, who asked me to write this work and very critically reviewed it. This chapter is dedicated to my wife Sophie, in memory of our common work on the development of therapeutic HSV vaccines and its consequence.

REFERENCES

Allen, E.M., Weir, J.P., Martin, S., Mercadal, C. and Rouse B.T. (1990). Role of coexpression of IL-2 and herpes simplex virus proteins in recombinant vaccinia virus vectors on levels of induced immunity. *Virol. Immunol.* **3**: 207

Altomare, G.F., Pigatto, P.D., Polenghi, M.M. and Germogli, R. (1986). Relapsed herpes simplex specific immunotherapy with killed virus. *Acta Toxicol. Ther.* **713**: 201–210

American Social Health Association (1998). Sexually transmitted diseases in America: how many cases and at what cost? Publication #1445 prepared for the Kaiser Foundation Family, Research Triangle Park, NC 27709

Anker-Peterslund, N.A. (1991). Herpes virus infection: an overview of the clinical manifestations. *Scand. J. Infect.* Suppl. **78**, 15–20

Arvin, A.M. (1991). Relationship between maternal immunity to herpes simplex virus and the risk of neonatal herpesvirus infection. *Rev. Infect. Dis.* **13** (Suppl. 11): S 953–956

Benson, C.A., Turyk, M.E., Wilbanks, C.D. *et al.* (1995). A placebo-controlled trial of vaccination with a mixed glycoprotein herpes simplex virus type 1 vaccine for the modulation of recurrent genital herpes. IDSA, 33rd Annual meeting, San Francisco, LA. Abstract #418

Berman, P.W., Gregory, T., Crase, D. and Lasky, L.A. (1984). Protection from genital herpes simplex virus type 2 infection by vaccination with cloned type 1 glycoprotein D. *Science* **227**: 1490–1492

Berman, P.W., Vogt, P.E., Gregory, T., Lasky, L.A. and Kerm, E.R. (1988). Efficacy of recombinant glycoprotein D subunit vaccines on the development of primary, recurrent and latent genital infections with herpes simplex virus type 2 in guinea pigs. *J. Infect. Dis.* **157** (5): 897–902

Bernstein, D.I., Harrison, C.J., Tepe, E.R., Shahwan, A. and Miller, R.L. (1995). Effect of imiquimod as an adjuvant for immunotherapy of genital HSV in guinea-pigs. *Vaccine* **13** (1): 72–76

Bernstein, D.I., Harrison, C.J., Jenski, L.J., Myers, M.G. and Stanberry, L.R. (1991). Cell-mediated immunologic responses and recurrent genital herpes in the guinea pig: effects of glycoprotein immunotherapy. *J. Immunol.* **146** (10): 3571–3577

Biberstein, H. and Jessner, M. (1935). Untersuchungen zur herpesfrage (immunbiologische beziehungen zwischen herpes und vaccine, diagnostische und therapeutische verwendbarkeit eines herpes antigens (herpin)). *Arch. F. Dermat. und Syph.* **173**: 48

Blank, H. and Haines, H.G. (1973). Experimental human reinfection with herpes simplex virus, *J. Invest. Dermatol.* **61**: 223–225

Brain, R.T. (1936). Biological therapy in virus diseases. *Br. J. Dermatol.* **48**: 21–26

Burke, R.L., Goldbeck, C., Ng, P., Stanberry, L., Ott, G., Van Nest, G. (1994). The influence of adjuvant on the therapeutic efficacy of a recombinant genital herpes vaccine. *J. Infect. Dis.* **170**: 1110–1119

Burke, R.L. (1993). In: 'The human herpes viruses' (ed. B. Roizman *et al.*), Current status of HSV vaccine development, pp. 367–380. Raven Press, New York

Burke, R.L. (1991). Development of a herpes simplex virus subunit glycoprotein vaccine for prophylactic and therapeutic use. *Rev. Infect. Dis.* **13** (Suppl. 11): S906–911

Cadoz, M., Micoud, M., Seigneurin, J.M. *et al.* (1992). Phase I trial of R7020: a live attenuated recombinant herpes simplex virus (HSV) candidate vaccine. (Abstract #341), 32nd ICAAC, Anaheim, CA

Cappel, R., Sprecher, S., Decuyper, F. and DeBraekeleer, J. (1985). Clinical efficacy of a herpes simplex subunit vaccine. *J. Med. Virol.* **16**: 137–145

Cheeseman, S.H., Rubin R.H., Stewart, J.A. *et al.* (1979). Controlled clinical trial of prophylactic human-leukocyte interferon in renal transplantation: effects on cytomegalovirus and herpes simplex virus infections., *N. Engl. J. Med.* **300**: 1345–1349

Cheever, F.S. and Daikos, G. (1950). Studies on the protective effect of gamma globulin against herpes simplex infections in mice. *J. Immunol.* **65**: 135–141

Cremer, K.J., Mackett, M., Wohlenberg, C., Notkins, A.L. and Moss, B. (1985). Vaccinia virus recombinant expressing herpes simplex virus type 1 glycoprotein D prevents latent herpes in mice. *Science* **228**: 737–740

Cunningham, A.L. and Merigan, T.C. (1983). Interferon-α production appears to predict time of recurrence of herpes labialis. *J. Immunol.* **144**: 307–312

Dix, R.D. (1987). Prospect for a vaccine against herpes simplex virus types 1 and 2. *Prog. Med. Virol.* **34**: 89–128

Douglas, J.M., Vontver, A., Stamm, W.M. *et al.* (1985). Ineffectiveness and toxicity of BCG vaccine for the prevention of recurrent genital herpes. *Antimicrob. Agents Chemother.* **27**: 203–206

Dundarov, S., Andonov, P., Bakalov, B., Nechev, K. and Tomov, C. (1982). Immunotherapy with inactivated polyvalent herpes vaccines. *Devel. Biol. Standard.* **52**: 351–358

Eing, B.R., Kühn, J.E. and Braun, R.W. (1989). Neutralizing activity of antibodies against the major herpes simplex virus type 1 glycoproteins. *J. Med. Virol.* **27**: 59–65

Eisenberg, R.J., Cerini, C.P., Heilman, C.J. *et al.* (1985). Synthetic glycoprotein D-related peptides protect mice against herpes simplex virus challenge. *J. Virol.* **56** (3): 1014–1017

Evans, C.A., Slavin, H.B. and Berry, G.P. (1946). Studies on herpetic infections in mice. IV. The effect of specific antibodies on the progression of the virus within the nervous system of young mice. *J. Exp. Med.* **84**: 429–447

Farrell, H., McLean, C.S., Harley, C., Efstathiou, S., Inglis, S.C. and Minson, A.C. (1994). Vaccine potential of a herpes simplex virus type 1 mutant with an essential glycoprotein deleted. *J. Virol.* **68**: 927–932

Forrester, A., Farrell, H., Wilkinson, G., Kaye, J., Davis-Poynter, N. and Minson, T. (1992). Construction and properties of a mutant herpes simplex virus type 1 with glycoprotein gH coding sequences deleted. *J. Virol.* **66** (1): 341–348

Foster, P.D. and Abshier, A.B. (1937). Smallpox vaccine in treatment of recurrent herpes. *Arch. Dermatol. Syphilis.* **36**: 294

Frank, S.B. (1938). Formalized herpes virus therapy and the neutralizing substance in herpes simplex. *J. Invest. Dermatol.* 267–282

Frenkel, L.M., Dillon, M., Garraty, E. *et al.* (1990). A randomized double blind, placebo-controlled phase I trial of a herpes simplex virus purified glycoprotein (gD1) vaccine. (Abstract 721, p. 206), 30th Interscience Conference on Antimicrobial Agents and Chemotherapy, Atlanta, GA

Freund, H. (1928). Beitrage zur kenntnis von herpes und zoster. *Arch. Dermatol. Syph.* **155**: 282

Gallichan, W.S. and Rosenthal, K.L. (1998). Long-term immunity and protection against herpes simplex virus type 2 in the murine female genital tract after mucosal but not systematic immunization. *J. Infect. Dis.* **177** (5): 1155–1161

Gallichan, W.S., Johnson, D.C., Graham, F.L. and Rosenthal, K.L. (1993). Mucosal immunity and protection after intranasal immunization with recombinant adenovirus expressing herpes simplex virus glycoprotein B. *J. Infect. Dis.* **168**: 622–629

Glorioso, J.C., DeLuca, N., Bender, M.A. and Fink, D.J. (1995). Gene therapy applications using herpes simplex virus vectors. *J. Cell Biochem. Suppl.* **21a**: 359. Abstract # C6–016

Green, J.A., Weiss, P.N., Yeh, T.J., Spruance, S.L. (1985). Immunospecific interferon production by peripheral blood mononuclear leukocytes from patients with primary and recurrent orolabiol herpes simplex virus infections. *J. Med. Virol.* **16**: 297–305

Harrison, C.J., Stanberry, L.R. and Bernstein, D.I. (1991). Effect of cytokines and R-837, a cytokine inducer, on UV-irradiation augmented recurrent herpes simplex in guinea pigs. *Antiviral Res* **15**: 315–322

Heng, M.C.Y., Heng, S.Y. and Allen, S.G. (1994). Co-infection and synergy of human immunodeficiency virus-1 and herpes simplex virus-1. *Lancet* **343**: 255–258

Hirsch, M.S., Zisman, B. and Allison, A.C. (1970). Macrophages and age-dependent resistance to herpes simplex virus in mice. *J. Immunol.* **104**: 1160–1165

Ho, R.J.Y., Burke, R.L. and Merigan, T.C. (1992). Liposome-formulated interleukin-2 as an adjuvant of recombinant HSV glycoprotein D for the treatment of recurrent genital HSV-2 in guinea pigs. *Vaccine* 10: 209–213

Ho, R.J.Y., Burke, R.L. and Merigan, T.C. (1989). Antigen-presenting liposomes are effective in treatment of recurrent herpes simplex virus genitalis in guinea pigs. *J. Virol.* 63 (7): 2951–2958

Hook, E.W., Cannon, R.O., Nahmias, A.J. *et al.* (1992). Herpes simplex virus infection as a risk factor for human immunodeficiency virus infection in heterosexuals. *J. Infect. Dis.* 165: 251–255

Hull, R.N. and Peck, F.B. (1966). Vaccination against herpesvirus infections. *Pan. Am. Health Org. Sci. Publ.* 147:266–275

Isfort, R.J., Witter, R. and Kung, H.J. (1994). Retrovirus insertion into herpesviruses. *Trends Microbiol.* 2(5): 174–177

Jennings, R. and Green Kinghorn, G.R. (1998). Herpes virus vaccines: an update. *Biodrugs* 10 (4): 257–264

Kern, A.B. and Schiff, B.L. (1964). Vaccine therapy in recurrent herpes simplex. *Arch. Dermatol.* 89: 844–845

Kern, A.B. and Schiff, B.L. (1959). Smallpox vaccinations in the management of recurrent herpes simplex: a controlled evaluation. *J. Invest. Dermatol.* 33: 99

Kinghorn, G.R. (1993). Genital herpes: Natural history and treatment of acute episodes. *J. Med. Virol.* 1(suppl.): 33–38

Kitagawa, K. (1973). Therapy of herpes simplex with heat inactivated herpes virus hominis type 1 and type 2. *Z. Hautkr.* 48 (13): 533–535

Knipe, D.M., Da Costa, X., Morrison, L.A., Bourne, N. and Stanberry, L.R. (1995). Immunization against genital herpes disease with a replication-defective mutant of HSV-2. In Vaccines 95: *Molecular Approaches to the Control of Infectious Diseases* pp. 369–373. Cold Spring Harbor Laboratory Press

Kohl, S. (1991). Role of antibody-dependent cellular cytotoxicity in defense against herpes simplex virus infections. *Rev. Infect. Dis.* 13: 108–114

Kohl, S. (1990). Protection against murine neonatal herpes simplex virus infection by lymphokine-treated human leukocytes. *J. Immunol;* 144: 307–312

Koutsky, L.A., Stevens, C.E., Holmes, K.K. *et al.* (1992). Underdiagnosis of genital herpes by current clinical and viral-isolation procedures. *N. Engl. J. Med.* 326: 1533–1539

Koutsky, L.A., Ashley, R.L., Holmes, K.K. *et al.* (1990). The frequency of unrecognized type 2 herpes simplex virus infection among women. *Sex. Transm. Dis.* 17 (2): 90–94

Kuhls, T.L., Sacher, J., Pineda, E. *et al.* (1986). Suppression of recurrent genital herpes simplex virus infection with recombinant alpha-2 interferon. *J. Infect. Dis.* 154: 437–442

Kutinova, L., Benda, R., Kalos, Z. *et al.* (1988). Placebo-controlled study with subunit herpes simplex virus vaccine in subjects suffering from frequent herpetic recurrences. *Vaccine* 6: 223–228

Lamontagne, J. (1991). General discussion and future directions. *Rev. Infect. Dis.* 13 (51): S 974–S 979

Laskey, L.A. (1990). From virus to vaccine: recombinant mammalian cell lines as substrates for the production of herpes simplex virus vaccines. *J. Med. Virol.* 31: 59–61

Laskey, L.A., Dowbenko, D., Simonsen, C. and Berman, P.W. (1984). Protection of mice from lethal herpes simplex virus infection by vaccination with a secreted form of cloned glycoprotein D. *Biotechnology* 2: 527

Lazar, M.P. (1956). Vaccination for recurrent herpes simplex infection: initiation of a new disease site following the use of unmodified material containing the live virus. *Arch. Dermatol.* 73: 70–71

Leroux-Roels, G., Moreau, E., Desombere, I. *et al.* (1994). Persistence of humoral and cellular immune response and booster effect following vaccination with herpes simplex (gD2t) candidate vaccine with MPL. (Abstract # H57), 34th ICAAC, Orlando, FL

Leroux-Roels, G., Moreau, E., Verhasselt, B. *et al.* (1993). Immunogenicity and reactogenicity of a recombinant HSV-2 glycoprotein D vaccine with or without mono-phosphoryl lipid A in HSV seronegative and seropositive subjects. (Abstract #1209), 33rd ICAAC, New Orleans, LA

Levin, M.J., Judson, F.N., Eron, L. *et al.* (1989). Comparison of recombinant leukocyte interferon-alpha (rIFN-2A) with topical aciclovir for the treatment of first episode herpes genitalis and prevention of subsequent recurrences. *Antimicrob. Agents Chemother.* **33**: 649–652

Luyet, F., Samra, D., Soneji, A. and Marks, M.I. (1975). Passive immunization in experimental herpesvirus hominis infection of newborn mice. *Infect. Immun.* **12**: 1258

Lyon, E. (1961). The problem of treatment of recurrent herpes simplex by antismallpox vaccination. *Israel Med. J.* **20** (3–4): 103–108

Manickan, E., Rouse, R.J.D., Yu, Z., Wire, W.S. and Rouse, B.T. (1995). Genetic immunization against herpes simplex virus. *J. Immunol.* **155**: 259–265

McKenzie, R. and Straus, S.E. (1996a). Vaccine therapy for herpes simplex virus infections: an historical perspective. *Rev. Med. Virol.* **6**: 85–96

McKenzie, R. and Straus, S.E. (1996b). Therapeutic immunization for recurrent herpes simplex virus infections. *Adv. Exp. Med. Biol. (United States)* **394**: 67–83

McLean, C.S., Ertirk, M., Jennings, R. *et al.* (1994). Protective vaccination against primary and recurrent disease caused by herpes simplex virus (HSV) type 2 using a genetically disabled HSV-1. *J. Infect. Dis.* **170**: 1100–1109

Meignier, B., Martin, B., Whitley, R.J. and Roizman B. (1990). *In vivo* behavior of genetically engineered herpes simplex viruses R7017 and R7020. II. Studies in immunocompetent and immunosuppressed owl monkeys (*Aotus trivirgatus*). *J. Infect. Dis.* **162**: 313–321

Meignier, B., Longnecker, R. and Roizman, B. (1988). *In vivo* behavior of genetically engineered herpes simplex viruses R7017 and R7020: construction and evaluation in rodents, *J. Infect. Dis.* **158**: 602–614

Meignier, B., Longnecker, R. and Roizman, B. (1987). In: 'Vaccines 87' (ed. Chanock R.M.), Construction and in vivo evaluation of two genetically engineered prototypes of live attenuated herpes simplex virus vaccines, pp 368–373. Cold Spring Harbor Laboratory, Cold Spring Harbor, NY

Melnick, J.L., Adam, E. and Rawls, W.E. (1974). The causative role of herpes virus type 2 in cervical cancer. *Cancer* **34**: 1375–1385

Mendelson, J., Clecner, B. and Eiley, S. (1986). Effect of recombinant interferon alpha 2 on clinical course of first episode genital herpes infection and subsequent recurrences. *Genitourinary Med.* **62**: 97–101

Metcalf, J.F., Chatterjee, S., Koga, J. and Whitley, R.J. (1988). Protection against herpetic ocular disease by immunotherapy with monoclonal antibodies to herpes simplex virus glycoproteins. *Intervirology* **29**: 39–49

Mindel, A. (1993). Long-term clinical and psychological management of genital herpes. *J. Med. Virol.* **s1**: 39–44

Morisson, L.A. and Knipe, D.M. (1994). Immunization with replication-defective mutants of herpes simplex virus type 1: sites of immune intervention in pathogenesis of challenge virus infection. *J. Virol.* **68** (2): 689–696

Myers, M.G., Bernstein, D.I., Harrisson C.J. and Stanberry L.R. (1988). Herpes simplex virus glycoprotein treatment of recurrent genital herpes reduces cervicovaginal virus shedding in guinea pigs. *Antiviral Res.* **10**: 83–88

Nahmias, A.J., Lee, F.K. and Beckman-Nahmias, S. (1990). Sero-epidemiological and -sociological patterns of herpes simplex virus infection in the world, *Scand. J. Infect. Dis. Suppl.* **69**: 19–36

Nakao, M., Hazama, M., Mayumi-Aono, A., Hinuma, S. and Fujisawa, Y. (1993). Immunotherapy of acute and recurrent herpes simplex virus type 2 infection with an adjuvant-free form of recombinant glycoprotein D-interleukin-2 fusion protein. *J. Infect. Dis.* **169**: 787–791

Nass, P.H., Elkins, K.L. and Weir, J.P. (1998). Antibody response and protective capacity of plasmid vaccines expressing three different herpes simplex virus glycoproteins. *J. Infect. Dis.* **178**: 611–617

Nesburn, A.B., Burke, R.L., Ghiasi, H., Slanina, S.M. and Wechsler, S.L. (1998). A therapeutic vaccine that reduces recurrent Herpes simplex Type 1 corneal disease. *Invest. Ophthalmol. Vis. Sci.* **39** (7): 1163–1170

Nesburn, A.B., Burke, R.L., Ghiasi, H., Slanina, S., Bahri, S., Wechsler, S.L. (1994). Vaccine therapy for ocular herpes simplex virus (HSV) infection: periocular vaccination reduces spontaneous ocular HSV type 1 shedding in latently infected rabbits. *J. Virol.* **68** (8): 5084–5092

Nguyen, L.H., Knipe, D.M. and Finberg, R.W. (1992). Replication-defective mutants of herpes simplex virus (HSV) induce cellular immunity and protect against lethal HSV infection. *J. Virol.* **66**: 7067–7072

Norrild, B. (1980). Immunochemistry of herpes simplex virus glycoproteins. *Curr. Topics Microbiol. Immunol.* **90**: 67–106

Oakes, J.E. and Launch, R.W. (1984). Monoclonal antibodies suppress replication of herpes simplex virus type 1 in trigeminal ganglia. *J. Virol.* **51**: 656

Pazin, G.J., Armstrong, J.A., Lam, M.T., Tarr, G.C., Janetta, P.J. and Ho, M. (1979). Prevention of reactivated herpes simplex infection by human leukocyte interferon after operation on the trigeminal root. *N. Engl. J. Med.* **301**: 225–230

Reeves, W.C., Corey, L., Adams, H.G., Vontver, L.A. and Holmes, K.K. (1981). Risk of recurrence after first episode of genital herpes. Relation to HSV type and antibody response. *N. Engl. J. Med.* **305**: 315–319

Richards, C.M., Shimeld, C., Williams, N.A. and Hill, T.J. (1998). Induction of mucosal immunity against herpes simplex type 1 in the mouse protects against occular infection and establishment of latency. *J. Infect. Dis.* **177**: 1451–1457

Roberts, P.L., Duncan, B.E., Raybould, T.J.G. and Watson, D.H. (1985). Purification of herpes simplex virus glycoproteins B and C using monoclonal antibodies and their ability to protect mice against lethal challenge. *J. Gen. Virol.* **66**: 1073–1085

Rollinson, E. (1998) Therapeutic vaccination: a novel approach to the management of genital herpes. *Int. Antiviral News* **6**: 125–127

Rooney, J.F., Wohlenberg, C., Cremer, K.J. and Notkins, A.L. (1989). Immunized mice challenged with herpes simplex virus by the intranasal route show protection against latent infection. *J. Infect. Dis.* **159**: 974–976

Rooney, J.F., Wohlenberg, C., Cremer, K.J., Moss, B. and Notkins, A.L. (1988). Immunization with a vaccinia virus recombinant expressing herpes simplex virus type 1 glycoprotein D: long-term protection and effect of revaccination. *J. Virol.* **62**: 1530–1534

Rouse, B.T., Miller, L.S., Turtinen, L. and Moore, R.N. (1985). Augmentation of immunity to herpes simplex virus by in vivo administration of interleukin 2. *J. Immunol.* **134**: 926–930

Sanchez-Pescador, L., Burke, R.L., Ou, G. and Van Nest, G. (1988). The effect of adjuvants on the efficacy of a recombinant herpes simplex virus glycoprotein vaccine. *J. Immunol.* **141**: 1720–1727

Schiff, B.L. and Kern, A.B. (1954). Multiple smallpox vaccinations in the treatment of recurrent herpes simplex. *Postgrad. Med.* **15**: 32–36

Schmersamz, P. and Rudiger, G. (1975). Behandlungshergebnisse mit dem Herpes simplex-Antigen Lupidon H. bzw Lupidon G, Z. Hautartzt **50**: 105

Scott Schmid, D. and Mawle, A.C. (1991). T cell responses to herpes simplex viruses in humans. *Rev. Infect. Dis.* **13** (11): S946–949

Simmons, A. and Nasch, A.A. (1985). Role of antibody in primary and recurrent herpes simplex virus infection. *J. Virol.* **53**: 944–948

Sin, J-I., Kim, J.J., Boyer, J.D., Ciccarelli, R.B., Higgins, T.J. and Weiner, D.B. (1999). In vivo modulation of vaccine-induced immune responses toward a Th1 phenotype increases potency and vaccine effectiveness in a herpes simplex virus type 2 mouse model. *J. Virol.* **73**(1): 501–509

Singh, M., Carlson, J.R., Briones, M. *et al.* (1998). A comparison of biodegradable microparticles and MF59 as systemic adjuvants for recombinant gD from HSV-2. *Vaccine* **16** (19): 1822–1827

Skinner, G.R.B., Fink, C.G., Durham, J., Hartley, C.E., Hallworth, J. and Buchan, A. (1989). In: 'Vaccines for Sexually Transmitted Diseases' (eds, Meheus, A., Spier, R.E.). Follow-up report on 101 subjects vaccinated with Skinner Herpes vaccine, pp. 202–207. Butterworths, London

Skinner, G.R.B., Woodman, C., Hartley, C. *et al.* (1982). Early experience with 'antigenoid' vaccine AcNFU1 (S-) MRC towards prevention or modification of herpes genitalis. *Devel. Biol. Standard.* **52**: 333–344

Spear, P.G. (1984). Glycoprotein specified by herpes simplex viruses. In: Roizman B., ed. *Herpesviruses.* New York: Plenum Press

Spector, F.C., Kern, E.R., Palmer, J. *et al.* (1998). Evaluation of a live attenuated recombinant virus RAV 9395 as a herpes simplex virus type 2 vaccine in guinea pigs. *J. Infect. Dis.* **177** (5): 1143–1154

Stanberry, L.R. (1996). Herpes immunization – on the threshold. *Review. J. Eur. Acad. Dermatol. Venereol.* **7**: 120–128

Stanberry, L.R. (1995). Herpes simplex virus vaccines as immunotherapeutic agents. *Trends Microbiol.* **3** (6): 244–247

Stanberry, L.R. (1993). Genital and neonatal herpes simplex virus infections: epidemiology, pathogenesis and prospects for control. *Rev. Med. Virol.* **3**: 37–46

Stanberry, L.R. (1990). In: 'Herpesviruses, the Immune System and AIDS' (ed. Aurelian), Subunit viral vaccines: prophylactic and therapeutic use, pp. 309–341. Kluwer Academic Press, Norwell, MA

Stanberry, L.R., Harrisson, C.J., Bernstein, D.I. *et al.* (1989). Herpes simplex virus glycoprotein immunotherapy of recurrent genital herpes: factors influencing efficacy. *Antiviral Res.* **11**: 203–214

Stanberry, L.R., Burke, R.L. and Myers, M.G. (1988). Herpes simplex virus glycoprotein treatment of recurrent genital herpes. *J. Infect. Dis.* **157** (1): 156–163

Stanberry, L.R., Bernstein, D.I., Burke, R.L., Pachl, C., Myers, M.G. (1987). Vaccination with recombinant herpes simplex virus glycoproteins: protection against initial and recurrent genital herpes. *J. Infect. Dis.* **155** (5): 914–920

Stanberry, L.R., Kern, E.R., Richards, J.T. and Overall, J.C.J.R. (1985). Recurrent genital herpes simplex virus infection in guinea pigs. *Intervirology* **24**: 226–231

Stanberry, L.R., Kern, E.R., Richards, J.T., Abbott, T.M. and Overall, J.C.J.R. (1982). Genital herpes in guinea pigs: pathogenesis of the primary infection and description of recurrent disease. *J. Infect. Dis.* **146**: 396–404

Straus, S.E., Wald, A., Kost, R.G. *et al.* (1997). Immunotherapy of recurrent genital Herpes with recombinant herpes simplex virus type 2 glycoproteins D and B: results of a placebo-controlled vaccine trial. *J. Infect. Dis.* **176**: 1129–1134

Straus, S.E., Corey, L., Burke, R.L. *et al.* (1994). Placebo-controlled trial of vaccination with recombinant glycoprotein D of herpes simplex virus type 2 for immunotherapy of genital Herpes. *Lancet* **343**: 1460–1463

Straus, S.E., Savarese, B., Tigges, M. *et al.* (1993). Induction and enhancement of immune responses to herpes simplex virus type 2 in humans by use of a recombinant glycoprotein D vaccine. *J. Infect. Dis.* **167**: 1045–1052

Wachsman, M., Aurelian, L., Smith, C.C., Perkus, M.E. and Paoletti, E. (1989). Regulation of expression of herpes simplex virus (HSV) glycoprotein D in vaccinia recombinants affects their ability to protect from cutaneous HSV-2 disease. *J. Infect. Dis.* **159**: 625–634

Wachsman, M., Aurelian, L., Smith, C.C., Lipinskas, B.R., Perkus, M.E. and Paoletti E. (1987). Protection of guinea pigs from primary and recurrent herpes simplex virus (HSV) type 2 cutaneous disease with vaccinia virus recombinants expressing HSV glycoprotein D. *J. Infect. Dis.* **155**: 1188–1197

Weinberg, A., Rasmussen, L. and Merigan, T. (1986). Acute genital infection in guinea pigs: effect of recombinant interleukin-2 on herpes simplex virus type 2. *J. Infect. Dis.* **154**: 134

Weitgasser, H. (1977). Kontrollierte klinische studie mit den Herpes Antigenen Lupidon®H und Lupidon®G.Z. Hautkr. **52** (11): 625–628

Whitley, R.J. (1993a). Prospects for vaccination against herpes simplex virus. *Pediatr. Ann.* **22**: 726–732

Whitley, R.J., Kern, E.R., Chatterjee, S., Chou, J., Roizman, B. (1993b). Replication, establishment of latency, and induced reactivation of herpes simplex virus γ_1 34.5 deletion mutants in rodent models. *J. Clin. Invest.* **91**: 2837–2843

Whitley, R.J. (1993c). Neonatal herpes simplex virus infections. *J. Med. Virol.* suppl. 1: 13–21

Wise, F. and Sulzberger, M.B. (1934). The 1934 year book of Dermatology and Syphilology, 426. The Year Book Publisher, Inc., Chicago

Woodman, C.B.J., Buchan, A., Fuller, A. *et al.* (1983). Efficacy of vaccine Ac Nfu1, (S-) MRC given after an initial clinical episode in the prevention of herpes genitalis. *Br. J. Vener. Dis.* **59**: 311–313

York, L.J., Giorgio, D.P. and Mishkin, E.M. (1995). Immunomodulatory effects of HSV2 glyco-protein D in HSV1 infected mice: implications for immunotherapy of recurrent HSV infection. *Vaccine* **13** (17): 1706–1712

10
CYTOMEGALOVIRUS

JOHN D. SHANLEY

Division of Infectious Diseases, Department of Medicine
University of Connecticut Health Center
Farmington, Connecticut, USA

CLINICAL MANIFESTATIONS

Infections due to cytomegalovirus (CMV) were initially recognized in the early part of the 20th century in the form of the devastating congenital condition termed congenital inclusion disease (CID). As infants affected by CID were rare, it was originally felt that infections due to CMV were uncommon. However, with the isolation of CMV in 1956 and 1957 (Rowe *et al.*, 1956; Smith, 1956; Weller *et al.*, 1957), our understanding of the extent to which this agent affects humanity changed radically (Weller, 1971). With the availability of appropriate diagnostic reagents to define the epidemiology of CMV, it rapidly became apparent that infections due to this agent were extremely common worldwide, approaching 100% in developing countries (Gold and Nankervis, 1989). Moreover, the vast majority were not clinically apparent. Clinical disease due to CMV has been found to occur in certain groups of patients, either congenitally or as infections in patients with abnormal cell-mediated immunity (CMI) (Griffiths and Emery, 1997) (Table 10.1). Infections in the normal host are generally asymptomatic, and when disease occurs it usually takes the form of a mononucleosis-like syndrome.

CMV remains an important cause of congenital disease and is the leading infectious cause of congenital abnormalities in the USA (Alford and Britt, 1993; Gehrz, 1991), where 0.3–2.2% of liveborne infants are infected with CMV. Of these, approximately 10% are severely affected with CID. This is characterized by hepatosplenomegaly, jaundice, 'blueberry muffin' rash and severe neurological damage, such as cerebral calcifications, microcephaly and anacephaly. The laboratory abnormalities in CID include thrombocytopenia, hemolytic anemia, atypical lymphocytosis, and abnormal liver function tests. Of infants

■ **TABLE 10.1 Clinical manifestations of human cytomegalovirus infection**

Congenital Infection
  Congenital inclusion disease (CID)
  Asymptomatic congenital infection

Neonatal infection
  Asymptomatic infection
  CMV of the neonate (pneumonitis, hepatitis)

Immunocompetent adults
  Asymptomatic
  Mononucleosis
  Fever of unknown etiology
  Miscellaneous

Immunocompromised
  Asymptomatic
  Fever, hepatitis
  Pneumonitis
  Retinitis
  Gastrointestinal (colitis, oral ulcers, esophagitis, gastritis)
  Adrenalitis
  Encephalitis

---

affected by CID, 4–37% die and the majority who survive have severe, long-term neurological sequelae. The remaining congenital infections are frequently asymptomatic at birth, but often lead to debilitating neurological abnormalities (Stagno *et al.*, 1977)

Individuals with abnormal cellular immune defenses are the other major group at risk for serious disease due to CMV (Gehrz, 1991; Zaia, 1994). In neonates, the majority of HCMV infections acquired perinatally are asymptomatic, although hepatitis and atypical lymphocytosis occur and up to 20% of infected infants develop radiological evidence of pneumonitis. Serious CMV disease occurs primarily in low-birthweight infants who acquire infection at birth (Yeager, 1974). Originally, CMV was often acquired via the administration of blood products (Yeager *et al.*, 1981). In this setting, acquisition of infection in the infants of CMV-naïve mothers was associated with disease characterized by hepatitis, pneumonitis and death. Passive maternal antibody from CMV-seropositive mothers appeared to be protective in this setting (Yeager *et al.*, 1981).

With the advent of solid organ transplantation and allogeneic bone marrow transplantation, new populations of patients affected by severe CMV infections were recognized (Apperly and Goldman, 1988; Betts and Hanshaw, 1977; Zaia, 1994). The predisposing factors in these populations appear to be the severity of suppression of cellular immunity and the presence of graft-vs-host disease. The nature of the disease in these patients varies with the nature of the transplant type (van der Meer *et al.*, 1996). In solid organ transplants, such as kidney and heart, these manifestations of CMV infection range from asymptomatic viral shedding to severe multiorgan failure (Betts 1982; Snydman *et al.*, 1993; Weisner *et al.*, 1993). Mild disease includes fever and hepatitis.

In lung and allogeneic bone marrow transplants CMV is often associated with fatal interstitial pneumonitis (Zaia, 1994). In allogeneic bone marrow transplants, CMV infection occurs in 30–70% of patients at risk (Neiman *et al.*, 1977; Winston *et al.*, 1979). Of patients with CMV infection, 10–30% develop pneumonitis, with a mortality rate of >80% (Zaia, 1986). The pathogenetic mechanisms leading to CMV pneumonitis remain obscure, but appear to be immunopathogenic in nature (Grundy *et al.*, 1987; Zaia, 1994). Despite advances in antiviral treatment and prophylaxis, CMV remains a serious source of diseases for these patients (van der Meer *et al.*, 1996).

With the advent of the HIV/AIDS epidemic, CMV was rapidly recognized as a common and serious opportunistic infection in these patients (Jacobson and Mills, 1988; Smith and Brennessel, 1994). In the setting of waning cellular immune function, reactivation of latent infection is a common event in patients with serological evidence of prior CMV infection. The majority of CMV infections are initially asymptomatic, and CMV can be recovered from multiple tissues in the absence of disease (Drew, 1988). In the setting of advanced HIV, CMV infection has been the cause of damage to a number of organ systems (Smith and Brennessel, 1994). Retinitis is the most common manifestation of CMV disease, leading to destruction of the retinal epithelium with necrosis and blindness (Bowen *et al.*, 1995). The gastrointestinal and hepatobiliary tracts are also affected by CMV, with a variety of manifestations, including oral ulcers, esophagitis, colitis and acalculous cholecystitis (Goodgame, 1993). CMV is also recognized in the central and peripheral nervous systems, associated with glial nodules, encephalitis and peripheral neuropathies (Kalayjian *et al.*, 1993; Katlama, 1993; Morgello *et al.*, 1987; Vinters *et al.*, 1989). CMV appears to induce disease in patients with AIDS by direct effects of virus replication in tissues, and many diseases are arrested by antiviral treatment.

EPIDEMIOLOGY/EXPECTED IMPACT OF AN EFFECTIVE VACCINE

The extent to which CMV affects humans was not appreciated until the availability of reagents to define its epidemiology (Weller, 1971). Subsequent studies following its isolation have demonstrated that CMV infection is extremely common worldwide (Krech, 1973). Humans are the only known reservoir for CMV, and transmission of infection requires intimate person-to-person contact (Gold and Nankervis, 1989). As with other herpes viruses, CMV will persist in the host after acute infection in a state of latency or chronic persistent viral shedding (Jordan, 1983). In previously infected patients CMV can be isolated from many different tissue sites (Table 10.2), leading to transmission by a variety of mechanisms (Onorato *et al.*, 1985). CMV has been isolated from oropharyngeal secretions, breast milk, semen and cervical secretions. CMV transmission has also been associated with the administration of blood products, most often associated with leukocytes (Winston *et al.*, 1980). Studies have shown that CMV can be transferred by solid organ grafts and bone marrow transplantation, indicating the presence of virus in these tissues (Apperly and Goldman, 1988; Ho *et al.*, 1975; Zaia, 1994). Thus, CMV utilizes multiple avenues of transmission to survive in nature.

▮ TABLE 10.2 Routes of HCMV transmission

Oral droplet

Cervical exposure

Breast milk

Intrauterine transmission

Venereal transmission

Blood and leukocyte transfusion

Organ graft transfer

Vertical transmission occurs in 30–40% of women with primary CMV infection in pregnancy, and occasionally in seropositive women who reactivate their latent virus (van der Meer *et al.*, 1996).

Horizontal transmission occurs by several mechanisms. At the time of birth, neonates can be infected by exposure to CMV in cervical secretions in the birth canal (Reynolds *et al.*, 1973). Cervical shedding of CMV has been shown to increase in frequency as pregnancy progresses (Numazaki *et al.*, 1970; Stagno *et al.*, 1975). Neonates can also be infected by breast milk or by postnatal administration of blood products (Stagno *et al.*, 1980; Yeager *et al.*, 1981). Oropharyngeal droplet infection appears to be important in settings such as daycare centers, or in families with CMV-infected infants (Pass *et al.*, 1986), After puberty, sexual transmission begins to play an increasingly important role in the transmission of CMV (Ho, 1992b). This route is extremely important in certain groups such as homosexual males (Smith and Brennessel, 1994). Studies have shown that the acquisition of CMV infection in homosexual males is more common than in the general population, approaching 80% in some settings (Drew, 1988). Moreover, infections with multiple strains of CMV have been documented (Drew *et al.*, 1984; Chandler *et al.*, 1987; Chou, 1989).

Infection via the administration of blood products was first recognized in the 1960s as the postperfusion syndrome (Lang and Hanshaw, 1969). The transmission of CMV by blood products is related to the presence of leukocytes. The use of blood products from CMV-seronegative donors or blood products that have been leukocyte depleted has significantly reduced the risks of CMV transmission (Bowden *et al.*, 1986). Similarly, CMV can be transferred to the recipients of either solid organs or bone marrow grafts (Zaia, 1990, 1994). In the setting of either solid organ or bone marrow transplantation, the transfer of CMV from a seropositive donor to a seronegative recipient is common, suggesting that the donor graft can act as a vehicle for CMV transmission.

Many active CMV infections are the result of reactivation of endogenous latent virus harbored by previously infected patients (Jordan, 1983). In situations of suppressed cellular immunity, whether iatrogenic or disease induced, CMV is able to reactivate to a state of active replication, with the potential to induce disease. Thus, CMV infections are common in patients undergoing intense immunosuppression for transplantation, and in patients who develop diseases that chronically suppress cellular immunity (Snydman *et al.*, 1993).

   Immunity to CMV infection following acute infection is insufficient to prevent reinfection (Collier *et al.*, 1989; Chandler *et al.*, 1987). Although there is evidence of some antigenic variation among CMV isolates, there is strong conservation of antigens relevant to host immunity, and variation is insufficient to describe distinct serotypes. It is clear that there are multiple genotypes of CMV that can be distinguished by restriction fragment polymorphism (RFP) or differences in genes amplified by polymerase chain reaction (PCR), but these differences are not sufficient to define distinct genotypes (Alford and Britt, 1993; Griffiths and Emery, 1997). The recovery of multiple strains of CMV from individual patients has been documented, establishing that host immunity after natural infection is incapable of preventing reinfection by CMV. These experiments of nature cast doubt upon the ability of a vaccine strategy to effectively prevent the transmission of CMV infection.

MICROBIOLOGY

Cytomegalovirus is one of eight herpes viruses known to infect humans (Mocarski, 1993). It is a β-herpes virus, which is characterized by restricted species range and a slow replication cycle. CMV is a double-stranded DNA virus with the large genome of the family, 2×10^8 Da or 230 kilobases (DeMarchi *et al.*, 1978). The genomic sequence of one of the laboratory prototypes, AD-169, has recently been defined and its genome shown to encode for at least 208 open reading frames (ORF) and more than 100 polypeptides (Chee *et al.*, 1990). Examination of several wild-type strains of CMV has indicated that at least 19 additional ORFs, not seen in AD-169, are present (Cha *et al.*, 1996). CMV possesses the morphology typical of the herpes group, with an icosahedral capsid and a dense core surrounded by an amorphous matrix (Figure 10.1). The virus possesses an envelope containing a number of glycoproteins thought to play a role in the attachment of CMV to the cell to initiate infection (Gretch *et al.*, 1988). The genome is organized into a long unique (UL) gene region and a short unique (US) region (Figure 10.2), and contains internal repeats and terminal repeats which allow the virus to exist in four isomeric forms (Mocarski, 1993). The function of this organization is unknown, but is probably related to the mechanism of DNA replication. Also, although there appears to be only a serotype of CMV, there are multiple genotypes that can be distinguished by RFP or PCR (Britt, 1996; Griffiths and Emery, 1997). This genotypic polymorphism is stable for each virus strain and has been useful in epidemological analysis of virus transmission (Chou 1990).
   The replication cycle of CMV is complex and not fully delineated. CMV appears to initiate infection by attachment to a specific cellular receptor (Cooper *et al.*, 1991). HCMV has been shown to bind to annexin, a 39.5 kDa cellular protein of the adhesin family (Wright *et al.*, 1994), and is then associated with an 89 kDa protein (Keay and Baldwin, 1992; Keay *et al.*, 1989). The nucleocapsid enters the cytoplasm of the cell by membrane fusion rather than endocytosis, and is then transported to the nucleus by as yet undefined mechanisms. The virus then undergoes its replication steps by an interlocking cascade termed α (immediate–early), β (early) and γ (late) (Mocarski, 1993). The replication of

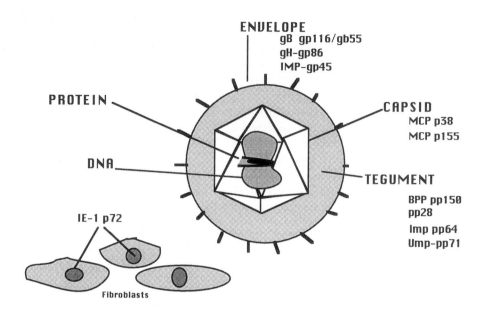

FIGURE 10.1 The localization of the major proteins of human cytomegalovirus in the virion and infected fibroblast cells.

viral DNA occurs between the early (β) and late (γ) stages, and is thought to occur by rolling circle replication. Concatameric molecules are synthesized and packaged in the capsid. The nucleocapsid is assembled in the nucleus and transported through the nuclear membrane and then the endoplasmic reticulum and Golgi, where its host-derived envelope appears to be modified by the addition of viral glycoproteins. A unique characteristic of HCMV replication is its interrelation and dependence on host cell functions (Mocarski, 1993). Unlike many other viruses, which shut off host cell metabolism, CMV requires certain ill-defined cellular functions for successful replication. For example, the immediate–early phase of replication can be influenced by such host cell factors as NFκb (Sambucetti *et al.*, 1989). Finally there is an interesting dichotomy for

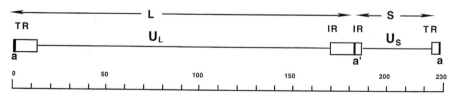

FIGURE 10.2 A schematic representation of the 230 kilobase genomic structure of human cytomegalovirus AD-169.

CMV replication: *in vitro*, CMV has been best studied in human fibroblasts, although it has been shown to replicate in a number of other cell types; *in vivo*, CMV replication appears to involve replication-cell types other than mesenchymal cells (Sinzger *et al.*, 1995).

There are two important characteristics of CMV biology that have a bearing on the development of any vaccination strategies. Concern has been raised by the observation that a number of the genes of CMV have been shown to be oncogenic *in vitro*, although CMV-related tumors have not been documented *in vivo* (Rapp and Li, 1975). Acute CMV infection has been shown to be followed by the development of latent infection and chronic viral shedding, raising concern that vaccine strains may induce latency (Jordan, 1983). The sites of latency, unlike the α herpes-viruses, are not restricted to neural structures but are known to involve multiple sites and cell types. As mentioned above, repetitive infection by different CMVs has been documented, making the ability of vaccines to prevent infection improbable. These issues must be addressed in the development of any candidate vaccine for CMV.

IMMUNOLOGY

In the immunocompetent host CMV infection is associated with both humoral and cellular immune responses (Gehrz, 1991; Zaia, 1994). The relative contributions of these responses in the prevention of infection and the resolution of disease are only now being defined. The understanding of the viral antigens that are relevant in immunity to CMV has improved dramatically in the last decade, but is still incomplete.

In acute CMV infection, antibodies are generated to a large number of structural and non-structural proteins (Gold *et al.*, 1988; Landini *et al.*, 1985, 1988; Pereira *et al.*, 1982; Schoppel *et al.*, 1997). Although CMV codes for more than 100 proteins (Chee *et al.*, 1990), few of these induce neutralizing antibodies, and the role of non-neutralizing antibodies in host defense is unknown (Britt, 1991). The envelope glycoproteins of CMV, gB and gH, both induce neutralizing antibodies (Britt, 1996), that by gB being somewhat more common than that by gH. The role neutralizing antibody in CMV infections is not known. In most situations, passive transfer of antibody to CMV has had little effect on the prevention of CMV infection, although it appears to lessen the frequency of CMV-associated diseases in certain transplant patients (Levin *et al.*, 1979; Snydman *et al.*, 1993). In view of the fact that reinfection with CMV has been documented, it is unlikely that antibody can prevent transmission. However, humoral immunity may play a significant role in modifying the development of disease. In situations such as the vertical transmission of virus from seropositive mothers to their fetuses, or in neonatal transmission in infants of seropositive mothers, antibody is thought to play a protective role in the occurrence of disease (Demmler, 1991).

The development of delayed-type hypersensitivity and cell-mediated immunity to CMV correlates with the prevention of disease and the resolution of infection (Quinnan *et al.*, 1984a; Rasmussen, 1990). In bone marrow transplants, the recovery of cytotoxic T cells (CTL) is correlated with fewer

CMV-related disease processes and improved survival (Quinnan *et al.*, 1981). Riddell and co-workers have shown that passive reconstitution of cellular immunity to CMV can provide protection (Riddell *et al.*, 1991; Riddell *et al.*, 1992). The relevant antigenic targets for cellular immunity to CMV are not completely understood. In the murine system, the non-structural α-protein of mouse CMV (MCMV), designated IE-1, has been shown to be the immuno-dominant antigen for cell-mediated protection of BALB/c mice (Koszinowski *et al.*, 1992). In contrast, one of the major viral matrix proteins, pp65, has been shown to be the immunodominant antigen for CTL responses in humans (Riddell *et al.*, 1993). The glycoprotein gB has also been shown to elicit a CTL response (Borysiewicz *et al.*, 1988). The potential importance of other viral antigens in CMI responses to CMV remain problematic.

VACCINES

As previously mentioned, several aspects of the natural history of infection have provided important insights into the potential role of vaccination for CMV. The observation that repeated infections with CMV occur naturally makes it unlikely that vaccines will prevent the transmission of HCMV infection. Never-theless, there are substantial data to encourage the development of vaccines for the prevention of disease due to CMV. In congenital infection, the majority of CMV infections resulting in disease production occur in CMV naive women undergoing primary infection during gestation (Ho, 1992a). Congenital infec-tion in infants from women previously infected with CMV has been described, but disease rarely occurs. Given the fact that CMV remains the most common infectious cause of congenital abnormalities, vaccination of women to prevent congenital disease is a worthwhile goal. Similarly, the presence of maternal antibody is correlated with absence of CMV disease in neonatal CMV infection of low-birthweight infants (Yeager, 1974; Yeager *et al.*, 1981). Again, vaccina-tion of women in the childbearing years may afford protection in this setting. Further support for this concept comes from animal studies using the guinea-pig model of congenital CMV infection (Harrison *et al.*, 1995, Bratcher *et al.*, 1995; Bia *et al.*, 1984).

Similarly, symptomatic disease in the majority of solid organ transplants occurs in patients undergoing primary CMV infection (Betts and Hanshaw, 1977; Ho *et al.*, 1975). Furthermore, studies described below suggest that vaccination, despite not preventing the transmission of infection, can lessen the occurrence of disease in renal transplant patients. Finally, recovery from CMV infection in bone marrow transplant patients correlates with the recovery of CMV-specific CTL. It may be possible to hasten the recovery of CMV-specific immunity in bone marrow recipients by vaccine modification of the marrow donor or recipient, thus improving outcome. All of these options are worthy of exploration.

The primary target populations for vaccination would appear to be seronegative women of childbearing age, individuals likely to undergo solid organ transplantation, and potentially, the donors or recipients of allogeneic bone marrow transplants. In contrast, patients with HIV who are seropositive are less likely to benefit from vaccination, as their infections arise from the reactivation of latent virus. It is possible that the presence of immunity to CMV, either natural or vaccine induced, may delay the onset of CMV disease in the face of declining host defenses. Thus, there may be a role for the vaccination of CMV-seronegative HIV patients who are at risk for primary CMV infection, although this is not clear at present.

Vaccine Strategies

Live virus vaccines

A variety of vaccine strategies for CMV have been explored in both human and animal model systems (Britt, 1996). Because CMVs are highly species specific and there are clear differences between animals in the immune mechanisms that are important to host protection for CMV, the relevance of studies using animal models to human vaccination and immunity must be interpreted cautiously. Nevertheless, studies in animals have provided important insights into the mechanism of CMV-induced diseases and the impetus to studies of vaccines in humans. However, vaccination in humans is the primary focus of this chapter.

The largest body of information relevant to human vaccination for CMV is related to the development of attenuated live-virus vaccines. In 1974 Elek and Stern reported on the use of tissue culture-adapted AD-169. The prototype CMV strain, originally isolated from adenoid tissues and designated AD-169, was passed 54 times in tissue culture. Oral administration of vaccine virus did not induce subsequent CMV infection. Intradermal inoculation was associated with the development of a local erythematous raised itchy papule at the inoculation site in seronegatives, and was not seen in seropositives. None of the subjects receiving intradermal inoculation developed antibody. Subcutaneous inoculation of human volunteers resulted in complement fixing (CF) antibody production in 96% of seronegative subjects. There were only minor local reactions, redness and swelling, and no evidence of secondary spread of vaccine strains. In 1984, an 8-year follow-up of these patients revealed that 50% of vaccines retained antibody or lymphocyte transformation responses (Stern, 1984). In 1979, in a subsequent study, AD-169 was given subcutaneously to 20 seronegative and four seropositive subjects (Neff *et al.*, 1979). All seronegatives developed CF and IFA antibodies, with 95% producing neutralizing antibody. There was no significant change in the antibody titers of the seropositive patients. Neither major side-effects nor virus shedding were observed. Stern reported his experience with 36 normal volunteers vaccinated with AD-169 (Stern, 1984). All developed CF and neutralization antibody by 1 month. At 1 and 3 years only 30.5% and 23.3% respectively retained CF antibody, whereas 100% possessed neutralizing antibody at both time points. In five patients awaiting renal transplantation, 100% developed antibody and lymphocyte proliferation responses to CMV antigen. Following transplantation three patients did not excrete vaccine virus, despite immunosuppressive therapy. One of these patients, who received a kidney from a seropositive donor, developed exogenous CMV infection and shed wild-type, but not vaccine, virus. No further studies using AD-169 have been reported.

The majority of the work on live-virus vaccine has been performed using the Towne strain of HCMV (Marshall and Plotkin, 1990; Plotkin *et al.*, 1990; Starr *et al.*, 1991). This virus was isolated in 1970 in WI-38 cells and underwent 125 passes and three limit dilutions prior to study (Plotkin, 1991). Although there are no clear markers of viral attenuation, subsequent studies indicated that this virus was less virulent than wild-type strain. Prior to use in humans this vaccine was examined *in vitro* for evidence of safety (Plotkin *et al.*, 1975). These studies detected no adventitial agents, no evidence of transformation by both *in vitro* and *in vivo* testing, and no aberrant effects in animals. The vaccine virus differed in increased resistance to trypsin, although it was not clear if this was a marker of attenuation.

Initial studies in volunteers were directed at establishing the safety and immunogenicity of this vaccine. Studies in normal patients have addressed both of these issues (Fleisher *et al.*, 1982; Just *et al.*, 1975; Plotkin *et al.*, 1976). There was no evidence of vaccine-induced infection following virus administration by either the oral or the intranasal route. Administration by the subcutaneous route consistently resulted in antibody and lymphoproliferative responses in seronegative subjects (Fleisher *et al.*, 1982; Just *et al.*, 1975; Plotkin *et al.*, 1976). The humoral responses included antibody detected by complement fixation, indirect immunofluorescence and neutralization. Whereas CF antibodies tended to deteriorate over time, neutralizing antibodies persisted for at least 3 years in all patients, and in 50% of patients at 8 years (Fleisher *et al.*, 1982). Characterization of the humoral responses by Western blot analysis revealed that antibodies are induced to multiple viral proteins (Gonczol *et al.*, 1989). Antibody to pp65 appeared rapidly in most subjects, whereas the appearance of neutralizating antibodies correlated with antibody to a 58 kDa protein, presumably a fragment of gB. Vaccine administration did not appear to induce antibody to the immediate–early proteins of the virus (Friedman *et al.*, 1982). The vaccine was also found to induce humoral and cellular responses in patients awaiting renal transplantation (Glazer *et al.*, 1979; Marker *et al.*, 1981). However, the responses in these patients were somewhat delayed and less intense than in normal subjects. Nevertheless these studies established the immunogenicity of Towne vaccine.

The safety of a live-virus vaccine is always of major concern. However, the studies of this vaccine have revealed no serious problems, even in patients undergoing intense immunosuppressive treatment (Plotkin *et al.*, 1984a). Consistently, only minor local reactions at the site of inoculation have been noted. These consisted of local pain, redness and swelling, which appeared several weeks after immunization and were thought to be a form of delayed hypersensitivity to the vaccine. There has been no evidence of shedding of vaccine virus, even in the face of intense immunosuppression. Secondary spread of vaccine virus has not been documented.

The development of natural CMV infection in adults is often manifested by fever, a mononucleosis syndrome, viremia, viruria, suppression of lymphocyte proliferation to lectins, and alteration in the CD4 and CD8 lymphocyte ratio (Carney *et al.*, 1981). Studies in volunteers of the Toledo strain of CMV, a low-passage virus isolate, revealed that as few as 10 plaque-forming units given subcutaneously would induce a vigorous immune response, accompanied by clinical symptoms and laboratory abnormalities, including hepatitis and viremia (Quinnan *et al.*, 1984b). In contrast, immunization with Towne strain vaccine induced no clinical symptoms or laboratory abnormalities, and there was no evidence of lymphocytosis, alteration in CD4 or CD8, or lectin-induced lymphoproliferation (Plotkin *et al.*, 1990). In the aggregate, these findings indicate that Towne strain vaccine is both safe, attenuated and immunogenic.

There have been several studies in normal volunteers that address the ability of the Towne vaccine to prevent virus transmission or modify CMV disease. Studies using the Toledo strain have shown that prior immunization with Towne can induce resistance to symptomatic infection comparable to naturally induced immunity (Plotkin *et al.*, 1990).

Recently, the ability of the Towne vaccine to reduce horizontal transmission of CMV from children to their mothers was examined (Adler *et al.*, 1995). Towne vaccine was shown to induce both humoral and cellular immunity in recipients, although the levels of neutralizing antibody were less than in wild-type infection. Although wild-type infection appeared to reduce horizontal transmission of infection from children to their mothers, similar reductions following vaccination were not seen. More recent studies using a new lot of Towne strain vaccine have shown that immunization induced levels of neutralizing antibody nearly equivalent to natural infection (Adler *et al.*, 1998). No studies are yet available to demonstrate the effects of vaccine-induced immunity on congenital transmission.

Patients undergoing solid organ transplantation are another population in whom immunization has been studied, but these studies are complicated by the fact that recipients can obtain their infections from a variety of sources, including endogenous reactivation, transfer of virus in the organ graft, and exogenous sources such as blood products (Betts 1982; van der Meer *et al.*, 1996; Zaia, 1994). A number of studies have addressed the use of vaccination in renal transplant candidates and recipients (Balfour, 1991; Balfour *et al.*, 1984; Brayman *et al.*, 1988; Marker *et al.*, 1981; Plotkin *et al.*, 1984b, 1991). There have been a number of important and confounding observations in these patients that influence vaccine studies in transplant recipients. It had previously been observed that symptomatic CMV infection was more common in transplant recipients undergoing primary, rather than secondary, infection. In addition, CMV infection could result from administration of blood products, be transferred in the grafted organ, or arise from reactivation of latent infection. Therefore, it is essential that vaccine studies in transplant patients control for the CMV serostatus of the donor–recipient pairs and the administration of blood products. The degree to which CMV vaccine induced an immune response in these patients was less than that seen in normal subjects. In general, the studies have been in agreement that vaccination of transplant recipients does not lessen the occurrence of CMV infection. However, a number of studies have shown that the severity of infection was ameliorated by vaccine-induced immunity. In one study there was a significant improvement in graft survival (Brayman *et al.*, 1988). Thus, the rationale of vaccination in this group of patients to the prevention of disease rather than prevention of infection seems to be sustained. Improving on the magnitude of the immune response induced by immunization may further enhance the observed protection.

Subunit Vaccines, including Vectored Vaccines

Recently, other strategies have been explored for the development of a vaccine for CMV. The efforts to develop a subunit vaccine strategy have the focused on the major envelope glycoprotein of CMV, gB, for a number of reasons (Adler, 1994; Britt, 1996). CMV gB is known to be immunogenic and is the dominant target of the virus neutralizing antibody response (Britt, 1991; Liu et al., 1988). It is also highly conserved among strains of CMV, and there is evidence that it elicits a CD8 CTL response (Liu et al., 1993). This glycoprotein is the gene product of the UL55 ORF of CMV (Spaete, 1991). The protein, composed of 907 amino acid residues, is synthesized as a 160 kDa precursor that undergoes N-linked and O-linked glycosylation. The glycoprotein then undergoes proteolysis to yield a heterodimer, composed of 55–58 kDa and 116–130 kDa proteins linked by a disulfide bridge. The mature protein can be found on the nuclear and cytoplasmic membranes and is plentiful in the viral envelope. CMV gB is thought to play a role in the process of viral attachment to the cell surface. Neutralizing epitopes are found throughout the protein, although many are clustered in the midportion. The gene sequence of gB has been expressed in yeast, CHO cells and baculovirus, and has been engineered in truncated forms that are secreted, as well as forms that resist proteolysis. Studies in animals have demonstrated that it is possible to induce neutralizing antibody response to the subunit. The character of the antibody response is influenced by the cell type producing the molecule, the degree of glycosylation and the adjuvant used in vaccination (Britt et al., 1995; Spaete, 1991). A form of gB has been used in human phase I trials demonstrating safety and immunogenicity (Adler, 1996).

The gene for CMV gB has also been inserted into a number of viral vectors as potential vaccine candidates. The gene of gB was inserted into adenovirus type 5 downstream of the E3 promoter (Marshall et al., 1990). A549 and MRC-5 cells, infected with this construct, expressed a 58 kDa protein and some probable degradation products. Intranasal inoculation of hamsters induced the production of neutralizing antibodies to gB. CMV gB has also been inserted into vaccinia virus and expressed in vitro in infected cells. Infection of mice with this construct elicited CTL responses only to the vaccinia vector. These constructs have not as yet been tried in human infections.

The membrane protein gH is also immunogenic, and this protein is thought to play a role in viral entry into the cell (Chou, 1992; Pachl et al., 1989). It has also been cloned and expressed, but no vaccine studies have been conducted.

Newer Vaccine Approaches

In 1994, Wolff and co-workers reported that the intramuscular injection of DNA expression vectors containing specific reporter genes, such as CAT, luciferase and β-galactosidase, resulted in the protracted expression of these reporter proteins in tissues at the site of injection (Wolff et al., 1990). Ulmer et al. (1993) demonstrated that an expression vector containing the nucleo-protein of influenza A virus also resulted in expression of this protein, and that this expression led to humoral and cellular immune responses in mice. The immune responses induced in this study resulted in protection when the mice were challenged with virus. These observations have ushered in new and exciting possibilities for vaccination, so-called DNA or polynucleotide vaccination. DNA-based vaccines have the advantage of being easy to engineer, inexpensive to produce, and more stable than conventional vaccines (Ulmer *et al.*, 1993, 1996). The expression of multiple microbial genes has been reported, including hepatitis B surface antigen, rabies, malaria, mycoplasma, and a number of herpes viruses. A limited number of preliminary studies in animals have addressed DNA vaccination for cytomegalovirus. DNA immunization of mice with a vector expressing IE-1 of murine CMV was shown to elicit humoral and CTL responses leading to protection (Gonzalez-Armas *et al.*, 1996). In our laboratory we have developed DNA vaccines using a truncated and a full-length form of gB murine CMV (unpublished observations). Immunization of BALB/c mice with either construct induces a humoral response to gB of MCMV. When vaccinated animals are challenged with a sublethal dose of MCMV, the full-length construct, but not the truncated construct, induced a reduction in viral titers in lung, liver and spleen. Pande *et al.* (1995) have reported the induction of humoral and CTL responses to the matrix protein, pp65, of HCMV in mice immunized with DNA expression vector. At the time of writing no other reports are available concerning the use of DNA-based vaccines in CMV infections. Nevertheless, the ability to induce an immune response to a specific gene product of CMV may open the door to new vaccination strategies.

FUTURE PROSPECTS

It is clear from the earlier studies that a modification of host immunity to CMV in specific patient populations in order to modify to occurrence of disease is a worthwhile and achievable goal. Studies of the natural history of CMV, and the early work using attenuated vaccines, indicate that although it may not be possible to prevent the transmission of CMV, it is certainly likely that disease can be prevented using this strategy. Despite the successes demonstrated with the studies using live attenuated CMV, there has been general reluctance to embrace this vaccine strategy. This probably stems from the theoretical concerns related to the potential oncogenicity of CMV and the uneasiness of using live-virus vaccines in the settings of pregnancy and immunosuppression. Nevertheless, these studies have clearly paved the way by demonstrating that it is possible to alter the pathogenesis of CMV disease by vaccine alteration of host immunity. The future focus is most likely to be on the use of defined segments of the virus which are known to be immunogenic. The elements required for success in this approach include a better understanding of the viral antigens which are relevant to humoral and cellular immunity, and further insight into the roles that both antibody and cellular immunity play in preventing disease. Understanding of these elements will permit the development of vaccine constructs that target the relevant immunogens and elicit a sufficient response to afford protection from disease.

REFERENCES

Adler, S.P. (1994). Development and clinical status of cytomegalovirus vaccines. *Clin. Immunother.* 2(1): 1–6

Adler, S.P. (1996). Current prospects for immunization against cytomegaloviral disease. *Infect. Agents Dis.* 5(1): 29–35

Adler S.P., Hempfling S.H., Starr S.E., Plotiin S.A. and Riddell S. (1998). Safety and immunogenicity of the Towne strain cytomegalovirus vaccine. *Pediatr. Infect. Dis. J.* 17: 200–206

Adler, S.P., Starr, S.E., Plotkin, S.A. *et al.* (1995). Immunity induced by primary human cytomegalovirus infection protects against secondary infection among women of childbearing age. *J. Infect. Dis.* 171(1): 26–32

Alford, C.A. and Britt, W.J. (1993). Cytomegalovirus. In: The Human Herpesviruses, B. Roizeman, R.J. Whitley and C. Lopez, eds., Raven Press, New York, 227–255

Apperly, J.F. and Goldman, J.M. (1988). Cytomegalovirus: biology, clinical features and methods for diagnosis. *Bone Marrow Transpl.*, 3: 253–264

Balfour, H.H.J. (1991). Prevention of cytomegalovirus disease in renal allograft recipients. *Scand. J. Infect. Dis. (-Suppl)*, 80: 88–93

Balfour, H.H.J., Sachs, G.W., Welo, P., Gehrz, R.C., Simmons, R.L. and Najarian, J.S. (1984). Cytomegalovirus vaccine in renal transplant candidates: progress report of a randomized, placebo-controlled, double-blind trial. *Birth Defects: Original Article Series* 20(1): 289–304

Betts, R.F. (1982). Cytomegalovirus infection in transplant patients. *Prog. Med. Virol.* 28: 44–64

Betts, R.F. and Hanshaw, J.B. (1977). Cytomegalovirus (CMV) in the compromised host(s). *Ann. Rev. Med.* 28: 103–110

Bia, F.J., Miller, S.A., Lucia, H.L. *et al.* (1984). Vaccination against transplacental cytomegalovirus transmission: vaccine reactivation and efficacy in guinea pigs. *J. Infect. Dis.* 149: 355–362

Borysiewicz, L.K., Graham, S., Hickling, J.K., Mason, P.D. and Sissons, J.G.P. (1988). Human cytomegalovirus-specific cytotoxic Tcells: their precursor frequency and stage specificity, *Eur. J. of Immunol.* 18: 269–275

Bowden, R.A., Sayers, M., Flournoy, N. *et al.* (1986). Cytomegalovirus immune globulin and seronegative blood products to prevent primary cytomegalovirus infection after marrow transplantation. *N. Engl. J. Med.* 314: 1006–1010

Bowen, E.F., Wilson, P., Atkins, M. *et al.* (1995). Natural history of untreated cytomegalovirus retinitis. *Lancet*, 346(8991–8992): 1671–1673

Bratcher, D.F., Bourne, N., Bravo, F.J. *et al.* (1995). Effect of passive antibody on congenital cytomegalovirus infection in guinea pigs. *J. Infect. Dis.* 172: 944–950

Brayman, K.L., Dafoe, D.C., Smythe, W.R. *et al.* (1988). Prophylaxis of serious cytomegalovirus infection in renal transplant candidates using live human cytomegalovirus vaccine. Interim results of a randomized controlled trial. *Arch. Surg.* 123(12): 1502–1508

Britt, W.J. (1991). Recent advances in the identification of significant human cytomegalovirus-encoded proteins. *Transplant. Proc.* 23: 64–69

Britt, W., Fay, J., Seals, J. and Kensil, C. (1995). Formulation of an immunogenic human cytomegalovirus vaccine: responses in mice, *J. Infect. Dis.* 171: 18–25

Britt, W.J. (1996). Vaccines against human cytomegalovirus: time to test. *Trends Microbiol.* 4(1): 34–38

Carney, W.P., Rubin, R.H., Hoffman, R.A., Hansen, W.P., Healey, K. and Hirsch, M.S. (1981). Analysis of T lymphocyte subsets in cytomegalovirus mononucleosis. *J. Immunol.* 126: 2114–2116

Cha, T.A., Tom, E., Kemble, G.W., Duke, G.M., Mocarski, E.S. and Spaete, R.R. (1996). Human cytomegalovirus clinical isolates carry at least 19 genes not found in laboratory strains. *J. Virol.* 70: 78–83

Chandler, S.H., Handsfield, H.H. and McDougall, J.K. (1987). Isolation of multiple strains of cytomegalovirus from women attending a clinic for sexually transmitted disease. *J. Infect. Dis.* 155(4): 655–60

Chee, M.S., Bankier, A.T., Beck, S. *et al.* (1990). Analysis of the protein-coding content of the sequence of human cytomegalovirus strain AD 169. *Curr. Topics Molec. Biol.* 154: 125–169

Chou, S.W. (1989). Reactivation and recombination of multiple cytomegalovirus strains from individual organ donors, *J. Infect. Dis.* 160: 11–5

Chou, S. (1990). Differentiation of cytomegalovirus strains by restriction analysis of DNA sequences amplified from clinical specimens, *J. Infect. Dis.* **162**: 738–742

DeMarchi, J.M., Blankship, M.L., Brown, G.D. and Kaplan, A.S. (1978). Size and complexity of human cytomegalovirus DNA. *Virology* **89**: 643–646

Chou, S. (1992). Molecular epidemiology of envelope glycoprotein H of human cytomegalovirus. *J. Infect. Dis.* **166**: 604–7

Collier, A.C., Chandler, S.H., Handsfield, H.H., Corey, L. and McDougall, J.K. (1989). Identification of multiple strains of cytomegalovirus in homosexual men. *J. Infect. Dis.* **159**(1): 123–6

Cooper, N.R., Nowlin, D., Taylor, H.P. and Compton, T. (1991). Cellular receptor for human cytomegalovirus, *Transplant. Proc.* **23**(S): 56S–59S

DeMarchi, J.M., Blankship, M.L., Brown, G.D. and Kaplan, A.S. (1978). Size and complexity of human cytomegalovirus DNA. *Virology* **89**: 643–646

Demmler, G.J. (1991). Summary of a Workshop on Surveillance for Congenital Cytomegalovirus Infection, *Rev. Infect. Dis.* **13**: 315–329

Drew, W.L. (1988). Cytomegalovirus infection in patients with AIDS. *J. Infect. Dis.* **158**(2): 449–456

Drew, W.L., Sweet, E.S., Miner, R.C. and Mocarski, E.S. (1984). Multiple infections by cytomegalovirus in patients with acquired immunodeficiency syndrome: documentation by Southern blot hybridization. *J. Infect. Dis.* **150**: 952–953

Elek, S.D. and Stern, H. (1974). Development of a vaccine against mental retardation caused by cytomegalovirus infection in utero. *Lancet* **1**(845): 1–5

Fleisher, G.R., Starr, S.E., Friedman, H.M. and Plotkin, S.A. (1982). Vaccination of pediatric nurses with live attenuated cytomegalovirus. *Am. J. Dis. Child.* **136**(4): 294–296

Friedman, A.D., Michelson, S. and Plotkin, S.A. (1982). Detection of antibodies to pre-early nuclear antigen and immediate-early antigens in patients immunized with cytomegalovirus vaccine. *Infect. Immun.* **38**(3): 1068–72

Gehrz, R.C. (1991). Human cytomegalovirus: biology and clinical perspectives. *Adv. Pediatr.* **38**: 203–232

Glazer, J.P., Friedman, H.M., Grossman, R.A. *et al.* (1979). Live cytomegalovirus vaccination of renal transplant candidates. A preliminary trial. *Ann. Intern. Med.,* **91**(5): 676–683

Gold, D., Ashley, R., Handsfield, H.H. *et al.* (1988). Immunoblot analysis of the humoral immune response in primary cytomegalovirus infection. *J. Infect. Dis.* **157**(2): 319–326

Gold, E. and Nankervis, G.A. (1989). Cytomegalovirus. In: Viral Infections of Humans, A.E. Evans, ed., Plenum Press, New York, 169–189

Gonczol, E., Ianacone, J., Furlini, G., Ho, W. and Plotkin, S.A. (1989). Humoral immune response to cytomegalovirus Towne vaccine strain and to Toledo low-passage strain. *J. Infect. Dis.* **159**(5): 851–859

Gonzalez-Armas, J.C., Morello, C.S., Cranmer, L.D. and Spector, D.H. (1996). DNA immunization confers protection against murine cytomegalovirus infection. *J. Virol.* **70**(11): 7921–7928

Goodgame, R.W. (1993). Gasterointestinal cytomegalovirus disease. *Ann. Intern. Med.* **119**: 924–935

Gretch, D.R., Kari, B., Rasmussen, L., Gehrz, R.C. and Stinski, M.F. (1988). Identification and characterization of three distinct families of glycoprotein complexes in the envelopes of human cytomegalovirus. *J. Virol.* **62**: 875–881

Griffiths, P.D. and Emery, V.C. (1997). Cytomegalovirus. In: Clinical Virology, D.D. Richman, R.J. Whitely and F.G. Hayden, eds., Churchill Livingstone, New York, 445–470

Grundy, J.E., Shanley, J.D. and Griffiths, P. (1987). Cytomegalovirus pneumonitis in Transplant Patients – A Hypothesis. *Lancet* **2**: 996–999

Harrison, C.J., Britt, W.J., Chapman, N.M. *et al.* (1995). Reduced congenital cytomegalovirus (CMV) infection after maternal immunization with a guinea pig CMV glycoprotein before gestational primary CMV infection in the guinea pig model. *J. Infect. Dis.* **172**: 1212–1220

Ho, M. (1992a). Congenital and perinatal human cytomegalovirus infections. In: Cytomegalovirus: Biology and Infection, M. Ho, ed., Plenum Medical Book Company, New York, 205–227

Ho, M. (1992b). Epidemiology of cytomegalovirus infection in man. In: Cytomegalovirus: Biology and Infection, M. Ho, ed., Plenum Medical Book Company, New York, 155–188

Ho, M., Suwansirikul, S., Dowling, J.N., Youngblood, L.A. and Armstrong, J.A. (1975). The transplanted kidney as a source of cytomegalovirus infection. *N. Engl. J. Med.* **293**(22): 1109–1112

Jacobson, M.A. and Mills, J. (1988). Serious cytomegalovirus disease in the acquired immunodeficiency syndrome (AIDS). *Ann. Intern. Med.*, **108**: 585–594

Jordan, M.C. (1983). The latent and elusive cytomegalovirus. *Rev. Infect. Dis.* **5**: 205–215

Just, M., Buergin-Wolff, A., Emoedi, G. and Hernandez, R. (1975). Immunization trials with live attenuated cytomegalovirus TOWNE 125. *Infection* **3**(2): 111–114

Kalayjian, R.C., Cohen, M.L., Bonomo, R.A. and Flanigan, T.P. (1993) Cytomegalovirus ventriculoencephalitis in AIDS. A syndrome with distinct clinical pathologic features, *Medicine* **72**: 67–77

Katlama, C. (1993). Cytomegalovirus infection in acquired immunodeficiency syndrome, *J. Med. Virol.* **1S**: 128–133

Keay, S. and Baldwin, B. (1992). The human fibroblast receptor of gp86 of human cytomegalovirus is a phosphorylated glycoprotein. *J. Virol.*, **66**: 4834–4838

Keay, S., Merigan, T.C. and Rasmussen, L. (1989). Identification of cell surface receptors for the 86-kilodalton glycoprotein of human cytomegalovirus. *Proc. Natl. Acad. Sci. USA* **86**: 10100–10103

Koszinowski, U.H., Reddehase, M.J. and Del-Val, M. (1992). Principles of cytomegalovirus antigen presentation in vitro and in vivo. *Semin. Immunol.* **4**(2): 71–79

Krech, U. (1973). Complement-fixing antibodies against cytomegalovirus in different parts of the world. *Bull. WHO* **49**: 103–106

Landini, M.P., Mirolo, G. and La Placa, M. (1985). Human immune response to cytomegalovirus structural polypeptides studied by immunoblotting. *J. Med. Virol.* **17**: 303–311

Landini, M.P., Rossier, E. and Schmitz, H. (1988). Antibodies to human cytomegalovirus structural polypeptides during primary infection. *J. Virol. Meth.* **22**: 309–317

Lang, D.J. and Hanshaw, J.B. (1969). Cytomegalovirus and the post-perfusion syndrome-(Recognition of primary infections in four patients). *N. Engl. J. Med.* **280**: 1145–1149

Levin, M.J., Rinaldo, C.R., Leary, P.L., Zaia, J.A. and Hirsch, M.S. (1979). Immune response to herpesvirus antigens in adults with acute cytomegalovirus mononucleosis. *J. Infect. Dis.* **140**: 851–855

Liu, Y.N., Curtsinger, J., Donahue, P.R. *et al.* (1993). Molecular analysis of the immune response to human cytomegalovirus glycoprotein B.I. Mapping of HLA-restricted helper T cell epitopes on gp93. *J. Gen. Virol.* **74**(10): 2207–2214

Liu, Y.-N. C., Kari, B. and Gehrz, R.C. (1988). Human immune responses to major human cytomegalovirus glycoprotein complexes. *J. Virol.* **62**: 1066–1070

Marker, S.C., Simmons, R.L. and Balfour, H.H.J. (1981). Cytomegalovirus vaccine in renal allograft recipients. *Transplant. Proc.* **13**(1 Pt 1): 117–119

Marshall, G.S. and Plotkin, S.A. (1990). Progress toward developing a cytomegalovirus vaccine. *Infect. Dis. Clin. N. Ame.* **4**(2): 283–98

Marshall, G.S., Ricciardi, R.P., Rando, R.F. *et al.* (1990). An adenovirus recombinant that expresses the human cytomegalovirus major envelope glycoprotein and induces neutralizing antibodies. *J. Infect. Dis.* **162**(5): 1177–1181

Mocarski, E.S. (1993). Cytomegalovirus Biology and Replication. In: The Human Herpesviruses, B. Roizeman, R.J. Whitley and C. Lopez, eds., Raven Press, Ltd, New York, 173–226

Morgello, S., Cho, E.-S., Nielsen, S., Devinsky, O. and Petito, C.K. (1987). Cytomegalovirus encephalitis in patients with acquired immunodeficiency syndrome: An autopsy study of 30 cases and a review of the literature, *Hum. Pathol.* **18**: 289–297

Neff, B.J., Weibel, R.E., Buynak, E.B., McLean, A.A. and Hilleman, M.R. (1979). Clinical and laboratory studies of live cytomegalovirus vaccine Ad-169. *Proc. Soc. Exp. Biol. Med.*, **160**(1): 32–37

Neiman, P.E., Reeves, W., Ray, G. *et al.* (1977). A prospective analysis of interstitial pneumonia and opportunistic viral infection among recipients of allogeneic bone marrow grafts. *J. Infect. Dis.* **136**: 754–767

Numazaki, Y., Yano, N., Morizuka, T., Takai, S. and Ishida, N. (1970). Primary infection with human cytomegalovirus: virus isolation from healthy infants and pregnant women. *Am. J. Epidemiol.* **91**(4): 410

Onorato, I.M., Morens, D.M., Martone, W.J. and Stansfield, S.K. (1985). Epidemiology of cytomegalovirus infections: recommendations for prevention and control. *Rev. Infect. Dis.* 7: 479–497

Pachl, C., Probert, W.S., Hermsen, K.M. *et al.* (1989). The human cytomegalovirus strain Towne glycoprotein H gene encodes glycoprotein p86. *Virology* **169**(2): 418–426

Pande, H., Campo, K., Tanamachi, B., Forman, S.J. and Zaia, J.A. (1995). Direct DNA immunization of mice with plasmid DNA encoding the tegument protein pp65 (ppUL83) of human cytomegalovirus induces high levels of circulating antibody to the encoded protein. *Scand. J. Infect. Dis. – (Suppl)* **99**: 117–120

Pass, R.F., Hutto, S.C., Ricks, R. and Cloud, G.A. (1986). Increased rate of cytomegalovirus infection among parents of children attending day care center. *N. Engl. J. Med.* **314**: 1414–1418

Pereira, L., Hoffman, M. and Cremer, N. (1982). Electrophoretic analysis of polypeptides immune precipitated from cytomegalovirus infected cell extracts with human sera. *Infect. Immun.* **36**(3): 933–942

Plotkin, S.A. (1991). Cytomegalovirus vaccine development – past and present. *Transplant. Proc.* **23**(3S): 85–89

Plotkin, S.A., Farquhar, J. and Horberger, E. (1976). Clinical trials of immunization with the Towne 125 strain of human cytomegalovirus. *J. Infect. Dis.* **134**(5): 470–475

Plotkin, S.A., Furukawa, T., Zygraich, N. and Huyglen, C. (1975). Candidate cytomegalovirus strain for human vaccination. *Infect. Immun.* **12**: 521–527

Plotkin, S.A., Smiley, M.L., Friedman, H.M. *et al.* (1984a). Prevention of cytomegalovirus disease by Towne strain live attenuated vaccine. *Birth Defects: Original Article Series* **20**(1): 271–287

Plotkin, S.A., Smiley, M.L., Friedman, H.M. *et al.* (1984b). Towne-vaccine-induced prevention of cytomegalovirus disease after renal transplants. *Lancet* **1**(8376): 528–530

Plotkin, S.A., Starr, S.E., Friedman, H.M. *et al.* (1991). Effect of Towne live virus vaccine on cytomegalovirus disease after renal transplant. A controlled trial. *Ann. Intern. Med.* **114**(7): 525–531

Plotkin, S.A., Starr, S.E., Friedman, H.M., Gonczol, E. and Brayman, K. (1990). Vaccines for the prevention of human cytomegalovirus infection. *Rev. Infect. Dis.* **12** Suppl(7): S827–838

Quinnan, G.V., Burns, W.H., Kirmani, N. *et al.* (1984a). HLA-restricted cytotoxic T lymphocyte are an early immune response and important defense mechanism in cytomegalovirus infection. *Rev. Infect. Dis.* **6**: 156–163

Quinnan, G.V., Delery, M., Rook, A.H. *et al.* (1984b). Comparative virulence and immunogenicity of the Towne strain and a nonattenuated strain of cytomegalovirus. *Ann. Intern. Med.* **101**: 478–483

Quinnan, G.V., Kirmani, N., Esber, E. *et al.* (1981). HLA-restricted cytotoxic T lymphocyte and nonthymic cytotoxic lymphocyte responses to cytomegalovirus infection of bone marrow transplant recipients. *J. Immunol.* **126**: 2036–2041

Rapp, F. and Li, J.L. (1975). Demonstration of the oncogenic potential of herpes simplex viruses and human cytomegalovirus. *Cold Spring Harbor Symposia on Quantitative Biology* **39**(2): 747–63

Rasmussen, I. (1990). Immune response to human cytomegalovirus. *Curr. Topics Microbiol. Immunol.* **154**: 221–254

Reynolds, D.W., Stagno, S., Hosty, T.S., Tiller, M. and Alford, C.A., Jr. (1973). Maternal cytomegalovirus excretion and perinatal infection. *N. Engl. J. Med.* **289**: 1–5

Riddell, S., Reusser, P. and Greenberg, P.D. (1991). Cytotoxic T cells specific for cytomegalovirus: A potential therapy for immunocompromised patients. *Rev. Infect. Dis.* **113**(S): 966–973

Riddell, S.R., Gilbert, M.J., Cheng-rong, L., Walter, B.A. and Greenberg, P.D. (1993). Reconstitution of protective CD8+ cytotoxic T lymphocyte responses to human cytomegalovirus in immunodeficient humans by adoptive transfer of T cell clones. In: Multidisciplinary Approach to Understanding Cytomegalovirus Disease, S. Michelson and S.A. Plotkin, eds., Elsevier Science, Amsterdam, 155–164

Riddell, S.R., Watanabe, K.S., Goodrich, J.M., Li, C.R., Agha, M.E. and Greenberg, P.D. (1992). Restoration of viral immunity in immunodeficient humans by the adoptive transfer of T cell clones. *Science* **257**: 238–241

Rowe, W.P., Hartley, J.W., Waterman, S., Turner, H.C. and Huebner, R.J. (1956). Cytopathic agent resembling human salivary gland virus recovered from tissue cultures of human adenoids. *Proc. Soc. Exp. Biol. Med.* **92**: 418–424

Sambucetti, L.C., Cherrington, J.M., Wilkinson, G.W.G. and Mocarski, E.S. (1989). NF-kB activation of the cytomegalovirus enhancer is mediated by a viral transactivator and by T cell stimulation. *EMBO J.* **8**: 4251–4258

Schoppel, K., Kropff, B., Schmidt, C., Vornhagen, R. and Mach, M. (1997). The humoral immune response against human cytomegalovirus is characterized by a delayed synthesis of glycoprotein specific antibodies. *J. Infect. Dis.* **175**: 533–544

Sinzger, C., Grefte, A., Plachter, B., Gouw, A.S., The, H.T. and Jahn, G. (1995). Fibroblasts, epithelial cells, endothelial cells and smooth muscle cells are major targets of human cytomegalovirus infection in lung and gastrointestinal tissues, *J. Gen. Virol.* **76**: 741–750

Smith, M.A. and Brennessel, D.A. (1994). Cytomegalovirus. *Infect. Dis. Clin. N. Am.* **8**(2): 427–438

Smith, M.G. (1956). Propagation in tissue culture of a cytopathic virus from human salivary gland disease. *Proc. Soc. Exp. Biol. Med.* **92**: 424–423

Snydman, D.R., Rubin, R.H. and Werner, B.G. (1993). New developments in cytomegalovirus prevention and management. *Am. J. Kidney Dis.* **21**(2): 217–228

Spaete, R.R. (1991). A recombinant subunit vaccine approach to HCMV vaccine development. *Transplant. Proc.* **23**(3S): 90–96

Stagno, S., Reynolds, D., Tsiantos, A. *et al.*, (1975). Cervical cytomegalovirus excretion in pregnant and nonpregnant women: suppression in early gestation. *J. Infect. Dis.* **131**: 522–527

Smith, M.A. and Brennessel, D.A. (1994). Cytomegalovirus. *Infectious Disease Clinics of North America* **8**(2): 427–38

Stagno, S., Reynolds, D.W., Huang, E.S., Thames, S.D., Smith, R.J. and Alford, C.A. (1977). Congenital cytomegalovirus infection. *N. Engl. J. Med.* **296**(22): 1254–1258

Stagno, S., Reynolds, D.W., Pass, R.F. and Alford, C.A. (1980). Breast milk and the risk of cytomegalovirus infection. *N. Engl. J. Med.* **302**(19): 1073–1076

Starr, S.E., Friedman, H.M. and Plotkin, S.A. (1991). The status of cytomegalovirus vaccine. *Rev. Infect. Dis.* **13S**(11): S964–965

Stern, H. (1984). Live cytomegalovirus vaccination of healthy volunteers: eight-year follow-up studies. *Birth Defect: Original Article Series* **20**(1): 263–269

Ulmer, J.B., Donnelly, J.J., Parker, S.E. *et al.* (1993). Heterologous protection against influenza by injection of DNA encoding a viral protein. *Science* **259**: 1745–1749

Ulmer, J.B., Sadoff, J.C. and Liu, M.A. (1996). DNA vaccines, *Curr. Opin. Immunol.* **8**: 531–536

van der Meer, J.T., Drew, W.L., Bowden, R.A. *et al.* (1996). Summary of the International Consensus Symposium on Advances in the Diagnosis, Treatment and Prophylaxis and Cytomegalovirus Infection. *Antiviral Res.* **32**(3): 119–140

Vinters, H.V., Kwok, M.K., Ho, H.W. *et al.* (1989). Cytomegalovirus in the nervous system of patients with acquired immunodeficiency syndrome, *Brain* **112**: 245–268

Weisner, R.H., Marin, E., Porayko, M.K., Steers, J.L., Krom, R.A.F. and Paya, C.V. (1993). Advances in the diagnosis, treatment, and prevention of cytomegalovirus infection after liver transplantation. *Adv. Liver Transplant.* **22**(2): 351–366

Weller, T.H. (1971). The cytomegaloviruses: ubiquitous agents with protean clinical manifestations. *N. Engl. J. Med.* **285**: 203–214, 267–274

Weller, T.H., Macauley, J.C., Craig, J.M. and Wirth, P. (1957). Isolation of intranuclear inclusion producing agents from infants illnesses resembling cytomegalic inclusion disease. *Proc. Soc. Exp. Biol. Med.* **94**: 4–12

Winston, D.J., Gale, R.P., Meyers, D.V. and Young, L.S. (1979). Infectious complications of human bone marrow transplantation. *Medicine* **58**: 1–31

Winston, D.J., Ho, W.G., Howell, C.L. *et al.* (1980). Cytomegalovirus infections associated with leukocyte transfusions. *Ann. Intern. Med.* **93**(5): 671–675

Wolff, J.A., Malone, R.W., Williams, P. *et al.* (1990). Direct gene transfer into mouse muscle in vivo. *Science* **247**: 1465–1468

Wright, J.F., Kurosky, A. and Wasi, S. (1994). An endothelial cell-surface form of annexin II binds human cytomegalovirus. *Biochem. Biophys. Res. Commun.* **198**(3): 983–989

Yeager, A.S. (1974). Transfusion-associated cytomegalovirus infection in newborn infants. *Am. J. Dis. Child.* **128**: 478–483

Yeager, A.S., Grumet, F.C., Hafleigh, E.B., Arvin, A.M., Bradley, J.E. and Prober, C.G. (1981). Prevention of transfusion-acquired cytomegalovirus infection in newborn infants. *J. Pediatr.* **988**: 281–287

Zaia, J.A. (1986). The biology of human cytomegalovirus infection after bone marrow transplantation. *Int. J. Cell Cloning* **4**: 135–154

Zaia, J.A. (1990). Viral infections associated with bone marrow transplantation. *Hematol./Oncol. Clin. N. Am.* **4**: 603–662

Zaia, J.A. (1994). Cytomegalovirus Infection. In: Bone Marrow Transplantation, S.J. Forman, K.G. Blume and E.D. Thomas, eds., Blackwell Science London, 376–403

11
GENITAL PAPILLOMAVIRUS INFECTION

JESSICA L. SEVERSON,*
KARL R. BEUTNER†
AND
STEPHEN K. TYRING‡

* Departments of Internal Medicine and Microbiology/Immunology,
University of Texas Medical Branch, Children's Hospital, Galveston, Texas, USA

† Solano Dermatology Associates, Vallejo, California, USA and Department of Dermatology,
University of California, San Francisco, California, USA

‡ University of Texas Medical Branch, Departments of Dermatology, Internal Medicine,
and Microbiology/Immunology, Galveston, Texas, USA

CLINICAL DISEASE

Most genital human papillomavirus (HPV) infections are benign: in fact, sub-clinical genital infections with HPV are extremely common. Genital HPV infection causes a spectrum of disease, ranging from asymptomatic infection to invasive cervical cancer. HPV types have been divided into low and high risk, based on their association with cervical cancer. When low-risk HPV infections are clinically expressed, they produce external genital warts (Beutner *et al.*, 1998a). There is also a rare verrucous carcinoma of the external genital area (giant condyloma of Buschke–Löwenstein) caused by HPV type 6 (and sometimes HPV 11). The incubation period from the time of infection to development of warts may be weeks, months or years. Once clinically apparent, they may persist, spread, grow, spontaneously regress and/or recur. Juvenile laryngeal papillomatosis (JLP) is a rare but devastating condition caused by perinatally acquired laryngeal low-risk HPV infection. HPV 6 and 11 are most commonly found in laryngeal papillomas.

When recognized, the high-risk HPV types cause abnormal pap smears, low-grade squamous intraepithelial lesions (LSIL), high-grade squamous intraepithelial lesions (HSIL), atypical squamous cells of uncertain significance (ASCUS) and atypical glandular cells of uncertain significance (AGUS) or, on biopsy, cervical intraepithelial neoplasia, carcinoma in situ and invasive cervical cancer.

There is a strong association between infection with certain HPV types and some anogenital cancers. HPV DNA can be detected in most invasive cervical cancers and cervical intraepithelial neoplasia (CIN). Latency from infection to the appearance of genital warts, cervical dysplasia, and cervical cancer may range from weeks to decades.

Although HPV is clearly the etiologic agent of cervical cancer, and is required for the cancer to develop, other as yet unknown factors are also needed, as the vast majority of those infected with high-risk HPVs never develop cervical cancer. Although the cofactors are not currently known, it is believed that if the acquisition of HPV could be prevented cervical cancer could be greatly reduced, if not eliminated.

When the high-risk HPV types are clinically recognized on the external genital area they produce papular lesions (Bowenoid papulosis) or erythematous scaly patches (Bowen's disease of the genital area), which histologically have been classified as squamous cell carcinoma in situ or vulvar intraepithelial neoplasia, or penile intraepithelial neoplasia; or, in the vagina, as vaginal intraepithelial neoplasia. Invasive squamous cell cancer of the penis, vulva or vagina is rare.

In addition to the cervix, the oncogenic potential of the high-risk HPV types is frequently found in the anus. Interestingly, both the cervix and the anus histologically have transformation zones where columnar epithelium changes to stratified squamous epithelium. To date, anal HPV infection has best been characterized in men, and its natural history is the subject of ongoing studies. Current evidence strongly suggests that HPV is the etiologic agent of many anal cancers.

EPIDEMIOLOGY/EXPECTED IMPACT OF AN EFFECTIVE VACCINE

It has been estimated that 1% of sexually active persons in the United States have visible genital warts at any given time (Koutsky *et al.*, 1988). The majority (80–90%) of genital warts are caused by HPV 6. Thus an effective monovalent vaccine is feasible for these low-risk infections. Although rarely associated with significant mortality or morbidity, cases of genital warts consume health-care resources and are emotionally troubling for the patients. It is also conceivable that genital warts, particularly those that are friable, may facilitate the transmission or acquisition of bloodborne pathogens, such as hepatitis viruses and human immunodeficiency virus. For these reasons, immunization is an important goal.

About 50% of young sexually active women are infected with HPV (Bauer *et al.*, 1991; Koutsky *et al.*, 1992; Ho, 1998). Cervical cancer is the leading cause of cancer death in women under the age of 50 worldwide (Rowen and Lacey, 1998). The impact of HPV infection is greatest in countries where screening is unavailable or underused. The strong association between sexually acquired HPV and cervical cancer makes vaccination against HPV an attractive goal. Relatively few HPV types account for the majority of cervical carcinomas, and animal models have consistently shown that both prophylactic and therapeutic vaccination is feasible. Both preventive and therapeutic vaccines have worldwide public health implications, not only to save lives, but to decrease the cost of screening and treating premalignant cervical disease. The Institute of Medicine has recently estimated the annual cost of HPV-related disease to be $10 billion (Eng and Butler, 1996).

MICROBIOLOGY

Human papillomavirus is a small, non-enveloped icosahedral virus with a diameter of about 55 nm. It contains 8–10 genes that are circular, double-stranded DNA. The HPV genome contains three functional regions. The early region contains open reading frames, the products of which control viral replication, transcription and cellular transformation, and encode for E6 and E7 oncoproteins. The late region encodes for structural proteins, two capsid proteins (L1 the major and L2 the minor) and proteins expressed inside virally infected cells. The long-control region contains transcription enhancer genes and promoter elements. Each of these proteins is immunogenic.

Over 80 types of papillomavirus infecting humans have been identified. HPV types 16, 18, 31, 33, 35, 45, 51, 52, 56, 58 and 66 are considered high risk as they may cause squamous cell carcinoma and premalignant cervical lesions (Lorincz *et al.*, 1992; Kiviat and Koutsky, 1993; Schiffman *et al.*, 1993; Bosch *et al.*, 1995). HPV types 16 and 18 cause up to 90% of cervical carcinomas (Gissmann *et al.*, 1983; Brown *et al.*, 1993; Walboomers *et al.*, 1994). About 80% of anal carcinomas contain HPV DNA, type 16 being the most common (Zaki *et al.*, 1992). HPV types 6 and 11 are considered low risk as they cause >90% of condyloma acuminata.

IMMUNOLOGY/IMMUNOBIOLOGY OF INFECTION

HPV infects stratified squamous epithelial cells. Viral DNA amplification occurs via E1 helicase and E2 transactivator. Next, the viral capsid proteins (L1 and L2) are produced in the presence of differentiating epithelium. Instead of cell lysis, human papillomavirus-containing epithelial cells are shed from the surface of the skin. The progeny of a single HPV-infected stem cell will extend over a relatively large area of skin, probably in the order of 2–3 mm^2 (Steele and Gallimore, 1990). No cell lysis results in no release of viral proteins and no antigen presentation to the immune system. However, there is production of antibody to both capsid proteins of the virus during the course of HPV infection (Christensen *et al.*, 1992). This natural immunity occurs on average 6 months after infection and may be protective, but is not likely to be therapeutic.

Host immune responses to HPV appear to be important in preventing the progression of HPV infection to clinical disease. Papillomaviruses (PV) elicit both humoral and cell-mediated immune responses. Cellular immunity, particularly the T-cell system, has a crucial role in modulating the effects of HPV, such as lesion persistence and spontaneous regression. This is evidenced by the fact that immunosuppressed individuals (transplant, lymphoma, HIV disease) have enhanced PV proliferation and an increased frequency of HPV infection and associated disease (Kast *et al.*, 1996). Patients with established warts frequently have a decreased cellular immunity (Kienzler *et al.*, 1983). In addition, dense infiltrates of T lymphocytes and macrophages are found in regressing warts, and reduced numbers of Langerhans' cells are found in CIN (Sherman *et al.*, 1998).

Viral DNA in benign lesions remains extrachromosomal. In most cervical cancers the viral DNA is incorporated into the host chromosome. Cervical cancers consistently retain and express two of the viral genes, E6 and E7. In cultured cells these same HPV types show transforming activity (Schlegel *et al.*, 1988). The transforming proteins are needed for continued growth and act as tumor rejection antigens.

HISTORY/BACKGROUND OF VACCINE RESEARCH AND OTHER STRATEGIES FOR CONTROL

Vaccine development has been hampered by the inability to culture papillomavirus and host tropism for the virus. Over 60 years ago, cottontail rabbit papillomavirus (CRPV) studies established that antibodies elicited by the injection of intact virions protect against experimental challenge by the homologous viral type (Shope, 1935). Further study discovered that intact virions displaying immunodominant epitopes are needed to induce protective antibodies (Shope, 1935; Kidd, 1938; Pilacinski *et al.*, 1986; Jin *et al.*, 1990; Ghim *et al.*, 1991). Advances in technology demonstrated that the L1 major capsid protein of papillomaviruses expressed in eukaryotic cells self-assembles into virus-like particles (VLPs). These resemble authentic virions without the viral genome and its transforming genes. Antibodies generated against neutralizing, confirmational epitopes found on the surface of native virions are sufficient to prevent infection both in vitro and in animal models.

Another strategy for HPV control is immunotherapy. Adoptive cellular transfer consists of collecting cells involved in host defenses from patients with cervical cancer or a histocompatible donor, and then growing these cells and activating them ex vivo by cytokines such as recombinant IL-2 and transferring them back to the cancer patient as therapy (Hines *et al.*, 1998). In mouse models, some protection against HPV-16 and 18 E6 and E7-positive tumors is obtained with this approach (Boursnell *et al.*, 1996; Krul *et al.*, 1996).

VACCINE STRATEGIES

Prophylactic Vaccines

HPV subunit vaccines induce protective anti-HPV antibodies to prevent infection. Most prophylactic vaccine studies use L1 and L2 as antigenic targets. These vaccines have been prepared using fusion proteins, vaccinia virus recombinants, plasmids and virus-like particles (VLPs). VLPs are spherical 50 nm structures resembling hollow viral capsids produced using molecular techniques (Rose *et al.*, 1993). They possess structurally intact viral capsid proteins, lack oncogenic DNA, and may be used in enzyme-linked immunosorbent and hemagglutination assays to detect humoral responses to HPV (Sherman *et al.*, 1998).

Therapeutic Vaccines

The goals of therapeutic vaccination include the elimination of residual cancer, regression of existing CIN or warts, and preventing the progression of infection or low-grade disease to higher-risk lesions. HPV E6 and E7 epitope peptides selectively maintained and expressed during malignant progression are targets in the development of therapeutic vaccines. *In vitro* studies show that human leukocyte antigen (HLA)-specific, human cytotoxic T-lymphocytes (CTLs) induced against E6 and E7 peptides will cause lysis of HLA-specific HPV-positive cervical carcinoma cell lines (Feltkamp *et al.*, 1993; Alexander *et al.*, 1996).

ANIMAL MODELS

HPV does not cause disease in animals, so that extrapolating data to humans must be done with caution. However, encouraging results have been obtained using species-specific papillomaviruses. Immunization with protein produced in bacteria or immunization with vaccinia vectors that express L1 and/or L2 prevents experimental infection *in vivo*, although low levels of neutralizing antibodies are induced (Pilacinski *et al.*, 1986; Jarrett *et al.*, 1991; Lin *et al.*, 1992). VLP vaccines, on the other hand, are strongly immunogenic. Infection by PVs specific to beagle dogs (COPV) (Bell *et al.*, 1994; Suzich *et al.*, 1995), cottontail rabbits (CRPV) (Donnelly *et al.*, 1991; Lin, *et al.*, 1993; Breitburd *et al.*, 1995; Christensen *et al.*, 1996) and calves (BPV) (Kirnbauer *et al.*, 1996) can be prevented by immunization with PV VLPs followed by experimental challenge with native virus. Protection against experimental viral challenge is both species and type specific. It is also dependent on confirmational epitopes present on the surface of VLP.

Protection of a mucosal surface against the natural transmission of PV has been achieved by systemic vaccination in animal models. Immunization with formalin-inactivated virions of COPV wart vaccine completely protected against COPV on oral mucosa by experimental challenge or natural infection (Bell *et al.*, 1994). Lowe *et al.* (1997) showed that HPV-11 L1 VLP-specific IgG present in the cervicovaginal secretions of monkeys parenterally immunized with HPV-11 L1 VLPs is sufficient to neutralize HPV-11 in the athymic mouse xenograft system.

Therapeutic vaccines target E6 and E7 proteins. Rats vaccinated with vaccinia-transfected HPV 16 E7 had regression of tumors derived from HPV 16-transformed oncogenic cell lines (Meneguzzi *et al.*, 1991). Both prevention and regression of CRPV tumors was achieved using a vaccine derived from recombinant *Listeria monocytogenes* (Jensen *et al.*, 1997). HPV-16 L1 VLPs also induce T-cell responses, providing a mechanism to eliminate cells undergoing productive viral infection (Dupuy *et al.*, 1997; Luxton *et al.*, 1997).

Despite these encouraging results, studies with animal models have not evaluated authentic routes of PV infection, the longevity of immune responses, or established that tumors can be eliminated with vaccination.

CLINICAL STUDIES

Immunogenicity and Efficacy Studies

Numerous clinical studies are currently under way throughout the world. A few have obtained data as described below.

A vaccine specific for HPV type 11 was found in a phase I study with 65 healthy volunteers to be both safe and immunogenic (Reichman *et al.*, 1998). The gene for the major capsid protein, L1, was placed in a recombinant virus and then expressed in insect cells. Virus-like particles carrying the L1 protein on their surface were produced. The volunteers, aged 18–45, were seronegative for HPV-11. Vaccine composed of 3, 9, 30 or 100 μg of virus-like particles or placebo was given at 0, 4 and 16 weeks. Anti-HPV-11 neutralizing antibody titers of 1:1000 or greater were achieved in 7 of 10 subjects who received the 3 μg dose, 9 of 10 who received the 9 μg dose, all 12 who received the 30 μg dose and all 10 who received the 100 μg dose. All doses were well tolerated.

A phase I/II study in advanced cervical carcinoma using a recombinant vaccinia virus expressing E6 and E7 epitope peptides of HPV 16 and 18 resulted in no significant toxicities, an antivaccinia antibody response in all eight participants, and HPV-specific antibody response in three of eight participants (Borysiewicz *et al.*, 1996).

Open-label studies with an L2 E7 fusion protein vaccine, incorporating an alum adjuvant, combined with conventional therapy for the treatment of anogenital warts in men, produced good humoral responses, although lympho-proliferative responses were variable (Lacey *et al.*, 1997; Rowen *et al.*, 1997).

Vaccine Development

There are a number of vaccines in various stages of development (Table 11.1). Although modern pivotal efficacy trials with HPV have not been completed, there are a number of observations that would seem to indicate that effective HPV vaccination is feasible. These include the following:

1. Immunosuppression results in more persistent clinical expression of HPV infection. Patients who are immunosuppressed more frequently express their HPV infection, which also tends to be more difficult to treat;
2. Favorable results from vaccine studies with animal papillomavirus systems;
3. The observation that many infected humans never express their infection;
4. Observed spontaneous regression of both external genital warts, non-genital warts and HPV-related cervical lesions;
5. The observation that CD4 lymphocytes are frequently found in the infiltrates of regressing warts;
6. Reports of the successful use of autologous vaccines in humans.

There are also hints that a genital HPV vaccine might not be effective. Factors that make HPV vaccines particularly challenging include the following:

1. Protective immunity has not been defined. Although there is good reason to believe that the proper immune response could be protective and/or therapeutic, what response to which antigen has not been defined. In animal systems antibodies appear to protect against infection, and cell-mediated immune responses are required for wart regression or protection against the development of warts.

■ **TABLE 11.1 Candidate HPV vaccines**

Virus-like particles
  HPV 11 L1   Reichman *et al.*, 1999
  HPV 6 L1   Zhang *et al.*, 1999
  HPV 16 L1   Da Silva *et al.*, 1999

Protein-based vaccines
  HPV 16 E6-E7   Mallarios *et al.*, 1999
  HPV 16 E7 + BCG H5/65   Zhu *et al.*, 1995
  HPV 16 E7   Hariharan *et al.*, 1998
  HPV 16 E2   Heinemann *et al.*, 1999;
                                               Hibma *et al.*, 1999
  HPV 16 E4 with HbeAg   El Mehdaoui *et al.*, 1999

Peptide vaccines
  HLA-A0201 HPV-16 lipopeptides   Steller and Schiller, 1996
  HPV 16 E7 PADRE   Feltkamp *et al.*, 1993
                                               Brandt *et al.*, 1999
  HPV 16 E7 'ISCAR' conjugate   Tindle *et al.*, 1995

Viral vector vaccines
  Vaccinia HPV 16 L1   Cooney *et al.*, 1991; Gao *et al.*, 1994
  Vaccinia HPV 16-E7   Lin *et al.*, 1996, 1993, 1992
  Vaccinia HPV 16+18 E6 + E7   Borysiewicz *et al.*, 1996

Bacterial vectors
  Salmonella typhimuriun   HPV16 E6 and E7   Krul *et al.*, 1996; London *et al.*, 1996
  Salmonella typhimurium   HPV16 L1   Nardelli-Haflinger *et al.*, 1997
  Streptococcus gordovii   HPV16 E7   Jensen *et al.*, 1997

2. In animal systems PV immunization can protect against mucosal infection. This has not been demonstrated in humans.

3. The route and frequency with which the immunized subject is exposed to or naturally challenged may be more chronic and repetitive with HPV and other sexually transmitted diseases than with other viral infections for which we have effective vaccines. What level of immunity is required to protect against frequent genital exposure? For the most part, viral infections for which we currently have vaccines are not characterized by frequent (potentially daily) exposure for years.

The two most obvious vaccine strategies are prophylactic and therapeutic (Table 11.2). Of these, prophylactic is the most appealing, based on experience with animal PVs and in humans, with a large number of viral vaccines currently in use.

Although a number of autologous studies appear to have achieved HPV therapeutic vaccination, these vaccines in all likelihood contained autologous host cellular antigens as well as viral antigens. There has yet to be a virion-based therapeutic vaccine for any human viral infection.

It is difficult to conceive of a vaccine to control genital HPV that would not involve immunization of both men and women. We have a good understanding of how to detect HPV in women, the attack rate of HPV in women, and the natural history of HPV in women. The information about infectivity, attack

TABLE 11.2 Potential clinical trial end point

Low-risk genital HPVs
    Prophylactic
        Acquisition of infection
        Development of genital warts
    Therapeutic
    Enhancing current therapies
    Prevention of recurrence after treatment

High-risk genital HPVs
    Prophylactic
        Acquisition of infection
        Shortening duration of infection
        Decrease in frequency of cytologic abnormalities
        Decrease in duration of cytologic abnormalities
    Therapeutic
        Resolution of cytologic abnormalities
        Regression of cancer

rate and natural history of HPV in men is poorly understood, particularly as it relates to the high-risk HPV types. Without this information, developing a vaccination strategy and designing clinical trials is very difficult.

Another obstacle to HPV vaccine development, study, design and implementation is the lack of a readily available standardized and validated serologic test to determine who is 'immune' and who is susceptible. This currently necessitates the detection of viral DNA as the only way to characterize patients at entry, and as the only virologic endpoint.

In additional to the scientific challenges there are major psychosocial challenges. Is the public ready to vaccinate their children for sexually transmitted diseases? There are small but vocal groups that question the value of current childhood vaccinations. The public has very little awareness of HPV, their risk of acquiring HPV and the reality that cervical cancer is a sexually transmitted disease.

POTENTIAL VACCINES

Virus-like Particles

Unfortunately, HPV cannot be cultivated *in vitro*, which prevents the development of traditionally killed or live attenuated vaccines. Fortunately, HPV viral capsid proteins L1 and L2, produced by recombinant DNA technology in yeast, bacteria or insect cells, will assemble into particles that look like HPV, and have been termed virus-like particles (VLP). In this form the VLP have the potential to present confirmational epitopes comparable to native virus. VLP-based vaccines have proved safe in animal systems, and early studies are under way in humans.

VLP vaccines can be not only L1 or L1 and L2, either of which can also be combined with early proteins to form chimeric VLPs. VLP vaccines in clinical trials include an HPV 11 VLP (Reichman, 1998, 1999); an HPV6 L1 VLP (Zhang *et al.*, 1999) and an HPV16 L1 VLP (Da Silva *et al.*, 1999).

Proteins

Biotechnology has allowed for the production of fusion protein of all HPV gene products, but which of these or which combinations should be advanced to clinical trials is not known. Experiments in animals indicate that these proteins have the greatest potential for therapeutic vaccines. There is no information on the role of human immunoresponse to the various antigens on the natural history of HPV infection, but there are *in vitro* hints. In the rabbit system, CRPV E1 and E2-based vaccine results in the regression of warts. In humans an HPV6L2-E7 fusion protein has been shown to be immunogenic, and perhaps therapeutic, in the treatment of genital warts.

Peptide Vaccines

Peptide vaccines, particularly if capable of binding to HLA and generating cytotoxic T lymphocytes, have obvious therapeutic appeal. This approach has been used to immunize mice against HPV-16-induced tumor cells and to kill transfected tumor cell targets. At present peptide vaccines are in early trials as candidate therapeutic vaccines for cervical cancer.

Viral Vectors

In essence, viral vector vaccine allows for the manufacture of something potentially equivalent to a live attenuated HPV vaccine. Such a vaccine could potentially generate humoral and CTL responses and be polyvalent. Vaccine-based viral vector vaccines are in early clinical trials in human cervical cancer patients. The vaccine being studied is a vaccinia vaccine expressing modified forms of E6 and E7 from HPV 16 and 18 (Borysiewicz *et al.*, 1996). This appears to be immunogenic. A similar vaccine of vaccinia vaccine big E7/LAMP-1 has been studied in mice. Other candidate protein-based vaccines include HPV-16 E6-E7 protein (Mallarios *et al.*, 1999); HPV-16 E6-E7 + BCG HS/65 (Chu *et al.*, 1999); HPV-16 E7 (Hariharan *et al.*, 1998); HPV-16-E2 protein (Heinemann *et al.*, 1999; Hibma *et al.*, 1999); and chimeric HPV16 E4 with hepatitis B core antigen (E1 Mehdaoui *et al.*, 1999).

Other Approaches

Effecting virus-specific immune responses utilizing direct introduction into the host of viral DNA coding for viral antigens is theoretically possible. To date, this approach has only been investigated in the rabbit system. Other approaches currently in early animal studies include the use of bacterial vector-based vaccines with *Salmonella* (Krul *et al.*, 1996; Londono *et al.*, 1996), *Listeria monocytogenes* (Nardelli-Haflinger *et al.*, 1997) and streptococcus gordoviik

(Jensen *et al.*, 1997). The use of dendritic cells to determine HPV antigen is also being studied.

Vaccine Trial Design

Once candidate vaccines are available, the next obstacle is the design of trials with clinically meaningful yet achievable endpoints.

Low-risk Types

One would hope that vaccination with low-risk HPV types would prevent the acquisition of infection, or at least the development of external genital warts. Although the presence or absence of external genital warts would represent a clear clinical endpoint, there is a lack of precise knowledge about infectivity, susceptibility, attack rate and time from infection to the development of external genital warts. Without this information, designing clinical trials in terms of sample size and duration of follow-up is at best challenging.

Alternatively, a low-risk HPV vaccine could be therapeutic, either by accelerating the resolution of external genital warts, enhancing the efficacy of current therapies, or preventing wart recurrence after current therapy has produced a wart-free state. Whereas 'spontaneous' regression of genital warts has been observed, its precise frequency and factors that favor or prevent resolution are poorly understood. Based on the placebo arm of controlled clinical trials, complete resolution of genital warts over a few months appears to be in the range of 0–20% (Beutner *et al.*, 1998b).

It is possible to enhance the effectiveness of current therapies, but most have about a 50% complete response rate, and so to prove that supplementing current treatment with a vaccine and increasing complete responses from 50% to 80% would require a trial using hundreds of patients.

Similar challenges exist if one hopes to decrease recurrences after conventional therapy. Although published recurrence rates with most modalities appear to be high, there have been no studies prospectively powered and designed to establish the frequency of or time to recurrence following treatment of genital warts with any modality. This information will be required to design vaccine trials adequately.

High-risk HPV Types

The obvious aim of vaccination against high-risk HPV types would be the prevention of cervical cancer. However, this is not a practical endpoint for human vaccine trials. Histologic endpoints are not feasible because the biopsy may alter the natural history. One is then left with several potential surrogate endpoints: the acquisition of infection; the virologic persistence; and a decrease in proportion of subjects who experience cytologic abnormalities.

Preventing the acquisition of high-risk HPV infection is in theory the surest way to prevent cervical cancer. Achieving this may also be the greatest challenge. One of the many problems with this endpoint is again the lack of a serologic test to accurately identify those who are either susceptible or immune. Obviously, true virgins would be susceptible, but limiting enrolment to virgins has some logistic limitations. The presence or absence of HPV DNA in genital and/or cervical samples at the time of enrolment may or may not identify those susceptible.

One current theory as to why a minority of those infected develop cervical cancer is that the persistence of HPV, not just the acquisition of a short-lived infection, is a critical step in the pathway to cancer. If this is true, a vaccine could decrease the proportion of the population with persistent HPV, thereby potentially preventing cervical cancer. As to cytologic endpoints, the use of a screening cytologic test as a clinical endpoint has inherent limitations in terms of interpretation and reproducibility.

Considering that HPV can not be grown *in vitro* and that until very recently there have been no serologic tests for HPV, we have come a long way on the path to a vaccine. With our current understanding of the natural history of genital HPV infection and the availability of candidate vaccines, we now have the challenge of designing and executing large-scale clinical trials. It is difficult to know whether an effective genital HPV vaccine is just around the corner or far away.

REFERENCES

Alexander, M., Salgaller, M., Celis, E. *et al.* (1996). Generation of tumor-specific cytolytic T lymphocytes from peripheral blood of cervical cancer patients by in vitro stimulation with a synthetic human papillomavirus type 16 E7 epitope. *Am. J. Obstet. Gynecol.* 175: 1586–1593

Bauer, H., Ting, Y. and Greer, C. (1991). Genital human papillomavirus infection in female university students as determined by a PCR-based method. *JAMA* 265: 472–477

Bell, J., Sundberg, J., Ghim, S., Newsome, J., Jenson, A. and Schlegel, R. (1994). A formalin-inactivated vaccine protects against mucosal papillomavirus infection: a canine model. *Pathobiology* 62: 194–198

Beutner, K., Wiley, D., Douglas, J. and Tyring, S. (1998a). Genital warts and their treatment. *Clin. Infect. Dis.* 28(Suppl 1): S37–56

Beutner, K., Richwald, G., Wiley, D., Reitano, M., and the AMA Expert Panel on External Genital Warts (1998b). External genital warts: report of the American Medical Association Consensus Conference. *Clin. Infect. Dis.* 27: 796–806

Borysiewicz, L., Fiander, A., Nimako, M., Winkinson, G., Westmoreland, D. and Evans, A. (1996). A recombinant vaccinia virus encoding human papillomavirus types 16 and 18, E6 and E7 proteins as immunotherapy for cervical cancer. *Lancet* 347: 1523–1527

Bosch, F., Manos, M., Munoz, N. *et al.* (1995). Prevalence of human papillomavirus in cervical cancer: a worldwide perspective. *J. Natl. Cancer Inst.* 7: 796–802

Boursnell, M., Rutherford, E., Hickling, J., Rollinson, E., Munro, A. and Rolley, N. (1996). Construction and characterization of a recombinant vaccinia virus expressing human papillomavirus proteins for immunotherapy of cervical cancer. *Vaccine* 14: 1485–1494

Brandt R.M.P., Ressing M.E., de Jong J.H. *et al.* (1999). The HPV16 E7 peptide based vaccine trial in end stage cervical carcinoma: final report. Proceedings of the 17th International Papillomavirus Conference January 9–15, Charleston, SC

Breitburd, F., Kirnbauer, R., Hubbert, N. *et al.* (1995). Immunization with virus-like particles from cottontail rabbit papillomavirus (CRPV) can protect against experimental CRPV infection. *J. Virol.* 69: 3959–3963

Brown, D., Bryan, J., Cramer, H. and Fife, K. (1993). Analysis of human papillomavirus types in exophytic condylomata acuminata by hybrid capture and Southern blot techniques. *J. Clin. Microbiol.* 31: 2667–2673

Christensen, N., Kreider, J., Shah, K. and Rando, R. (1992). Detection of human serum antibodies that neutralize infectious human papillomavirus type 11 virions. *J. Gen. Virol.* **73**: 1261–67

Christensen, N., Reed, C., Cladel, N., Han, R. and Kreider, J. (1996). Immunization with viruslike particles induces long-term protection of rabbits against challenge with cottontail rabbit papillomavirus. *J. Virol.* **70**: 960–965

Chu, N.R., Wu, B., Bantkoch, S. *et al.* (1999). Immunotherapy of human papillomavirus type 16 associated carcinoma using adjurant-free fusion protein encoding M. Boris BCG HSP65 and HPV16 E7. Proceedings the 17th International Papillomavirus Conference, January 9–16, Charleston, SC, USA

Cooney, E.L., A.C. Collier, P.D., Greenberg, R.W. *et al.* (1991). Safety of and immunological response to a recombinant vaccinia virus vaccine expressing HIV envelope glycoprotein. *Lancet* **337**: 567–72

Da Silva, D.M., Nieland, J.D., Greenstone, H.L., Schiller, J.T. and Kast, W.M. (1999). Chimeric papillomavirus virus-like particles induce antigen-specific therapeutic immunity against tumours expressing the HPV-16 E7 protein. Proceedings of the 17th International Papillomavirus Conference, January 9–15, Charleston, SC, USA

Donnelly, J., Martinez, D., Jansen, K., Ellin, R., Montgomery, D. and Liu, M. (1991). Protection against papillomavirus with a polynucleotide vaccine. *J. Infect. Dis.* **713**: 314–320

Dupuy, C., Buzoni-Gatel, D., Touze, A., Cann, P., Bout, D. and Coursaget, P. (1997). Cell mediated immunity induced in mice by HPV 16 L1 virus-like particles. *Microb. Pathogen.* **22**: 219–225

El Mehdaoui S., Touzé A. and Coursaget P. (1999). Production of chimeric human papillomavirus 16 E4 – hepatitis B core recombinant particles. Proceedings of the 17th International Papillomavirus Conference January 9–15, Charleston, SC, USA

Eng T. and Butler W. (1996). The hidden epidemic: confronting sexually transmitted diseases. Committee on Prevention and Control of Sexually Transmitted Diseases. Institute of Medicine, National Academy Press.

Feltkamp, M., Smits, H. and Vierboom, M. (1993). Vaccination with cytotoxic T lymphocyte epitope-containing peptide protects against a tumor induced by human papillomavirus type 16-transformed cells. *Eur. J. Immunol.* **23**: 2242–2249

Feltkamp, M.C.W., Vreugdenhil, G.R., Vierboom, M.P.M. *et al.* (1995). Cytotoxic T lymphocytes raised against a subdominant epitope offered as a synthetic peptide eradicate human papillomavirus type 16-induced tumors. *Eur. J. Immunol.* **25**: 2638–42

Gao, L., Chain, B., Sinclair, C. *et al.* (1994). Immune response to human papillomavirus type-16 E6 gene in a live vaccina vector. *J. Gen. Virol.* **75**: 157–64

Ghim, S., Christensen, N., Kreider, J. and Jenson, A. (1991). Comparison of neutralization of BPV-1 infection of C127 cells and bovine fetal skin xenografts. *Int. J. Cancer.* **49**: 285–289

Gissmann, L., Wolnik, L., Ikenberg, H., Koldovsky, U., Schnurch, H. and zur-Hausen, H. (1983). Human papillomavirus types 6 and 11 DNA sequences in genital and laryngeal papillomas and in some cervical cancers. *Proc. Natl. Acad. Sci. USA* **80**: 560–563

Hariharan, K., Braslawsky, G., Barnett, R.S. *et al.* (1998). Tumor regression in mice following vaccination with human papillomovirus E7 recombinant protein in Provax™. *Int. J. Oncol.* **12**: 1229–1235

Heineman, L., Amer, M., Van Nest, G. and Hibma, M.A. (1999). Immune response to human papillomavirus type 16 early protein E2 following immunization with adjuvant MF 59 in a mouse model. Proceedings of the 17th International Papillomavirus Conference January 9–15, Charleston, SC, USA

Hibma, M.H., Amer, M., Heinemann, L. and Van Nest, G. (1999). Immune responses to the human papillomavirus type 16 early protein E2 following several methods of immunization. Proceedings of the 17th International Papillomavirus Conference January 9–15, Charleston, SC, USA

Hines, J., Ghim, S. and Jenson, A. (1998). Prospects for human papillomavirus vaccine development: emerging HPV vaccines. *Curr. Opin. Obstet. Gynecol.* **10**: 15–19

Ho, G. (1998). Natural history of cervicovaginal papillomavirus infection in young women. *N. Engl. J. Med.* **338**: 423–8

Jarrett, W., Smith, K., O'Neil, B. *et al.* (1991). Studies on vaccination against papillomaviruses: prophylactic and therapeutic vaccination with recombinant structural proteins. *Virology* **184**: 33–42

Jensen, E., Selvakumar, R. and Shen, H. (1997). Recombinant *Listeria monocytogenes* vaccination eliminates papilloma virus induced tumors and prevents papilloma formation from viral DNA. *J. Virol.* **71**: 8467–8474

Jin, X., Cowsert, L., Marchall, D. *et al.* (1990). Bovine serological response to a recombinant BPV-1 major capsid protein vaccine. *Intervirology* **31**: 345–354

Kast, W., Feltkamp, M., Ressing, M., Vierboom, M., Brandt, R. and Melief, C. (1996). Cellular immunity against human papillomavirus associated cervical cancer. *Semin. Virol.* **7**: 117–123

Kidd, J. (1938). The course of virus-induced rabbit papillomas as determined by virus, cells, and host. *J. Exp. Med.* **67**: 551–574

Kienzler, J., Lemoine, M., Orth, G. *et al.* (1983). Humoral and cell-mediated immunity to human papillomavirus type 1 (HPV-1) in human warts. *Br. J. Dermatol.* **108**: 665–666

Kirnbauer, R., Chandachud, L., O'Neil, B. *et al.* (1996). Virus-like particles of bovine papillomavirus type 4 in prophylactic and therapeutic immunization. *Virology* **219**: 37–44

Kiviat, N. and Koutsky, L. (1993). Specific human papillomavirus types as the causal agents of most cervical intraepithelial neoplasia: implications for current views and treatment. *J. Natl. Cancer Inst.* **85**: 934–935

Koutsky, L., Holmes, K. and Critchlow, C. (1992). A cohort study of the risk of cervical intraepithelial neoplasia grade 2 or 3 in relation to papillomavirus infection. *N. Engl. J. Med.* **327**: 1272–1278

Koutsky, L.A., Galloway, D.A. and Holmes, K.K. (1988). Epidemiology of genital human papillomavirus infection. *Epidemiol. Rev.* **10**: 122–63

Krul, M., Tijhaar, E., Kleijne, J., Van Loon, A., Nievers, M. and Schipper, H. (1996). Induction of an antibody response in mice against human papillomavirus (HPV) type 16 after immunization with HPV recombinant *Salmonella* strains. *Cancer Immunol. Immunother.* **43**: 44–48

Lacey, C., Monteiro, E. and Thompson, H. (1997). *A phase IIa study of a therapeutic vaccine for genital warts.* Medical Society for the Study of Venereal Disease Spring Meeting, Oxford, UK

Lin, Y., Borenstein, L., Ahmed, R. and Wettstein, F. (1993). Cottontail rabbit papillomavirus L1 protein-based vaccines: protection is achieved only with a full-length, nondenatured product. *J. Virol.* **67**: 4154–4162

Lin, Y.-L., Borenstein, L., Selvakumar, R., Ahmed, R. and Wettstein, F. (1992). Effective vaccination against papilloma development by immunization with L1 or L2 structural protein of cottontail rabbit papillomavirus. *Virology* **187**: 612–619

Lin, K.Y., Guarnieri, F.G., Staveley-O'Carroll, K.F. *et al.* (1996). Treatment of established tumors with a novel vaccine that enhances major histocompatibility class II presentation of tumor antigen. *Cancer Res.* **56**: 21–6

Londono, L.P., Chatfield, S., Tindle, R.W. *et al.* (1996). Immunization of mice using *Salmonella typhimurium* expressing human papillomavirus type-16 E7 epitopes inserted into hepatitis-B virus core antigen. *Vaccine* **14**: 545–52

Lorincz, A., Jenson, R., Greenberg, M., Lancaster, W. and Kurman, R. (1992). Human papillomavirus infection of the cervix: relative risk associations of 15 common anogenital types. *Obstet. Gynecol.* **79**: 328–337

Lowe, R., Brown, D., Bryan, J. *et al.* (1997). Human papillomavirus type 11 (HPV-11) neutralizing antibodies in the serum and genital mucosal secretions of African green monkeys immunized with HPV-11 virus-like particles expressed in yeast. *J. Infect. Dis.* **176**: 1141–1145

Luxton, J., Rose, R., Coletart, T., Wilson, P. and Shepherd, P. (1997). Serological and T-helper cell responses to human papillomavirus type 16 L1 in women with cervical dysplasia or cervical carcinoma and in healthy controls. *J. Gen. Virol.* **78**: 917–923

Mallarios J., Quinn C., Arnold F. and Macfarlan R. (1999). Cell-mediated immune responses to recombinant HPV-16 E6E7hh protein generated by immunization with chelating Iscomatrix[TM] adjuvant. Proceedings of the 17th International Papillomavirus Conference January 9–15, Charleston, SC, USA

Meneguzzi, G., Cerni, C. and Kieny, M. (1991). Immunization against human papillomavirus type 16 tumor cells with recombinant vaccinia viruses expressing E6 and E7. *Virology* **181**: 62–69

Nardelli-Haflinger, D., Roden, R.B.S., Benyacoub, J. *et al.* (1997). Human papillomavirus type 16 virus-like particles expressed in attenuated *Salmonella typhimurium* elicit mucosal and systemic neutralizing antibodies in mice. *Infect. Immun.* **65**: 3328–36

Pilacinski, W., Glassman, D., Glassman, K. *et al.* (1986). Immunization against bovine papilloma-virus infection. *Ciba Foundation Symposium* **120**: 136–156

Reichman R., Balsley J., Carlin D. *et al.* (1999). Evaluation of the safety and immunogenicity of a recombinant HPV-11 L1 virus like particle vaccine in healthy adult volunteers. Proceedings of the 17th International Papillomavirus Conference January 9–15, Charleston, SC, USA

Reichman, R., Bonnez, W., O'Brien, D. *et al.* (September 24–27, 1998). A phase I study of a recombinant virus-like particle vaccine against human papillomavirus type 11 in healthy adult volunteers. 38th Interscience Conference on Antimicrobial Agents and Chemotherapy, San Diego, California

Rose, R., Bonnez, W., Reichman, R. and Garcea, R. (1993). Expression of human papillomavirus type 11 L1 protein in insect cells: in vivo and in vitro assembly of viruslike particles. *J. Virol.* **67**: 1936–1944

Rowen, D. and Lacey, C. (1998). Toward a human papillomavirus vaccine. *Dermatology Clinics* **16**(4): 835–838

Rowen, D., Thompson, H. and O'Neill (1997). A phase IIa, open study to evaluate the safety, immunogenicity and clinical response to TA-GW given in combination with cryotherapy to men with genital warts. Medical Society for the Study of Venereal Disease Spring Meeting, Oxford, UK

Schiffman, M., Bauer, H., Hoover, R. *et al.* (1993). Epidemiologic evidence showing that human papillomavirus infection causes most cervical intraepithelial neoplasia. *J. Natl. Cancer Inst.* **85**: 958–964

Schlegel, R., Phelps, W., Zhang, Y.-L. and Barbosa, M. (1988). Quantitative keratinocyte assay detects two biological activities of human papillomavirus DNA and identifies viral types associated with cervical carcinoma. *EMBO J* **7**: 3181–3187

Sherman, M., Schiffman, M., Strickler, H. and Hildesheim, A. (1998). Prospects for a prophylactic HPV vaccine: rationale and future implications for cervical cancer screening. *Diagn Cytopathol* **18**: 5–9

Shope, R. (1935). Immunization of rabbits to infectious papillomatosis. *J Exp Med* **65**: 219–231

Steele, J. and Gallimore, P. (1990). Humoral assays of human sera to disrupted and nondisrupted epitopes of human papillomavirus type 1. *Virology* **174**: 388–398

Steller M.A., Gurski K.J., Murakami M. *et al.* (1998). Cell-mediated immunological responses in cervical and vaginal cancer patients immunized with a lipidated epitope of human papilloma-virus type 16 E7. *Clin. Cancer Res.* **4**(9): 2103–2109

Steller M.A., Schiller J.T. (1996). Human papillomavirus immunology and vaccine prospects. *Monogr. Natl. Cancer Inst.* **21**: 145–148

Suzich, J., Ghim, S. and Palmer-Hill, F. (1995). Systemic immunization with papillomavirus L1 protein completely prevents the development of viral mucosal papillomas. *Proc. Natl. Acad. Sci. USA* **92**: 11553–11557

Tindle R.W., Croft S., Herd K. *et al.* (1995). A vaccine conjugate of ISCAR immunocarrier and peptide epitopes of the E7 cervical cancer – associated protein of human papillomavirus type 16 elicits specific Th1 – and Th2 – type responses in immunized mice in the absence of oil – based adjuvants. *Clin. Exp. Immunol.* **101**: 265–271

Walboomers, J., De Roda Husman, A., Van Den Brule, A., Snijders, P. and Meijer, C. (1994). Detection of genital human papillomavirus infections: critical review of methods and prevalence studies in relation to cervical cancer. Oxford, UK, Oxford University Press

Zaki, S., Judd, R., Coffield, L., Greer, P., Rolston, F. and Evatt, B. (1992). Human papillomavirus infection and anal carcinoma. Retrospective analysis by in situ hybridization and the polymerase chain reaction. *Am. J. Pathol.* **140**: 1345–1355

Zhang L.F., Zhou J., Shao C. *et al.* (1999). A phase 1 trial of HPV 6 B virus like particles as immunotherapy for genital warts. Proceedings of the 17th International Papillomavirus Conference January 9–15, Charleston, SC

Zhu, X., Tommasino, M., Vousden, K. *et al.* (1995). Both immunization with protein and recombinant vaccinia virus can stimulate CTL specific for the E7 protein of human papilloma-virus 16 in H-2(D) mice. *Scand. J. Immunol.* **42**: 557–63

12
THERAPEUTIC VACCINES FOR CONTROL OF HUMAN PAPILLOMAVIRUS CHRONIC INFECTIONS

PIERRE VANDEPAPELIÈRE

SmithKline Beecham Biologicals, Rixensart, Belgium

CLINICAL ASPECTS AND EPIDEMIOLOGY

Introduction

Human papillomavirus (HPV) infections are among the most common chronic viral infections in humans and are considered to be the major causative agent of cervical carcinoma (zur Hauzen, 1991; Kjaer *et al.*, 1996). They are principally sexually transmitted and cause various diseases in the anogenital area. The prevalence of HPV infections ranges between 20% and 80% in sexually active adults, depending on the various risk factors. The majority remain asymptomatic (Koutsky *et al.*, 1988). It is estimated that 1 million new cases of HPV disease are diagnosed annually in the United States (ASHA, 1998; CDC, 1983; US CDC, 1996; IMC PC of STD, 1997; Beutner *et al.*, 1998), and approximately 500 000 new cases of cervical carcinoma develop annually worldwide (Munoz and Bosch, 1989; Bosch *et al.*, 1995). Over 95% of these tumors contain DNA from high-risk HPV types (Munoz and Bosch, 1996; Walboomers and Meijer, 1997).

To date, more than 80 types of papillomavirus have been identified, and others are still being characterized (de Villiers, 1989, 1994). These viruses can be differentiated as they are both species and tissue specific, and some are benign whereas others lead to the development of cancer. In humans, HPV can therefore be subdivided into low-risk HPV, which do not lead to neoplastic transformation, and high-risk HPV, which are associated with the development of cancers, mainly in the genital area, and also according to the site of infection, i.e. either cutaneous or mucosal (Table 12.1).

TABLE 12.1 Most common human HPV types and associated diseases

Group	Virus types		Associated diseases
	Major	**Minor**	
Low risk			
Cutaneous	HPV1		Verruca plantaris
	HPV2		Verruca vulgaris
	HPV3		Verruca plana
Mucosal	HPV6		Genital warts
			(papular or flat)
	HPV11		Laryngeal papillomatosis
			Tumor of Buschke-Loewenstein
High risk			
Cutaneous	HPV8		Epidermodysplasia
			verruciformis ($\rightarrow$ carcinoma)
Muscosal	HPV16–18	31-33-45-	Flat warts
		35-39-51-	Cervical dysplasia and carcinoma
		52-56-68	Carcinoma of penis, vulva, vagina, anus

Low-risk HPV

HPV6 and HPV11 are the most frequent low-risk HPV in humans and are mainly responsible for the development of genital warts (Stone, 1995). HPV6 is the causative agent in 80% of the cases in immunocompetent people, whereas in immunodepressed subjects HPV11 has been reported to be predominant (Brown *et al.*, 1999). These benign tumors develop on the internal and external anogenitalia and present either as cauliflower-like warts, which are named condylomata acuminata, or else as flat warts, usually when the infection develops internally. They are generally asymptomatic but are sometimes accompanied by itching, burning or bleeding. However, they cause major psychosocial disturbances and are highly infectious. Their evolution is variable, leading either to growth, persistence or spontaneous regression, with a mean regression rate of 2–11% in 3 months (Beutner and Ferenczy, 1997). Newly developed warts are said to be more prone to regress spontaneously but definite proof is still needed.

A rare but severe clinical presentation is laryngeal papillomatosis, which causes significant morbidity and is potentially life threatening (Kashima *et al.*, 1996). Laryngeal papillomatosis can be acquired at birth by babies from infected mothers, or during adulthood. Warts develop in the upper respiratory tract, usually in the larynx and on vocal cords, and can lead to obstruction of airways and suffocation. The only effective therapy is laser surgery, under general anesthesia (Bauman and Smith, 1996). Relapse is almost a rule, sometimes with invasion of the lower respiratory tract with a severe outcome (Rimell, 1999). The mean number of surgical interventions is five per year for neonatally acquired infection and two per year when the disease is contracted during adulthood.

High-risk HPV

Infection by high-risk HPVs can evolve to low-grade preneoplastic lesions (CIN I), to a high-grade precancerous stage (CIN III), to cancer in situ (CIS), and ultimately to invasive cervical carcinoma (CC) (see Richart, 1987; Figure 12.1). Only a small percentage will progress from one step to the next and this process usually takes several years. On the other hand, Critchlow and Kiviat support the hypothesis that some high-risk HPV infections are CIN III from the onset (Critchlow and Kiviat, 1999; Kiviat *et al.*, 1992). Persistent infection by high-risk HPV is a prerequisite for the development and maintenance of CIN III (Meijer *et al.*, 1999), and a high viral load has been associated with persistent HPV infection and increased risk of cervical dysplasia (Caballero *et al.*, 1999; Franco, 1999), but this is controversial. Also, at each stage, with the exception of CIS and CC, which are not spontaneously reversible, the possibility of reversion to the previous stage is important, including in some cases total clearance of lesions and infection (Nasiell *et al.*, 1986). Progression from HPV infection to CIN I is detected in approximately 3–5% of cases per year. In a 15-year follow-up study in Finland the rate of progression remained remarkably stable, between 8 and 17%, whatever the duration of follow-up (Syrjänen, 1996). However, the rate of regression increased with time, with a parallel decrease in the number of persistent cases. A long-term follow-up study in the Netherlands of women with initially abnormal or normal cytology with or without high-risk HPV infection at entry (Walboomers and Meijer, 1999) revealed that women with abnormal cytology and high-risk HPV infection were 29 times more at risk of developing CIN III than HPV-negative women. If the cytology was normal, the presence of high-risk HPV increased the possibility of developing CIN III lesions by 116 times. In the case of persistent HPV infection, progression to CIN III occurred at 7% and 14% per year for patients with normal or abnormal cytology at entry, respectively. Sixteen per cent of untreated mild dysplasia (CIN I) will progress to carcinoma in situ (CIS) over a period of 84–96 months.

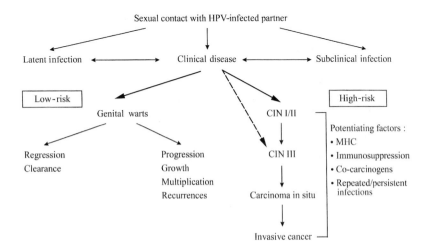

FIGURE 12.1 Human papillomavirus and disease cycles

HPV Transmission

Transmission usually occurs through sexual contact. HPV is resistant to heat and dessication and is therefore very stable. Up to 60% of individuals will develop the disease after a single sexual contact with an infected partner: the number of sexual partners appears to be the most important risk factor for the acquisition of HPV (Burk *et al.*, 1996). The virus is thought to penetrate through microlesions in the skin or the mucosa: semen probably plays a major role in transmission and the prostate could be an important reservoir. There is a variable interval between exposure and the development of HPV lesions, ranging from 8 weeks to 8 months.

Vertical transmission from infected mothers to neonates probably occurs very frequently, but rarely leads to the development of disease. However, if it does occur the infection may persist for many years in the mouths of these children (Lacey, 1996; Puranen *et al.*, 1996; Rice *et al.*, 1999). Neonatal infection by HPV6, or more often HPV11, may lead to the development of laryngeal papillomatosis in a small percentage of babies (Kashima *et al.*, 1996; Bauman and Smith, 1996). High-risk HPV may also be detected in the mouths of babies for several months or even years after delivery (Fredricks *et al.*, 1993; Cason *et al.*, 1995).

Finally, transmission could also occur without physical contact owing to the high stability of the virus. Transmission has been described through contaminated fomites (Ferenczy *et al.*, 1989), infected towels or clothes (Bergeron *et al.*, 1990), surgical instruments or as a result of a baby bathing with an infected mother (Gibson *et al.*, 1990). It has also been detected in the plume generated by laser vaporization of warts (Kashima *et al.*, 1991). In the latter case, if the vaporized HPV DNA are inhaled this could (in theory) lead to

TABLE 12.2 Main therapies for low-risk HPV-associated warts

Physical ablation
Surgical excision
Electrocautery
Laser
Cryotherapy

Caustic cytodestruction
Podophyllotoxin
Podophyllin
Bichloroacetic/trichloroacetic acid

Antiviral/immunomodulatory chemotherapies
Cidofovir (in development)
Imiquimod
Interferon
Bleomycin
5-Fluorouracil

Others
Phototherapy (laryngeal papillomatosis)
Retinoids

TABLE 12.3 Therapies for cervical dysplasia

Electrocautery

Cryotherapy

CO_2 laser vaporization

LEEP (loop electrosurgical excisional procedures)

Electroconization

Laser conization

Cold knife conization

laryngeal HPV infection of the attending medical personnel. These anecdotal reports need proper confirmation.

Current Therapies

Low-risk HPV

There are numerous therapies for genital warts (see Table 12.2), the efficacy of which varies from study to study. However, in general the recurrence rate appears to remain similar whatever the therapy used, with an average of 30% recurrence within 3 months post therapy (Beutner and Frerenczy, 1997; Baker and Tyring, 1997).

High-risk HPV (see Table 12.3)

High-grade cervical dysplasia is usually treated by ablation or excision with a high cure rate of between 92% and 96% (Barasso *et al.*, 1997). There is no consensus on the necessity to treat low-grade lesions, as most of these will regress spontaneously. A close follow-up is usually recommended, especially if it is an infection by high-risk HPV.

PATHOGENESIS: MECHANISMS OF PERSISTENCE

Pathogenesis (Figure 12.1).

Papillomaviruses are made up of circular double-stranded DNA enclosed in an isocahedral 550 nm nucleocapsid. They have no envelope. The genome is divided into two regions: an early region which usually contains six (E1, E2, E4, E5, E6 and E7) open reading frames (ORF), coding for viral replication and transformation, and a late region, containing two open reading frames (L1 and L2) which code for nucleocapsid proteins.

Among the early region ORF, E1 and E2 are mainly involved in replication and transcription, whereas E6 and E7 are responsible for transformation. In particular in oncogenic HPV, E6 binds with tumor suppressor p53 and E7 with tumor suppressor Rb, blocking their action. These proteins are continuously expressed within tumor cells and are necessary to maintain the transformed state (von Knebel *et al.*, 1994; zur Hauzen, 1995). L1 codes for the major nucleocaspid protein and L2 for the minor nucleocapsid protein.

HPV infections are usually limited to their host cells, which are the keratinocytes, and do not pass the stratum basale. There is therefore no evidence of a viremic phase and little if any exposure to the systemic immune system.

Infection of the cervix usually occurs at the squamocolumnar junction of the ectocervix and endocervix, or the transformation zone. Limited replication occurs in the suprabasal layers of the epithelium, with expression of the early ORF proteins. As infected cells mature and move towards the upper layers of the epithelium, the late proteins L1 and L2 are expressed to form capsids. Viral shedding does not occur within the epithelium and mature viruses are only shed externally. In the case of oncogenic viruses, E6 and E7 proteins transform the cells leading to CIS and CC. Benign tumor cells contain episomal DNA and capsid proteins, whereas CIN lesions and cancer also contain HPV DNA integrated into host DNA. Once integrated, HPV16 is non-replicating and non-infectious. E6 and E7 are expressed at high levels in cervical cancer.

Mechanisms of persistence

The facts that the usual infection remains confined to the epithelium, together with the absence of inflammation and (most likely) some viral effects on local Langerhans cells and on antigen presentation are probably associated with the systemic immune system's degree of ignorance of HPV lesions, which prevents the mounting of an adequate immune response. However, some expression to the immune system must occur, as antibodies are detected in most patients with clinical disease, when sensitive detection methods are used (Foster *et al.*, 1997). These antibodies disappear rapidly after resolution of the HPV lesions. Therefore, the persistence and the relatively limited growth of warts support the theory that either the virus-infected cells are able to escape the antiviral immune response, or that this response is only slightly or not at all induced.

The cell-mediated immune response appears to play a major role in regression of the warts (Tagami *et al.*, 1974), and is probably downregulated or modified by HPV viruses (Frazer, 1996a). Women with multiple condylomata may have a higher percentage of suppressor/cytotoxic T cells, a lower percentage of T-helper cells, and the ratio T-helper/T-suppressor cells can be decreased (Carson *et al.*, 1986; Cauda *et al.*, 1987). In the tissue of condyloma acuminata and of CIN lesions there is a depletion of T-helper cells, with a reversal of T4/T8 ratio to less than 1. The local production of IL-2 and γ-IFN is also decreased (Tay *et al.*, 1987). On the other hand, spontaneously regressing warts are massively infiltrated with CD4+ T cells and an increased level of IL-12 expression is observed (Coleman *et al.*, 1994). Therefore, it seems that persistence correlates with a local depletion of CD4+ T-cell activity, which is restored in case of spontaneous regression.

Host factors also seem to play a role in influencing the outcome of HPV infection. An association has been found between some HLA class II genes and the development of cervical cancer following HPV16 infection (Wank *et al.*, 1991, 1992). On the other hand, data suggesting an association between some MHC class II phenotypes and regression of PV-associated lesions have recently been reported in rabbits (Frazer, 1996a).

Finally, as opposed to hepatitis B virus infection, immune depression or suppression favors the progression of warts, and also significantly increases the risk of carcinogenesis (Benton *et al.*, 1992; Palefsky, 1991).

In conclusion, chronic HPV infections probably result from multiple mechanisms, including limited exposure to the immune system, bias of cellular immune responses and genetic host factors.

Animal Models and Therapeutic Vaccines

Because of the large variety of papillomaviruses (PV) infecting various animals, numerous animals can be used as experimental models for HPV infection. Although they cannot be infected by the human PV because papillomaviruses are species specific, some of these animal PV share characteristics with HPV infections and diseases, for example the tissue specificity for the skin or mucosa, and the development of malignancies in some cases (Brandsma, 1994).

The two most studied models are papillomaviruses infecting the cottontail rabbit (CRPV) and those infecting cattle (BPV-1, 2 and 4). CRPV was the first PV identified, by Shope in 1933, with the association between PV infection and carcinoma observed shortly thereafter (Rous and Beard, 1935). Following the observation of spontaneous regression of warts, therapeutic vaccination using autologous or heterologous warts was shown to increase the frequency of regression (Evans *et al.*, 1962). More recent data demonstrated that vaccination with purified E1 and E2 Trip E fusion proteins induced regression of papillomavirus-induced lesions (Selvakumar *et al.*, 1995a), but this efficacy was not improved by adjuvantation with the Th1 inducer Ribi adjuvant. It was also shown in CRPV that vaccination with the E6 but not the E7 protein induced regression of warts. Tumor regression is associated with specific antibody and lymphoproliferative responses to E2 in the cottontail rabbit model (Selvakumar, 1995b).

Bovine papillomavirus (BPV) infections differ from HPV in that there is a dermal involvement and no neoplastic progression (Campo, 1994). On the other hand, some BPV (BPV-1 and 2) cause cutaneous lesions and others affect mucosal sites, such as BPV-4. In this bovine model, regression of lesions caused by BPV-4 has been induced by the E7 protein (Campo *et al.*, 1993) and not by E2. In addition, the development of lesions was not prevented by vaccination with L2, but this did induce an early regression of the tumor. L2 could therefore be considered a valid candidate therapeutic antigen (Brandsma, 1996; Kirnbauer *et al.*, 1996). Similarly, in BPV-2 infection, L2 administered prior to challenge does not protect against the development of lesions, but these warts are massively infiltrated by T cells and rapidly regress (Campo *et al.*, 1994).

In both animal models the structural protein L1 is not able to induce regression, but can produce large amounts of neutralizing antibodies that protect against challenge (Jarret *et al.*, 1991; Christensen *et al.*, 1996; Kirnbauer *et al.*, 1996).

Two other animal models of interest have been less intensively evaluated. The canine oral papillomavirus (COPV) induces the development of buccal, lingual and gingival papillomas, but there is no malignant progression. COPV has been used as a model for the prevention of oral mucosal warts in dogs

(Suzich *et al.*, 1995). The Rhesus monkey papillomavirus (Rh PV1) could be an excellent model for evaluation of HPV-associated cervical carcinogenesis, but infectious virus is difficult to obtain.

An HPV-induced tumor model in mice has recently been developed and is now extensively used for testing of prophylactic and therapeutic vaccines (Feltkamp *et al.*, 1993; Lin *et al.*, 1996; Greenstone *et al.*, 1998; De Bruijn *et al.*, 1999).

In conclusion, regression of papillomavirus-induced lesions can be obtained in at least two different animal models. This regression is accompanied by both strong humoral and cellular immune responses.

RATIONALE FOR THERAPEUTIC VACCINES

There are numerous facts that support the rationale for therapeutic vaccination as potential therapy for HPV-induced diseases. They can be summarized as follows:

1. Spontaneous regression of lesions is observed in animals and also in humans (Tagami *et al.*, 1974; Coleman *et al.*, 1994).
2. The spontaneous regression is accompanied by a vigourous humoral and lymphoproliferative immune response (Stanley *et al.*, 1994; Aiba *et al.*, 1986).
3. Examination of the cellular infiltrate in spontaneously regressing warts shows a massive T-cell invasion, mainly of the CD4+ type (Coleman *et al.*, 1994).
4. This is in correlation with the observation of a reverse T-helper/ T-suppressor (T4/T8) ratio in infected subjects. The T-helper response is restored in the case of spontaneous regression.
5. In animal models, early ORF proteins and sometimes structural proteins are able to induce regression of lesions.
6. Immunosuppression increases the prevalence and the severity of HPV infections, as well as the risk of carcinogenesis in high-risk HPV-induced lesions (Lutzner, 1985; Benton *et al.*, 1992; Fruchter *et al.*, 1996).
7. Numerous clinical trials of autogenous vaccines in humans have claimed positive results in reducing the recurrence rate and in inducing regression of warts (Powell *et al.*, 1970).

Therefore, there is a strong rationale for developing therapeutic vaccines for the treatment of HPV-induced lesions and to prevent their recurrence.

IMMUNOTHERAPEUTIC APPROACHES

Therapeutic vaccination for HPV-induced lesions is an old concept, with the first attempt being described in 1924. Early attempts were, however, limited to autogenous vaccines, and it was not until recently that research expanded in this field (Hagensee, 1997; Vonka, 1996). This explains why, besides the experience with autogenous vaccines, clinical evaluation with more recent compounds is so far limited to phase I or II trials.

Active Specific Immunotherapy

Autogenous Vaccines

The first reports of immunization therapy for HPV were described in 1924 (Biberstein, 1924). In 1970, Powell *et al.* published the results of therapy for condyloma acuminata resistant to conventional treatment but which responded to autogenous vaccination (Powell *et al.*, 1970). This vaccine was prepared from sterilized heat-inactivated wart extracts and, after a series of weekly injections, warts regressed completely in 20 out of 24 subjects. Several other reports of similar vaccination from wart homogenate were described with excellent success rates (Biberstein, 1944; Abcarian *et al.*, 1976; Abcarian and Sharon, 1977, 1982; Eftaiha *et al.*, 1982). However, all these studies present pitfalls in methodology such as absence of control group and small number subjects. In a double-blind, randomized, placebo-controlled cross-over design, Malison *et al.* failed to show a significant difference between autogenous vaccine and placebo (Malison *et al.*, 1982). Another vaccinal approach was recently reported as autogenous vaccination by Wiltz *et al.* (Wiltz *et al.*, 1995). Homogenized warts were planted in four types of bacterial media plates. Organisms that grew on these plates – mainly *Escherichia coli*, *Klebsiella* and *Enterococcus* species – were then recultivated several times and inactivated. When these bacteria stopped growing they were placed in solution, which formed the vaccine. Patients were injected subcutaneously with the vaccine at increasing doses, from 0.1 to 1.0 ml, three times a week for up to 12 weeks. The total follow-up period ranged from 3 months to 7 years. The recurrence rate after excision plus vaccine was 4.6%, compared to 50% following excision alone. It is difficult to interpret these data since no HPV material should normally remain in the injected vaccine after that preparation. The reported efficacy would then be related to a non-specific adjuvant effect.

Current Therapeutic Vaccine Approaches

More recent therapeutic vaccine approaches are based on specific immune stimulation against various well defined HPV antigens. The immune stimulation must be T-cell mediated, inducing both CD4+ and CD8+ types. Initially, the antigens selected were mainly early proteins, especially E6 and E7 (Chen *et al.*, 1991, 1992; Ho *et al.*, 1999; Murakami *et al.*, 1999), but recently L1, under the presentation of chimeric virus-like particles (VLPs), has also gained interest in the therapeutic area.

Live Vector Vaccines

In order to be able to destroy tumor-infected cells an immune response against intracellularly expressed proteins should primarily involve cytotoxic T-lymphocyte activity (CTL). One way to induce CTL is to use viral vectors such as vaccinia.

A recombinant vaccinia vaccine expressing modified forms of E6 and E7 from both HPV16 and HPV18 has been developed by Cantab Pharmaceuticals Research Ltd. E6 and E7 were modified in order to suppress the oncogenic potential by deleting the regions involved in binding with the two cellular proteins that control cell division, p53 and Rb (Boursnell *et al.*, 1996). This vaccine was successful in inducing CTLs in mice.

Phase I results with this vaccine have recently been reported (Borysiewicz *et al.*, 1996). A single dose of vaccine was injected into eight women with late-stage cervical cancer. As well as exhibiting an excellent safety profile, three patients mounted a HPV-specific antibody response, and HPV-specific cytotoxic T lymphocytes could be detected in one patient. This study was too limited to draw final conclusions, but none-the-less this vaccine deserves further investigation. Another group has also developed a recombinant vaccinia vaccine (Sig/E7/LAMP-1) which has a significant antitumor effect in an HPV16 E6/E7-expressing murine tumor model, TC-1, and which induces E7-specific CD4+ and CD8+ T-cell responses (Che, 1999a,b).

Protein-Based Vaccines

For the reasons explained above, the initial proteins of choice for therapeutic vaccines against high-risk HPV types are the early proteins, especially E6 and E7. Several groups are currently developing such protein-based vaccines, usually in combination with adjuvants and/or vehicles in order to induce a stronger cellular immune response. Some of these are as follows:

1. HPV16 E6/E7 protein (fused on hexa-his tag) with chelating Iscomatrix, developed by CLS, Australia. Results: induction of CTLs specific for E7 epitopes in mice (Mallarios *et al.*, 1999);
2. HPV16 E7 + BCG HS/65 by Stress Gene Biotechnologies (Chu *et al.*, 1999);
3. HPV16 E7 protein combined with an adjuvant formulation, Provax, a microfluidized emulsion, induces a significant reduction of tumor growth in mice (Hariharan *et al.*, 1998).
4. HPV16 – E2 protein with MF59 adjuvant developed by Chiron Pharmaceuticals: it induces a strong Th2 response in mice (Heineman *et al.*, 1999; Hibma *et al.*, 1999).
5. Chimeric HPV16 E4 with hepatitis B core antigen (El Mehdaoui *et al.*, 1999).

A more advanced vaccine development program is that described by Cantab Pharmaceuticals for the treatment of genital warts. The vaccine is an L2E7 fusion protein produced in *E. coli* and combined with aluminium salts. Three doses of 300 µg vaccine induced a trend towards fewer recurrences in patients with recurrent condyloma acuminata (Thompson *et al.*, 1999; Lacey *et al.*, 1999). However, this uncontrolled study needs confirmation. The addition of an adjuvant, monophosphoryl lipid A, increased antigen-specific *in-vitro* T-cell proliferative response, γ-interferon secretion and *in-vivo* delayed-type hypersensitivity response (Thompson *et al.*, 1998). This antigen is now currently licensed to SmithKline Beecham Biologicals Immunotherapeutics, Rixensart, Belgium, and is being clinically evaluated using more potent adjuvant systems.

Cantab has also generated results in mice with an HPV16 L2 protein, showing some cross-neutralization between HPV16 L2 and HPV18 L2, although VLP L1/L2 from the same virus types do not cross-neutralize (Roden *et al.*, 1999).

The capsid proteins L1 and L2 were initially developed for prophylactic use only. L1 in particular, when expressed recombinantly in various systems such as yeast, insect cells or tissue culture, self-assembles into virus-like particles L1 VLP (Kirnbauer *et al.*, 1992; Rose *et al.*, 1993). These are very immunogenic in animals and produce high levels of neutralizing antibodies. For this reason they are primary candidates for prophylactic vaccines (Frazer, 1996b). L1 VLPs also induce strong CD4+ and CD8+ cellular immune responses both in mice (De Bruijn *et al.*, 1998; Rudolf *et al.*, 1999) and in other animal models (Jansen *et al.*, 1995). Chimeric VPLs have been tested therapeutically in the mouse tumor regression model, with very positive results (Schäfer *et al.*, 1999; Da Silva *et al.*, 1999; Velders *et al.*, 1999). Those could therefore also be interesting therapeutic candidates. This is further supported by additional interesting properties of L1 VLPs:

1. Chimeric L1 VLPs can also be constructed which include L2 alone or L2 plus early proteins. In such constructions L1 VLPs serve as immunostimulants and as carriers of other antigens, while also presenting their own epitopes (Da Silva *et al.*, 1999; Neeper *et al.*, 1999).
2. L1 VLPS can be administered via the mucosal route, either nasally or orally, and this induces a good mucosal immune response (Liu *et al.*, 1999; Rose *et al.*, 1999; Nardelli-Haefliger *et al.*, 1999; Hagensee *et al.*, 1993).
3. VLPS can serve as carriers of plasmid DNA (Touze and Coursaget, 1999; Kawana *et al.*, 1999), of costimulatory proteins, for example B7.2 (Shi *et al.*, 1999), or of foreign antigen such as HIV1 gP41 (Slupetzky *et al.*, 1999).

At least three L1 VLPs have reached the clinical stage:

1. A phase I study has been conducted with the HPV11 L1 VLP on Al(OH)$_3$ from Med Immune (Reichman *et al.*, 1999). Four doses of 3, 9, 30 and 100 µg were injected into healthy volunteers following a 0–1–4-month schedule. The safety profile was very good and the three higher doses produced equivalent levels of neutralizing and ELISA antibodies.
2. An HPV6 L1 VLP without aluminium has been evaluated in subjects with genital warts in China: 33 volunteers received three injections at 4-week intervals of either 1, 5 or 10 µg of VLP L1 HPV6 B. At week 20, 26 of 33 subjects had cleared their warts, 29/33 had low antibody titers, and 23/24 had a positive DTH skin test (Zhang *et al.*, 1999).
3. Another dose comparison study of an HPV16 L1 VLP produced in baculovirus (10 and 50 µg doses with or without aluminium) was presented by an NIH group during the 17th International Papillomavirus Conference (1999). After two injections the antibody levels were significantly higher with the higher dose.

In conclusion, several protein-based vaccines are in early clinical development and appear very promising as potential therapeutic vaccines for HPV-induced lesions.

Peptide Vaccines

Another approach aimed at specific induction of cytotoxic T lymphocytes (CTLs) is the administration of peptides associated with class I MHC molecules (van Driel et al, 1996).

Several groups are developing HLA-A0201 restricted HPV16 lipopeptide vaccines. Steller *et al.* (1998) published results from a phase I study in women with cervical carcinoma, using HLA H2 restricted CTL HPV16 E7 peptide linked to PADRE, a non-specific helper peptide. All women mounted a cellular immune response and 4 of 10 evaluable subjects elicited specific CTLs. There was no toxicity, but also no clinical response.

Another group has developed a vaccine composed of two CTL-specific peptides. HPV16 E7 peptides are combined with the same T-helper peptide, PADRE, and with a mineral oil-based adjuvant Montanide ISA 51 (Feltkamp *et al.*, 1993). The final results were presented recently (Brandt *et al.*, 1999). Fifteen women who were HLA-A0201, HPV16 positive with recurrent cervical cancer were enrolled into one of three groups to receive four subcutaneous injections of 100, 300 or 1000 μg vaccine at 3-week intervals. No CTL response could be detected either before or after injection in any of the subjects. The lymphoproliferative response was also very weak compared with healthy subjects. This, as well as a decreased CTL response against influenza, is indicative of altered immunocompetence in women with cervical cancer. In terms of clinical efficacy, in the 100 μg group two of five had stable disease for more than 1 year, two progressed and one died; in the 300 μg group one out of five had a stable disease, three progressed and one died; and in the 1000 μg group all five had progressive disease.

In Australia, researchers at the University of Queensland are working on an HPV16 E7 peptide conjugated with ISCAR (immunostimulatory carrier) which stimulates the induction of peptide-specific antibodies, the T-cell proliferative response and specific cytotoxic T lymphocytes (Tindle *et al.*, 1995).

Other Approaches

As previously mentioned, research in the field of HPV vaccines is literally blossoming. Numerous other approaches are also being considered, such as:

1. DNA vaccination: DNA plasmids are constructed expressing various early proteins and/or structural proteins (Kell *et al.*, 1999; Ji *et al.*, 1999). In order to improve immunogenicity, plasmid DNA can be combined with gold microparticles and injected subcutaneously (Han *et al.*, 1999).
2. Dendritic cells have also recently been identified as potent inducers of CD4+ cellular response. Dendritic cells transfected with HPV16 E7 exhibit a good antitumor effect in mice (Wang *et al.*, 1999).

These techniques are still at an experimental stage but they all aim at inducing a strong Th1-type immune response.

Passive Specific Immunotherapy

Adoptive therapy is sometimes used in the treatment of HPV-positive cervical carcinoma. The patient's lymphocytes are stimulated *in vitro* with immunologically relevant peptides; after expansion in vitro using interleukin-2, these lymphocytes are transfused back in the patient (Steller and Schiller, 1996). This technique is of course limited to specific cases of disease.

Passive, Non-specific Immunotherapy

Interferons (IFN) have been extensively evaluated in the treatment of genital warts (Rockley and Tyring, 1995). Used alone, α-IFN injected intralesionally was more effective than when injected subcutaneously, with a cure rate ranging from 30% to 70% and a recurrence rate similar to other therapies. Topical creams have been disappointing. However, there is some indication that α-IFN could be valuable as an adjuvant to conventional therapy. Several studies have shown a higher clearance rate, and sometimes a decreased recurrence rate, when it is used in combination with or prior to cryotherapy, laser therapy or 5-fluorouracil therapy. However, the side-effects caused by both intralesional and systemic administration of α-IFN are substantial, leading to discontinuation of treatment or reduction of the dose in a significant proportion of patients. Therefore, as the efficacy is not significantly better than other therapies, the frequency and severity of side-effects and the costs of α-IFN therapies preclude their systematic use in patients with HPV-induced lesions.

Another, more recent topical immunomodulator is Imiquimod (Aldara 3M Pharmaceuticals), a heterocyclic amine that is a potent inducer of α-IFN and which stimulates monocytes and macrophages and the secretion of various cytokines (Testerman *et al.*, 1995). When added *in vitro* to PBMC, Imiquimod induces the secretion of interleukin-1 (IL1), IL6, IL8, tumor necrosis factor (TNF) and IFN (Tyring *et al.*, 1998). Imiquimod used as a 5% cream three times a week for up to 16 weeks induces clearance in 50% of patients vs 11% in placebo recipients. However, it does not reduce the recurrence rate (Edwards *et al.*, 1998). Significantly higher clearance rates were observed in females (72%) than in males (37%), which is supposedly due to differences in the keratinization of wart tissue. Local reactions, usually mild to moderate, are the most frequently reported reactions, mainly erythema (60%) and erosion (30%). Imiquimod has the advantage over other therapies of being self-applicable by the patient, and safe as well as resulting in relatively good clearance rates. Therefore, despite its higher cost, it is a direct competitor for podophyllotoxin as first-line therapy. However, the main disadvantage is its limited effect on the recurrence rate.

FUTURE PROSPECTS AND CONCLUSIONS

Therapeutic vaccination of patients with low-risk HPV has relatively straightforward objectives. The current therapies are quite effective in clearing the warts, but the major clinical problem lies in the high recurrence rate, whatever the therapy used. Therefore, the first objective of therapeutic vaccination will be to prevent or at least to decrease the frequency of recurrence. Whether this goal is achieved using a vaccine alone or combined with ablative or topical therapies is not a major question. Additional but not mandatory objectives could be the improvement of the efficacy of some available therapies, for example by shortening the duration of application of topical creams or lotions, or reducing the number of applications of cryotherapy. In fact, the combination of a therapeutic vaccine with conventional treatment is particularly appealing, as the local destruction of warts by ablation, caustic agents or topical immuno-modulators would cause local inflammation and presentation of viral epitopes to the immune response concomitantly enhanced by therapeutic vaccination.

Immunotherapy with vaccine alone also deserves consideration, provided there is a short and relatively safe course of administration and also good patient compliance, as we know from the animal models that regression of the warts is likely to be a progressive and slow process that can take a weeks or months. In any case, studies will need a blinded placebo control group to be evaluable, and adequate stratifications between relevant subgroups are man-datory. Finally, because HPV6 and HPV11 are very closely related, share many antigenic determinants and cross-neutralize, it is likely that a vaccine based on one virus would also be efficacious against the other. Another more limited indication is laryngeal papillomatosis. This disease is very incapacitating, potentially life threatening, and can only be treated effectively by laser surgery. Recurrences are also very frequent. These patients would therefore benefit from a therapeutic vaccine that would reduce or eliminate recurrences. However, an additional complication for clinical development in this instance would be that such a vaccine would have to be safe enough for pediatric use, as most patients with laryngeal papillomatosis are children.

Immunotherapy of diseases caused by high-risk HPV is less straightforward, for several reasons. First, several diseases with totally different prognoses are caused by these viruses, ranging from local dysplasia to invasive cervical cancer. Second, both infection and all phases of cervical dysplasia (CIN I–II–III) are spontaneously reversible in a high percentage of cases. Third, the rather long intervals between the different stages of disease render the evaluation of thera-peutic approaches difficult. Fourth, these diseases are caused by numerous HPV types, and although HPV16 and 18 are currently detected in more than 80% of CIN III and cervical cancers, therapeutic vaccination against these two viruses only could select the emergence of currently less prevalent viruses. Fifth, HPV infection usually occurs very soon after the first sexual experience, and ther-apeutic vaccination would then have to be applied to young women of high childbearing potential. Sixth, current approaches combining systematic detec-tion by pap smears, and more recently, HPV detection, together with very efficacious therapeutic protocols for cervix conization in CIN III, puts the value of a therapeutic vaccine for the early stages of infection into question. Also, a therapeutic vaccine that would eliminate HPV infection in the early stages of disease should ideally be able to prevent reinfection by the same subtype for several years at least. Therefore, to be acceptable as immunotherapy for the early stage of disease, a therapeutic vaccine should be composed of the most prevalent HPV subtypes, should be able to eliminate HPV infection in close to 100% of the treated women after only one or two injections, and should have its efficacy demonstrated in very large, long-term studies. Finally, it is difficult to draw a line between prevention and therapy, as a therapeutic vaccine that would eliminate asymptomatic infection or CIN I would in fact ultimately prevent the development of cancer. However, this last point is rather semantic. In conclusion, therapeutic HPV vaccines seem highly feasible and the main difficulties will probably lie in the selection of adequate methods of clinical evaluation.

ACKNOWLEDGEMENTS

My sincere thanks go to Charles Lacey for his thorough review and his very helpful comments.

I also want to thank my wife, Sophie, and my children, for their support and patience.

REFERENCES

Abcarian, H. and Sharon, N. (1982). Long-term effectiveness of the immunotherapy of anal condyloma acuminata. *Dis. Colon Rectum* **25**: 648–651

Abcarian, H. and Sharon, N. (1977). The effectiveness of immunotherapy in the treatment of anal condyloma acuminatum. *J. Surg. Res.* **22**: 231–236

Abcarian, H., Smith, D. and Sharon, N. (1976). The immunotherapy of anal condyloma acuminatum. *Dis. Colon Rectum* **19**: 237–244

Ablin, Z.J. and Curtis, W.W. (1974). Immunotherapeutic treatment of condyloma acuminata. *Gynecol. Oncol.* **2**: 446–450

Aiba, S., Rokugo, M. and Tamani, H. (1986). Immunohistological analysis of the phenomenon of spontaneous regression of numerous flat warts. *Cancer* **58**: 1246–1251

American Social Health Association (1998). Sexually transmitted diseases in America: how many cases and at what cost? Publication #1445 prepared for the Kaiser Foundation Family, Research Triangle Park, NC 27709

Baker, G.E. and Tyring, S.K. (1997). Therapeutic approach to papillomavirus infections. *Dermatol. Clin.* **15**(2): 331–340

Barrasso, R., Zanardi, C., Huynh, B. and Ferenczy, A. (1997). In: Human papillomavirus infection. A clinical atlas (Ed. Gross-Barrasso). Cervix and vagina treatment. Chapter 7, 277–290. Ullstein Mosby Gmbh & Co. KG, Berlin/Wiesbaden

Bauman, N.M. and Smith, J.H. (1996). Recurrent respiratory papillomatosis. *Pediatr. Clin. N. Am.* **43**(6): 1385–1401

Benton, C., Shahidullah, H. and Hunter, J.A.A. (1992). Human papillomavirus in the immunosuppressed. *Papillomavirus Report* **3**(2): 23–26

Bergeron, C., Ferenczy, A. and Richart, R. (1990). Underwear: contamination by human papillomavirus infection. *Epidemiol. Rev.* **10**: 122–163

Beutner, K.R., Wiley, D.J., Douglas, J.M. *et al.* (1999). Genital warts and their treatment. *Clin. Infect. Dis.* **28**(Suppl. 1): S37–56

Beutner, K.R., Reitano, M.V., Richwald, G.A. and Wiley, D.J. and the AMA Expert Panel on external genital warts (1998). External genital warts: Report of the American Medical Association Concensus Conference. *Clin. Infect. Dis.* **27**: 796–806

Beutner, K.R. and Wiley, D.J. (1997). Recurrent external genital warts: a literature review. *Papillomavirus Report* **8**(3): 69–74

Beutner, K.R. and Ferenczy, A. (1997). Therapeutic approaches to genital warts. *Am. J. Med.* **105**(5A): 28–37

Biberstein, H. (1944). Immunization therapy of warts. *Arch. Dermatol.* **50**: 12–22

Borysiewicz, L.K., Fiander, A., Nimako, M. *et al.* (1996). A recombinant vaccinia virus encoding human papillomavirus types 16 and 18, E6 and E7 proteins or immunotherapy for cervical cancer. *Lancet* **347**: 1523–1527

Bosch, F.X., Manos, M.M., Munoz, N. *et al.* (1995). Prevalence of human papillomavirus in cervical cancer: a worldwide perspective. *J. Natl. Cancer Inst.* **87**: 796–802

Boursnell, M.E.G., Rutherford, E., Hickling, J.K. *et al.* (1996). Construction and characterisation of a recombinant vaccinia virus expressing human papillomavirus protein for immunotherapy of cervical cancer. *Vaccine* **14**(16): 1485–1494

Brandsma, J.L. (1996). Animal models for human papillomavirus vaccine development. 69–78. Papillomavirus Reviews: current research on papillomaviruses. Ed: Ch. Lacey. Publisher: Leeds University Press

Brandsma, J.L. (1994). Animal models for HPV vaccine development. *Papillomavirus Report* **5**: 105–111

Brandt, R.M.P., Ressing, M.E., de Jong, J.H. *et al.* (1999). The HPV16 E7 peptide based vaccine trial in end stage cervical carcinoma: final report. Proceedings of the 17th International Papillomavirus Conference January 9–15, Charleston, SC

Brown, D.R., Schroeder, J.M., Brian, J.T., Stoler, M.H. and Fife, K.H. (1999). Detection of multiple HPV types in condyloma acuminata lesions from normal and immunosuppressed patients. *J. Clin. Microbiol.*, in press

Burk, R.D., Ho, G.Y.F., Beardsley, L., Lempa, M., Peters, M. and Bierman, R. (1996). Several behavior and partner characteristics are the predominant risk factors for genital human papillomavirus infection in young women. *J. Infect. Dis.* **174**: 679–689

Caballero, O.L., Trevisan, A., Villa, L.L., Ferenczy, A. and Franco, E.L. (1999). High viral load is associated with persistent HPV infection and risk of cervical dysplasia. Proceedings of the 17th International Papillomavirus Conference January 9–15, Charleston, SC

Campo, M.S. (1994). Vaccination against papillomavirus in cattle. *Curr. Topics Microbiol. Immunol.* **186**: 255–266

Campo, M.S., O'Neil, B.W., Barron, R.J. and Jarrett, W.F.H. (1994). Experimental reproduction of the papilloma–carcinoma complex of the alimentary canal in cattle. *Carcinogenesis* **15**: 1597–1601

Campo, M.S., Grindlay, G.J., O'Neil, B.W., Chandrachud, L.M., McGarvie, G.M. and Jarrett, W.F.H. (1993). Prophylactic and therapeutic vaccination against a mucosal papillomavirus. *J. Gen. Virol.* **74**: 945–953

Carson, L.F., Twiggs, L.B., Fukushima, M., Ostrow, R.S., Faras, A.J. and Okagaki, T. (1986). Human genital papilloma infections: an evaluation of immunologic competence in the genital neoplasia-papilloma syndrome. *Am. J. Obstet. Gynecol.* **155** 784–789

Cason, J., Kaye, J.N., Jewers, R.J. *et al.* (1995). Perinatal infection and persistence of human papillomavirus types 16 and 18 in infants. *J. Med. Virol.* **47**: 209–218

Cauda, R., Tyring, S.K., Grossi, C.E. *et al.* (1987). Patients with condyloma acuminatum exhibit decreased interleukin-2 and interferon gamma production and depressed natural killer activity. *J. Clin. Immunol.* **7**: 304–311

Centers for Disease Control and Prevention (1983). Condyloma accuminatum – United States (1966–1981). MMWR **32**: 306–308

Che, M., Lin, K.Y., Pardoll, D.M., Kurman, R.J. and Wu, T.C. (1999a). Determination of E7-specific CD8⁺ and CD4⁺ T-cell precursors following vaccination with SIG/E7/LAMP-1 vaccinia using elispot assays. Proceedings of the 17th International Papillomavirus Conference January 9–15, Charleston, SC

Che, M., Lin, K.Y., Pardoll, D.M., Kurman, R.J. and Wu, T.C. (1999b). Variation of antitumor immunity in C57BL/6 and BALB/C mice after vaccination with SIG/E7/LAMP-1 vaccinia. Proceedings of the 17th International Papillomavirus Conference January 9–15, Charleston, SC

Chen, L., Mizuno, M.T., Singhal, M.C. *et al.* (1992). Induction of cytotoxic T lymphocytes specific for a syngeneic tumor expressing the E6 oncoprotein of human papillomavirus type 16. *J. Immunol.* **148**: 2617–2621

Chen, L., Thomas, E.K., Hu, S.L., Hellström, I. and Hellström, K.E. (1991). Human papillomavirus type 16 nucleoprotein E7 is a tumor rejection antigen. *Proc. Natl. Acad. Sci. USA* **88**: 110–114

Christensen, N.D., Reed, C.A., Cladel, N.H., Han, R. and Kreider, J.W. (1996). Immunization with virus like particles induces long-term protection of rabbits against challenge with cottontail rabbit papillomavirus. *J. Virol.* **70** (2): 960–965

Chu, N.R., Wu, B., Bantkoch, S. *et al.* (1999). Immunotherapy of human papillomavirus type 16 associated carcinoma using adjuvant-free fusion protein encoding *M. bovis* BCG HSP65 and HPV16 E7. Proceedings of the 17th International Papillomavirus Conference January 9–15, Charleston, SC

Coleman, N., Birley, H.D.L., Renton, A.M. *et al.* (1994). Immunological events in regressing genital warts. *Ann. J. Clin. Pathol.* **102**: 768–774

Critchlow, C.W. and Kiviat, N.B. (1999). Old and new issues in cervical cancer control. *J. Natl Cancer Inst.* **91**: 200–201

Da Silva, D.M., Nieland, J.D., Greenstone, H.L., Schiller, J.T. and Kast, W.M. (1999). Chimeric papillomavirus virus-like particles induce antigen-specific therapeutic immunity against tumors expressing the HPV-16 E7 protein. Proceedings of the 17th International Papillomavirus Conference January 9–15, Charleston, SC

De Bruijn, M.L.M., Greenstone, H.L., Vermeulen, H. *et al.* (1998). L1-specific protection from tumor challenge elicited by HPV16 virus-like particles. *Virology* 250 (2), 371–376

De Villiers, E-M. (1994). Human pathogenic papillomavirus types : an update. *Curr. Topics Microbiol Immunol* 186: 1–12

De Villiers, E-M. (1989). Minireview: heterogenicity of the human papillomavirus group. *J. Virol.* 63 (11), 4898–4903

Edwards, L., Ferenczy, A., Eron, L. *et al.* and the HPV study group (1998). Self-administered topical 5% Imiquimod cream for external ano-genital warts. *Arch. Dermatol.* 134: 25–30

Eftaiha, M.S., Amshel, A.L., Shonberg, I.L. and Batshon B. (1982). Giant and recurrent condyloma acuminata: appraisal of immunotherapy. *Dis. Colon Rectum* 25: 136–138

El Mehdaoui, S., Touzé, A. and Coursaget, P. (1999). Production of chimeric human papillomavirus 16 E4 – hepatitis B core recombinant particles. Proceedings of the 17th International Papillomavirus Conference January 9–15, Charleston, SC

Evans, C.A., Gormann, L.R., Ito, Y. and Weiser, R.J. (1962a). A vaccination procedure which increases the frequency of regressions of Antitumor immunity in Shope papilloma of rabbits. *Nature* 193: 288–289

Evans, C.A., Gormann, L.R. and Ito, Y. and Weiser, R.S. (1962b) Antitumor immunity in the Shope papilloma–carcinoma complex of rabbits. I. Papilloma regression induced by homologous and antologous tissue vaccines. *J. Natl Cancer Inst.* 29: 277–285

Feltkamp, M.C.W., Smith, H.L., Vierboom, M.P.C. *et al.* (1993). Vaccination with cytotoxic T lymphocyte epitope-containing peptide protects against a tumor induced by human papillomavirus type 16-induced cells. *Eur. J. Immunol.* 23: 2242–2249

Ferenczy, A., Bergeron, C. and Richart, R.M. (1989). Human papillomavirus DNA in fomites on objects used for the management of patients with genital human papillomavirus infections. *Obstet. Gynecol.* 74: 950–954

Foster, C.A., Hall, K.S. and Smith, L.H. (1997). Detection of neutralizing antibodies to papillomaviruses: applications for vaccine study. *Papillomavirus Report* 8(5): 127–131

Franco, E.L. (1999). Epidemiology and natural history of HPV induced disease. Proceedings of the 17th International Papillomavirus Conference January 9–15, Charleston, SC

Frazer, I.H. (1996a). Immunology of papillomavirus infection. *Curr. Opin. Immunol.* 8(4): 484–491

Frazer, I.H. (1996b). The role of vaccines in the control of STD's: HPV vaccines. *Genitourinary Med* 72: 398–403

Fredricks, B.D., Balkin, A., Daniel, H.W., Schonrock, J., Ward, B. and Frazer, I.H. (1993). Transmission of human papillomaviruses from mother to child. *Aust. NZ J. Obstet. Gynaecol.* 33: 30–32

Fruchter, R.G., Maiman, M., Sedlis, A., Bartley, L., Camilien, L. and Arrastia, C. (1996). Multiple recurrences of cervical intraepithelial neoplasia in women with human immunodeficiency virus. *Obstet. Gynecol.* 87(3): 338–344

Gibson, P.E., Gardner, S.D. and Best, S.J. (1990). Human papillomavirus types in anogenital warts of children. *J. Med. Virol.* 30: 142–145

Greenstone, H.L., Nieland, J.D., De Visser, K.E. *et al.* (1998). Chimeric papillomavirus virus-likw particles elicit antitumor immunity against E7 oncoprotein in an HPV-16 tumor model. *Proc Natl. Acad. Sci.* USA **95**: 1800–1805

Hagensee, M.E. (1997). Progression in the development of HPV vaccines. *Infect. Med.* **14**(7): 555–564

Hagensee, M.E., Yaegashi, N. and Galloway, D.A. (1993). Self-assembly of human papillomavirus type I capsids by expression of the L1 protein alone or by coexpression of the L1 and L2 capsid proteins. *J. Virol.* **67**: 315–322

Han, R., Cladel, N.M., Reed, C.A., Peng, X. and Christensen, N.D. (1999). Intracutaneous genetic vaccination with a combination of cottontail rabbit papillomavirus E1, E2, E6 and E7 genes induces protective anti-papillomavirus immunity in rabbits. Proceedings of the 17th International Papillomavirus Conference January 9–15, Charleston, SC

Hariharan, K., Braslawsky, G., Barnett, R.S. *et al.* (1998). Tumor regression in mice following vaccination with human papillomovirus E7 recombinant protein in Provax^tm. *Int. J. Oncol.* **12**: 1229–1235

Heineman, L., Amer, M., Van Nest, G. and Hibma, M.A. (1999). Immune response to human papillomavirus type 16 early protein E2 following immunization with adjuvant MF 59 in a mouse model. Proceedings of the 17th International Papillomavirus Conference January 9–15, Charleston, SC

Hibma, M.H., Amer, M., Heinemann, L. and Van Nest G. (1999). Immune responses to the human papillomavirus type 16 early protein E2 following several methods of immunization. Proceedings of the 17th International Papillomavirus Conference January 9–15, Charleston, SC

Ho, G.Y.F., Timmins, P., Wang, Y.X. *et al.* (1999). Lymphoproliferation cells mediated immune responses to specific HPV16 E6 and E7 peptides are associated with resolution of genital HPV infection and for regression of CIN. Proceedings of the 17th International Papillomavirus Conference January 9–15, Charleston, SC

IMPC of STD: Institute of Medicine Committee on Prevention and Control of Sexually Transmitted Diseases (1997). The hidden epidemic: confronting sexually transmitted diseases. National Academic Press, Washington DC

Jansen, K.U., Rosolowsky, M., Schultz, L.D. *et al.* (1995). Vaccination with yeast-expressed cottontail rabbit papillomavirus (CRPV) virus-like particles protects rabbits from CRPV-induced papilloma formation. *Vaccine* **13**(16): 1509–1514

Jarrett, W.F., Smith, K.T., O'Neil, B.W. *et al.* (1991). Studies on vaccination against papillomaviruses: prophylactic and therapeutic vaccination with recombinant structural proteins. *Virology* **184**: 33–42

Ji, H., Chen, H., Kurman, R.J., Pardoll, D.M. and Wu, T.C. (1999). Both CD-8^+ T-cells and CD-4^+ T-cells are essential in mediating potent anti-tumor immunity of DNA vaccines expressing cytosolic antigens. Proceedings of the 17th International Papillomavirus Conference January 9–15, Charleston, SC

Kashima, H.K., Mounts, P. and Shah, K. (1996). Recurrent respiratory papillomatosis. *Obstet. Gynecol. Clin. N. Am.* **23**(3): 699–706

Kashima, H.K., Kessis, T., Mounts, P. and Shah, K. (1991). Polymerase chain reaction identification of human papillomavirus DNA in CO_2 laser plume from recurrent respiratory papillomatosis. *Otolaryngol. Head Neck Surg.* **104**: 191–195

Kawana, K., Yoshikawa, H. and Taketani, Y. (1999). In vitro construction of pseudovirions of human papillomavirus type 16: incorporation of plasmid DNA into reassembled L1/L2 capsids. Proceedings of the 17th International Papillomavirus Conference January 9–15, Charleston, SC

Kell, B., Mant, C., Best, J., Cason, J. and Shepherd, P. (1999). Development of DNA vaccines against HPV-16 L1. Proceedings of the 17th International Papillomavirus Conference January 9–15, Charleston, SC

Kirnbauer, R., Chandrachud, L.M., O'Neil, B.W. *et al.* (1996). Virus-like particles of bovine papillomavirus type 4 in prophylactic and therapeutic immunization. *Virology* **219**: 37–44

Kirnbauer, R.V., Booy, F., Cheng, N., Lowy, D.R. and Schiller, J.T. (1992). Papillomavirus L1 major capsid protein self-assembles into virus-like particles that are highly immunogenic. *Proc. Natl. Acad. Sci.* USA **89**: 12180–12184

Kiviat, N.B., Critchlow, C.W. and Kurman, R.J. (1992). Reassessment of morphological continuum of cervical intrapithelial lesions: does it reflect different stages in the progression to cervical carcinoma? *IARC Sci. Publ.* pp. 59–66

Kjaer, S.K., Van den Brule, A.J.C., Bock, J.E. *et al.* (1996). Human papillomavirus – the most significant risk determinant of cervical intraepithelial neoplasia. *Int. J. Cancer* **65**: 601–606

Koutsky, L.A., Galloway, D.A., Holmes, M.K. (1988). Epidemiology of genital human papilloma-virus infection. *Epidemiol. Rev.* **10**: 122–163

Lacey, C.J.N. (1996). Genital warts in children. Papillomavirus Report. **77**: 83–87

Lacey, C.J.N., Thompson, H.S.G., Monteiro, E.F., O'Neill, T., Davies, M.L., Holding, F.P., Fallon, R.E. and Roberts, JStC (1999). Phase IIa safety and immunogenicity of a therapeutic vaccine, TA-GW, in persons with genital warts. *J. Infect. Dis.* **179**: 612–618

Lin, K.Y., Guarnieri, F.G., Stavely-O'Carroll, K.F. *et al.* (1996). Treatment of established tumors with a novel vaccine that enhances major histocompatability class II presentation of tumor antigen. *Cancer Res.* **56**: 21–26

Liu, X.S., Abdul-Jabbar, I., Qui, Y.M., Frazer, I.H. and Zhou, J. (1999). Mucosal immunization with papillomavirus-like particles elicits and mucosal immunity in mice. Proceedings of the 17th International Papillomavirus Conference January 9–15, Charleston, SC

Lutzner, M.A. (1985). Papillomavirus lesions in immunodepression and immunosuppression. *Clin. Dermatol.* **3**: 165–169

Malison, M.D., Morris, R. and Jones, L.W. Autogenous vaccine therapy for condyloma acuminatum. A double-blind controlled study. *Br. J. Vener. Dis.* **58**(1): 62–5

Mallarios, J., Quinn, C., Arnold, F. and Macfarlan, R. (1999). Cell-mediated immune responses to recombinant HPV-16 E6E7hh protein generated by immunization with chelating Iscomatrix[tm] adjuvant. Proceedings of the 17th International Papillomavirus Conference January 9–15, Charleston, SC

Meijer, C.J.L.M., Nobbenhuis, M.A.E., Helmerhorst, T.J.M. *et al.* (1999). Human papillomavirus and cervical lesions in a prospectus study of 353 women with abnormal cytology. Proceedings of the 17th International Papillomavirus Conference January 9–15, Charleston, SC

Munoz, N. and Bosch, F.X. (1996). Current vaccines on the epidemiology of HPV and cervical cancer. In: Lacey C. (ed) Papilloma reviews: current research on papillomaviruses. Leeds University Press, 227–238

Munoz, N. and Bosch, F.X. (1989). Epidemiology of cervical cancer. In: Munoz, N., Bosch, F.X., Jensen, O.K. (eds) Human papillomaviruses and cervical cancer. Lyon IARC, Scientific Publications, pp. 9–40

Murakami, M., Gurski, K.J. and Steller, M.A. (1999). Human papillomavirus vaccines for cervical cancer. *J. Immunother.* **22**(3): 212–218

Nardelli-Haefliger, D., Roden, R., Potts, A., Balmelli, C., Schiller, J. and De Grandi, P. (1999). Only mucosal immunization with human papillomavirus type 16 (HPV-16) virus-like particles (VLP) elicits a continuous induction of HPV-16 neutralizing antibodies along the estrous cycle of mice. Proceedings of the 17th International Papillomavirus Conference January 9–15, Charleston, SC

Nasiell, K., Roger, V. and Nasiell, M. (1986). Behavior of mild cervical dysplasia during long-term follow-up. *Obstet. Gynecol.* **67**: 665–669

Neeper, M.P., Ling, J.C., Wang, X.M. and Cook, J.C. (1999). HPV 16 virus-like particles containing nonstrucutural HPV 16 proteins. Proceedings of the 17th International Papillomavirus Conference January 9–15, Charleston, SC

Palefsky, J. (1991). Human papillomavirus infection among HIV-infected individuals. *Hematol. Oncol. Clin. North. Am.* **5**: 357–370

Powell, L.C. Jr., Pollard, M. and Jinkins, J.L. Sr. (1970). Treatment of condyloma acuminata by antigenous vaccine. *South. Med. J.* **63**: 202–205

Puranen, M., Yliskoski, M., Saarikoski, S., Syrjänen, K. and Syrjänen, S. (1996). Vertical transmission of human papillomavirus from infected mothers to their newborn babies and persistence of the virus in childhood. *Am. J. Obstet. Gynecol.* **174**(2): 694–699

Reichman, R., Balsley, J., Carlin, D. *et al.* (1999). Evaluation of the safety and immunogenicity of a recombinant HPV-11 L1 virus like particle vaccine in healthy adult volunteers. Proceedings of the 17th International Papillomavirus Conference January 9–15, Charleston, SC

Rice, P.S., Cason, J., Best, J.M. and Banatvala, J.E. (1999). High risk genital papillomavirus infections are spread vertically. *Rev. Med. Virol.* **9**: 15–21

Richart, R.M. (1987). Causes and management of cervical intraepithelial neoplasia. *Cancer* **60**: 1951–1959

Rimell, F.L. (1999). Metastatic pulmonary disease by recurrent respiratory papillomatosis. Proceedings of the 17th International Papillomavirus Conference January 9–15, Charleston, SC

Rockley, P.F. and Tyring, S.K. (1995). Interferons, alpha, beta and gamma therapy of anogenital human papillomavirus infections. *Pharmac. Ther.* **65**: 265–287

Roden, R., Thompson, S., Inglis, S., Lowy, D. and Schiller, J. (1999). Detection of cross-neutralizing epitopes in the L2 minor capsid protein of genital human papillomaviruses. Proceedings of the 17th International Papillomavirus Conference January 9–15, Charleston, SC

Rose, R.C., Lane, C., Wilson, S., Suzich, J.A., Rybicki, E. and Williamson, A.L. (1999). Oral administration of human papillomavirus (HPV) virus-like particles (VLPs) induces serum neutralizing antibodies in mice. Proceedings of the 17th International Papillomavirus Conference January 9–15, Charleston, SC

Rose, R.C., Bonnez, W., Reichman, R.C. and Garcea, R.L. (1993). Expression of human papillomavirus type 11L1 protein in insect cells: in vivo and in vitro assembly of viruslike particles. *J. Virol.* **67**(4): 1936–1944

Rous, P. and Beard, J.W. (1935). The progression to carcinoma of virus induced rabbit papillomas (Shope). *J. Exp. Med.* **62**: 523–549

Rudolf, M.P., Nieland, J.D., Velders, M.P. *et al.* (1999). Induction of HPV16 L1/L2 virus-like particle specific human T-cell responses in vitro. Proceedings of the 17th International Papillomavirus Conference January 9–15, Charleston, SC

Schäfer, K., Faath, S., Gabelsberger, J., Müller, M., Gissmann, L. and Jochmus, I. (1999). Induction of tumor protection and regression by HPV-16 L1E7 chimeric virus-like particles. Proceedings of the 17th International Papillomavirus Conference January 9–15, Charleston, SC

Selvakumar, R., Borenstein, L.A., Lin, Y.L., Ahmed, R. and Wettstein, F.O. (1995a). Immunization with nonstructural proteins E1 and E2 of cottontail rabbit papillomavirus stimulates regression of virus-induced papillomas. *J. Virol.* **69**(1): 602–605

Selvakumar, R., Ahmed, R. and Wettstein, F.O. (1995b). Tumor repression in associated with a specific immune response to the E2 proteins of Cottontail Rabbit Papillomavirus. *Virology* **208**: 298–302

Shi, W., Zdinak, L., Paintsil, J. and Qiao, L. (1999). BPV pseudovirions deliver human B7.2 gene into cancer cells and enhance their immunogenicity. Proceedings of the 17th International Papillomavirus Conference January 9–15, Charleston, SC

Shope, R.E. and Hurst, E.W. (1933). Infections papillomatosis of rabbits. *J. Exp. Med.* **58**: 607–623

Slupetzky, K., Brandt, S. and Kirnbauer, R. (1999). Chimeric papillomavirus-like particles as an antigen carrier system: insertion of the HIV1 GP41–Eldkwas epitope into a putative L1–capsid surface loop. Proceedings of the 17th International Papillomavirus Conference January 9–15, Charleston, SC

Stanley, M., Coleman, N. and Chambers, M. (1994). The host response to lesions induced by human papillomavirus. In: Vaccines against virally induced tumors. Ciba Foundation Symposium. No 187, London, 21–31

Steller, M.A., Gurski, K.J., Murakami, M. *et al.* (1998). Cell-mediated immunological responses in cervical and vaginal cancer patients immunized with a lipidated epitope of human papillomavirus type 16 E7. *Clin. Cancer Res.* **4**(9): 2103–2109

Steller, M.A. and Schiller, J.T. (1996). Human papillomavirus Immunology and vaccine prospects. *Monogr. Natl. Cancer Inst.* **21**: 145–148

Stone, K.M. (1995). Human papillomavirus infection and genital warts: update on epidemiology and treatment. *Clin Infect Dis* **20**: S91–S97

Suzich, J.A., Ghim, S.J., Palmer Hill, F.J. *et al.* (1995). Systematic immunization with papillomavirus L1 protein completely presents the development of viral mucosal papillomas. *Proc. Natl. Acad. Sci. USA* **92**: 11553–11557

Syrjänen, K.J. (1996). Natural history of genital human papillomavirus infections pp. 183–206. Papillomavirus Reviews: current research on papillomaviruses. Ed: Ch. Lacey. Leeds University Press

Tagami, H., Ogino, A., Takigawa, M., Imamura S. and Ofuji S. (1974). Regression of plane warts following spontaneous inflammation: an histopathological study. *Br. J. Dermatol.* **90**: 147–154

Tay, S.K., Jenkins, D., Maddox, P. and Singer, A. (1987). Lymphocyte phenotypes in cervical intraepithelial neoplasia and human papillomavirus infection. *Br. J. Obstet. Gynaecol.* **94**: 16–21

Testerman, T.L., Gerster, J.F., Imbertson, L.M. *et al.* (1995). Cytokine induction by the immuno-modulator Imiquimod and S-27609. *J. Leukocyte Biol.* **58**: 365–372

Thompson, H.S.G., Davies, M.L., Holding, F.P. *et al.* (1999). Phase I safety and antigenicity of TA-GW: a recombinant HPV6 L2E7 vaccine for the treatment of genital warts. *Vaccine* 17: 40–49

Thompson, H.S.G., Davies, M.L., Watts, M.J. *et al.* (1998). Enhanced immunogenicity of a recombinant genital warts vaccine adjuvanted with monophosphoryl lipid A. *Vaccine* 16(20): 1993–1999

Tindle, R.W., Croft, S., Herd, K. *et al.* (1995). A vaccine conjugate of 'ISCAR' immunocarrier and peptide epitopes of the E7 cervical cancer-associated protein of human papillomavirus type 16 elicits specific Th1- and Th2-type responses in immunized mice in the absence of oil-based adjuvants. *Clin. Exp. Immunol.* 101: 265–271

Touze, A. and Coursaget, P. (1999). Gene transfer with human papillomavirus virus-like particle. Application to DNA immunization. Proceedings of the 17th International Papillomavirus Conference January 9–15, Charleston, SC

Tyring, K.S., Arany, I., Stanley, M.A. *et al.* (1998). A randomized controlled, molecular study of condyloma acuminata clearance during treatment with Imiquimod. *J. Infect. Dis.* 178: 551–555

US Centers for Disease Control, Division of STD Prevention (1996). Sexually transmitted diseases surveillance (1995) Atlanta: centers for disease control and prevention, US Department of Health Services, Public Health Services, 1996.

Van Driel, W.J., Ressing, M.E., Brandt, R.M.P. *et al.* (1996). The current status of therapeutic HPV vaccine. *Ann. Med.* 28: 471–477

Velders, M.P., Nieland, J.D., Da Silva, D.M., De Visser, K.E., Muller, M. and Kast, W.M. (1999). Prevention of peptide induced tolerance and induction of protective and therapeutic immune responses by chimeric human papillomavirus like particle. Proceedings of the 17th International Papillomavirus Conference January 9–15, Charleston, SC

Vonka, V. Human papillomavirus vaccines: review (1996). Folia Biologica (Praha) 42: 73–78

Von Knebel Doeberitz, M., Ritmuller, C., Aengeneyndt, P., Jansendurr, D., Spitovsky, D. (1994). Reversible regression of papillomavirus oncogene expression in several carcinoma cells. Consequences for the phenotype and E6–p53 and E7–pRb interactions. *J. Virol.* 68: 2811–2821

Walboomers, J.M.M., Meijer, C.J.L.M. (1999). Testing for HPV in the Netherlands: latest results. Proceedings of the 17th International Papillomavirus Conference January 9–15, Charleston, SC

Walboomers, J.M.M. and Meijer, C.J.L.M. (1997). Editorial: Do HPV-negative cervical carcinomas exist? *J. Pathol.* 181: 253–254

Wang T-L., Shih I-M., Lu, Z.B., Chen, C.H. and Kurman, R.J. (1999). Vaccination with dendritic cells transfected with human papillomavirus type 16 (HPV-16) oncogene-E7 protects against E7-expressing tumor cells. Proceedings of the 17th International Papillomavirus Conference January 9–15, Charleston, SC

Wank, R., Schendel, D.J. and Thomssen, C. (1992). HLA antigens and cervical carcinoma. *Nature* 356: 22–23

Wank, R. and Thomssen, C. (1991). High risk of squamous cell carcinoma of the cervix for women with HLA-DQw3. *Nature* 352: 723–725

Wiltz, O.H., Torregrosa, M. and Wiltz, O. (1995). Autogenous vaccine: the best therapy for perianal condyloma acuminata? *Dis. Colo Rectum* 38: 838–884

Zhang, L.F., Zhou, J., Shao, C. *et al.* (1999). A phase 1 trial of HPV 6 B virus like particles as immunotherapy for genital warts. Proceedings of the 17th International Papillomavirus Conference January 9–15, Charleston, SC

Zur Hausen, H. (1995). Are human papillomavirus infections not necessary or sufficient causal factors for invasive cancer of the cervix? *Int. J. Cancer* 63(2): 315–316

Zur Hausen, H. (1991). Human papillomaviruses in the pathogenesis of anogenital cancer. *Virology* 184: 9–13

13
■ HEPATITIS B VIRUS INFECTION

JANE N. ZUCKERMAN*
AND
ARIE J. ZUCKERMAN†

Academic Unit of Travel Medicine and Vaccines;
Royal Free and University College Medical School, University of London, London, UK

†*WHO Collaborating Centre for Reference and Research on Viral Diseases,*
Royal Free and University College Medical School, University of London, London, UK

INTRODUCTION

Hepatitis B virus is responsible for the most common form of parenterally transmitted viral hepatitis, referred to originally as 'homologous serum jaundice' or serum hepatitis, and it is an important cause of acute and chronic liver disease. The incubation period is variable, with a range of 6–8 weeks. Clinically, acute infection resembles that of the other viral hepatitides, hepatitis A, C, D and E. Acute hepatitis B is frequently anicteric and asymptomatic, although a severe illness with jaundice can occur, and occasionally acute liver failure (fulminant hepatitis) may develop.

More than a third of the world's population has been infected with hepatitis B virus, and the World Health Organization estimates that hepatitis B virus results in 1–2 million deaths every year.

The virus persists in approximately 5–10% of immunocompetent adults, and in as many as 90% of infants infected perinatally. Persistent carriage of hepatitis B, defined by the presence of hepatitis B surface antigen (HBsAg) in the serum for more than 6 months, has been estimated to affect about 350 million people worldwide. The pathology is mediated by the responses of the cellular immune response of the host to the infected hepatocytes. Long-term continuing virus replication may lead to progression to chronic liver disease and cirrhosis in approximately 25% of carriers, and to hepatocellular carcinoma.

Hepatocellular carcinoma is one of the 10 most common cancers in the world, and it is particularly common among men in Southeast Asia, the Western Pacific region, sub-Saharan Africa and a number of other developing regions. Up to 80% of hepatocellular cancers are attributable to hepatitis B virus, which is thus second only to tobacco among the known human carcinogens (WHO, 1983).

STRUCTURE OF HEPATITIS B VIRUS

The hepatitis B virion is a 42 nm particle comprising an electron-dense nucleo-capsid 27 nm in diameter, surrounded by an outer envelope of the surface protein (HBsAg) embedded in membraneous lipid derived from the host cell. The surface antigen (originally referred to as Australia antigen) is produced in excess by the infected hepatocytes and is secreted in the form of 22 nm particles and tubular structures of the same diameter (Figure 13.1).

The 22 nm particles are composed of the major surface protein in both non-glycosylated (p 24) and glycosylated (gp 27) form, in approximately equimolar amounts, together with a minority component of the so-called middle proteins (gp 33 and gp 36) which contain the pre-S2 domain, a glycosylated 55 amino acid N-terminal extension. The surface of the virion has a similar composition, but also contains the large surface proteins (p 39 and gp 42) which include both the pre-S1 and pre-S2 regions.

These large surface proteins are not found in the 22 nm spherical particles, they are immunogenic, and of importance for the development of third-generation vaccines. The domain that binds to the specific HBV receptor on the hepatocyte is believed to reside within the pre-S1 region.

The nucleocapsid of the virion consists of the viral genome surrounded by the core antigen (HBcAg). The genome, which is approximately 3.2 kilobases in length, has an unusual structure and is composed of two linear strands of DNA held in a circular configuration by base-pairing at the 5' ends. One of the strands is incomplete and the 3' end is associated with a DNA polymerase molecule which, when supplied with deoxynucleoside triphosphates, is able to complete that strand.

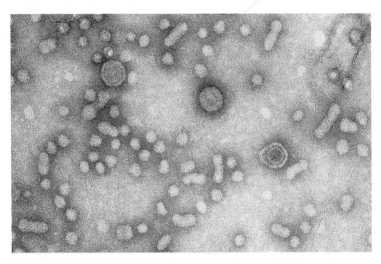

FIGURE 13.1 Double-shelled hepatitis B virus particles measuring 42 nm in diameter, surface antigen particles and tubular forms approximately 22 nm in diameter. Negative staining × 250 000. (Electron micrograph from a series by A.J. Zuckerman et al.).

The genomes of more than a dozen isolates of hepatitis B virus have been cloned and the complete nucleotide sequences determined. Analysis of the coding potential of the genome reveals four open reading frames (ORFs), which are conserved between all of these isolates (Figure 13.2).

IMMUNE RESPONSES AND SEROLOGICAL DIAGNOSIS

Antibody and cell-mediated immune responses to various hepatitis B viral antigens are induced during the infection; however, not all these are protective and, in some instances, may cause autoimmune phenomena that contribute to disease pathogenesis. The immune response to infection is directed towards at least three antigens: hepatitis B surface antigen, the core antigen, and the *e* antigen. There is evidence that the pathogenesis of liver damage in the course of hepatitis B infection is related to the immune response by the host.

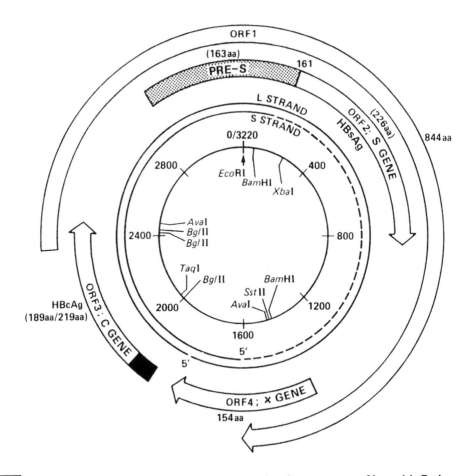

FIGURE 13.2 Schematic diagram of the molecular structure of hepatitis B virus.

The surface antigen (HBsAg) appears in the sera of most patients during the incubation period, 2–8 weeks before biochemical evidence of liver damage or the onset of jaundice. The antigen persists during the acute illness and usually clears from the circulation during convalescence. Next to appear in the circulation is the virus-associated DNA polymerase, which correlates in time with damage to liver cells, as indicated by elevated serum transaminases. The polymerase activity persists for days or weeks in acute cases, and for months or years in some persistent carriers. Antibody of the IgM class to the core antigen is found in the serum 2–10 weeks after the surface antigen appears, and persists during replication of the virus. Core antibody of the IgG class is detectable for many years after recovery. Finally, antibody to the surface antigen component, anti-HBs, appears.

During the incubation period and during the acute phase of the illness, surface antigen–antibody complexes may be found in the sera of some patients. Immune complexes have been found in the sera of all patients with fulminant hepatitis, but are seen only infrequently in non-fulminant infection.

Immune complexes are also important in the pathogenesis of chronic liver disease, and of other disease syndromes characterized by severe damage to blood vessels (for example polyarteritis nodosa, some forms of chronic glomerulonephritis, and infantile papular acrodermatitis).

EPIDEMIOLOGY

Although various body fluids (blood, saliva, menstrual and vaginal discharges, serous exudates, seminal fluid and breast milk) have been implicated in the spread of infection, infectivity appears to be especially related to blood and to body fluids contaminated with blood. The epidemiological propensities of this infection are therefore wide: they include infection by inadequately sterilized syringes and instruments, transmission by unscreened blood transfusion and blood products, by close contact, and by sexual contact. Antenatal (rare) and perinatal (frequent) transmission of hepatitis B infection from mother to child may take place, and in some parts of the world (Southeast Asia, China and Japan) perinatal transmission is very common.

Hepatitis B virus is spread in the community in the following principal ways:

- There is much evidence for the transmission of hepatitis B virus by intimate contact and by the sexual route, particularly among those who change sexual partners frequently. In the industrialized countries hepatitis B is primarily a disease of young adults, and most infections in those aged 20–35 years are linked to sexual activity: 41% were attributed to heterosexual activity and 14% to homosexual contact in the USA in 1990 (Alter *et al.*, 1990). In France, 30–35% of cases of hepatitis B were attributed to sexual transmission (Sepetjan, 1991).
- Mother to infant transmission is a principal mode of infection in Southeast Asia, China, the Western Pacific and in Japan (and particularly among certain ethnic groups, e.g. the Chinese). The risk of

infection (70–90%) is particularly high if the mother carries the *e* antigen. Persistent carriage is high (85–90%) in infants infected perinatally, compared to 10–15% following acute infection in adults.

• Intravenous drug abuse is a major risk among those using unsterile equipment or sharing needles.

It should be noted, however, that in as many as 30% of infected individuals the cause or route of infection is not known (Figure 13.3).

THE CARRIER STATE

The carrier state is defined as persistence of hepatitis B surface antigen in the circulation for more than 6 months. It may be lifelong and may be associated with liver damage, varying from minor changes in the nuclei of hepatocytes to chronic hepatitis, cirrhosis and primary liver cancer. Several risk factors have been identified in relation to the carrier state. It is more frequent in males, more likely to follow infections acquired in childhood than those acquired in adult life, and more likely to occur in patients with natural or acquired immune deficiencies. A carrier state becomes established in approximately 5–10% of infected adults. In countries where hepatitis B infection is common the highest prevalence of the surface antigen is found in young children, with steadily declining rates among older age groups. The *e* antigen (HBeAg) is more common in young than in adult carriers, whereas the prevalence of *e* antibody (anti-HBe) increases with age.

The survival of hepatitis B virus is ensured by the reservoir of carriers, estimated to number over 350 million worldwide. The prevalence of carriers,

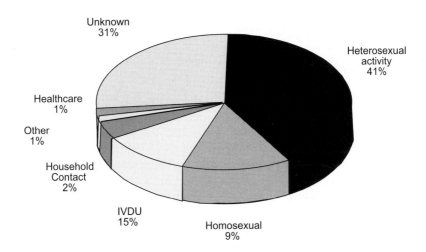

FIGURE 13.3 Risk factors associated with acute cases of hepatitis B in the USA, 1992–1993. Source: Centers for Disease Control, Atlanta.

particularly among blood donors, in Northern Europe, North America and Australia, is 0.1% or less; in Central and Eastern Europe up to 5%; in Southern Europe, the countries bordering the Mediterranean, and parts of Central and South America the frequency is even higher; and in some parts of Africa, Asia and the Pacific Region as many as 20% or more of the apparently healthy population may be carriers. There is an urgent need to introduce methods of interruption of transmission of the infection and establishment of the carrier state, particularly in endemic areas. The management of the carrier state is a complex issue with personal, social and economic implications.

DIAGNOSIS

Direct identification of the virus in serum samples is feasible by electron microscopy (not practical as a routine procedure) and by detecting virus-associated DNA polymerase, which has now been superseded by detection and assay of HBV DNA. In general laboratories, however, diagnosis is based on serological tests, which are widely available in the form of commercial kits.

Hepatitis B surface antigen (HBsAg) first appears in the circulation during the late stages of the incubation period and is easily detectable by sensitive and specific enzyme immunoassays. Radioimmunoassays are no longer used routinely. HBsAg persists during the acute phase of the disease and decreases sharply when antibody to the surface antigen (anti-HBs) becomes detectable. Antibody of the IgM class to the core antigen (anti-core IgM) is found in the serum after the onset of clinical symptoms, and declines slowly after recovery. Persistence of anti-core IgM in high titre suggests continuation of the infection. Core antibody of the IgG class persists for many years, and provides evidence of past infection.

PROTECTION AGAINST HEPATITIS B

The discovery of variation in the epitopes on the surface of the virions and subviral particles identified several subtypes of HBV which differ in their geographical distribution. All isolates of the virus share a common epitope, *a*, which is a domain of the major surface protein which is believed to protrude as a double loop from the surface of the particle. Two other pairs of mutually exclusive antigenic determinants, *d*, or *y* and *w* or *r*, are also present on the major surface protein. Four principal subtypes of HBV are recognized: *adw*, *adr*, *ayw* and *ayr*. Subtype *adw* predominates in Northern Europe, the Americas and Australasia, and is also found in Africa and Asia.

Subtype *ayw* is found in the Mediterranean region, Eastern Europe, Northern and Western Africa, the near East and the Indian subcontinent. In the Far East *adr* predominates, but the rarer *ayr* may be found occasionally in Japan and Papua New Guinea.

PREVENTION BY IMMUNIZATION

Passive Immunization

Hepatitis B immunoglobulin (HBIG) is prepared specifically from pooled plasma with a high titre of hepatitis B surface antibody, and may confer temporary passive immunity under certain defined conditions. The major indication for the administration of hepatitis B immunoglobulin is a single acute exposure to hepatitis B virus, such as occurs when blood containing surface antigen is inoculated, ingested or splashed on to mucous membranes and the conjunctiva. Doses in the range of 250–500 IU have been used effectively. It should be administered as early as possible after acute exposure, and preferably within 48 hours, usually 3 ml (containing 200 IU of anti-HBs per ml) in adults. It should not be administered 7 days or more following exposure. It is generally recommended that two doses be given 30 days apart.

Results with the use of hepatitis B immunoglobulin for prophylaxis in neonates at risk of infection with hepatitis B virus are good if the immunoglobulin is given as soon as possible after birth, or within 12 hours of birth, and the chance of the baby developing the persistent carrier state is reduced by about 70%. Studies using combined passive and active immunization indicate an efficacy approaching 90%. The dose of hepatitis B immunoglobulin recommended in the newborn is 1–2 ml (200 IU of anti-HBs per ml).

Active Immunization

The major response of recipients of hepatitis B vaccine is to the common *a* epitope, with consequent protection against all subtypes of the virus. First-generation vaccines were prepared from 22 nm HBsAg particles purified from plasma donations from asymptomatic chronic carriers. These preparations are safe and immunogenic, but have been superseded in many countries by recombinant vaccines produced by the expression of HBsAg in yeast cells. The expression plasmid contains only the 3′ portion of the HBV surface ORF, and only the major surface protein, without pre-S epitopes, is produced. Vaccines containing pre-S2 and pre-S1, as well as the major surface proteins expressed by recombinant DNA technology, have been developed and are undergoing clinical trials.

In many areas of the world with a high prevalence of HBsAg carriage, such as China and Southeast Asia, the predominant route of transmission is perinatal. Although HBV does not usually cross the placenta, the infants of viraemic mothers have a very high risk of infection at the time of birth.

Administration of a course of vaccine, with the first dose immediately after birth, is effective in preventing transmission from an HBeAg-positive mother in approximately 70% of cases, and this protective efficacy rate may be increased to more than 90% if the vaccine is accompanied by the simultaneous administration of hepatitis B immunoglobulin.

Immunization against hepatitis B is now recognized as a high priority in preventive medicine in all countries, and strategies for immunization are being revised. Universal vaccination of infants and adolescents is recommended.

More than 85 countries now offer hepatitis B vaccine to all children, including the United States, Canada, Italy, France, and most Western European countries.

There are three main approaches to developing new hepatitis B immunization strategies:

1. The introduction of universal antenatal screening to identify hepatitis B carrier mothers, and vaccination of their babies. It is important that any other strategies do not interfere with the delivery of vaccine to this group. Immunization of this group will have the greatest impact in reducing the number of new hepatitis B carriers. For children outside this group it is difficult to estimate the lifetime risk of acquiring a hepatitis infection.
2. Vaccinate all infants.
3. Vaccinate all adolescents. This approach delivers vaccination at a time close to the time when 'risk behaviour' would expose adolescents to infection. Vaccination could be delivered as part of a wider package on health education in general, to include sex education, the risk of AIDS, the dangers of drug abuse and smoking, and the benefits of a healthy diet and lifestyle.

The problems with this approach are as follows:

- Persuading parents to accept vaccination of the children against a sexually transmitted disease, a problem they may not wish to address at that time.
- Ensuring that a full course of three doses is given.
- Evaluating and monitoring vaccine coverage. The systems for monitoring uptake of vaccine in this age group may not operate efficiently.

Vaccination of Infants

The advantages of this approach are:

- It is now known that vaccination can be delivered to babies.
- Parents will accept vaccination against hepatitis B along with other childhood vaccinations, without reference to sexual behaviour.

The disadvantages are:

- It is not known whether immunity will last until exposure in later life. This may become less of a problem as more people are vaccinated and the chance of exposure to infection thereby reduced.
- The introduction of another childhood vaccination may reduce the uptake of existing childhood vaccinations. This problem would be avoided if hepatitis B could be delivered in a combined vaccine containing DPT (diphtheria, polio, tetanus), and such preparations have been developed.

Vaccination of infants is preferable to vaccination of adolescents, as there are sufficient mechanisms to ensure, monitor and evaluate cover. A booster dose could be given in early adolescence, combined with a health education package. A rolling programme could be introduced, giving priority to urban areas.

However, immunization against hepatitis B is at present recommended in a number of countries with a low prevalence only for groups at an increased risk of acquiring this infection. These include individuals requiring repeated transfusions of blood or blood products; prolonged inpatient treatment; patients who require frequent tissue penetration or need repeated circulatory access; patients with natural or acquired immune deficiency; and patients with malignant diseases. Viral hepatitis is an occupational hazard among health-care personnel and the staff of institutions for the mentally handicapped, and in some semi-closed institutions. High rates of infection with hepatitis B occur in narcotic drug addicts and intravenous drug abusers, sexually active male homosexuals and prostitutes. Individuals working in high endemic areas are, however, at an increased risk of infection and should be immunized (Zuckerman, 1984).

Young infants, children and susceptible persons (including travellers) living in certain tropical and subtropical areas where present socioeconomic conditions are poor and the prevalence of hepatitis B is high should also be immunized. It is emphasized that in about 30% of patients with hepatitis B the mode of infection is not known, and this is therefore a powerful argument for universal immunization.

Site of Injection

Hepatitis B vaccination should be given in the upper arm into the deltoid muscle or the anterolateral aspect of the thigh, and not in the buttock. There are over 100 reports of unexpectedly low antibody seroconversion rates after hepatitis B vaccination using injection into the buttock. In one centre in the USA a low antibody response was noted in 54% of healthy adult health-care personnel. Many studies have since shown that the antibody response rate was significantly higher in centres using deltoid injection than in those using the buttock. On the basis of antibody tests after vaccination, the Advisory Committee on Immunization Practices of the Centers of Disease Control, USA, recommended that the arm be used as the site of hepatitis B vaccination in adults, as has the Department of Health in the UK.

A study in the USA by Shaw et al. (1989) showed that participants who received the vaccine in the deltoid had antibody titres up to 17 times higher than those of subjects who received the injections in the buttock.

Furthermore, those who were injected in the buttock were 2–4 times more likely to fail to reach a minimum antibody level of 10 IU/l after vaccination. Recent reports have also implicated buttock injection as a possible factor in a failure of rabies postexposure prophylaxis using a human diploid cell rabies vaccine.

The injection of vaccine into deep fat in the buttocks is likely with needles shorter than 5 cm, and there is a lack of phagocytic or antigen-presenting cells in layers of fat. Another factor may involve the rapidity with which antigen becomes available to the circulation from deposition in fat, leading to a delay in processing by macrophages and eventually presentation to T and B cells. An additional factor may be denaturation by enzymes of antigen that has remained in fat for hours or days. The importance of these factors is supported by the findings that thicker skin folds are associated with a lowered antibody response.

These observations have important public health implications, well illustrated by the estimate that about 20% of subjects immunized against hepatitis B via the buttock in the USA by March 1985 (about 60 000 people) failed to attain a minimum level of antibody of 10 IU/l, and were therefore not protected.

Hepatitis B surface antibody titres should be measured in all individuals who have been immunized against hepatitis B by injection into the buttocks, and when this is not possible a complete course of three injections of vaccine should be administered into the deltoid muscle or the anterolateral aspect of the thigh, the only acceptable sites for hepatitis B immunization (Zuckerman *et al.*, 1992).

PRODUCTION OF HEPATITIS B VACCINES BY rDNA TECHNIQUES

Recombinant DNA techniques have been used for expressing hepatitis B surface antigen and core antigen in prokaryotic cells (*Escherichia coli* and *Bacillus subtilis*) and in eukaryotic cells, such as mutant mouse LM cells, HeLa cells, COS cells, CHO cells and yeast cells (*Saccharomyces cerevisiae*).

Recombinant yeast hepatitis B vaccines have undergone extensive evaluation by clinical trials, the results indicating that they are safe, immunogenic and free from side-effects (apart from minor local reactions in a proportion of recipients). The immunogenicity is similar, in general terms, to that of the plasma-derived vaccine. Recombinant yeast hepatitis B vaccines are now being used in many countries. A vaccine based on HBsAg expressed in mammalian (CHO) cells is in use in the People's Republic of China.

NON-RESPONSE TO HEPATITIS B VACCINES

All studies of the antibody response to currently licensed plasma-derived hepatitis B vaccines and hepatitis B vaccines prepared by recombinant DNA technology have shown that between 5% and 10% or more of healthy immunocompetent subjects do not mount an antibody response (anti-HBs) to the surface antigen component (HBsAg) present in these preparations (non-responders), or that they respond poorly (hyporesponders). The exact proportion depends partly on the definition of non-responsiveness or hyporesponsiveness: generally less than 10 IU/l or 100 IU/l of anti-HBs, respectively, against an international antibody standard (reviewed by Zuckerman, 1996).

Non-responders remain susceptible to infection with HBV. Several factors are known adversely to affect the antibody response to HBsAg, including the site and route of injection, gender, advancing age, body mass (overweight), immunosuppression and immunodeficiency. Other mechanisms underlying non-responsiveness to the S component of HBsAg in humans, however, remain largely unexplained, although evidence is accumulating that there is an association between different HLA-DR alleles and specific low responsiveness in different ethnic populations. Considerable experimental evidence is available to suggest that the ability to produce antibody in response to specific protein antigens is controlled by dominant autosomal class II genes of the major histocompatibility complex (MHC) in the murine model.

There is also evidence that the pre-S1 and pre-S2 domains have an important immunogenic role in augmenting anti-HBs responses, preventing the attachment of the virus to hepatocytes, eliciting antibodies which are effective in viral clearance, stimulating cellular immune responses, and circumventing genetic non-responsiveness to the S antigen. A number of studies indicate that the inclusion of pre-S components in recombinant or future synthetic vaccines should be developed. For example, the pre-S2 region is more immunogenic at the T- and B-cell levels than the S regions in the mouse model, as is the case with pre-S1 in the mouse and in humans, and circumvents S-region non-responsiveness at the level of antibody production!

Milich *et al.* (1986) demonstrated in the murine model that the independence of MHC-linked gene regulation of immune responses to pre-S1, pre-S2 and S regions of HBsAg would assure fewer genetic non-responders to a vaccine containing all three antigenic regions. Studies conducted in humans with experimental recombinant hepatitis B vaccines containing all three S components of the viral envelope polypeptides demonstrated the enhanced immunogenicity of such preparations compared to conventional yeast-derived vaccines (Yap *et al.*, 1995; Shouval *et al.*, 1994; Zuckerman *et al.*, 1997), although several earlier studies with vaccines containing the S and pre-S components revealed significant differences from preparations containing only the S antigen.

Zuckerman *et al.* (1997) evaluated the immunogenicity of a vaccine containing the S, pre-S1 and pre-S2 components of hepatitis B surface antigen produced in a continuous mammalian cell line, C127 mouse cells in a group of 86 persistent non-responder health-care personnel. A total of 55 of 86 subjects (64%) seroconverted following a single dose of the vaccine, with an overall rate of 66% (57/86) 4 months after the booster dose. A single dose of 20 µg of the vaccine was shown to be as effective as either two doses of 20 µg or 40 µg, in terms of seroconversion, seroprotection levels or geometric mean titres (seroconversion and seroprotection as defined by an antibody titre > 10 IU/l and > 100 IU/l, respectively, against an international antibody standard).

HEPATITIS B ANTIBODY ESCAPE MUTANTS

Production of antibodies to the group antigenic determinant *a* mediates cross-protection against all subtypes, as has been demonstrated by challenge with a second subtype of the virus following recovery from an initial experimental infection. The epitope *a* is located in the region of amino acids 124–148 of the major surface protein, and appears to have a double-loop conformation. A monoclonal antibody which recognizes a region within this *a* epitope is capable of neutralizing the infectivity of hepatitis B virus for chimpanzees, and competitive inhibition assays using the same monoclonal antibody demonstrate that equivalent antibodies are present in the sera of subjects immunized with either plasma-derived or recombinant hepatitis B vaccine.

During a study of the immunogenicity and efficacy of hepatitis B vaccines in Italy, a number of individuals who had mounted a successful immune response with antisurface antibody (anti-HBs) production later became infected with HBV (Carman *et al.*, 1990). These cases were characterized by the coexistence of non-complexed anti-HBs and HBsAg, and in 32 of 44 vaccinated subjects there were other markers of hepatitis B infection. Furthermore, analysis of the antigen using monoclonal antibodies suggested that the *a* epitope was either absent or masked by antibody. Subsequent sequence analysis of the virus from one of these and other cases revealed a mutation in the nucleotide sequence encoding the *a* epitope, the consequence of which was a substitution of arginine for glycine at amino acid position 145 (Carman *et al.*, 1990; Oon *et al.*, 1995, 1996).

There is now considerable evidence for a wide geographical distribution of this point mutation in hepatitis B virus (guanosine to adenosine at position 587), resulting in the amino acid substitution at position 145 from glycine to arginine in the highly antigenic group determinant *a* of the surface antigen. This stable mutation has been found in viral isolates from children several years later, and it has been described in Italy, Singapore, Japan, Brunei, the USA, China and elsewhere, and from liver transplant recipients with hepatitis B in the USA, Germany and the UK who had been treated with specific hepatitis B immuno-globulin or humanized hepatitis B monoclonal antibody. The region in which this mutation occurs is an important virus epitope to which vaccine-induced neutralizing antibody binds, and the mutant virus is not neutralized by antibody to this specificity. It can replicate as a competent virus, implying that the amino acid substitution does not alter the attachment of the virus to the liver cell.

Other surface antigen mutants have also been identified with amino acid substitutions within and outside the *a* determinant, but the 145 aa mutation appears to be the most frequent and the most stable. Hepatitis B virus was considered to be genetically stable, but it is now clear that mutations occur frequently, probably because the reverse transcriptase, which is essential in the replication cycle of HBV, is error-prone. Surveillance programmes are essential in order to monitor the epidemiology of HBV mutants. Because some of the most widely used serological tests for HBsAg fail to recognize certain HBsAg mutants, there is a critical need to improve existing techniques; polyclonal antibody-based detection assays are far more likely to identify such mutants than the monoclonal antibody-based assays which are widely used at present. Improved tests which can detect mutant forms of HBsAg are vital for blood donor screening programmes to prevent HBV mutants from entering the blood supply.

THE IMMUNE RESPONSE AND SOME UNRESOLVED ISSUES

Hepatitis B virus continues to pose many questions, even though it has been studied intensively for some 30 years. Some of these issues are as follows:

- Anti-*a* is generally considered to be the neutralizing antibody, but which particular sequences within anti-*a* or within the *dywr* locus are important for neutralization?
- What are the kinetics of antibody production to various antigenic determinants following the administration of hepatitis B vaccines?
- Very little is known about antigenic binding and antibody avidity.
- What is the composition of the ideal hepatitis B vaccine?
- Will recombinant vaccines incorporating pre-S1, pre-S2 and S components help circumvent infection by HBsAg mutants?

HBV PRECORE AND OTHER MUTANTS

The nucleotide sequence of the genome of a strain of HBV cloned from the serum of a naturally infected chimpanzee was reported in 1988. A surprising feature was a point mutation in the penultimate codon of the precore region which changed the tryptophan codon (TGG) to an amber termination codon (TAG). The nucleotide sequence of the HBV precore region from a number of anti-HBe-positive Greek patients and others was investigated by direct sequencing PCR-amplified HBV DNA from serum. An identical mutation of the penultimate codon of the precore region to a termination codon was found in seven of eight anti-HBe-positive patients who were positive for HBV DNA in serum by hybridization. In most cases there was an additional mutation in the preceding codon. Similar variants were found by amplification of HBV DNA from serum from anti-HBe-positive patients in other countries. In many cases precore variants have been described in patients with severe chronic liver disease and who may have failed to respond to therapy with interferon. This observation has raised the question of whether some mutants may be more pathogenic than the wild-type virus.

Mutations have also been described in the X gene of HBV and in the gene coding for the polymerase. More data on these mutants are anticipated.

TREATMENT OF HEPATITIS B

Specific antiviral therapy is not available for the treatment of acute viral hepatitis, nor for chronic infection associated with hepatitis B, C and D.

Although many antiviral compounds have been evaluated for the treatment of chronic hepatitis B, at present, only interferon-α remains the mainstay of treatment. Under optimal conditions and careful patient selection 30–50% respond at least transiently, and about 20% clear the virus, accompanied by an improvement in liver function. A comparison of prednisolone withdrawal followed by interferon, versus treatment with interferon only, shows no added benefit for the combined regimen. Combination therapy with nucleoside analogues and other drugs has shown a variable response. Lamivudine and famciclovir are promising but not yet licensed. Rapid progress is expected in the development of specific antiviral therapy.

REFERENCES

Alter, M.J., Hadler, S.C., Margolis, H.S. *et al.* (1990). The changing epidemiology of hepatitis B in the United States: need for alternative vaccine strategies. *J. Am. Med. Assoc.* **263**: 1218–1222

Carman, W.F., Zanetti, A.R., Karayiannis, P. *et al.* (1990). Vaccine-induced escape mutant of hepatitis B virus. *Lancet* **336**: 325–329

Milich, D.R., McLachlan, A., Chisari, F.V. *et al.* (1986). Immune response to the pre-S1 region of hepatitis B surface antigen (HBsAg): A pre-S1-specific T cell response can bypass nonresponsiveness to pre-S2 and S regions of the HBsAg. *J. Immunol.* **137**: 315–322

Oon, C.-J., Lim, G.-K., Ye, Z. *et al.* (1995). Molecular epidemiology of hepatitis B virus vaccine variants in Singapore. *Vaccine* **13**: 699–702

Oon, C.-J., Tan, K.-L., Harrison, T. and Zuckerman, A.J. (1996). Natural history of hepatitis B surface antigen mutants in children. *Lancet* **348**: 1524

Sepetjan, M. (1991). Hepatites Virales. Annual report of the laboratory of preventive medicine, public health and hygiene. Lyon, France

Shaw, F.E., Jr., Guess, I.J.A., Roets, J.M. *et al.* (1989). Effect of anatomic site, age and smoking on the immune response to hepatitis B vaccination. *Vaccine* **7**: 425–430

Shouval, D., Ilan, Y., Adler, R. *et al.* (1994). Improved immunogenicity in mice of a mammalian cell-derived recombinant hepatitis B vaccine containing pre-S1 and pre-S2 antigens as compared with conventional yeast-derived vaccines. *Vaccine*, 1453–1459

WHO (1983). Prevention of Liver Cancer. World Health Organization Technical Report Series **691**: 1–30

Yap, I., Guan, R. and Chan, S.H. (1995). Study on the comparative immunogenicity of a recombinant DNA hepatitis B vaccine containing pre-S components of the HBV coat protein with non pre-S containing vaccine. *J. Gastroenterol. Hepatol.* **10**: 51–55

Zuckerman, A.J. (1984). Who should be immunised against hepatitis B? *Br. Med. J.* **289**: 1243–1244

Zuckerman, J.N., Cockcroft, A. and Zuckerman, A.J. (1992). Site of injection for vaccination. *Br. Med. J.* **305**: 1158

Zuckerman, J.N. (1996). Nonresponse to hepatitis B vaccines and the kinetics of anti-HBs production. *J. Med. Virol.* **50**: 283–280

Zuckerman, J.N., Sabin, C., Craig, F.M., Williams, A. and Zuckerman, A.J. (1997). Immune response to a new hepatitis B vaccine in health care workers who had not responded to standard vaccine: randomised double blind dose-response study. *Br. Med. J.* **314**: 329–333

14
THERAPEUTIC VACCINES FOR CHRONIC HEPATITIS B INFECTION

PIERRE VANDEPAPELIÈRE

SmithKline Beecham Biologicals, Rixensart, Belgium

INTRODUCTION

Hepatitis B is not usually regarded as a major sexually transmitted disease, probably because there is no genital disease manifestation and transmission by other routes is more common worldwide. Hepatitis B virus (HBV) is, however, 8.6 times more infectious through the sexual route than is HIV in homosexuals (Kingsley *et al.*, 1990), and sexual transmission is the most frequent mode of transmission in the United States. A recent study by the American Sexual Health Association (ASHA, 1998) estimated that in the United States 750 000 chronic HBV carriers were infected through sexual transmission. Despite the availability of effective prophylactic vaccines, the worldwide prevalence and incidence of hepatitis B has not decreased globally (Coleman *et al.*, 1998), with the notable exceptions of countries such as Taiwan and Italy, where universal vaccination programs have been implemented.

Although the evaluation of therapeutic HBV vaccines is limited, this is an attractive approach that is receiving considerable attention. This chapter will review the current status of therapeutic HBV vaccines as well as the clinical and immunological aspects of the disease. Particular emphasis is placed on aspects relevant to the development of therapeutic vaccines.

EPIDEMIOLOGY AND CLINICAL ASPECTS

Epidemiology

Hepatitis B is a disease with worldwide implications. After infection, the disease can either resolve or evolve towards chronicity. Around 350 million people worldwide are chronic carriers (Kane *et al.*, 1989). The WHO divides the world into regions of low endemicity (<2% carrier rate, i.e. Western Europe, North America, Australia), intermediate endemicity (2–8%: Eastern Europe, Middle East, Japan) and high endemicity (>8%: South East Asia, Far East, Africa, China); 75% of all HBV carriers live in Asia. Clinical manifestations associated with chronic HBV carriage are highly variable, ranging from the absence of hepatic lesions to rapidly evolving destruction of the liver, leading to cirrhosis and hepatic insufficiency (Dienstag, 1984; Hoofnagle, 1990). HBV chronic carriage increases the risk of developing hepatocellular carcinoma (HCC) by a factor of 100, and chronic hepatitis B (CHB) infection is the leading cause of HCC in the world (Beasley *et al.*, 1981a,b; 1988).

Hepatitis B can be transmitted through vertical, sexual, parenteral or horizontal routes (Davis *et al.*, 1989; Beasley and Hwang, 1983c). Although vertical transmission (i.e. from mother to neonate) remains the most frequent route of infection, particularly in areas of high endemicity, sexual activity, both heterosexual and homosexual, is a major route for transmission in areas of low prevalence (Alter and Margolis, 1990).

Chronic hepatitis B Patient Subgroups: Subdivision Based on Clinical, Virological, Serological and Immunological Criteria

Chronic hepatitis B is a lifelong infection which evolves in response to changes in the balance between the immune response and viral activity. Therefore, the same infected individual can present over time with different clinical, virological and immunological aspects of disease (Figure 14.1).

This is perhaps best illustrated in Asians, who are usually infected at birth from their infected mothers (Lok, 1992). If the mother is HBeAg positive, the

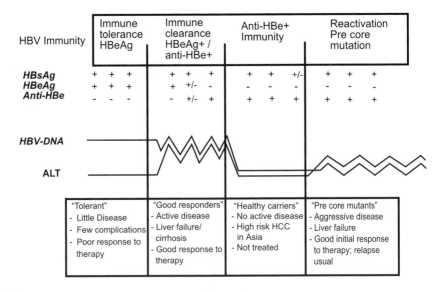

FIGURE 14.1 CHB patient subgroups

baby has a greater than 90% chance of developing a chronic infection (Stevens *et al.*, 1975), but there is usually little, if any, clinical disease during childhood or early adolescence. This is the so-called 'tolerant' phase, as the immune system is almost fully tolerized to the virus. In this phase, HBV-DNA levels are high, the aspartate aminotransferase (ALT) levels are usually normal, and individuals are HBeAg positive. These patients respond very poorly to α-interferon (IFN) therapies. In a small percentage there are occasional flares of transient acute hepatitis with ALT increases, most likely corresponding to attempts by the immune system to clear the infected hepatocytes. If these flares of immune activity are very frequent, it can lead to early cirrhosis. This is, however, a rare phenomenon.

Between 15 and 25 years of age the serological and biochemical patterns change, with frequent flares of increased ALT levels, sometimes accompanied by clinical manifestations of acute hepatitis, and progressive decreases of HBV-DNA levels (Lok *et al.*, 1987, 1990). During this period, patients respond well to α-IFN therapy, especially when the ALT levels are high and the HBV-DNA level is low.

This phase, called the 'immune clearance phase', will lead in most cases to the loss of HBeAg. However, if it is too long and too aggressive it can also induce cirrhosis or hepatic decompensation (Sheen *et al.*, 1985; Liaw and Tsai, 1997). The phenomenon of spontaneous seroconversion generally occurs at a rate ranging from 10% to 30% per year, depending upon the population (Evans *et al.*, 1997).

When HBeAg disappears, it is usually accompanied by normalization of ALT and decreases in HBV-DNA levels to below the detection limits of hybridization techniques. At this time anti-HBe antibodies eventually appear. This is a phase of anti-HBe immmunity, and the only markers of infection are serological (HBsAg+, anti-HBe+, anti-HBc+). The inflammatory disease is decreased, although it is unclear whether the development of HCC is also slowed down. What is clear is that such subjects in Asia are still at a much higher risk of developing HCC than the non-infected population. Even the loss of HBsAg in this population does not necessarily imply a good prognosis (Huo *et al.*, 1998). Thus, in Asia such subjects are not referred to as 'healthy carriers' as they are in western countries, but simply as 'asymptomatic carriers' (Hoofnagle *et al.*, 1987; de Franchis *et al.*, 1993; Lok and Lai, 1988).

Another possible evolution from the immune clearance phase, which may be immediate or some years after anti-HBe immunity, is reactivation of the disease despite sustained loss of HBeAg (Fattovitch *et al.*, 1986). This is usually associated with mutation in the precore and/or core regions of the viral genome which prevents the secretion of HBeAg (Carman *et al.*, 1989, 1992; Bozkaya *et al.*, 1996). These viruses may be referred to as 'precore mutants', even though a mutation in the precore region cannot always be detected. The disease may be more severe, with rapid evolution to cirrhosis, HCC and/or liver failure (Brunetto *et al.*, 1991; Omata *et al.*, 1991). As opposed to HBeAg-positive CHB, transmission of the HBeAg negative virus (precore mutant) from the mother to her baby can result in fulminant hepatitis, as if the tolerogenic mechanisms were lost. These patients respond well to IFN therapy, but virtually all will relapse at the end of the treatment (Hadziyannis, 1995).

In western countries, infection usually occurs through parenteral or sexual transmission, during adolescence or adulthood, and only 5% evolve to chronicity. This suggests that establishment of tolerance is less frequent. However, all the subpopulation patterns described above can be found with similar characteristics and similar responses to therapy.

Currently Available Therapies

Interferon

Interferons (α and β) are the only widely approved drugs for the treatment of chronic hepatitis B. They have antiviral and immunomodulatory activities, and have been shown to limit virus spread from focal sites of infection, to shorten the duration of viremia, and to decrease the severity of acute systemic infection. They also have a complex immune-enhancing capacity, affecting multiple cell types, mainly through increasing cytotoxic T-lymphocyte (CTL) expression and natural killer cell activity (Peeters, 1989). Treatment consists of subcutaneous administration of 6–10 million units, three times a week, for 3–6 months (Bailly and Trepo, 1997), but sometimes for up to 24 months (Lampertico et al., 1997).

Interferon therapy is indicated only in a fraction of CHB patients, i.e. preferably those with HBeAg-positive hepatitis, elevated ALTs and limited viral replication (Perillo, 1993; Lok et al., 1998). It has been shown that these patients also have a higher natural tendency for spontaneous seroconversion (Lok et al., 1989; Krogsgaard et al., 1994). However, even in this patient subgroup sustained efficacy is limited to approximately 20% of treated subjects (Wong et al., 1993; Thomas et al., 1994). The rate of response is similar in children (Vajro et al., 1996; Sokal et al., 1998).

Long-term follow-up has demonstrated that interferons are so far the only drugs that increase survival time (Niederau et al., 1996) and decrease the incidence of HCC (Ikeda et al., 1998). It has been repeatedly claimed that Asian patients are poor responders to IFN therapy (Lok et al., 1989, 1993). This has been associated with vertical transmission, a higher tolerance, a longer duration of disease, and genetic factors. However, if the same selection criteria are used, especially with regard to ALT and HBV-DNA levels, the response rate is similar in Asian and western subjects (Lau et al., 1991; Wu et al., 1992; Lok et al., 1998). Long-term sustained viral clearance, however, does occur less frequently in Asian patients than in Caucasians (Lok et al, 1993; Lau et al., 1997).

Patients with HBe antigen-negative CHB have a good initial response to IFN therapy but almost always relapse during follow-up (Alberti and Fattowitch, 1994; Brunetto et al., 1995). High initial ALT and IgM anti-HBc levels (Lobello et al., 1998) and a long duration of therapy may improve the sustained response rate (Lampertico et al., 1997).

Asymptomatic 'healthy' carriers do not benefit from IFN therapy; on the contrary, it can cause a reactivation of the disease (Rodriguez-Iningo et al., 1997).

Interferon treatment also has many side-effects that lead to dose reduction or discontinuation of the therapy in approximately 20% of patients (Bailly and Trepo, 1997). Most of the adverse reactions resolve spontaneously, but some are more severe or indeed irreversible (Yoshikawa *et al.*, 1995; Manesis *et al.*, 1998). In rare cases hepatic failure (Marcellin *et al.*, 1991; Perillo *et al.*, 1988), or more frequently induction of precore mutations (Fattovitch *et al.*, 1995; Lampertico *et al.*, 1995), have been described.

In conclusion, interferons were until recently the only drugs available for the treatment of chronic hepatitis B. Their limited efficacy and the poor safety profile emphasize the need for the development of better alternatives.

Antiviral Drugs

Research into the development of antiviral drugs for the treatment of CHB is very active. The most advanced is lamivudine (3TC, Glaxo Wellcome), which was initially developed and licensed for the treatment of HIV infection (Dusheiko, 1998). Large phase III studies in CHB patients are now completed, and the drug was recently approved in the US for treatment of HBV, with registration ongoing in many countries. Treatment with lamivudine, 100 μg od, causes a rapid drop in HBV-DNA levels, accompanied by normalization of transaminases (Jaeckel and Manns, 1997). After 1 year of treatment, it induces HBeAg seroconversion in approximately 15% of patients, both in western (Dienstag *et al.*, 1998; Schiff *et al.*, 1998) and Asian populations (Lai *et al.*, 1998). Histological improvements are seen in 50% of patients, including slowing of the fibrosis process. As with interferon therapy, a complete response, evidenced by HBeAg loss and appearance of anti-HBe antibodies, is almost exclusively seen in patients with initially elevated transaminases (Lok, 1998, Personal communication).

Unfortunately, there are two pitfalls associated with lamivudine therapy. First, at the end of therapy there is a frequent relapse of disease with a rebound of ALT that, in rare cases, results in liver necrosis (Honkoop *et al.*, 1995). HBeAg seroconversions, which are rare also, are not always sustained. This implies that patients responding to lamivudine should in theory be treated for many years, if not for life. Second, lamivudine rapidly induces several mutations that confer a reduced sensitivity to the treatment. These are mainly in the YMDD motif of the viral polymerase: 14% of patients after 1 year of therapy and 38% after 2 years (Chayama *et al.*, 1998; Allen *et al.*, 1998; Niesters *et al.*, 1998). Wild-type virus takes over from the mutated virus when lamivudine treatment is stopped.

Several other antivirals are in earlier phases of development (Table 14.1). Despite claimed improvements, it is unlikely that any will be devoid of the two major problems described above. Therefore, antiviral drugs are probably not the ultimate treatment for chronic hepatitis B. Combinations of antivirals with interferon (Marinos *et al.*, 1996; Schiff *et al.*, 1998), with thymosin α_1 or with other antivirals, is currently being evaluated.

PATHOGENESIS OF THE DISEASE: MECHANISMS OF VIRAL PERSISTENCE

There is a large body of evidence suggesting that HBV is not cytopathic in itself, but that the necroinflammatory process in the liver cells is caused by the immune response against infected hepatocytes (Alberti *et al.*, 1984). An

TABLE 14.1 Antiviral drugs in development for therapy of chronic HBV

Drug	Company	Development Phase
Lamivudine (Epivir)	Glaxo Wellcome	III–IV
Famciclovir (Famvir)	SmithKline Beecham Pharmaceuticals	III
Lobucavir	Bristol Myers Squibb	III
Adefovir	Gilead	II
BMS 200 475	Bristol Myers Squibb	I–II

inadequate immune response probably leads to viral persistence and chronicity of the disease (Tsai and Huang, 1997). If the cellular immune response plays a major role in the pathogenesis of hepatocellular damage in CHB, both cellular and humoral responses are probably needed for viral clearance (Milich, 1997c).

Immune Response in Acute and Chronic HBV Diseases

In acute self-limiting HBV infection there is a vigorous, heterogeneous CD4+ T-helper cell response specific to HBeAg and HBcAg (Jung *et al.*, 1995), whereas the response to envelope antigens is much weaker (Ferrari *et al.*, 1990; Jung *et al.*, 1991). On the other hand, CTLs specific for multiple epitopes within the envelope, polymerase, HBc and HBeAg antigens have been detected (Penna *et al.*, 1991; Nayersina *et al.*, 1993; Reherman *et al.*, 1995). More CTL epitopes have been identified on polymerase and surface antigens than on the core antigen (Ferrari *et al.*, 1991).

During chronic hepatitis B the CD4+ T-helper response is predominantly directed against the core antigen and HBeAg, to a much lesser extent against envelope antigens (Inoue, 1989; Vento *et al.*, 1987; Löhr *et al.*, 1995; Marinos *et al.* 1995). In fact, the core antigen is 100 times more immunogenic than HBsAg, at both the T-cell and B-cell levels (Milich, 1997a). However, the cellular immune response to core and HBeAg is much weaker in CHB patients than in acute hepatitis. The CTL response to all HBV proteins is undetectable or very weak during chronic hepatitis B (Löhr *et al.*, 1993). The chronic but moderate inflammatory destruction of liver cells during chronic HBV, however, is most probably due to this CTL activity, which is too weak to eliminate the virus (Milich *et al.*, 1993).

CD4+ T-helper cells specific for HBc/HBeAg play a major role in viral clearance. This response also correlates both with acute disease (Tsai *et al.*, 1992; Jung *et al.*, 1995) and with clearance of HBsAg and HBV DNA (Löhr *et al.*, 1995). The T-helper response is of the Th1 type and is characterized by secretion of IL-2, γ-IFN and TNF-α (Fukuda *et al.*, 1995; Maruyama *et al.*, 1993). Serum levels of IL-2, IL-12 and γ-IFN have also been shown to correlate with flares of hepatic enzymes (ALT). However, serum IL-12 is increased only when these flares are followed by loss of HBeAg and the appearance of anti-HBe antibodies, i.e. HBeAb seroconversion (Rossol, 1997).

CTLs also play a major role in viral clearance (Missale *et al.*, 1993; Nayersina *et al.*, 1993), and although CTLs specific for surface antigens are predominant during acute infection, HBc/HBeAg-specific T-cell responses may be more important during chronic HBV infection. The strength of the CTL response is a determinant for clearance or persistence of infection (Chisari, 1995).

In conclusion, a strong CD4+ response against the core Ag seems mandatory to initiate the process of clearance, but polyclonal, multispecific humoral and cellular immune stimulations are most likely needed for a full clearance.

Mechanisms of Persistence

Numerous factors have been identified to explain the mechanisms of persistence of HBV viral infection. Milich separates mechanisms of induction from those responsible for maintenance of persistence (Chisari, 1995; Milich, 1997c). In both cases HBeAg seems to play a major role (Table 14.2).

Factors Associated with Induction of Persistence

In the majority of cases, neonates born to HBeAg carrier mothers will become chronic carriers of HBV (Okada *et al.*, 1976; Beasley *et al.*, 1977). This is most probably due to transplacental passage of HBeAg, which is then identified as a self-antigen and renders T cells (but not B cells) tolerant to HBeAg. Because of the cross-reactivity between HBeAg and HBcAg, this tolerance is extended to HBc (Milich *et al.*, 1990). In contrast, neonates can respond to HBsAg, as evidenced by the good immune response and protection induced by prophylactic surface antigen-based vaccines when administered soon after birth (Beasley *et al.*, 1983a; Poovorawan *et al.*, 1989). The in-utero exposure to HBeAg is supported by the fact that HBeAg, but not HBsAg, can be detected in the cord blood of neonates from HBeAg-positive mothers (Hsu *et al.*, 1992). Infection in utero is, however, very rare and only occurs following placental leakage (Li *et al.*, 1986; Lee *et al.*, 1989). The role of HBeAg in the induction of

---

TABLE 14.2 Mechanisms of HBV persistence

Factors associated with induction of persistence
Neonatal induction
Transplacental passage of HbeAg in utero
Immaturity of the immune system

Adult infection:
Immunodepression at the time of infection
MHC genetic characteristics

Factors associated with maintenance of persistence
HBV viral protein in excess in serum
HBeAg
HBsAg

Virus mutants and variants

Modulation of immune system by viral proteins

HBV infection of extrahepatic sites

---

tolerance at birth is further supported by case reports of neonates born to HBeAg-negative anti-HBe antibody-positive CHB mothers who developed fulminant hepatitis (Terazawa *et al.*, 1991; Carman *et al.*, 1991; Liang *et al.*, 1991). One possible explanation in these cases is that the absence of HBeAg prevents tolerance towards the virus. Some cases of fulminant hepatitis have also been reported after sexual transmission from anti-HBe-positive CHB patients (Fagan *et al.*, 1986; Tassopoulos *et al.*, 1986).

Only 5–10% of infected adults become chronic carriers of the virus, but this proportion is dramatically increased when the subject is immunocompromised at the time of infection. This suggest that immune responses in immunologically normal hosts can control the acute HBV infection and prevent persistent infection. MHC genes also seem to play a role in protecting against or facilitating the establishment of chronic infection (Thursz *et al.*, 1995; Thomas and Thursz, 1997).

Factors Associated with the Maintenance of Persistence

The balance between the two subsets of regulatory CD4+ T-helper (Th) cells, namely Th1 and Th2, most likely plays an important role in the maintenance or clearance of chronic HBV (Milich, 1997b). The Th1 cells, which promote cellular immunity with secretion of IL-2, γ-IFN and TNF-α, are predominant during the clearance phase. On the other hand, the Th2 cells that help B cells to produce antibodies and induce secretion of IL-4, IL-5, IL-6, IL-10 and IL-13 (Paul and Seder, 1994), clearly represent the major T-helper subset during the tolerant phase of chronic HBV (Ferrari *et al.*, 1990; Milich *et al* 1997b). Therefore, the predominant Th1 or Th2 pattern most likely drives the immune system towards either induction of clearance or maintenance of tolerance, respectively. It has been shown in mice that HBeAg may play a major role in the down-regulation of antiviral clearance mechanisms by eliciting predominant induction of Th2 cells with secretion of Th2 cytokines (Milich *et al.*, 1997). In the same experiment, HBcAg preferentially induced stimulation of Th1 cells. The secreted HBeAg, circulating in excess in the serum, could thus be a factor involved in promotion of the Th2 response (Milich *et al.*, 1993, 1998).

This hypothesis is also supported by the development of fulminant hepatitis in adults infected by anti-HBe positive CHB patients (Carman *et al.*, 1991; Liang *et al.*, 1991), and also by the aggravation of the disease when there is a shift from infection by the wild-type virus to infection by the mutated precore virus (Brunetto *et al.*, 1991). The excess of HBsAg could also play an important role in virulence, through absorption of the neutralizing anti-HBs antibodies, which could prevent viral spread. In addition, it has been shown that HBsAg of T-cell origin may have a role in suppressing anti-HBsAb production (Nagaraju *et al.*, 1997).

Other mechanisms of increased pathogenesis include the induction of HBV variants, which are not recognized by neutralizing antibodies or by the cellular immune response. Variants with mutations in a determinant region of the HBsAg have been described in neonates following prophylactic HBV vaccination. However, the most frequent mutation is a point mutation in the precore region (codon 28), which prevents translation of the precore protein and liver secretion of HBeAg (Carman *et al.*, 1989; Brunetto, 1991a,b). A mutation in core antigen may appear illogical, as HBeAg is a major factor in persistence. A possible explanation is that after a long chronic immune response against the virus HBeAg becomes a target for CTLs during an acute necroinflammatory phase, and thus becomes deleterious to virus survival (Milich *et al.*, 1988). This phenomenon could be explained by mutations in the CTL epitopes which interfere with recognition of the wild-type epitope, and therefore act as T-cell receptor antagonists (Bertoletti *et al.*, 1994).

Other factors, such as modulation of the immune system by viral proteins, mainly HBeAg (Whitten *et al.*, 1991) and polymerase (Foster *et al.*, 1991), have been described. In addition, the virus can locate in privileged extrahepatic sites, i.e. sites which cannot be reached by CTLs (Ando *et al.*, 1994). Hepatitis B virus has been detected in mesenteric lymph nodes, spleen, kidney, pancreas, brain, and some endocrine tissues such as testis, ovary, adrenal and thyroid glands (Yoffe *et al.*, 1990).

ANIMAL MODELS

The animal models used in the study of HBV can be divided into three categories: animals that can be infected by human HBV; animals infected by species-specific viruses from the hepatitis virus family (Hepadnaviridae); and transgenic models.

The chimpanzee is probably one of the best animal models to evaluate therapeutic vaccines, as it can be infected by human HBV. Chimps develop chronic infection and disease and are similar to man with respect to the antibody titers required for protection (Davis, 1996). However, their cost and scarcity render them difficult to use for large-scale testing of multiple immuno-therapeutic approaches.

Recently another animal model with susceptibility to human HBV infection has been described. A primitive monkey, *Tupaia belangeri*, develops acute and chronic diseases similar to those observed in humans, and could prove to be very useful in the future (Köck *et al.*, 1998; Brown *et al.*, 1998).

In the search for other animal viruses that cause hepatocellular carcinoma several other viruses of the Hepadnaviridae family have been identified. The two most extensively described are the Woodchuck and the Peking duck hepatic viruses, which have been and still are very useful for the evaluation of antiviral drugs.

The Woodchuck hepatitis virus (WHV) is the most attractive, as it shares considerable similarities in morphology, genomic organization and protein composition (Tennant and Gerin, 1994; Michalak, 1998). Chronically WHV-infected woodchucks develop liver disease with chronic sequelae, similar to those observed in HBV-infected humans, and virtually 100% of these wood-chucks eventually develop hepatocellular carcinoma. Their relatively low cost and availability compared to chimpanzees make woodchucks an ideal model for the evaluation of anti-HBV therapeutic compounds (Hervas-Stubbs *et al.*, 1997). However, their use in the immunotherapeutic area has so far been limited by the lack of recognized markers of woodchuck lymphocyte subsets, established cell lines, immunoreagents and purified recombinant WHV antigens. These limitations will probably soon be overcome thanks to the rapid development of the necessary assays, methods and reagents. Although cheaper, the Peking duck model suffers from the same limitations, plus the difference between the mammalian and the avian immune systems, which renders it less attractive.

The most extensively used animal models are transgenic mice, which constitutively express HBV antigens in the liver even before birth. Although this model is not fully representative of a real HBV infection and there is no liver pathology, transgenic mice are very useful for evaluating the role of the various antigens in the different subsets of the immune response, and in the mechanisms of immune tolerance (Chisari, 1995a). They can also be used for testing therapeutic vaccines for their capacity to induce HBsAg and/or HBe seroconversion (Fazle Akbar *et al.*, 1997; Davis *et al.*, 1997b).

RATIONALE FOR A THERAPEUTIC VACCINE TO TREAT CHRONIC HBV

There are many arguments in favor of therapeutic vaccination as a rational treatment for chronic hepatitis B. Studies of the pathogenesis of the disease unanimously suggest that chronic necroinflammatory destruction of hepatocytes is due to the host immune response and not to a direct cytopathic viral effect. Initiation and maintenance of the persistence of viral infection is dependent upon a complex interaction between virus and host, with the immune response to the viral antigens believed to be a dominant factor associated with the chronic carrier state.

However, this state of partial or total immune tolerance can be broken and the immunological effector mechanisms may be corrected during acute exacerbations of the disease, which can lead to immunoclearance. This natural phenomenon of seroconversion is therefore one of the main arguments in favor of the feasibility of active immunotherapy, even if there are still many immunological and viral aspects of its mechanism that we do not understand.

In addition, unlike herpes viruses, the hepatitis B virus is mainly located in cells that can be reached by CTLs and cytokines. Furthermore, massive numbers of hepatocytes can be destroyed without rapidly impairing hepatic function, and with a high probability of rapid recovery thanks to the great regenerative capacity of the liver. Indeed, acute liver failure following either seroconversion (natural or induced) or necroinflammatory flares usually only occurs in the presence of pre-existing severely impaired hepatic function, for example in advanced liver cirrhosis (Marcellin *et al.*, 1991; Honkoop *et al.*, 1995).

It is also important to note that certain therapies with immunomodulatory properties, such interferon, thymosin-α and/or some candidate therapeutic vaccines, have demonstrated a sustained efficacy, i.e. seroconversions that persist for years following the end of treatment. This efficacy is correlated with the level of pre-existing necroinflammatory activity, as expressed by the levels of transaminases. Also pure antiviral drugs, which are very efficient at decreasing viral replication and viral load, can induce HBe seroconversion mainly in patients with initially elevated ALTs. The reason why these seroconversions are not always sustained after stopping antiviral therapy is not clearly understood.

However, even in the case of sustained HBe seroconversion, normalization of liver enzymes and loss of HBV-DNA (measured by hybridization assays), the virus is probably never completely eliminated from the body. Indeed, HBV-DNA can still be detected by sensitive PCR techniques years after immunoclearance, and CTLs specific for hepatitis antigens can be measured up to 15 years after successfully resolving acute hepatitis (Reherman *et al.*, 1996). Therefore, it is probable that low levels of viral replication coexist with minimal specific CTL activity for the lifetime of the infected individual, despite the absence of any obvious sign of liver disease or marker of viral activity. It is not clear whether this silent persistence has any negative impact on the subject's health. It has, however, been demonstrated that HBe and HBs seroconversion are both correlated with improvement in liver histology, and even in survival (Niederau *et al.*, 1996). This silent infection can be transmitted under specific circumstances, such as blood transfusion. Virus can also be reactivated by immunosuppression and immunodeficiency, further proving the importance of immune control.

In conclusion, the development of a therapeutic vaccine that would induce an adequate immune response able to revert the tolerant state, to destroy the infected hepatocytes and to control viral replication and spread, is in theory a logical approach. It is, however, unlikely that any kind of therapy would ever be able to completely eradicate the virus, but it is also unlikely that this is really necessary.

THERAPEUTIC VACCINES

Introduction

The experience of immunotherapy in the field of chronic hepatitis B is limited, probably because of the relatively recent discovery of the virus (Blumberg *et al.*, 1967; Dane *et al.*, 1970).

Initial efforts focused on prophylactic vaccine development. A safe and effective vaccine has been available since the early 1980s, and since 1986 the initial plasma-derived vaccines have progressively been replaced by recombinant vaccines (André and Zuckerman, 1994). The fact that vaccines based on surface antigens (HBsAg) alone were very protective has limited the extent of research on other antigens. Also at that time chronic hepatitis B was poorly understood and classifications were based on histological criteria only, with chronic active hepatitis and chronic persistent hepatitis recognized as the two main forms of hepatitis B (Schmid *et al.*, 1994).

Prophylactic vaccines were the first to be evaluated for the treatment of CHB, in an attempt to induce anti-HBs antibodies which could neutralize HBsAg and then overcome the persistence of the virus. As these first attempts were unsuccessful, further efforts were for several years concentrated on the evaluation of interferon. The disappointing therapeutic profile of IFN, and a better understanding of the virus and of the mechanisms of the disease, have led to the current extensive work on research and development of antiviral drugs and, to a lesser extent, on immunotherapeutic approaches.

Immunotherapeutic approaches can be divided into four categories, depending on whether they are active or passive, specific or non-specific (Table 14.3).

Active Specific Immunotherapy

The goal of active specific immunotherapy or therapeutic vaccination is to induce a specific immune response against target antigens of the hepatitis B virus, with the ultimate objective of breaking the immune tolerance, destroying infected hepatocytes, and preventing viral spread and infection of adjacent cells.

Until now, only HBs-derived protein-based vaccines or core peptide vaccines have reached the clinical evaluation stage in humans.

Protein-based Vaccines

The first approach to be evaluated, and hence the most advanced, is the use of the classic vaccine concept of viral proteins as antigens. The first attempt of active immunotherapy was published by Dienstag *et al.*, in 1982, using a prophylactic plasma-derived HBsAg vaccine (HEPTAVAX-B 40 µg/mL Merck Sharpe & Dohme, West Point, Pennsylvania). The rationale was to try to break the immunological tolerance by exposing the tolerant host to antigens that differed slightly from the tolerated antigen. To achieve this goal, the 22 nm spherical forms of HBsAg, treated with digestive enzymes and formalin, were absorbed on alum. It was hypothesized that the injection of HBsAg of a different subtype could help break the tolerance. Sixteen HBsAg chronic carriers, all of the *adw* subtypes, received six monthly injections of 40 µg vaccine, of either the *ayw* or the *adw* subtype. The primary objective was the induction of anti-HBs antibodies and clearance of HBsAg. To summarize the results, the vaccine was considered safe but ineffective. ALT flares were induced after two to four injections in two subjects, leading to cessation of therapy as these flares were

---

TABLE 14.3 Immunotherapeutic strategies

Active specific immunotherapy = therapeutic vaccines

Protein based
  HBsAg ± preS antigens
  Adjuvants
HBsAg anti-HBs complex
Peptide vaccine: CTL-specific peptide
DNA vaccines
Viral vectors
HBcAg as epitope Carrier

Active non-specific vaccines
BCG
HDV
MTH68/B

Passive specific Immunotherapy
Adoptive transfer of immunity
HB immunoglobulins

Passive non-specific immunotherapy
Cytokines
Others

---

believed to be sporadic non-A, non-B hepatitis caused by a contaminated vaccine lot. Consequently, after these two flares, subsequent doses of the same lot were not injected into other volunteers. Therefore, only five of the 16 subjects received the full course of six doses. Another subject lost HBeAg and developed anti-HBe antibodies. No subject lost HBsAg or developed anti-HBs antibodies.

Sixteen years later, these data can be reconsidered in the light of our better understanding of HBV pathogenesis. The objective of HBs seroconversion would now be considered as too stringent a criterion, and ALT flares as beneficial. Unfortunately, the negative conclusions of this study prevented further research for many years.

In 1994, Brechot and Pol reported promising results from a pilot study using another existing prophylactic vaccine (GENHEVAC B, Pasteur-Mérieux sérums et vaccins, Marnes-la-Coquette, France), produced recombinantly on CHO cells and containing HBsAg and PreS2 (Pol *et al.*, 1994). Thirty-two HBsAg carriers, 29 of whom were HBeAg positive, were given three doses of the vaccine at 1-month intervals. The objective was to induce loss of HBV-DNA. Six months after the first injection HBV-DNA became undetectable in 12 patients (37.5%) and a significant decrease was observed in three others; 21 patients were subsequently given interferon-α therapy for 4 months, and over the total follow-up period ranging from 12 to 32 months HBV-DNA disappeared in 18 of the 32 initial patients (53.1%). Anti-HBe antibodies developed in 15 patients, sometimes up to 2 years after the disappearance of HBV-DNA.

A transaminase flare preceded the disappearance or decrease of HBV-DNA in 13 of the 18 (72.2%) who lost HBV-DNA. These results were compared to historical control groups, with a mean spontaneous clearance of 1.7% over 6 months (Pol, 1995).

Another group in Turkey has evaluated varying doses of the same vaccine in 22 CHB patients who were either HBeAg positive or anti-HBe antibody positive (Senturk *et al.*, 1998). Preliminary results indicate HBV-DNA loss in 24%, with three of the nine HBeAg-positive patients also losing HBeAg.

A large multicentre, placebo-controlled trial was initiated to confirm these results, using a vaccine containing only HBsAg, recombinantly produced in yeast, as a control (Recombivax, Merck, Sharpe & Dohme, West Point, Pennsylvania): 152 HBV-DNA positive subjects were randomized between three groups, to receive six doses of one of the two vaccines or the placebo. Three doses were given at monthly intervals, followed by three doses at 3-months intervals over a 12-month period (Pol *et al.*, 1998). The primary end-point was again HBV-DNA loss. Although there was some difference at month 6 (3%, 20% and 22% in the placebo, preS2-S vaccine and S vaccine groups, respectively), there was no difference between the three groups at month 12 (30%, 25%, 35% respectively). A lymphoproliferative response was detected in vaccinated subjects.

In conclusion, no clear clinical benefit could be demonstrated and the preS2 antigen present in this vaccine did not seem to have any additional effect. However, the two different expression systems (mammalian vs yeast) render direct comparison hazardous.

These results have prompted the therapeutic evaluation of another prophylactic vaccine also produced in mammalian cells. Hepagene (Medeva) contains all the HBV envelope antigens, i.e. HBsAg, part of preS1 and the whole preS2, combined with aluminum hydroxide as the adjuvant. Antigens are recombinantly produced in mouse cells. Vaccination was reported to induce higher anti-HBs antibody titers than classic recombinant yeast-derived hepatitis vaccines in healthy subjects (Poma et al., 1990; Thoma et al., 1990) and in subjects who are non-responders to classic vaccination (Williams and Young, 1998; Zuckerman et al., 1997). This vaccine also induces a strong cellular immune response in chimpanzees (Pride et al., 1998).

In a pilot trial, 24 chronic hepatitis B patients, all initially positive for HBV-DNA and HBeAg, were given eight 20 µg doses of Hepagene in two series of 4-monthly injections with an interval of 5 months (Carman et al., 1998). At the end of administration, eight of the 22 patients who completed the study had a sustained loss of HBV-DNA and seven lost HBeAg. Factors predictive of response included raised ALT and non-Asian ethnicity, but not the pretreatment HBV-DNA levels. The majority of the responses were elicited during the second course of vaccinations. Most HBeAg losses were preceded by a temporary elevation of transaminases. This limited uncontrolled trial is encouraging and supports the concept of therapeutic vaccination, and warrants confirmation in larger, controlled trials.

Recently, Medeva reported preliminary results from a larger ongoing phase II study in Indonesia: 108 HBeAg-positive CHB patients were randomized to receive either four doses of Hepagene at monthly intervals or placebo, and stratified according to initial ALT levels (ALT less than or at least 1.5 times the upper limit of normal). Nine months after the last immunization, 6.7% and 38.1% of the patients had lost serum HBV-DNA, in the low-ALT and high-ALT vaccine groups, respectively, versus 0% and 18.2% in the low-ALT and high-ALT placebo groups, respectively. Five of the 10 patients who lost HBV-DNA in the vaccine groups also underwent HBe seroconversion.

These approaches use aluminum as adjuvant. Because aluminum is known to induce a Th2-type immune response, it is possibly not the best adjuvant for HBV immunotherapy. Another group (SmithKline Beecham Biologicals Immunotherapeutics, Rixensart, Belgium) has initiated the development of therapeutic vaccines following extensive work on specific immunostimulants. The therapeutic vaccine formulation contains S, preS1 and preS2 antigens, namely SL* antigens, and a specific adjuvant system (SBAS2). SL* is a recombinant yeast-derived vaccine containing the small protein (S) and a modified large protein (L*) which contains HBsAg (S-antigen), and selected parts of preS1 (aa 12–52) and pre S2 (aas 133–145) (Cabezon et al., 1990). Unlike Hepagene, this preS-containing vaccine does not increase anti-HBs antibody response when compared to classic HBsAg recombinant vaccines (Leroux-Roels, 1997).

For therapeutic use, SL* is combined with strong and specific immunostimulants to induce Th1 and CTL responses. This adjuvant system, called SBAS2, contains MPL (monophosphoryl lipid A), QS21 (*Quillaria saponaria*) and an oil-in-water emulsion. When injected into HBsAg-expressing transgenic mice the formulation induces a complete and rapid disappearance of HBsAg and the appearance of high levels of anti-HBs antibodies (Wettendorf et al., 1996). In healthy adult volunteers, two doses induced very high levels of anti-HBs antibodies, a strong lymphoproliferative response and high interferon-γ secretion (Vandepapelière *et al.*, 1996). A phase II evaluation in CHB patients is ongoing.

Recently a plasma-derived vaccine containing HBs, preS1 and preS2 antigens has been marketed in Pakistan as being the first prophylactic and therapeutic vaccine, but no data could be found in the scientific literature (Hepavar, Korean Green Cross).

HBsAg–anti-HBs Complex

A different approach is based on the combination of HBsAg and anti-HBs antibodies, which form immune complexes to induce humoral and cytotoxic T-cell immune responses. A solid matrix antigen–antibody complex (SMAA) with Duck hepatitis B antigens and antibodies induced clearance of viremia and antigenemia in tolerant DHBV-infected ducks (Wen *et al.*, 1994).

For human testing, a complex formed from plasma-derived HBsAg hepatitis vaccine and anti-HBs human hepatitis B immoglobulins was evaluated in 14 patients with HBeAg-positive CHB. Nine patients (64.3%) became HBV-DNA negative and six lost HBeAg after three immunizations at 3-week intervals. HBV-DNA loss was accompanied in all cases by at least one ALT elevation, but without any safety complications (Wen *et al.*, 1995). The addition of plasmid DNA encoding for HBsAg or for core protein of hepatitis C improved the anti-HBs antibody response of the HBs antigen–antibody complex (Qu *et al.*, 1998).

Peptide Vaccines

A lipopeptide-based therapeutic vaccine has been developed by Cytel Corporation, San Diego, California. This approach is based on the evidence that major histocompatibility complex class I restricted cytotoxic T lymphocytes can play a major role in the prevention, control and cure of infectious diseases in general, and of hepatitis virus infection in particular. During chronic hepatitis B infection the CTL response is weak or undetectable, and a vaccine that could induce a strong CTL response against hepatitis B antigen(s) should in theory be able to eradicate the viral infection.

The vaccine is composed of three parts, namely an HLAA2.1 restricted HBV core peptide 18–27 as the CTL epitope, a tetanus toxoid peptide 830–843 as a T-helper peptide, and two palmitic acid molecules as lipids (Bertoletti *et al.*, 1997). Studies in healthy volunteers showed dose-dependent induction of CTLs to core peptide, reaching levels similar to those observed during acute hepatitis B infection (Livingston *et al.*, 1997; Vitiello *et al.*, 1995). This vaccine was then tested in patients with HBeAg-positive chronic hepatitis B. Again, CTLs were detected with the higher doses of vaccine, but only in patients with ALT initially elevated to $\geqslant 2 \times$ the upper normal limit. Despite ALT flares in the same subjects, no clear clinical efficacy could be demonstrated (Heathcote *et al.*, 1996).

DNA Vaccines

From a theoretical point of view DNA immunization presents many advantages. In mice, a simple intramuscular injection of DNA is efficient at inducing an immune response via an *in vivo* synthesis of the encoded antigens and a potent self-adjuvantation, probably through the CpG motifs. The type of immune response is very similar to that induced by natural infection in terms of kinetics, antibody isotypes and the specificity of the humoral response. More importantly, it induces a very strong cellular immune response, mainly of the Th1 and CTL types. This response is sustained for up to 17 months post vaccination (Davis, 1996), with a peak around 4–8 weeks after injection. DNA vaccines have other interesting advantages, such as low cost of production, ease of quality control, and the ability to be stored and transported at ambient temperatures.

Plasmids carrying genes encoding various HBV envelope proteins have been widely evaluated in animals (Davis *et al.*, 1997a). They induce high antibody titers and a strong cellular response after only one injection, in both mice and chimpanzees (Michel, 1995). In HBsAg-expressing transgenic mice, a single intramuscular injection of plasmid DNA encoding for surface antigens could induce the complete disappearance of HBsAg and generate anti-HBs antibodies (Davis *et al.*, 1997b). Vaccination could even provoke a decrease or a complete disappearance of the HBV messenger RNA from the liver, almost without cytopathic effect (Mancini *et al.*, 1996). This anti-HBs antibody response mimics a natural response, with an initial increase of IgM followed by IgG, mainly of the Th1-related isotypes (Michel *et al.*, 1995). The decrease of HBsAg is probably provoked by neutralizing antibodies and linked to the mRNA disappearance in the liver. There was no transaminase increase or liver cirrhosis, which implies that the downregulation of HBsAg expression was mediated by non-cytolytic cytokines (Davis, 1996).

Recently Powderject Pharmaceuticals, a UK biotechnology company, has announced positive results from a phase I study in humans, using DNA coding for HBsAg, coated on gold particles and injected transdermally using a jet gun. Other groups are working on core-expressing DNA immunization (Kuhober *et al.*, 1996) or a combination of HBc and envelope antigens (Geissler *et al.*, 1997). These constructions can induce strong antibody response to S, preS1, preS2 and core, as well as a potent Th1 cytokine secretion and CTLs against the large protein and HBcAg (Geissler *et al.*, 1997, 1998).

The combination of HBV antigen with cytokines such as IL-2 (Chow *et al.*, 1997) or IL-12 is also under development, as are modes of administration other than the classic intramuscular injection (Widera, 1997; Lee *et al.*, 1996).

Codelivery of DNA vaccines with cytokines can modulate the T-helper response in mice towards predominantly Th1 or Th2, depending upon the type of cytokines (Chow *et al.*, 1998). In such experiments, IL-12 or γ-IFN strongly enhanced Th1 cells and CTL activity. IL-2 or GM-CSF also stimulated Th1 cells but not CTLs, whereas IL-4 significantly enhanced Th2 cells. This strategy could be of major importance in the development of therapeutic vaccines. Another group has constructed gene sequences encoding for both human B7-1 and HBsAg, and induced a strong CTL response in mice with this (He *et al.*, 1996).

The two major pitfalls of DNA vaccination are the weaker immunogenicity in large animals, in particular primates, compared to mice and, more importantly, a safety concern because of the possibility of integration of the plasmid DNA into the host genome, with a potential risk of mutagenesis. If proven safe and immunogenic in humans, this approach is promising as it could induce the adequate immune response against the selected antigens and then succeed in overcoming the HBV tolerance.

Viral Vectors

Gene transfer technology using replication-defective viral vectors can efficiently deliver genes to mammalian cells and indirectly introduce immunogenic proteins for stimulation of the immune system (Michel, 1997). A retroviral vector expressing a fusion HBV core/neo protein efficiently induced humoral and CTL responses in mice and Rhesus monkeys (Ronlov *et al.*, 1996). It had a limited efficacy in three chronically HBV-infected chimpanzees.

Another approach has used recombinant adenovirus vectors coexpressing HBsAg and the costimulatory factor B7-I protein (CD80). This also induced strong antibody and CTL responses in mice (He *et al.*, 1996).

HBcAg as Epitope Carrier

The HBc particle has been extensively studied as an epitope carrier because it presents several useful characteristics, such as an increased immunogenicity at B-cell, T-cell and CTL levels in primates and rodents; a high level of production and correct self-assembly after expression in various systems, such as bacteria, yeast and various mammalian cells; and also a high acceptance of foreign insertions (Milich *et al.*, 1988; Pumpens *et al.*, 1995; Schödel *et al.*, 1996). The position of insertion of these foreign epitopes influences their immunogenicity as well as that of HBcAg (Schödel *et al.*, 1992). Constructs made fusing preS1 and preS2 epitopes to various sites in the core antigen appear very promising, as they display HBcAg and envelope epitopes with induction of a strong cellular and humoral immune response, without special external adjuvantation. This approach is currently being evaluated by Medeva.

Active Non-specific Immunotherapy

Non-HBV specific vaccinations have also been evaluated in CHB patients in an attempt to induce strong, non-HBV specific, cellular T-helper responses. The first was probably the Bacillus Calmette–Guérin vaccine (BCG), with promising initial results (Bassendine *et al.*, 1980), although no benefit could be detected in a later well controlled trial (McGilchrist and Follett, 1987). Another approach is the use of a modified hepatitis Delta virus. The concept is to introduce sequences into the HDV genome which could interfere with the replication of HBV (Taylor *et al.*, 1995).

Recently, Hungarian researchers have developed a vaccine based on Bursal disease virus, MTH-68/B, which should be harmless to humans, and it is claimed that this vaccine has prevented progression to chronicity after acute infection by both HBV and hepatitic C virus (Csatary *et al.*, 1998). The results need confirmation.

Passive Specific Immunotherapy

Adoptive transfer of Immunity

Adoptive transfer of HBV immunity to immunosuppressed patients through transplantation of bone marrow (BM) from HBV-immune bone marrow donors was first developed to confer immunity against hepatitis B (Shouval and Ilan, 1995a). The same procedure has since been applied to HBsAg carriers and led to the clearance of HBV infection (Ilan *et al.*, 1993; Shouval and Ilan, 1995b). The first report was an HBsAg+/HBV-DNA+ asymptomatic carrier with leukemia who received BMT from his naturally immune brother. These results were later confirmed in a larger study, where only patients who received anti-HBs positive bone marrow had a sustained clearance of HBsAg. It was further demonstrated that only bone marrow from donors who had recovered from acute hepatitis, i.e. who were naturally immune, could elicit a clearance of infection, and not bone marrow from donors immunized with HBsAg vaccine (Lau G.K.K., 1997, 1998). This suggests that transfer of memory cells against HBsAg only would not be sufficient to allow the clearance of chronic HBV infection, at least in the case of HBeAg-positive active disease, as was the case in this study. Interestingly, though, a case of HBsAg clearance after BMT from a donor immunized with one dose of recombinant HBsAg vaccine was recently reported (Brugger *et al.*, 1997). However, the recipient was an asymptomatic HBsAg carrier and the possibility of a pre-existing silent anti-HBc immunity in the donor had not been excluded. These results further confirm the possibility of control of HBV infection through an adequate immunity in situations where tolerance mechanisms have been abolished, in this case by immunosuppression. However, it is clear that this form of therapy cannot be used in otherwise healthy HBV carriers.

Immunoglobulins

Hepatitis B immunoglobulins have been used for many years for the prevention of infection in neonates from HBV carrier mothers (Beasley *et al.*, 1983b; Stevens *et al.*, 1987), for postexposure prophylaxis, and also for the prevention of reinfection in immunosuppressed patients, especially in infected liver transplant patients (Terrault *et al.*, 1996). In the case of neonatal protection the excess circulating antibodies prevent infection of hepatocytes, but this protection wanes over time. Only coadministration with a vaccine can confer long-term protection (Beasley, 1983a). In liver transplant patients continuous treatment with high doses of HBIg is necessary to prevent infection of the donor organ, and the disease usually reactivates when the HBIg is stopped.

Such therapy shows the usefulness of HBV-neutralizing antibodies to control viral spread, but it is not able to induce full seroconversion. Furthermore, the cost and inconvenience limits its use to specific cases.

Passive Non-specific Immunotherapy

Cytokines

The importance of the T-helper cell activity in the pathogenesis of liver cell damage in chronic hepatitis B, more specifically of the Th1 cell subset in the exacerbations of liver inflammation, that sometimes lead to seroconversion, is now well described. Particular attention has therefore been given to the cytokines secreted by the Th1 cells, in particular IL-2, γ-IFN and, more recently, IL-12, which promotes Th1 function. Although the clinical evaluations of IL-2

(Artillo *et al.*, 1998) and of γ-interferon (Lau *et al.*, 1991) as therapeutic agents have not been successful, IL-12 could yet play a more prominent role and be of major therapeutic interest (Naoumov and Rossol, 1997; Carreno and Quiroga, 1997; Gately *et al.*, 1992; Berg *et al.*, 1998).

Interleukin-12 is a dominant cytokine which specifically promotes Th1 cell differentiation and depresses Th2 activity (Trinchieri, 1995). In transgenic mice IL-12 inhibits viral replication in the liver and in extrahepatic sites (Cavanaugh *et al.*, 1997), and it probably plays a major role in viral clearance through its ability to induce γ-IFN. In HBeAg-positive transgenic mice it can modulate the HBeAg-specific Th response in favor of Th1 cells (Milich *et al.*, 1995b). This leads to massive anti-HBe Ab production, and hence, to HBeAg disappearance.

In patients treated with interferon, serum levels of IL-12, IL-2 and γ-IFN have been followed by sensitive, specific ELISAs. All three cytokines were elevated only in patients in whom ALT flares were followed by HBe seroconversion, and not in non-responders to α-IFN, despite similar ALT flares. This peak of IL-12, γ-IFN and, to a lesser extent, IL-2, occurred on average 9 weeks after the transaminase peak. This further supports the importance of a multispecific response. If CTLs are important to initiate the viral clearance process by killing infected hepatocytes, Th1 cytokines and IL-12 are most likely required for a sustained immune control over HBV replication, leading to HBe seroconversion. Further, these results tend to demonstrate dissociation in time between the immune-mediated lysis of hepatocytes and viral clearance.

Apart from the potential interest of IL-12 as an antiviral and immuno-modulatory therapeutic agent, these results are also important for the under-standing of the processes of viral clearance and control, and could give some indication for the development of adequate immunotherapeutic approaches. Human granulocyte–macrophage colony-stimulating factor (GM-CSF) was evaluated in a pilot trial in nine patients, either in combination with or without α-IFN therapy. It significantly reduced the HBV-DNA levels, but its use is probably limited to CHB patients with concomitant leukopenia and suscepti-bility to bacterial infections (Hasan *et al.*, 1998; Martin *et al.*, 1993).

Others

Several other strategies have been evaluated, usually in small uncontrolled studies. In Germany, 14 patients received intravenous injections of a non-specific immunomodulator of the monocyte–macrophage system consisting of isolated cell walls of *Proprionibacterium granulosum*, strain KP-45, and fol-lowed for up to 5 years. Complete remission was observed in 10 of the patients, either during or after the therapy (Gil *et al.*, 1992).

O-Palmitoyl-3 (+)-cotechin, or cyanidanol, has also been evaluated in CHB patients because of its immunostimulatory properties, but without success (Scully *et al.*, 1988).

Finally, Chinese herbal medicines should be mentioned here, as they probably represent the most widely used therapeutic compounds in the world because of their broad use in China. Some have been shown to inhibit secretion of HBsAg in cell culture (Huang *et al.*, 1996). Despite the usual scepticism of western scientists, the role of these ancient medicines should probably not be ignored as potential cofactors in clinical studies conducted in China.

CONCLUSIONS AND FUTURE AVENUES

One major complication of HBV infection is its persistence, leading to different patterns of chronic disease. In particular, chronic HBV carriage is the leading cause of hepatocellular carcinoma. Prophylactic vaccines are very effective in preventing acute and chronic infections, but universal vaccination is not yet globally established and there are more than 350 million HBV carriers worldwide, 75% of whom are in Asia.

Chronic hepatitis B should not be addressed as a single static disease caused by one single virus: it should rather be considered as various diseases caused by different viruses in several types of patient populations, and one should look at virological, immunological, genetic, epidemiological and clinical aspects to form separate syndromes which should then be addressed individually. For example, one young Chinese child who is HBeAg positive with very high HBV-DNA levels but with no liver disease cannot be approached in the same way as a 40-year-old Mediterranean adult with anti-HBe antibody-positive, HBeAg-negative active chronic hepatitis. Similarly, a young Caucasian adult with HBsAg and anti-HBc Ab as the only markers of infection, with no disease, infected through sexual contacts, has a very good prognosis compared to an Asian subject infected at birth with the same serological pattern, who is at high risk of developing hepatocellular carcinoma.

The mechanisms of viral persistence and chronic diseases are not fully understood, but it is widely accepted that immune mechanisms play a major role, most likely induced by various viral mechanisms. The cellular responses in particular, both the CD4+ Th1 response and the cytotoxic T-lymphocyte activity, are modified and downregulated during chronic infection, and strongly activated during viral clearance or acute disease. These responses seem to be directed against several HBV antigens, and it is difficult to identify the major player.

Currently available therapies, especially interferon and antivirals, are unsatisfactory, mainly because of limited sustained efficacy. There is clearly a need for improved therapies, and therapeutic vaccines appear to be the most rational approach for the treatment of chronic hepatitis B. A few therapeutic vaccines are currently being clinically evaluated. Most of them are based on HBV envelope antigens, and even though some efficacy was reported with one of these vaccines, these results need confirmation in a well-controlled trial.

Research has focused recently on immunological mechanisms of viral persistence and liver cytopathology, after several years during which serology and histology were the only two criteria for evaluation and follow-up of the disease. But even if therapeutic vaccination appears logical, the ideal design of a therapeutic vaccine is not clear in terms of antigen selection, type of immune stimulation, target population, schedule or endpoints.

For the selection of suitable antigens and induction of a specific immune response, the history of natural seroconversion tells us that a strong initial CD4+ Th1 and CTL response to nucleocapsid antigens and to preS antigens is crucial for induction of the process of clearance, breaking the immune tolerance, and destruction of infected hepatocytes, although a late induction of neutralizing anti-HBs antibodies would be useful to stop viral spread and prevent reinfection.

This would mean that the inclusion of preS antigens or HBcAg is essential for induction of the clearance process, together with a predominance of a strong Th1 CD4+ response.

However, other antigens could in theory have relevant cellular epitopes to induce the necessary immune response. HBsAg has been included in most tested vaccines as opposed to other antigens, partly because of its direct availability. DNA vaccine technology could offer new opportunities (Zu Putlitz, 1998a,b). Unfortunately, animal models which are very useful for the evaluation of antiviral drugs are of more limited support for therapeutic vaccines, and therefore the final answers can only come from human clinical trials.

It is possible that future therapeutic vaccines will have to induce polyclonal immune responses against various HBV antigens administered at regular intervals for many years, or with booster courses. Combination therapies are an attractive approach, especially the combination of an antiviral with any therapeutic compound. The rationale is to downregulate viral activity and break immune tolerance, in order to facilitate the vaccine-induced immune stimulations. The fact that T-cell responsiveness could be restored by lamivudine therapy supports this strategy (Boni, 1998). However, it could also be that the disappearance of the viral epitopes and the suppression of the inflammatory process induced by the antiviral would prevent the development of an adequate immune response by the immunotherapy. Here again the respective schedules of administration of both compounds, as well as the target population, could be critical.

Many questions remain unsolved. For example, we cannot explain why only a small percentage of patients with the same serological and virological characteristics will respond to interferon, antivirals or therapeutic vaccine; or why, after many years of harmless cohabitation during the immune tolerance phase, does the immune system at a defined age start to try and clear the viral infection? The answers to these questions could help in the designing of suitable therapeutic compounds. Alternatively, therapeutic trials using different strategies could bring useful information to help our understanding of these disease mechanisms. Clinical endpoints, schedules of administrations and study design should also be adapted according to the type of patient population, their specific medical needs and the therapeutic approach.

In conclusion, chronic hepatitis B is one of the chronic viral sexually transmitted diseases for which therapeutic vaccination appears to be a logical and highly feasible approach.

ACKNOWLEDGEMENTS

I dedicate this chapter and all the work and time spent on it to my wife, Sophie. I also wish to thank Geert Leroux-Roels and Tineke Rutgers for their critical review and their pertinent comments.

REFERENCES

Alberti, A. and Fattovich, G. (1994). Interferon therapy for the anti-Hbe positive form of chronic hepatitis B. *Antiviral Res.* 24: 145–153

Alberti, A., Trevisan, A., Fattovich, G. and Realdi, G. (1984). In: 'Advances in Hepatitis Research' (ed. Chisari, F.V.). The role of hepatitis B virus replication and hepatocyte membrane expression in the pathogenesis of HBV related hepatic damage, pp. 134–143. Masson Publishing, New York

Allen, M.I., Deslauriers, M., Andrews, C.W. et al. (1998). Identification and characterization of mutations in Hepatitis B virus resistant to Lamivudine. *Hepatology* 27: 1670–1677

Alter, M.J. and Margolis, H.S. (1990). The emergence of Hepatitis B as a sexually transmitted disease. *Med. Clin. N. Am.* 74 (6): 1529–1541

American Social Health Association (1998). Sexually Transmitted Diseases in America: How Many Cases and at What Cost? Publication #1445 prepared for the Kaiser Foundation Family, Research Triangle Park, NC 27709

Ando, K., Guidotti, L.G., Cerny, A., Ishikawa, T. and Chisari, F.V. (1994). Access to antigen restricts cytotoxic T-lymphocyte function *in vivo*. *J. Immunol.* 153: 482–488

André, F.E., Zuckerman, A.J. (1994). Review: protective efficacy of hepatitis B vaccines in neonates. *J. Med. Virol.* 44: 144–151

Artillo, S., Pastore, G., Alberti, A. et al. (1998). Double-blind, randomized controlled trial of interleukin-2 treatment of chronic hepatitis B. *J. Med. Virol.* 54: 167–172

Bailly, F. and Trepo, C. (1997). Interferon treatment of viral hepatitis. Practical recommendations. *Biodrugs* 8(1): 16–23

Bassendine, M.F., Weller, I.V.D., Murray, A., Summer, J., Thomas, H.C. and Sherlock, S. (1980). Treatment of HBsAg positive chronic liver disease with Bacillus Calmette–Guérin (BCG). *Gut* 21: A915

Beasley, R.P. (1988). Hepatitis B virus. The major etiology of hepatocellular carcinoma. *Cancer* 61: 1942–1956

Beasley, R.P., Hwang, K.Y., Lee, G.L.Y. et al. (1983a). Prevention of perinatally transmitted hepatitis B virus infections with hepatitis B immunoglobulin and hepatitis B vaccine. *Lancet* ii (8359): 1099–1102

Beasley, R.P., Hwang, K.Y., Stevens, C.E. et al. (1983b). Efficacy of Hepatitis B immune globulin for prevention of perinatal transmission of the hepatitis B virus carrier state: final report of a randomized double-bind, placebo-controlled trial. *Hepatology* 3(2): 135–141

Beasley, R.P. and Hwang, L. (1983c). Postnatal infectivity of hepatitis B surface antigen-carrier mothers. *J. Infect. Dis.* 147: 185–190

Beasley, R.P., Hwang, L.Y., Lin, C.C. and Chien, C. (1981a). Hepatocellular carcinoma and hepatitis B virus. *Lancet* ii: 1129–1133

Beasley, R.P., Lin, C.C., Hwang, L.Y. and Chien, C.S. (1981b). Hepatocellular carcinoma and hepatitis B virus: a prospective study of 22707 men in Taiwan. *Lancet* ii: 1129–1133

Beasley, R.P., Trepo, C., Stevens, C.E. and Szmuness, W. (1977). The e antigen and vertical transmission of hepatitis B surface antigen. *Am. J. Epidemiol.* 105 (2): 94–98

Berg, T., Teuber, G., Rödig, S., Naumann, U., Zeuzem, S. and Hopf, U. (1998). Treatment of chronic hepatitis B with Interleukin-12 (IL-12). Hepatology 28 (4): AASLD Abstracts Nr. 1702, 588A

Bertoletti, A., Southwood, S., Chesnut, R. et al. (1997). Molecular features of the hepatitis B virus nucleocapsid T-cell epitope 18–27: Interaction with HLA and T-cell receptor. *Hepatology* 26: 1027–1034

Bertoletti, A., Sette, A., Scaccaggia, P. et al. (1994). Natural variants of cytotoxic epitopes are T-cell receptor antagonists for antiviral cytotoxic T-cells. *Nature* 369: 407–410

Blumberg, D.S., Gerslev, B.S.J., Hungerford, D.A., London, W.T. and Sutnick, A.J. (1967). A serum antigen (Australian antigen) in Down's syndrome, leukemia and hepatitis. *Ann. Intern. Med.* 66: 924–931

Boni, C., Bertoletti, A., Penna, A. et al. (1998). Lamivudine treatment can restore T cell responsiveness in chronic hepatitis B. *J. Clin. Invest.* 102 (5): 1–8

Bozkaya, H., Ayola, B. and Lok, A.S.F. (1996). High rate of mutations in the hepatitis B core gene during the immune clearance phase of chronic hepatitis B virus infection. *Hepatology* 24: 32–37

Brown, J.J., Parashar, B., Kadakol, A., Rabbani, E., Engelhardt, D. and Chowdhury, J.R. (1998). Serological and pathomorphological characteristics of human hepatitis B infection in a small primate, *Tupaia belangyeri*. *Hepatology* 28(4) AASLD abstract #937: 397A

Brugger, S.A., Oesterreicher, C., Hofmann, H., Kalhs, P., Greinix, H.T. and Müller, C. (1997). Hepatitis B virus clearance by transplantation of bone marrow from hepatitis B immunized donor. *Lancet* **349**: 996–997

Brunetto, M.R., Oliveri, F., Colombatto, P., Capalbo, M., Barbera, C. and Bonino, F. (1995). Treatment of chronic anti-HBe positive hepatitis B with Interferon-alpha. *J. Hepatol.* **22** (S1): 42–44

Brunetto, M.R., Giarin, M.M., Oliveri, F. *et al.* (1991a). Wildtype and e antigen-minus hepatitis B viruses and course of chronic hepatitis. *Proc. Natl. Acad. Sci. USA.* **88**: 4186–4190

Brunetto, M., Giarin, M.M., Oliveri, F., Saracco, G., Barbera, C. and Parrella, T. (1991b). 'e' antigen defective hepatitis B virus and course of chronic infection. *J. Hepatol.* **13**: S82

Cabezon, T., Rutgers, T., Biemans, R., Vanderbrugge, D., Voet, P. and De Wilde, M. (1990). A new hepatitis B vaccine containing pre-S1 and pre-S2 epitopes from *Saccharomyces cerevisiae.* Vaccines 90, Modern approaches to new vaccines including prevention of AIDS, 199–203

Carman, W.F., Tucker, T., Song, E. *et al.* (1998). Efficacy of a third generation preS1/preS2 containing HBV carrier (Hepagene) as immunotherapy for HBeAg positive chronic hepatitis. *J. Hepatol.* **28**(S1), 190

Carman, W.F., Ferrao, M., Lok, A.S.F., Ma, O.C.K., Lai, C.L. and Thomas, H.C. (1992). Precore sequence variation in Chinese isolates of hepatitis B virus. *J. Infect. Dis.* **165**: 127–133

Carman, W.F., Fagan, E.A., Hadziyannis, S. *et al.* (1991). Association of a precore genomic variant of hepatitis B virus with fulminant hepatitis. *Hepatology* **14**: 219–222

Carman, W.F., Jacyna, M.R., Hadziyannis, S. *et al.* (1989). Mutations preventing formation of 'hepatitis B e antigen' in patients with chronic hepatitis B infection. *Lancet* **ii**: 588–591

Carreno, V. and Quiroga, J.A. (1997). Biological properties of interleukin-12 and its therapeutic use in persistent hepatitis B virus and hepatitis C virus infection. *J. Viral Hepatitis* **4**(suppl. 2), 83–86

Cavanaugh, V.J., Guidotti, L.G., Chisari, F.V. (1997). Interleukin-12 inhibits hepatitis B virus replication in transgenic mice. *J. Virol.* **71** (4): 3236–3243

Chayama, K., Suzuki, Y., Kobayashi M.as. *et al.* (1998). Emergence and take over of YMDD motif mutant hepatitis B virus during long term lamivudine therapy and re-takeover by wild type after cessation of therapy. *Hepatology* **27**: 1711–1716

Chisari, F. (1995a). Hepatitis B virus transgenic mice: insights into the virus and the disease. *Hepatology* **22**: 1316–1325

Chisari, F.V. and Ferrari, C. (1995b). Hepatitis B Virus Immunopathogenesis. *Annu. Rev. Immunol.* **13**: 29–60

Chow Y-H, Chiang B-L., Lee Y-L. *et al.* (1998). Development of Th1 and Th2 populations and the nature of immune responses to hepatitis B virus DNA vaccines can be modulated by codelivery of various cytokine genes. *J. Immunol.* **160** (3): 1320–1329

Chow Y-H., Huang W-L., Chi W-K., Chu Y-D., Tao M-H. (1997). Improvement of hepatitis B virus DNA vaccines by plasmids coexpressing hepatitis B surface antigen and interleukin-2. *J. Virol.* **71** (1): 169–178

Coleman, P.J., McQuillan, G.M., Moyer, L.A., Lambert, S.B. and Margolis, H.S. (1998). Incidence of hepatitis B virus infection in the United States 1976–1994: estimates from the National Health and Nutrition Examination Surveys. *J. Infect. Dis.* **178**: 954–959

Csatary, L.K., Telegdy, L., Gergely, P., Bodey, B. and Bakacs, T. (1998). Preliminary report of a controlled trial of MTH-68/B virus vaccine treatment in acute B and C hepatitis: a phase II study. United Cancer Research Institute, Alexandria, VA, USA

Dane, D.S., Cameron, C.H. and Briggs, M. (1970). Virus-like particles in serum of patients with Australian antigen-associated hepatitis. *Lancet* **i**: 695–698

Davis, H.L. and Brazolot-Millan, C.L. (1997a). DNA-based immunization against hepatitis B virus. Springer Semin. *Immunopathol.* **19** (2): 195–204

Davis, H.L., Brazolot-Millan, C.L., Mancini, M. *et al.* (1997b). DNA-based immunization against hepatitis B surface antigen (HBsAg) in normal and HBsAg-transgenic mice. *Vaccine* **15** (8): 849–852

Davis, H.L. (1996). DNA-based vaccination against hepatitis B virus. *Adv. Drug Deliv. Rev.* **21** (1): 33–47

Davis, L.G., Weber, D.J. and Lemon, S.M. (1989). Horizontal transmission of Hepatitis B virus. *Lancet* **i**: 889–893 de Franchis, R., Meucci, G., Vecchi, M. *et al.* (1993). The natural history of asymptomatic hepatitis B surface antigen carriers. *Ann. Intern. Med.* **118**: 191–194

Dienstag, J., Schiff, E., Wright, T. *et al.* and the US Lamivudine Investigator Group (1998). Lamivudine treatment for our year in previously untreated U.S. hepatitis B patients: histological improvement and hepatitis B e-antigen (HBeAg) seroconversion. The Digestive Disease Week (1998). May 16–22 New Orleans, LA

Dienstag, J.L. (1984). The Epidemiology of Hepatitis B, the virus, the disease and the vaccine, pp 55–65. Plenum Press, New York

Dienstag, J.L., Stevens, C.E., Bhan, A.K. and Szmuness, W. (1982). Hepatitis B vaccine administered to chronic carriers of hepatitis B surface antigen. *Ann. Intern. Med.* **96**: 575–579

Dusheiko, G. (1998). Lamivudine treatment of chronic hepatitis B. *Rev. Med. Virol.* **8**: 153–159

Evans, A.A., Fine, M. and Thomas London, W. (1997). Spontaneous seroconversion in hepatitis B e antigen-positive chronic hepatitis B: implications for Interferon therapy. *J. Infect. Dis.* **176**: 845–850

Fagan, E.A., Smith, P.M., Davison, F. and Williams, R. (1986). Fulminant Hepatitis B in successive female sexual partners of two anti-HBe-positive males. *Lancet* **ii**: 538–540

Fattovitch, G., Mc Intyre, G., Thursz, M. *et al.* (1995). Hepatitis B virus precore/core variation and interferon therapy. *Hepatology* **22**: 1355–1362

Fattovitch, G., Rugge, M., Brollo, L. *et al.* (1986). Clinical, virologic and histologic outcome following seroconversion from HBeAg to anti-HBe in chronic hepatitis type B. *Hepatology* **6**: 167–172

Fazle Akbar, S.M., Kajino, K., Tanimoto, K. *et al.* (1997). Placebo-controlled trial of vaccination with hepatitis B virus surface antigen in hepatitis B virus transgenic mice. *Hepatology* **26**: 131–137

Ferrari, C., Bertoletti, A., Penna, A. *et al.* (1991). Identification of immunodominant T cell epitopes of the hepatitis B virus nucleocapsid antigen. *J. Clin. Invest.* **88** i: 214–220

Ferrari, C., Penna, A., Bertoletti, A. *et al.* (1990). Cellular immune response to hepatitis B virus-encoded antigens in acute and chronic hepatitis B virus infection. *J. Immunol.* **145**: 3442–3449

Foster, G.R., Ackrill, A.M., Goldin, R.D., Kerr, I.M., Thomas, H.C. and Stark, G.R. (1991). Expression of the terminal protein region of hepatitis B virus inhibits cellular response to interferon alpha and gamma and double-stranded RNA. *Proc. Natl. Acad. Sci. USA.* **88**: 2888–2892

Fukuda, R., Ishimura, N., Nguyen, T. *et al.* (1995). The expression of IL-2, IL-4 and interferon-gamma (IFN-γ) mRNA using liver biopsies at different phases of acute exacerbation of chronic hepatitis B. *Clin. Exp. Immunol.* **100**: 446–451

Gately, M.K., Wolitzky, A.G., Quinn, P.M. and Chizzonite, R. (1992). Regulation of human cytolytic lymphocyte responses by interleukin-12. *Cell. Immunol.* **143**: 127–142

Geissler, M., Schirmbeck, R., Reimann, J., Blum, H.E. and Wands, J.R. (1998). Cytokine and hepatitis B virus DNA co-immunizations enhance cellular and humoral immune responses to the middle but not to the large hepatitis B virus surface antigen in mice. *Hepatology* **28**: 202–210

Geissler, M., Tokushige, K., Chante, C.C., Zurawski, V.R. and Wands, J.R. (1997). Cellular and humoral immune response to hepatitis B virus structural proteins in mice after DNA-based immunization, *Gastroenterology* **112** (4): 1307–1320

Gil, J., Szmigielski, S., Jeljaszewicz, J. and Pulverer, G. (1992). Immunotherapy of chronic active viral hepatitis B with propionibacterium granulosum KP-45 (a 5-year follow-up report). *Hepatogastroenterol* **39**: 325–329

Hadziyannis, S.J. (1995). Hepatitis B e Antigen Negative Chronic hepatitis B: from clinical recognition to pathogenesis and treatment. *Rev. Viral Hep. Rev.* **1**(1): 7–36

Hasan, M.S., Agosti, J., Tanzman, E., Treanor, J., Andrews, J. and Evans, T.G. (1998). GM-CSF as an adjuvant for hepatitis B vaccination of healthy adults. *Clin. Infect. Dis* **27** (4): 1057

He, X.S., Chen, H.S., Chu, K., Rivkina, K. and Robinson, W.S. (1996). Costimulatory protein B7-1 enhances the cytotoxic T cell response and antibody response to Hepatitis B surface antigen. *Proc. Natl. Acad. Sci. USA* **93**: 7274–7278

Heathcote, J., McHutchison, J., Benner, K. *et al.* (1996) CY-1899: A therapeutic vaccine for chronic hepatitis B. Hepatology 24(4) Pt2, 283A, AASLD abstract #628

Hervas-Stubbs, S., Lasarte, J.J., Sarobe, P. *et al.* (1997). Therapeutic vaccination of woodchucks against chronic Woodchuck hepatitis virus infection. Hepatology 27, 726–737

Honkoop, P., de Man, R.A., Heijtink, R.A. and Schalm, S.W. (1995). Hepatitis B reactivation after lamivudine. *Lancet* **346**: 1156–1157

Hoofnagle, J.H. (1990). Chronic hepatitis B. *N. Engl. J. Med.* **323** (5): 337–339

Hoofnagle, J.H., Shafritz, D.A. and Popper, H. (1987). Chronic Type B hepatitis and the 'healthy' HBsAg carrier state. *Hepatology* **7** (4): 758–763

Hsu, H.Y., Chang, M.H., Hsieh, K.H. *et al*. (1992). Cellular immune response to HBcAg in mother-to-infants transmission of hepatitis B Virus. *Hepatology* **15**: 770–776

Huang R-L., Chen C-C., Huang Y-L. *et al*. (1996). Osthole increases glycosylation of hepatitis B surface antigen and suppresses the secretion of hepatitis B virus in vitro. *Hepatology* **24** (3): 508–515

Huo, T.I., Wu, J.C., Lee, P.C. *et al*. (1998). Seroclearance of hepatitis B surface antigen in chronic carriers does not necessarily imply a good prognosis. *Hepatology* **28**: 231–236

Ikeda, K., Saitoh, S., Suzuki, Y. *et al*. (1998). Interferon decreases hepatocellular carcinogenesis in patients with cirrhosis caused by the hepatitis B virus: a pilot study. *Cancer* **82** (5): 827–835

Ilan, Y., Nagler, A., Adler, R., Tur-Kaspa, R., Slavin S. and Shouval D (1993). Ablation of persistent hepatitis B by bone marrow transplantation from a hepatitis B immune donor: A case report. *Gastroenterology* **104**: 1818–1821

Inoue, M., Kakumu, S., Yoshioka, K., Tsutsumi, Y., Wakita, T. and Arao, M. (1989). Hepatitis B core antigen-specific IFN-γ production of peripheral blood mononuclear cells in patients with chronic hepatitis B infection. *J. Immunol.* **142** (11): 4006–4011

Jaeckel, E. and Manns, M.P. (1997). Experience with lamivudine against hepatitis B virus. *Intervirology* **40**: 322–336

Jung, M.C., Diepolder, H.M., Spengler, U. *et al*. (1995). Activation of a heterogeneous hepatitis B (HB) core and e antigen-specific CD4+ T-cell population during seroconversion to anti-HBe and anti-HBs in hepatitis B virus infection. *J. Virol.* **69** (6): 3358–3368

Jung, M.C., Spengler, U., Schraut, W. *et al*. (1991), Hepatitis B virus antigen-specific T cell activation in patients with acute and chronic hepatitis B. *J. Hepatol.* **13**: 310–317

Kane, M.A., Alter, M.J., Hadler, S.C. and Margolis, H.S. (1989). Hepatitis B infection in the United States – recent trends and future strategies for control. *Am. J. Med.* **87**(suppl. 3A): 11S–13S

Kingsley, L.A., Rinaldo, C.R., Lyter, D.W., Valdiserri, R.O., Belle, S.H. and Ho, M. (1990). Sexual transmission efficiency of hepatitis B virus and human immunodeficiency virus among homosexual men. *JAMA* **264** (2): 230–234

Köck, J., Blum, H.E. and von Weizsäcker, F. (1998). Hepatitis B virus infection of primary *Tupaya* hepatocytes. Hepatology **28** (4): *AASLD Abstracts Nr.* **1682**: 583A

Krogsgaard, K., Bindslev, N., Christensen, E. *et al*. and the European Concerted Action of Viral Hepatitis (EUROHEP) (1994). The treatment effect of alpha-interferon in chronic hepatitis B is independent of pretreatment variables. Results based on individual patient data from 10 clinical controlled trials. *J. Hepatol.* **21**: 646–655

Kuhober, A., Pudollek, H.P., Reifenberg, K. *et al*. (1996). DNA immunization induces antibody and cytotoxic T cell responses to hepatitis B core antigen in H-2 (b) mice. *J. Immunol.* **156** (10): 3687–3695

Lai, C.L., Chien, R.N., Leung, N.W.Y. *et al*. (1998). A one-year trial of lamivudine for chronic hepatitis B. *N. Engl. J. Med.* **339** (2): 61–68

Lampertico, P., Del Ninno, E., Manzin, A. *et al*. (1997). A randomized controlled trial of a 24-month course of interferon-alpha 2b in patients with chronic hepatitis B who had hepatitis B virus DNA without hepatitis Be antigen in serum. *Hepatology* **26**: 1621–1625

Lampertico, P., Manzin, A., Rumi, M.G. *et al*. (1995). Hepatitis B virus precore mutants in HBeAg carriers with chronic hepatitis treated with Interferon. *J. Viral Hepatitis* **2**: 251–256

Lau, D.T.Y., Everhart, J., Kleiner, D.E. *et al*. (1997). Long term follow-up of patients with chronic hepatitis B treated with interferon alpha. *Gastroenterology* **113** (5): 1660–1667

Lau, G.K.K., Liang, Z., Lee, C.K. *et al*. (1998). Clearance of persistent hepatitis B virus infection in Chinese bone marrow transplant recipients whose donors were anti-hepatitis B case and anti-hepatitis B surface antibody-positive. *J. Infect. Dis.* **178**: 1585–1591

Lau, G.K.K., Lok, A.S.F., Liang, R.H.S. *et al*. (1997). Clearance of hepatitis B surface antigen after bone marrow transplantation: role of adoptive immunity transfer. *Hepatology* **25**: 1497–1501

Lau, J.Y.N., Lai, C.L., Wu, P.C., Chung, H.T., Lok, A.S.F. and Lin, H.I. (1991). A randomized controlled trial of recombinant interferon-γ in Chinese patients with chronic hepatitis B virus infection. *J. Med. Virol.* **34**: 184–187

Lee, J.H., Noh, E.W., Lee, H.H., Kim, G.C., Noh, E.R. and Youn, Y. (1996). Construction of a vector for the expression of hepatitis B virus therapeutic vaccine gene in plants. *Research Report of the Forest Genetics Research Institute* **0** (32): 80–87

Lee, S.D., Lo, K.J., Tsai, Y.T. and Wu J-C. (1989). Maternal hepatitis B virus DNA in mother–infant transmission. *Lancet* **i**: 719–724

Leroux-Roels, G., Desombere, I., De Tollenaere, G. *et al.* (1997). Hepatitis B vaccine containing surface antigen and selected pre-S1 and pre-S2 sequences. 1. Safety and immunogenicity in young, healthy adults. *Vaccine* **15**(16): 1724–1731

Li, L., Sheng, M.H., Tong, S.P., Chen, H.Z. and Wen, Y.M. (1986). Transplacental transmission of hepatitis B virus. *Lancet* **i**: 872–873

Liang, T.J., Hasegawa, K., Rimon, N., Wands, J.R. and Ben-Porath, E. (1991). A hepatitis B virus mutant associated with an epidemic of fulminant hepatitis. *N. Engl. J. Med.* **324**: 1705–1709

Liaw, Y.F. and Tsai, S.L. (1997). Pathogenesis and clinical significance of spontaneous exacerbations and remissions in chronic hepatitis B virus infection. *Viral Hepatitis Rev.* **3**(3): 143–154

Livingston, B.D., Crimi, C., Grey, H. *et al.* (1997). The hepatitis B virus-specific CTL responses induced in humans by lipopeptide vaccination are comparable to those elicited by acute viral infection. *J. Immunol.* **159**: 1383–1392

Lobello, S., Lorenzoni, U., Vian, A., Floreani, A., Brunetto, M.R. and Chiaramonte, M. (1998). Interferon treatment in hepatitis B surface antigen-positive hepatitis Be antibody-positive chronic hepatitis B: role of hepatitis B core antibody Ig M titre in patient selection and treatment monitoring. *J. Viral Hepatitis* **5**: 61–66

Löhr, H.F., Weber, W., Schlaak, J., Goergen, B., Meyer zum Büschenfelde, K.H. and Gerken, G. (1995). Proliferative response of CD4+ T cells and hepatitis B virus clearance in chronic hepatitis with or without hepatitis B e-minus hepatitis B virus mutants. *Hepatology* **22**: 61–68

Löhr, H.F., Gerken, G., Schlicht, H.J., Meyer Zum Büschenfelde, K.H. and Fleischer, B. (1993). Low frequency of cytotoxic liver-infiltrating T-lymphocytes specific for endogenous processed surface and core proteins in chronic hepatitis B. *J. Infect. Dis.* **168**: 1133–1139

Lok, A.S.F., Ghany, M.G., Watson, G. and Ayola, B. (1998). Predictive value of aminotransferase and hepatitis B virus DNA levels on response to interferon therapy for chronic hepatitis B. *J. Viral Hepatitis* **5**: 171–178

Lok, A.S.F., Chung, H.T., Liv, V.W.S. and Ma, O.C.K. (1993). Long-term follow-up of chronic hepatitis B patients treated with interferon-alpha. *Gastroenterology* **105**: 1833–1838

Lok, A.S.F. (1992). Natural history and control of perinatally acquired hepatitis B virus infection. *Dig. Dis.* **10**: 46–52

Lok, A.S.F. and Lai, C.L. (1990). Acute exacerbations in Chinese patients with chronic hepatitis B virus (HBV) infection. Incidence, predisposing factors and etiology. *J. Hepatol.* **10**: 29–34

Lok, A.S.F., Lai, C.L., Wu, P.C., Lau, J.Y.N., Leung, E.K.Y. and Wong, L.S.K. (1989). Treatment of chronic hepatitis B with Interferon: experience in Asian patients. *Semin. Liver Dis.* **9**(4): 249–253

Lok, A.S.F. and Lai, C.L. (1988). A longitudinal follow-up of asymptomatic hepatitis B surface antigen positive Chinese children. *Hepatology* **8**(5): 1130–1133

Lok, A.S.F., Lai, C.L., Wu, P.C., Leung, E.K.Y. and Lam, T.S. (1987). Spontaneous hepatitis B Antigen to antibody seroconversion and reversion in chinese patients with chronic hepatitis B virus infection. *Gastroenterology* **92**: 1839–1843

McGilchrist, A.J. and Follett, E. (1987). In: 'Viral Hepatitis and Liver disease'. Bacillus Calmette-Guérin vaccine as immunotherapy in chronic hepatitis B. pp. 963–964. Ed., A.J. Zuckerman. New-York, Alan R- Liss

Mancini, M., Hadchouel, M., Davis, H.L., Whalen, R.G., Tiollais, P. and Michel, M.L. (1996), DNA-mediated immunization in a transgenic mouse model of the hepatitis B surface antigen chronic carrier state. *Proc. Natl. Acad. Sci.* USA **93**: 12496–12501

Manesis, E.K., Moschos, M., Brouzas, D. *et al.* (1998). Neurovisual impairment: a frequent complication of alpha-interferon treatment in chronic viral hepatitis. *Hepatology* **27**: 1421–1427

Marcellin P, Collin, J.F., Boyer, N. *et al.* (1991). Fatal exacerbation of chronic hepatitis B induced by recombinant alpha-interferon. *Lancet* **338**: 828

Marinos, G., Naoumov, N.V. and Williams, R. (1996). Impact of complete inhibition of viral replication on the cellular immune response in chronic hepatitis B virus infection. *Hepatology* 24: 991–995

Marinos, G., Torre, F., Chokshi Sh. *et al.* (1995). Induction of T-helper cell response to hepatitis B core antigen in chronic hepatitis B: a major factor in activation of the host immune response to the hepatitis B virus. *Hepatology* 22: 1040–1049

Martin, J., Bosch, O., Moraleda, G., Bartolome, J., Quiroga, J.A. and Carreno, V. (1993). Pilot study of recombinant human granulocyte-macrophage colony-stimulating factor in the treatment of chronic hepatitis B. *Hepatology* 18 (4): 775–780

Maruyama, T., Iino, S., Koile, K., Yasuda, K. and Milich, D.R. (1993). Serology of acute exacerbations in chronic hepatitis B virus infection. *Gastroenterology* 105: 1141–1151

Michalak, T.I. (1998). The woodchuck animal model for hepatitis B. *Viral Hepatitis Rev.* 4(3): 139–165

Michel, M.L. (1997). Prospects for active immunotherapies for hepatitis B virus chronic carriers. *Res. Virol.* 148: 95–99

Michel, M.L. (1995). DNA-mediated immunization: prospects for hepatitis B vaccination. *Res. Virol.* 146: 261–265

Michel, M.L., Davis, H.L., Schleef, M., Mancini, M., Mollais, P. and Whalen, R.G. (1995). DNA-mediated immunization to hepatitis B surface antigen in mice: aspects of the humoral response mimic hepatitis B viral infection in humans, *Proc. Natl. Acad. Sci USA* 92: 5307–5311

Milich, D.R., Chen, M.K., Hughes, J.L. and Jones, J.E. (1998). The secreted hepatitis B precore antigen can modulate the immune response to the nucleocapsid: a mechanism for persistence. *J. Immunol.* 160 (4): 2013–2021

Milich, D.R. (1997a). Review: immune response to the hepatitis B virus: infection, animal models, vaccination. *Viral Hepatitis Rev.* 3(2): 63–103

Milich, D.R. (1997b). Influence of T-helper cell subsets and crossregulation in hepatitis B virus infection. *J. Viral Hepatitis* 4(suppl.2), 48–59

Milich, D.R. (1997c). Pathobiology of acute and chronic hepatitis B virus infection: an introduction. *J. Viral Hepatitis* 4(suppl.2): 25–30

Milich, D.R., Schödel, F., Hughes, J.L., Jones, J.E. and Peterson, D.L. (1997). The hepatitis B virus core and e antigens elicit different Th cell subsets: antigen structure can affect Th cell phenotype. *J. Virol.* 71 (3): 2192–2201

Milich, D.R., Peterson, D.L., Schödel, F., Jones, J.E. and Hughes, J.L. (1995a). Preferential recognition of hepatitis B nucleocapsid antigens by Th1 or Th2 cells is epitope and major histocompatibility complex dependent. *J. Virol.* 69 (5): 2776–2785

Milich, D.R., Wolf, S.F., Hughes, J.L. and Jones, J.E. (1995b). Interleukin-12 suppresses autoantibody production by reversing helper T-cell phenotype in hepatitis B e antigen transgenic mice. *Proc. Natl. Acad. Sci. USA* 92: 6847–6851

Milich, D.R., Jones, J.E., Hughes, J.L. and Maruyama, T. (1993). Role of T-cell tolerance in the persistence of hepatitis B virus infection. *J. Immunother.* 14: 226–233

Milich, D.R., Jones, J.E., Hughes, J.L., Price, J., Raney, A.K. and McLachlan, A. (1990). Is a function of the secreted hepatitis B e antigen to induce immunologic tolerance *in utero*? *Proc. Natl. Acad. Sci. USA* 87: 6599–6603

Milich, D.R., Hughes, J.L., McLachlan, A., Thornton, G.B. and Moriarty, A. (1988). Hepatitis B synthetic immunogen comprised of nucleocapsid T-cell sites and envelope B-cell epitopes. *Proc. Natl. Acad. Sci. USA.* 85: 1610–1614

Missale, G., Redeker, A., Person, J. *et al.* (1993). HLA-A31 and HLA-AW68 restricted cytotoxic T cell responses to a single hepatitis B virus nucleocapsid epitope during acute hepatitis. *J. Exp. Med.* 177: 751–762

Nagaraju, K., Naik, S.R. and Naik, S. (1997). Functional implications of Hepatitis B surface antigen (HBsAg) in the T cells of chronic HBV carriers. *J. Viral Hepatitis* 4: 221–230

Naoumov, N.V., Rossol, S. (1997). Studies of interleukin-12 in chronic hepatitis B virus infection. *J. Viral Hepatitis* 4(suppl. 2): 87–91

Nayersina, R., Fowler, P., Guilhot, S. *et al.* (1993). HLA-A2 restricted cytotoxic T-lymphocyte responses to multiple hepatitis B surface antigen epitopes during hepatitis B virus infection. *J. Immunol.* 150: 4659–4671

Niederau, C., Heintges, T., Lange, S. *et al.* (1996). Long term follow-up of HbeAg positive patients treated with interferon-alpha for chronic hepatitis B. N. *Engl. J. Med.* 334: 1422–1427

Niesters, H.G.M., Honkoop, P., Haagsma, E.B., de Man, R.A., Schalm, S.W. and Osterhaus, A.D.M.E. (1998). Identification of more than one mutation in the hepatitis B virus polymerase gene arising during prolonged lamivudine treatment. *J. Infect. Dis.* 177: 1382–1385

Okada, K., Kamiyama, I., Inomata, M., Miyakawa, Y. and Mayumi, M. (1976). E antigen and anti-e in the serum of asymptomatic carrier mothers as indicators of positive and negative transmission of hepatitis B virus to their infants. N. *Engl. J. Med.*, 294 (4): 746–749

Omata, M., Ehata, T., Yokosuka, O., Hosoda, K. and Ohto, M. (1991). Mutations in the precore region of hepatitis B virus DNA in patients with fulminant and severe hepatitis. N. *Engl. J. Med.* 324 (24): 1699–1704

Paul, W.E. and Seder, R.A. (1994). Lymphocyte responses and cytokines. *Cell* 76: 241–251

Penna, A., Chisari, F.V., Bertoletti, A. *et al.* (1991). Cytotoxic T-lymphocytes recognize an HLA-A2 restricted epitope within the hepatitis B virus nucleocapsid antigen. *J. Exp. Med.* 174: 1565–1570

Perillo, R.P. (1993). Interferon in the management of chronic hepatitis B. *Dig. Dis. Sci.* 38 (4): 577–593

Perillo, R.P., Regenstein, F.G., Peters, M.G. *et al.* (1988). Prednisone withdrawal followed by recombinant alpha interferon in the treatment of chronic type B hepatitis. *Ann. Intern. Med.* 109: 95–100

Peeters, M. (1989). Mechanisms of action of Interferons. *Semin. Liver Dis.* 9: 235–239

Pol, S., Driss, I., Couillin, I. *et al.* (1998). A controlled study of anti-HBV vaccine therapy in chronic hepatitis B infection. Hepatology 28 (4): AASLD abstract #1301; 488A

Pol, S. (1995). Immunotherapy of chronic hepatitis B by anti HBV vaccine, *Biomed. Pharmacother.* 49: 105–109

Pol, S., Driss, F., Michel, M.L., Malpas, B., Berthelot, P. and Brechot, C. (1994). Specific vaccine therapy in chronic hepatitis B infection (Letter). *Lancet* 342

Poma, D., Hötzinger, H., Wischnik, A. and Thoma, H. (1990). The superior immunogenicity of a preS1 containing HBV vaccine compared to a S-vaccine in comparative clinical trials. Proceedings of the 1990 International Symposium on Viral Hepatitis and Liver Diseases, April 4–8, 1990, Houston: 118 abstract #288

Poovorawan, Y., Sanpavat, S., Pongpunlert, W., Chumdermpadetsuk, S., Sentrakul, P. and Safary, A. (1989). Protective efficacy of a recombinant DNA hepatitis B vaccine in neonates of HBe antigen-positive mothers. *JAMA* 261: 3278–3281

Pride, M.W., Bailey, C.R., Muchmore, E. and Thanavala, Y. (1998). Evaluation of B and T cell responses in chimpanzees immunized with Hepagene, a hepatitis B vaccine containing preS1, preS2 and S gene products. *Vaccine* 16 (6): 543–550

Pumpens, P., Borisova, G.P., Crowther, R.A. and Grens, E. (1995). Hepatitis B virus core particles as epitope carriers. *Intervirology* 38: 63–74

Qu, D., Yuan, Z.H., He, L.F., Yang, L., Li, G.D. and Wen, Y.M. (1998). Effect of plasmid DNA on immunogenicity of HBsAg – antiHBs complex. *Virol. Immunol.* 11 (2): 65–72

Rehermann, B., Ferrari, C., Pasquinelli, C. and Chisari, F.V. (1996). The hepatitis B virus persists for decades after patients' recovery from acute viral hepatitis despite active maintenance of a cytotoxic T-lymphocyte response. *Nature Med.* 2 (10): 1104–1108

Rehermann, B., Fowler, P., Sydney, J. *et al.* (1995). The cytotoxic T-lymphocyte response to multiple hepatitis B virus polymerase epitopes during an acute hepatitis B infection. *J. Exp. Med.* 181 (3): 1047–1058

Rodriguez-Iningo, E., Bartolome, J., Lopez-Alcorocho, J.M., Cotonat, T., Oliva, H. and Carreno, V. (1997). Activation of liver disease in healthy hepatitis B surface antigen carriers during interferon-alpha treatment. *J. Med. Virol.* 53: 76–80

Ronlov, G., Anderson, C.G., Towsend, G. *et al.* (1996). In: 'Proceedings of the 1996 Meeting on Molecular Biology of Hepatitis B viruses'. Direct injection of a recombinant retroviral vector expressing HBV core in chronic carrier chimpanzees. Cold Spring Harbor, 18–22 September 1996, p.164

Rossol, S., Marinos, G., Carucci, P., Singer, M.V., Williams, R. and Naoumov, N.V. (1997). Interleukin-12 induction of Th1 cytokines is important for viral clearance in chronic hepatitis B. *J. Clin. Invest.* 99(12): 3025–3033

Schiff, E., Karayalcin, S., Grimm, I. *et al.* and the International Lamivudine Investigator Group (1998). A placebo controlled study of lamivudine and interferon α2b in patients with chronic hepatitis B who previously failed interferon therapy. AASLD abstract #901, p. 388A

Schmid, M., Flury, R., Bühler, H., Havelka, J., Grob, P.J. and Heitz, P.U. (1994). Chronic viral hepatitis B and C: an argument against the conventional classification of chronic hepatitis. *Virchows Archiv.* **425**: 221–228

Schödel, F., Peterson, D. and Milich, D. (1996) Hepatitis B virus core and e antigen: immunorecognition and use as a vaccine carrier moiety. *Intervirology* **39**: 104–110

Schödel, F., Moriarty, A.M., Peterson, D.I. *et al.* (1992). The position of heterologous epitopes inserted in hepatitis B core particles determines their immunogenicity. *J. Virol.* **6**(1): 106–114

Scully, L.J., Lloyd, C., P. Karayiannis, P. and Thomas, H.C. (1988). Clinical and immunologic response of chronic hepatitis B virus carriers to O-palmitoyl-3(+)-catechin. *Hepatology* **8** (5): AASLD Abstracts, abstract Nr. 415

Senturk, H., Tabak, F., Akdogan, M. *et al.* (1998). Therapeutic vaccination with a pre-S2 containing vaccine in chronic hepatitis B: a promising approach. *Hepatology* **28** (4): AASLD abstract #1701, 588A

Sheen, I.S., Liaw, Y.F., Tai, D.I. and Chu, C.M. (1985). Hepatic decompensation associated with hepatitis B e antigen clearance in chronic type B hepatitis. *Gastroenterology* **89**: 732–735

Shouval, D. and Ilan, Y. (1995a). Immunization against hepatitis B through adoptive transfer of immunity. *Intervirology* **38**: 41–46

Shouval, D. and Ilan, Y. (1995b). Adoptive transfer of immunity to hepatitis B: development of the concept and evolution of the research project for targeted immunotherapy against hepatic cells which express a viral antigen, Adv. *Drug Del. Rev.* **17**: 317–320

Sokal, E.M., Conjeevaram, H.S., Roberts, E.A. *et al.* (1998). Interferon-alpha therapy for chronic hepatitis B in children: a multinational randomized controlled trial. *Gastroenterology* **114**: 988–995

Stevens, C.E., Taylor, P.E., Tong, M.J. *et al.* (1987). Yeast recombinant hepatitis B vaccine. Efficacy with hepatitis B immune globulin in prevention of perinatal hepatitis B virus transmission. *JAMA* **257** (19): 2612–2616

Stevens, C.E., Beasley, R.P., Tsui, J. and Lee, W.C. (1975). Vertical transmission of hepatitis B antigen in Taiwan. *N. Engl. J. Med.* **292**: 771–774

Taylor, J.M., Netter, H.J. and Wu, T.T. (1995). Potential use of modified hepatitis Delta virus in the therapy of chronic hepatitis B virus infection. *Viral Hepatitis Rev.* **1**(1): 47–52

Tassopoulos, N.E., Papaevangelou, G.J. and Roumeliotou-Karayannis, A. (1986). Heterosexual transmission of hepatitis B virus from symptomless HBsAg carriers positive for antiHBe. *Lancet* **ii** (4): 972

Tennant, B.C., Gerin, J.L. (1994). The Woodchuck model of hepatitis B virus infection. The liver: biology and pathobiology, 3rd edn. Raven Press, New York, 1455–1466

Terazawa, S., Kojima, M., Yamanaka, T. *et al.* (1991). Hepatitis B virus mutants with precore-region defects in two babies with fulminant hepatitis and their mothers positive for antibody to hepatitis B e antigen, *Pediatr. Res.* **29**: 5–9

Terrault, N.A., Zhou, S., Combs, C. *et al.* (1996). Prophylaxis in liver transplant recipients using a fixed dosing schedule of hepatitis B immunoglobulin. *Hepatology* **24** (6): 1327–1333

Thoma, H.A., Hemmeling, A.E. and Hötzinger, H. (1990). In: 'Progress in Hepatitis B Immunization' (ed. P. Coursaget). Evaluation of immune response in a third generation hepatitis B vaccine containing pre-S proteins in comparative trials. *MJTONG* **194**: 35–42

Thomas, H.C. and Thursz, M.R. (1997). Immunogenetics of hepatitis B virus infection, J. Viral. *Hepatitis* **4**(suppl 2): 98–100

Thomas, H.C., Lok, A.S.F., Carreno, V. *et al.* and the International Hepatitis trial Group (1994). Comparative study of three doses of interferon- α-2 in chronic active hepatitis B. J. Viral. *Hepatitis* **1**: 139–148

Thursz, M.R., Kwiatowski, D., Allsop, C.E., Greenwood, B.M., Thomas, H.C. and Hill, A.V.S. (1995). Association between a MHC class II allele and clearance of hepatitis B virus in the Gambia. *N. Engl. J. Med.* **332** (16): 1065–1069

Trinchieri, G. (1995). Interleukin-12: a pro-inflammatory cytokine with immunoregulatory functions that bridge innate resistance and antigen-specific adoptive immunity. *Annu. Rev. Immunol.* **13**: 251–276

Tsai, S.L. and Huang, S.N. (1997). T cell mechanisms in the immunopathogenesis of viral hepatitis B and C. J. Gastroenterol. *Hepatol.* **12** (suppl.): S227–S235

Tsai, S.L., Cheno, P.J., Lai, M.Y. *et al.* (1992). Acute exacerbations of chronic type B hepatitis are accompanied by increased T cell responses to hepatitis B core and e antigens, *J. Clin. Invest.* **89**: 87–96

Vajro, P., Tedesco, M., Fontanella, A. *et al.* (1996). Prolonged and high dose recombinant interferon alpha 2b alone or after prednisone priming accelerates termination of active viral replication in children with chronic B infection. *Pediatr. Infect. Dis. J.* **15**: 223–231

Vandepapelière, P., Slaoui, M., Lebacq, E., Leroux-Roels, G. and Delchambre, M. (1996). In: 'Proceedings of the 36th Interscience Conference on Antimicrobial Agents and Chemotherapy'. Vaccine therapy for chronic hepatitis B: Immunogenicity and safety of two candidate vaccines in human healthy volunteers. September 15–19, New Orleans, LA. Abstract # H 20

Vento, S., Chen, S.H., Giuliani-Piccari, G. *et al.* (1987). Cellular immunity to nucleocapsid and pre-S determinants in asymptomatic carriers of hepatitis B virus. *Immunology* **62**: 593–598

Vitiello, A., Ishioka, G., Grey, H.M. *et al.* (1995). Development of a lipopeptide-based therapeutic vaccine to treat chronic HBV infection: induction of a primary cytotoxic T lymphocyte response in humans. *J. Clin. Invest.* **95**: 341–349

Wen, Y.M., Wu, X.H., Hu, D.C., Zhang, Q.S. and Guo, S.Q. (1995). Hepatitis B vaccine and antiHBs complex as approach for vaccine therapy. *Lancet* **345**: 1575–1576

Wen, Y.M., Xiong, S.D. and Zhang, W. (1994). Solid matrix–antibody–antigen complex can clear viraemia and antigenaemia in persistent Duck hepatitis B virus infection. *J. Gen. Virol.* **75**: 335–339

Wettendorff, M., Voet, P., Garcon, N., Thiriart, C. and Slaoui, M. (1996). In: 'Proceedings of the 36th Interscience Conference on Antimicrobial Agents and Chemotherapy'. Immunogenicity Study of Hepatitis B Therapeutic Vaccine Formulations in HBsAg Transgenic Mice. September 15–19, New Orleans, LO. Abstract # H 19

Whitten, T.M., Quets, A.T. and Schloemer, R.A. (1991). Identification of the Hepatitis B virus factor that inhibits expression of the beta interferon genome. *J. Virol.* **65**: 4699–4704

Widera, G. (1997). In: 'Proceedings of IBC, Third Annual Industry Congress, Hepatitis Latest Therapeutic Developments for hepatitis B and C'. Particle-mediated DNA immunization with HBV antigens: preclinical efficacy in small and larger animals. January 30–31, Washington DC

Williams, A. and Young, M. (1998). A multicenter, double blind, parallel-group comparative study to assess the efficacy and safety of Hepagene and Engerix B in subjects with a proven inadequate response to prior vaccination. *J. Hepatol.* **28**(S1): 118

Wong, D.K.H., Cheung, A.M., O'Rourke, K., Naylor, C.D., Detsky, A.J. and Heathcote, J. (1993). Effect of alpha-interferon treatment in patients with hepatitis B e antigen-positive chronic hepatitis B. A meta-analysis. *Ann. Intern. Med.* **119**: 312–323

Wu, P.C., Lok, A.S.F., Lau, J.Y.N., Lauder, I.J. and Lai, C.L. (1992). Histological changes in Chinese patients with chronic hepatitis B virus infection after interferon-α therapy, *Am. J. Clin. Pathol.* **98**: 402–407

Yoffe, B., Burns, D.K., Bhatt, H.S. and Combes, B. (1990). Extrahepatic hepatitis B virus DNA sequences in patient with acute hepatitis B infection. *Hepatology* **12**: 187–192

Yoshikawa, M., Fukui, H. and Tsujii, T. (1995), Immunological adverse effects of interferon treatment. *Clin. Immunother.* **4**(5): 361–375

Zu Putlitz, J., Encke, J. and Wands, J.R. (1998a). DNA-based immunization generates cellular immune responses against hepatitis B virus polymerase. Hepatology **28** (4): AASLD abstract # 1676, 581A

Zu Putlitz, J., Lanford, R.E., Carlson, R.I., Notvall, L., De La Monte, S.M. and Wands, J.R., (1998b). Properties of hepatitis B virus polymerase revealed by monoclonal antibodies. Hepatology **28** (4): AASLD abstract #1672, 580A

Zuckerman, J.N., Sabin, C., Craig, F.M., Williams, A. and Zuckerman, A.J. (1997). Immune response to a new hepatitis B vaccine in healthcare workers who had not responded to standard vaccine: randomized double blind dose-response study. *Br. Med. J.* **314**(7077): 329–333

15
CHLAMYDIA

ROBERT C. BRUNHAM*
AND
GRANT McCLARTY[†]

Centre for Disease Control,
University of British Columbia,
Vancouver, Canada

[†] *Department of Medical Microbiology*
University of Manitoba,
Winnipeg, Manitoba, Canada

INTRODUCTION

Chlamydiae are obligate intracellular bacteria that infect eukaryotic cells (Moulder, 1991). Their reduced biosynthetic capacity causes them to be absolutely dependent on the eukaryotic cell for many nutrients required for growth and replication (McClarty, 1994). Of the four recognized chlamydial species, two (*Chlamydia trachomatis* and *C. pneumoniae*) are primary human pathogens; this chapter will focus on *C. trachomatis*. Chlamydiae are among the most common of all human infectious agents and produce much disability, although little acute mortality. This chapter highlights selected aspects of *C. trachomatis* infection, including clinical features of disease, microbiology, immunobiology, and approaches to vaccine development. A vaccine to prevent *C. trachomatis* infection has been a long-sought goal of chlamydia researchers, and progress in elucidating the molecular and mechanistic basis for chlamydia immunity, together with new knowledge on how to deliver the vaccine immunogen, now makes the feasibility of a chlamydial vaccine more likely than ever.

CLINICAL ASPECTS

C. trachomatis is classifiable into two distinct groups of biological variants or biovars, which produce distinctively different diseases (Schachter, 1978). These are the trachoma biovar and the lymphogranuloma venereum biovar. As if to confuse the uninitiated, these designations are also the names for two important chlamydial diseases. The trachoma biovar is temperature sensitive, with its growth restricted to 35°C, which perhaps explains why it only produces

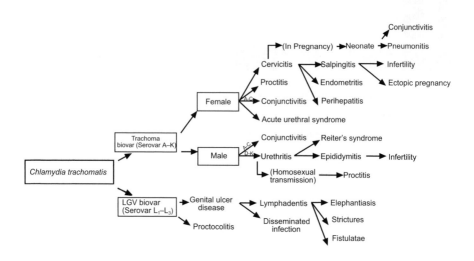

FIGURE 15.1 Shown in schematic form are the major chlamydial diseases caused by either biovar of *Chlamydia trachomatis*.

infection on the cooler mucosal surfaces of the body. The lymphogranuloma venereum (LGV) biovar is capable of growth at body temperature (37°C), and during human infection disseminates throughout the body, principally within the lymphoid system. Figure 15.1 illustrates the most frequent chlamydial diseases caused by infection with the two biovars of *C. trachomatis*.

Within each of the two biovars, *C. trachomatis* is serologically divisible into serovariants or serovars, with the major diseases among trachoma biovar organisms produced by serovars A to K, and among the LGV biovar organisms caused by serovars L_1, L_2 and L_3 (Grayston and Wang, 1975).

Transmission of trachoma biovar can be either by contaminated fomites, as in the case of serovars A–C, or by sexual and perinatal contact, as in the case of serovars D–K. LGV serovars are transmitted only by sexual contact. Geographically, the diseases trachoma and lymphogranuloma venereum are essentially restricted to developing areas of the world, whereas sexually and perinatally transmitted chlamydial infection are distributed globally.

Trachoma

Trachoma is a characteristic ocular disease caused by infection with *C. trachomatis* serovars A, B and C (Grayston and Wang, 1975; Jones, 1975). An estimated 500 million people worldwide are afflicted with trachoma, most of whom are young children. Trachoma is especially common in poor areas of sub-Saharan Africa, where it is a major public health problem because 1–5% of infected individuals later develop scarring which deforms the eyelid and causes inward turning of the eyelashes (entropion) and corneal abrasion (trichiasis). The consequent corneal opacification results in blindness. Trachoma is the most common infectious cause of blindness in the world, and an estimated 7 million people are currently blind as a result of trachoma. Most of these individuals are middle-aged and elderly adults. Active trachoma often has its onset within the first 1–2 years of life. Recurrences are common during childhood, but cease spontaneously by 10–15 years of age, presumably because of acquired immunity (Mabey and Bailey, 1996). Important risk factors for trachoma susceptibility include the frequency of face washing, access to water, sharing a sleeping room with an infected individual, and the intensity of exposure to eye-seeking flies. Active trachoma can also occur in adults, especially mothers caring for young children with the disease. Trichiasis is related to repeated episodes of trachoma occuring during childhood, is more common in women than in men, and seems to cluster in families, suggesting that there may be genetic determinants for trachoma scarring.

The *C. trachomatis* serovars that produce trachoma are spread by direct contact with contaminated secretion carried by fomites such as facecloths or eye-seeking flies. Perinatal exposure to *C. trachomatis* from maternal genital tract infection is not an important route for trachoma transmission.

Trachoma is a chronic follicular conjunctivitis which causes macroscopically visible lymphoid follicles to form in the submucosa of the conjunctiva. CD4 and CD8 T cells, as well as B cells, are numerous in areas of trachomatous inflammation (El-Asrar *et al.*, 1989; Reacher *et al.*, 1991). Infiltration with lymphocytes is especially apparent along the upper tarsal plate, and the bulbar conjunctiva is minimally involved. The cornea may also be involved, with superficial vascularization and lymphocytic infiltration (pannus).

Sexually/Perinatally Transmitted Chlamydial Infections

At present *C. trachomatis* is the most prevalent sexually transmitted bacterial infection in most developed countries (Hillis and Wasserheit, 1996). More than 4 million chlamydial infections occur annually in the United States, and prevalence rates are highest (>10%) among sexually active adolescent females. Prevalence is higher in inner-city areas, among lower socioeconomic status individuals, and among minority ethnic groups such as African-Americans in the United States and Native Americans in Canada. In the United States the direct and indirect costs of sexually transmitted chlamydial disease exceed $2.4 billion annually. From a global perspective, sexually transmitted chlamydial infections are a major cause of total disease burden and healthy life years lost, because of the effects they produce on the reproductive health of women and because chlamydial infection facilitates the transmission of human immunodeficiency virus (HIV) (Plummer *et al.*, 1991).

Urethritis

C. trachomatis causes 30–40% of cases of non-gonococcal urethritis (NGU) in men, and an estimated 40–60% of urethral chlamydial infections are symptomatic with NGU (Holmes *et al.*, 1975). NGU is characterized by complaints of mild urethral discharge, urethral discomfort and mild dysuria. Often this is only observed in the morning prior to voiding.

C. trachomatis urethral infection also occurs in women, where the urethritis it produces is called the acute urethral syndrome. In such cases the woman complains of dysuria; pyuria is found on urinalysis, but culture for uropatho-gens is negative. Urinary frequency and urgency are usually absent. A mild urethral exudate may be observed during pelvic examination when the urethra is compressed against the pubic ramus.

Epididymitis

In an estimated 1–3% of men with urethral chlamydial infection, infection spreads from the urethra to the epididymis (Berger *et al.*, 1978). This results in unilateral testicular pain, scrotal erythema and tenderness, or swelling over the epididymis. Epididymitis associated with urethritis is most commonly due to C. *trachomatis* or *Neisseria gonorrhoeae*, and among men <35 yeas of age C. *trachomatis* is the principal cause of epididymitis. Untreated, bilateral epidi-dymitis can cause scar formation in the epididymis and thus be a cause of male infertility.

Reiter's Syndrome

Reactive arthritis can complicate genital chlamydial infection (Martin *et al.*, 1984). This is more common in men than in women, and is strongly associated with the major histocompatibility antigen HLA-B27. Reiter's syndrome is characterized by the triad of asymmetric oligoarthritis of the large joints, inflammation at sites of tendon insertion (ensethopathy), urethritis and con-junctivitis. Painless oral ulceration, circinate balanitis or vulvitis, and kerato-dermia blenorrhagia are less common clinical findings. Up to 50% of men with Reiter's syndrome have urethral C. *trachomatis* infection, and it is estimated that approximately 1% of men with chlamydial urethritis develop Reiter's syndrome.

Mucopurulent Cervicitis

Mucopurulent cervicitis in women is the epidemiologic counterpart of non-gonococcal urethritis in men (Brunham *et al.*, 1984). As with non-gonococcal urethritis, C. *trachomatis* causes 40–50% of cases of mucopurulent cervicitis. Only about 20–50% of women with cervical chlamydial infection develop clinically apparent mucopurulent cervicitis, with the remainder being sub-clinically infected. Women with mucopurulent cervicitis may complain of a mucoidy vaginal discharge. Unless there is concurrent infection with other pathogens, the vaginal discharge lacks odour and vulvar pruritis does not occur. Mucopurulent cervicitis is best recognized during vaginal speculum examina-tion, when the cervix is fully exposed and well illuminated. A yellow or cloudy mucoid discharge issues from the cervix, with the colour of cervical mucus, and is better appreciated on the tip of a cotton swab than *in situ*. Often, a red area of columnar epithelium is visible on the face of the cervix (ectopy). The area is erythematous, edematous, and when infected with C. *trachomatis* bleeds easily when touched with a cotton-tipped swab. Cytologic examination of Papani-colaou-stained cervical smears demonstrates that the host response to chlamy-dial cervicitis is characterized by recruitment of neutrophils and macrophages, together with large numbers of activated and dividing lymphocytes, suggestive of strong cell-mediated immune responses (Kiviat *et al.*, 1985).

Endometritis and Salpingitis

C. trachomatis infection can spread from the cervix to the endometrium, producing endometritis, and to the fallopian tubes, producing salpingitis (Mardh *et al.*, 1976). Upper genital tract spread occurs in 10–40% of women with cervical chlamydial infection. *C. trachomatis* spread to the upper reproductive tract can occur either from subclinical cervical infection or from mucopurulent cervicitis. Spread to the endometrium can also occur after therapeutic abortion or following vaginal delivery, causing late-onset postabortal or postpartum endometritis. *C. trachomatis* endometritis and salpingitis can include the following features: subacute onset of low abdominal pain during menses or during the first 2 weeks of the menstrual cycle; pain on sexual intercourse (dyspareunia); and prolonged menses or intermenstrual vaginal bleeding. Fever is not a feature of *C. trachomatis* endometritis or salpingitis, although abnormal levels of acute-phase reactants, such as raised ESR and CRP, are commonly found. During bimanual vaginal examination, cervical motion tenderness, uterine compression tenderness and bilateral adnexal tenderness support the diagnosis of salpingitis. Some cases of chlamydial salpingitis are associated with transabdominal spread to produce perihepatitis, or the Fitz–Hugh–Curtis syndrome (Wolner-Hanssen *et al.*, 1980).

Infant Inclusion Conjunctivitis and Pneumonia

Perinatally transmitted *C. trachomatis* infection is an important health problem for infants born to infected mothers. Approximately two out of three infants born to a mother with *C. trachomatis* cervical infection acquire infection. Inclusion conjunctivitis of the newborn develops in one in three exposed infants (Laga *et al.*, 1988), and a distinctive pneumonia syndrome in about one in six (Beem and Saxon, 1977; Harrison *et al.*, 1978). Since an estimated 5–20% of pregnant women in the United States are infected with *C. trachomatis*, the morbidity due to perinatally transmitted chlamydial infection is substantial.

Inclusion conjunctivitis of the newborn usually has its onset between 1 and 3 weeks after birth. Onset is subacute, with moderate mucopurulent ocular discharge. In distinction to trachoma, no conjunctival lymphoid follicles are found in the neonate, and both upper and lower tarsal conjunctival surfaces are equally involved.

The distinctive pneumonia syndrome also has a subacute onset in infants, at between 1 and 4 months of age, possibly arising at this time as a consequence of the decline in maternally transferred antibody. The natural history of illness is protracted and, importantly fever is absent. The cardinal clinical characteristic is a distinctive staccato cough reminiscent of pertussis, but without the whoop or post-tussive vomiting. Auscultation discloses bilateral crackles. Chest X-ray shows hyperinflation with diffuse interstitial and patchy alveolar infiltrates. Hematologic examination consistently shows eosinophilia and hypergammaglobulinemia (Beem and Saxon, 1977). These hematologic changes strongly suggest the cytokine effects of a polarized T_{H2} type of immune response to *C. trachomatis*. As discussed later, polarized T_{H2} immune response during trachoma has been associated with delayed clearance of *C. trachomatis*, and thus a polarized T_{H2} immune response during neonatal chlamydial infection may underlie the spread of infection to the lungs.

Lymphogranuloma Venereum

Sexually transmitted infection with *C. trachomatis* serovars L_1, L_2 or L_3 can result in the disease lymphogranuloma venereum (Grayston and Wang, 1975). This is a systemic *C. trachomatis* infection with prominent involvement of lymphoid tissue (Bernstein *et al.*, 1984). In the United States in 1992 only 302 cases of lymphogranuloma venereum were reported to the CDC, but in the developing world, especially sub-Saharan Africa, lymphogranuloma venereum is a much more common disease, although accurate statistics are lacking. In a recent study of commercial sex workers in Nairobi, Kenya, 21% of 179 genotyped chlamydial infections were identified as being due to lymphogranuloma venereum strains, although clinical cases of lymphogranuloma venereum were not commonly observed (Brunham *et al.* 1996). Many individuals with lymphogranuloma venereum in developed countries often report exposure to a sexual partner from a developing country, usually a commercial sex worker. Like chancroid, lymphogranuloma venereum can present as a sexually acquired infection in a returned traveler.

The *C. trachomatis* serovars L_1, L_2 and L_3 which produce lymphogranuloma venereum are much more invasive than are the other *C. trachomatis* serovars, for reasons which are not understood but may relate to differences in temperature tolerance as well as other undefined differences.

Among heterosexuals, primary lymphogranuloma venereum infection produces an evanescent and rarely observed genital ulcer 2 or 3 weeks following exposure. The ulcer heals spontaneously and 2–4 weeks later painful bilateral inguinal lymphadenopathy develops, often associated with signs of systemic infection such as fever, headache, arthralgias, leukocytosis and hypergammaglobulinemia. Involved lymph nodes are sometimes biopsied to exclude malignant lymphoma, and show stellate abscesses surrounded by activated histiocytes reminiscent of necrotizing granulomas.

In women and homosexual men, rectal infection with *C. trachomatis* L_1, L_2 or L_3 strains produces a severe febrile protocolitic illness (Quinn *et al.*, 1981). Patients complain of frequent painful defecation (tenesmus), with urgency and, less commonly, mucopurulent bloody discharge in the stool. Biopsy of the rectal mucosa shows submucosal granulomas, crypt abscesses and diffuse mononuclear cell inflammation similar to changes seen with Crohn's disease.

In the absence of treatment, lymphogranuloma venereum can heal spontaneously or result in long-term persistent infection, sometimes leaving lymphatic scarring. Late fibrotic complications of lymphogranuloma venereum include genital elephantiasis, strictures, and fistulas of the penis, urethra and rectum.

MICROBIOLOGY

Chlamydiae are an unique monophyletic bacterial phylum as defined by 16S rRNA sequences (Moulder, 1991). They have an extremely ancient origin within the eubacterial kingdom, perhaps as ancient as the origin of the early eukaryote. The phylum is currently recognized to be composed of four species: *Chlamydia psittaci*, *C. pecorum*, *C. pneumoniae* and *C. trachomatis*. Other species probably exist. No free-living species are known, and all chlamydiae are entirely dependent on a eukaryotic host cell for growth and replication. Table 15.1 shows the classification of species, biovars and serovars within the genus chlamydiae.

C. *pneumoniae* and C. *trachomatis* are obligatory human pathogens, whereas C. *psittaci* primarily infects mammals and birds and produces human infection only sporadically; C. *pecorum* infects only ruminants such as cattle and sheep, and is not known to infect humans. In addition to displaying host specificity, chlamydiae also display cell trophism. For instance, the trachoma biovar infects only columnar cells, whereas the lymphogranuloma venereum biovar infects mononuclear cells. The molecular basis for the cellular trophism of chlamydiae remains to be defined.

The chlamydial cell has a Gram-negative cell wall structure consisting of an outer membrane and an inner cytoplasmic membrane. Despite being sensitive to penicillin and containing penicillin-binding proteins, chlamydiae lack measurable muramic acid, a structure found exclusively in peptidoglycan. In addition, recent genomic sequencing has shown that the chlamydial genome encodes the necessary information for peptidoglycan biosynthesis. The simplest explanation for these anomalies is that chlamydiae synthesize an analytically undetectable amount of peptidoglycan, perhaps restricted to a certain stage of the development cycle.

The outer membrane is extremely protein rich, comprising a single major outer membrane protein (MOMP 40 kDa) (Caldwell *et al.*, 1981). This varies substantially in amino acid sequence, both among and within each chlamydial species (Zhang *et al.*, 1993). Within the C. *trachomatis* species, MOMP variation determines the serologic types that define the serovar designation (Stephens *et al.*, 1986, 1987). The genetic basis for MOMP antigenic diversification has been well characterized. The single copy gene, *omp1*, shows four regions of sequence heterogeneity, called variable regions or variable domains, interspersed among five sequence-constant regions. At the protein level, the four variable segments are thought to be arrayed as surface exposed loops (Baehr *et al.*, 1988) and to have undergone sequence divergence by mutational and recombinational mechanisms, followed by immune selection (Brunham *et al.*, 1993). The MOMP is known to be the dominant antigen against which neutralizing antibodies are directed (Maclean *et al.*, 1988) (Figure 15.2).

TABLE 15.1 Classification of biological variants (biovars) and serological variants (serovars) of the genus chlamydiae

	Biovars	Serovars
C. *trachomatis*	Trachoma	15
	lymphogranuloma venereum	3
	Mouse pneumonitis	1
C. *pneumoniae*	TWAR	1
C. *psittaci*	Avian, mammals	Unknown, multiple
C. *pecorum*	Ruminants	Unknown, multiple

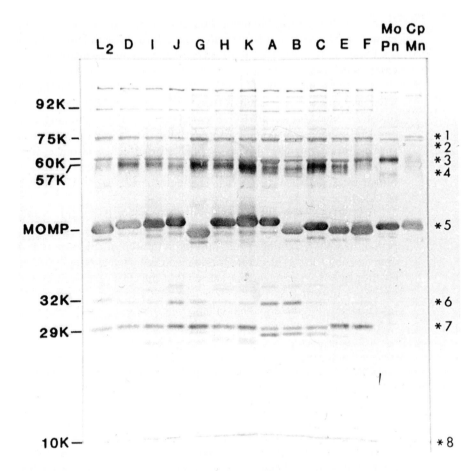

FIGURE 15.2 Immunoblot (using serovar J immune sera at 1:500) of 12 *C. trachomatis* serovars, the MoPn strain, and a *C. psittaci* strain separated on a 10% polyacrylamide gel and transferred to nitrocellulose paper. Molecular mass estimates (in kDa) of the more prominent antigens are given on the left, with the starred numbers 1–8 on the right designating the antigens to which monoclonal antibodies (Mab) were raised. Only Mabs to the MOMP were neutralizing in tissue culture (modified from Maclean et al., 1988).

As with other Gram-negative bacteria, the chlamydial outer membrane also contains lipopolysaccharide (LPS). Chlamydial LPS is a rough type without O-saccharides and is composed of a trisaccharide of 3-deoxy-D-manno-octulo-sonic acid (KDO). Although the core KDO sequences are shared by LPS from many other Gram-negative bacteria, the chlamydial LPS is unique because two of the three KDOs are bonded through a unique 2.8 instead of a 2.4 linkage. Thus, antibodies to chlamydial LPS are specific to the chlamydiae genus. Antibodies to chlamydial LPS are not protective against infection, either in tissue culture neutralization assays or in passive protection animal experiments. Furthermore, chlamydial LPS is a weak inducer of an inflammatory cytokine response (Ingallis et al., 1995).

All chlamydiae share a common and distinctive development cycle. The organisms are dimorphic and undergo developmental change during the growth cycle. Elementary bodies (EBs) are metabolically inactive cells and are found in the extracellular environment. They function as the transmission stage. Two features characterize EBs: the cysteine-rich outer membrane proteins (60 kDa and 12.5 kDa) are extensively cross-linked through inter- and intradisulfide bond formation, which confers structural rigidity on the chlamydial cell (Hatch *et al.*, 1984), and the DNA genome is highly condensed into a nucleoid structure by two major DNA compacting proteins, histone 1 (*hc1*) and histone 2 (*hc2*) (Barry *et al.*, 1992). The EB genome is transcriptionally inactive. Reticulate bodies (RBs) are found only within vacuoles in the perinuclear region of the eukaryotic cell. RBs are capable of DNA replication and binary fission. In RBs MOMP exists as monomers without extensive disulfide bond formation, and the two minor cysteine-rich proteins are absent. The RB chromosome is relaxed and the EB-specific histone proteins are absent. The RB genome is very transcriptionally active.

The size of the *C. trachomatis* genome is small, at 1045 kbp (Birkelund and Stephens, 1992), and has recently been determined to contain 920 open reading frames. Most strains of chlamydiae also contain a 7 kb cryptic plasmid. As will be discussed later, the *C. trachomatis* serovar D genome has been sequenced in its entirety and is available on the World Wide Web (http://chlamydia-www. berkeley.edu:4231/).

Chlamydial attachment, endocytosis and avoidance of lysosomal fusion are constituitive properties of the EB. Although chlamydiae infect eukaryotic cells through multiple attachment mechanisms, one important *C. trachomatis* attachment mechanism appears to involve a trimolecular complex, with a secreted heparan sulfate-like glycosaminoglycan synthesized by chlamydia acting as a bridge between ligands on the chlamydial and mammalian cell surfaces (Zhang and Stephens, 1992). Different mechanisms probably exist, and may explain the characteristic host and cellular trophism displayed by these bacteria.

After attachment, chlamydial EBs enter eukaryotic cells by parasite-directed endocytosis. Often the chlamydiae ladden endosome is associated with clathrin cages during the endocytic stage (Reynolds and Pearce, 1991; Taraska *et al.*, 1996). Approximately 2–8 hours after ingestion, the chlamydial EB organizes into an RB and asynchronously replicates 8–12 times, with a doubling duration of 2–3 hours. During RB replication *C. trachomatis* synthesizes substantial amounts of glycogen, which is initially stored inside the RB but is subsequently released into the vacuolar space, perhaps during RB to EB differentiation. In tissue cell culture the developmental cycle takes 48–72 hours, although this is likely to be much longer in the host. At the conclusion of a growth cycle RBs differentiate back to EBs, with each inclusion yielding 100–1000 new infectious EBs.

Chlamydiae are absolutely dependent on host cells for the basic building blocks of intermediate metabolism, which they polymerize and assemble into chlamydial-specific forms (McClarty, 1994). Chlamydiae were once speculated to be obligate energy parasites of their host cells, and thus having no capacity for net ATP generation. However, recent genomic sequencing indicates that they do possess the enzymes of the glycolytic pathway, and are thus capable of substrate-level phosphorylation. Furthermore, chlamydiae encode a complete pentose phosphate pathway but have a truncated tricarboxylic acid cycle. To augment substrate-level phosphorylation for ATP generation, chlamydiae also acquire ATP directly from the host cell through an ATP:ADP translocase, similar to that found in mitochondria but working in the reverse direction (Hatch *et al.*, 1982). Other phosphorylated and non-phosphorylated intermediates, such as glucose-6-phosphate, pyruvate, most amino acids, vitamins and cofactors, are also taken up by chlamydiae from the eukaryotic cell, probably by unique membrane transporters which are then used in the biosynthesis of lipids, proteins, carbohydrates and nucleic acids. Most notably, chlamydiae are auxotrophic for three of the four ribonucleotides, ATP, GTP and UTP, needed for DNA and RNA synthesis (Tipples and McClarty, 1993). Compared to other bacteria, chlamydiae are virtually unique in being able to transport phosphorylated intermediates, and this undoubtedly represents their premier adaptation to the intracellular environment in a eukaryotic cell (Moulder 1991; Hackstadt *et al.*, 1997).

IMMUNOBIOLOGY

Chlamydiae produce intracellular infection of either macrophage or non-macrophage host cells. As previously described, macrophages and other lymphoid cells are the principal target cell for the LGV biovar, whereas columnar epithelial cells are the usual host cells for the trachoma biovar. Chlamydial infection elicits both humoral and cellular immune responses, both arms of which appear to be necessary for the full expression of chlamydial immunity and immunopathology (Pearce, 1983; Kunimoto and Brunham, 1985; Morrison *et al.*, 1992; Brunham and Peeling, 1994; Ward, 1995).

Immunity

C. trachomatis infection elicits secretory IgA and systemic IgM and IgG antibodies. Mucosal antibodies appear more important in chlamydial immunity to trachoma biovar infection than do circulatory antibodies, probably because the infection is restricted to mucosal surfaces. In one study, the prevalence of mucosal IgA antibodies to *C. trachomatis* in cervical mucus was inversely correlated with the quantitative recovery of chlamydia in tissue culture, suggesting that IgA antibody may inhibit the replication of *C. trachomatis* and consequently limit the extent of local infection (Brunham *et al.*, 1983).

C. trachomatis infection often elicits antibodies to the serovar specific epitopes on the MOMP, and such antibodies are able to neutralize the organism in cell culture (Morrison *et al.*, 1992; Brunham and Peeling, 1994). However, the *in vivo* role of antibodies to the MOMP in protection against *C. trachomatis* infection is less clearly defined. The observation that mucosal IgA antibodies are inversely correlated with the quantitative recovery of chlamydia may suggest that MOMP antibodies do neutralize the organism *in vivo* (Brunham *et al.*, 1983).

Because chlamydiae produce intracellular infection, T cell-mediated immune (CMI) responses are also likely to be important in host clearance of infection. Antigen-specific lymphocyte proliferation responses are commonly observed among individuals with chlamydial infection, although it has been difficult to directly correlate the magnitude of lymphoproliferative responses with aspects of immunity or pathology (Brunham *et al.*, 1981). One longitudinal study noted that individuals with depressed lymphoproliferative responses to chlamydial antigens were more likely to have persistent chlamydial infection and scarring trachoma than were individuals with strong antigen-specific lymphocyte proliferative responses (Holland *et al.*, 1993). These observations suggest that intact CMI responses are likely to be associated with healing trachoma.

In support of the notion that CMI responses are protective for *C. trachomatis* infection, HIV has also been observed to be a risk factor for chlamydia infection acquisition and disease development in a prospective study of commercial sex workers in Nairobi, Kenya (Kimani *et al.*, 1996) (Figure 15.3).

Thus, the infection rate, the reinfection rate and the same-strain reinfection rate were all increased among women with HIV infection. Furthermore, as CD4 T cells were lost as a result of HIV infection, the risk that a chlamydial infection would cause acute salpingitis rose dramatically.

Gamma-interferon (IFN-γ), the principal cytokine of T_{H1} cells, has been detected in individuals with *C. trachomatis* infection and strong IFN-γ responses have been directly correlated with healing trachoma (Holland *et al.*, 1996). In cell culture, IFN-γ inhibits the intracellular replication of *C. trachomatis*, possibly through induction of a host-cell enzyme, indoleamine 2,3-dioxygenase, which depletes intracellular tryptophan pools, thereby causing nutritional deprivation of an essential amino acid for chlamydia.

Immunoepidemiologic data collected from individuals with trachoma suggest that T_{H2} activation is associated with the tissue-damaging phases of chlamydial infection (Holland *et al.*, 1996; Bailey *et al.*, 1995). By comparing individuals with scarring trachoma to community controls, it was observed that individuals with scarring disease more often had high serum antibody titers, high IL-4 responses, and low IFN-γ responses from antigen-stimulated peripheral blood lymphocytes than did normal controls. These data strongly suggest that polarized T_{H2}-like immune responses allow for persistent chlamydial infection, and thus contribute to conjunctival scarring.

More generally, the data suggest that *C. trachomatis* infections may exist as a spectrum of disease states, as has been well established for leprosy and leishmaniasis. In this paradigm, T_{H1}-polarized responses are associated with healing and T_{H2}-polarized responses are associated with progressive disease. Based on this useful heuristic model, chlamydia immunity is predicted to be correlated with T_{H1} immune responses.

Recently, cytotoxic T cells have also been detected during chlamydial infection (Starnbach *et al.*, 1994; Beatty and Stephens, 1994; Starnbach *et al.*, 1995), but their role in the immunobiology of infection is not well described. Although *C. trachomatis*-infected cells can be lyzed by CTLs, it is only with great difficulty. Furthermore, the late chlamydial inclusion specifically alters

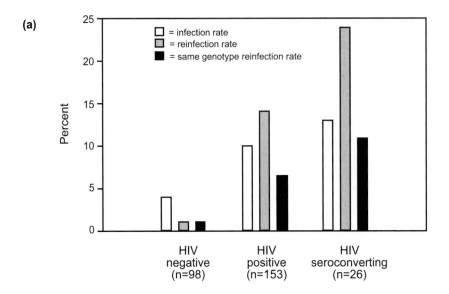

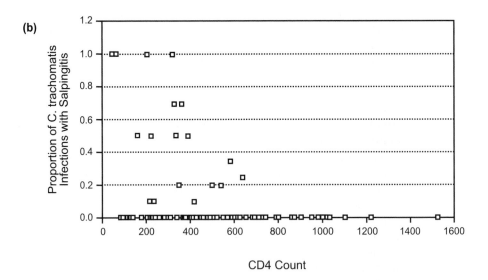

FIGURE 15.3 The effect of HIV on the natural history of *Chlamydia trachomatis* infection. (a) The *C. trachomatis* infection rate among HIV-negative, HIV-positive and HIV-seroconverting women (from Brunham *et al.*, 1996). (b) The proportion of *C. trachomatis* infections associated with acute salpingitis rose as a function of the CD4 T cell number among HIV infected women (from Kimani *et al.*, 1996).

the infected host cell to resist apoptosis (Fan *et al.*, 1998). Thus, CTLs may not be important effectors in chlamydia immunity. Also, several chlamydial diseases, such as Reiter's syndrome (HLA-B27), scarring trachoma (HLA-A28) and chlamydial salpingitis (HLA-A31), are associated with specific HLA class I antigens. It may in fact be that class I restricted CD8 T cells play a larger role in disease pathogenesis than in immunity.

Pathogenesis and Mechanism of Host Injury

Chronic infection or repeated episodes of reinfection appear to underlie the immune responses that cause host tissue injury (Grayston *et al.*, 1985), and not uncommonly evidence of non-cultivatable forms of chlamydiae is found in individuals with the inflammatory or scarring sequelae of chlamydial disease. Infection of an immunologically primed host results in an accelerated and intensified inflammatory response, and tissue destruction which appears to be directly correlated with the intensity of inflammation. This has been best elucidated for *C. trachomatis* ocular infection, where inflammatory and scarring trachoma are diseases of persistent infection and/or reinfection. Thus, the mechanism for host injury with *C. trachomatis* infection is due to immune-mediated tissue inflammation and fibrosis. As just described, a highly polarized T_{H2} cytokine response characterizes the immune responses in individuals with trachoma scarring. Downregulation of T_{H1} immune responses by T_{H2} cytokines such as IL-10 could cause persistence of chlamydia infection, and hence contribute to host injury (Yang *et al.*, 1996).

An additional distinctive immunologic feature that characterizes the late stages of *C. trachomatis* disease is the presence of high serum antibody titers to the chlamydial hsp60 protein (Brunham and Peeling, 1994). Thus, women with the reproductive sequelae of chlamydial infection, such as tubal infertility or ectopic pregnancy, often have prominent antibody responses to the chlamydial hsp60 protein. Furthermore, among women with laparoscopically verified chlamydial salpingitis, those with the highest titers of chlamydial hsp60 antibody had the most severe inflammatory disease. Asymptomatic women with high titers of chlamydial hsp60 antibodies were much more likely to develop chlamydial salpingitis when infected with chlamydia than were women with little or no hsp60 antibody. Because prokaryotic and eukaryotic heat shock proteins share extensive amino acid sequence identity, it has been suggested that autoimmunity to human hsp60 may be induced by chlamydial infection and that autoimmune inflammatory damage may contribute to the late fibrotic sequela of *C. trachomatis* infection. Autoimmune responses to hsp60 are readily induced in experimental animals by infection or immunization, although autoimmune responses to hsp60 have yet to be detected among humans with the late sequelae of chlamydia infection. Alternatively, antibodies to the hsp60 could signal the presence of persistent chlamydia infection, as 'persistent' chlamydial forms are known to express hsp60 disproportionately. This knowledge has proved useful in guiding vaccine design, as an ideal chlamydial vaccine immunogen should not contain the hsp60.

Rationale for and Expected Impact of a Vaccine

Public health efforts to control chlamydia transmission are required because of the high prevalence of infection, the major morbidity due to blindness, infertility and perinatal disease, and the observation that genital infection synergistically augments the transmission of HIV. Both vaccine and non-vaccine approaches to public health efforts for chlamydia control have been proposed. Because the interruption of chlamydia transmission can be successfully achieved by interventions other than a vaccine, it has been reasonably questioned whether control efforts should not be focused solely on these areas. For instance, the dynamics of sexually transmitted chlamydial infection can be substantially altered by changes in sexual behavior, and the incidence of ocular trachoma can be reduced by improved face-washing habits (West *et al.*, 1996). Also, chlamydiae are uniformly susceptible to inexpensive, widely available antibiotics such as the tetracyclines and macrolides. Thus, case detection and antibiotic treatment is an intervention which has the potential to dramatically alter chlamydia transmission dynamics (Scholes *et al.*, 1996). No doubt these strategies are and should be employed where they are feasible and affordable. However, the long-term control and eradication of C. *trachomatis* will require a vaccine approach. Vaccines are simply the most cost-effective public health tool available, and are the only approach that has a demonstrated capacity to eradicate an infectious disease. Thus, vaccine development for chlamydia should continue as a biomedical research priority.

As will be discussed in the next section, there are good reasons for being optimistic about the development of a chlamydial vaccine. This confidence has led to the development of mathematical models to estimate the impact of vaccination on the control of C. *trachomatis* genital infection. Modeling shows that even modestly effective vaccines ($\geqslant 50\%$ efficacy) that induce immunity for 10 or more years dramatically alter the transmission dynamics of C. *trachomatis* infection (de la Maza and de la Maza, 1995). This is because the risk for a sexually transmitted disease such as C. *trachomatis* is confined to a relatively narrow time frame in the lifecycle of an individual. Furthermore, the transmission of most infectious diseases, including STDs, is characterized by extreme heterogeneity, and therefore targeting prevention to risk groups can be highly cost-effective (Woolhouse *et al.*, 1997). In principle, the non-linear relationship between STD risk and vaccine immunity can be exploited in a successful vaccine campaign.

Vaccine Development

Overall, the major challenge for the science of vaccinology is to develop vaccines for infectious diseases that cause chronic persistent infection, which are capable of enduring in the host after the development of antigen-specific immunity. C. *trachomatis* presents such a challenge. The major immune evasion strategies exploited by chlamydiae include intracellular growth, as well as antigenic polymorphism of the major protective protein. C. *trachomatis* grows in epithelial cells that do not express major histocompatibility complex (MHC) class II molecules, and within the cell grows in an endocytic vacuole that does not intersect the MHC class I antigen processing and presentation pathway. The intracellular niche interferes with T-cell and antibody recognition of chlamydia. Within the extracellular environment, the *C. trachomatis* elementary body is able to interact with antibodies that can neutralize infectivity; however, *C. trachomatis* has diversified its major surface protein gene (*omp1*). As such, *C. trachomatis* antigenic diversification and intracellular growth are evasion strategies that allow *C. trachomatis* to persist in individual hosts and in populations.

Study of the natural history of chlamydial infection in a group of commercial sex workers in Nairobi, Kenya, graphically illustrates the consequences of antigenic diversification as an immune evasion strategy for chlamydia (Brunham *et al.*, 1996). Women were followed for over 2 years at monthly intervals, with nearly 3000 chlamydial tests done and over 300 infections identified. One of the strongest risk factors for chlamydial infection acquisition was the length of time that the individual had been a prostitute, with a breakpoint occurring at 10 years, suggesting a strong time dependence on the acquisition of immunity. An obvious question to ask is, 'why does it take so long to acquire immunity to *C. trachomatis*?' in these highly exposed individuals. Part of the answer is the large number of strain variants to which these women are exposed. Over 18 different chlamydia strains were identified in this population of sex workers (Table 15.2). The specific strain composition of the chlamydia population also fluctuated widely from year to year.

Fluctuations in the relative abundance of each chlamydial strain appeared to be due to strain-specific herd immunity. Thus, when women who had infection with one strain were evaluated for risk of reinfection with that strain, the risk was reduced (Brunham *et al.*, 1996). In aggregate, the data demonstrate that immunity to *C. trachomatis* is strain specific and takes a long time to acquire, partly because of the large diversity of strains. Thus, antigenic variation of the MOMP appears to underlie the slowly acquired and imperfect nature of host immunity to *C. trachomatis*.

None-the-less, vaccine development for the prevention of *C. trachomatis* is a feasible goal. Belief in this assertion derives from the early trachoma vaccine trials (Collier, 1961; Sowa *et al.*, 1969; Grayston and Wang, 1978). Shortly after the initial isolation of *C. trachomatis* from cases of trachoma in the late 1950s, several large-scale trials using whole inactivated bacterial cells injected intramuscularly were evaluated in humans for the prevention of trachoma. The best of these trials confidently established the proof-of-principle that vaccine prevention of chlamydial disease was possible (Grayston and Wang, 1978) (Table 15.3). However, it was also clear that vaccine-induced immunity was partial and short-lived. The absence of knowledge regarding the immunologic and antigenic determinants for a successful chlamydial vaccine at that time constituted a significant barrier to further human trials. In order to understand more about the characteristics of whole inactivated cell vaccines, further trials were subsequently performed in primates. These trials revealed two important principles: the first was that the protection elicited by vaccination was specific to the homologous strain (Table 15.4); the second was that the whole chlamydial cell contains antigens that accelerate inflammatory responses when breakthrough infection occurs. Thus, these trials established that chlamydiae have

TABLE 15.2 DNA sequence analysis of *C. trachomatis omp I* genotypes among 170 infections in Pumwani commercial sex workers. Data are no. (%). *omp I* allelic composition of population of chlamydial strains significantly changed during observation period ($p = 0.005$) (modified from Brunham et al., 1996)

omp I genotype	1991 (n = 24)	1992 (n = 51)	1993 (n = 104)
Prototype			
A			2 (2)
D			1 (1)
E	6 (25)	22 (43)	58 (56)
F	3 (12)	2 (4)	6 (6)
G			1 (1)
H		1 (2)	
I			1 (1)
L_2			6 (6)
Minor variant			
C	1 (4)		
D			7 (7)
K	3 (12)	1 (2)	1 (1)
L_1	1 (4)		
L_2		11 (22)	7 (7)
Major variant			
I/H	3 (12)	2 (4)	2 (2)
L_1/L_2	4 (16)	8 (16)	5 (5)
L_2/L_1		1 (2)	
L_3/H	2 (8)		1 (1)
New variant M			3 (3)
Mixed infection	1 (4)	3 (6)	3 (3)

two classes of antigens: a type-specific antigen that elicits protective immunity, and a cross-reactive antigen that causes a tissue-damaging immune-mediated hypersensitivity reaction (Schachter and Caldwell, 1980).

Based on these data, it was concluded that a successful chlamydial vaccine would require a subunit design that contained only the protective antigen and excluded the hypersensitivity antigen. It was also noted that the vaccine needed to be based on a sounder understanding of the exact immune effector mechanisms that characterize chlamydial immunity.

The subsequent two decades of research have shown that the type-specific antigen that engenders immunity is the MOMP of chlamydiae. Similar data have also suggested that the hypersensitivity antigen may be the chlamydia hsp60 protein. The precise immunologic mechanism that underpins chlamydia immunity still remains to be definitively established, although the data summarized

TABLE 15.3 Results of field trials of trachoma vaccine in pre-school-aged children (modified from Grayston and Wang, 1978).

Place and vaccine	No. of Subjects	Percentage of conversion to trachoma during follow-up		
		1 Yr	2 Yr	3 Yr
Taiwan				
Placebo	193	10	16	49
Vaccine	169	4	8	47
Effectiveness		*66%*	*47%*	
India				
Placebo	88	37	48	
Vaccine	90	19	42	
Effectiveness		*50%*	*12%*	
Vaccine	92	10	28	
Effectiveness		*73%*	*42%*	

in previous sections have led to the notion that C. *trachomatis* immunity is mediated by mucosal IgA antibodies and CD4 T cells that secrete IFN-γ. Considerable effort has therefore been devoted to developing a chlamydial vaccine based on the MOMP. Protein, peptide and live recombinant microbial vectors containing MOMP sequences from either C. *trachomatis* or C. *psittaci* strains have been evaluated in primate, mice, sheep and guinea pig models of infection (Taylor *et al.*, 1988; Campos *et al.*, 1995; Batteiger *et al.*, 1993; Tan *et al.*, 1990; Jones *et al.*, 1995; Pal *et al.*, 1997; Tuffrey *et al.*, 1992; Su *et al.*, 1995; Murdin *et al.*, 1993, 1995; Hayes *et al.*, 1991). Sadly, these direct approaches have failed to elicit protective immunity in most instances. The reasons for MOMP vaccine failure were not precisely defined in any of the studies, but part of the problem could have been because the MOMP

TABLE 15.4 Effect of highly purified bour strain vaccines in prevention of experimental infection of the monkey eye (modified from Grayston and Wang, 1978)

Vaccine	Bour strain infection		Disease score
	Rate of infection		
Control	16/27	(59%)	4.41 ± 1.17
Homologous vaccine	10/48	(21%)	5.73 ± 1.61
Heterologous vaccine	7/11	(64%)	7.69 ± 2.58

immunogens lacked structural information necessary for antibody-mediated neutralization, or because of their failure to prime antigen-specific memory lymphocytes for recall during challenge infection. In support of this conjecture, the best results were reported when the immunogen was prepared in a manner to preserve the conformational structure of the MOMP (Batteiger *et al.*, 1993). As *C. trachomatis* immunity appears to be strongly dependent on CD4 T lymphocytes that express T_{H1}-type cytokines such as interferon-γ, induction of this type of immune response may depend on a particular priming process. It is known that induction of CD4 T_{H1} immune responses by subunit vaccines in the absence of strong adjuvants such as Freund's complete adjuvant is notoriously difficult. Thus, the MOMP constructs may have failed in part because they failed to induce protective T-lymphocyte effector and memory cells. Collectively, the experiments suggest that delivery of the MOMP immunogen in a manner that elicits both T_{H1}-type immune responses and IgA antibodies to conformational determinants of the MOMP will probably be essential for a protective *C. trachomatis* vaccine.

Because of the failures with protein and peptide MOMP vaccines, we evaluated a naked DNA vaccine approach. DNA vaccines are known to be remarkably efficacious at inducing protective immunity to intracellular pathogens, and can induce strong CD8 CTL and CD4 T_{H1} responses (Donnelly *et al.*, 1997). Mucosal and serum antibody responses have also been produced by DNA vaccination.

DNA vaccinology grew out of the gene therapy field following the initial observation of Wolff that intramuscular injection of a eukaryotic expression plasmid containing β-galactoside resulted in 1–5% of myocytes persistently expressing the foreign protein (Wolff *et al.*, 1990). Shortly thereafter, Merck scientists led by Dr M. Liu demonstrated that intramuscular injection of DNA expressing influenza nucleoprotein produced protective immunity in mice against lethal influenza infection (Ulmer *et al.*, 1993).

The immunologic basis for DNA vaccination remains incompletely resolved. Available data show that immune responses are restricted by the major histocompatibility complex (MHC) antigens of bone-marrow derived antigen-presenting cells (APC) (Donnelly *et al.*, 1997). DNA transfection of APCs is not necessary, and it appears that transfected somatic cells such as myocytes produce proteins that cross-prime APCs for the induction of MHC class I and class II cellular immune responses (Fu *et al.*, 1997). The released protein probably also stimulates B lymphocytes for antibody production.

An additional unexpected benefit of naked DNA immunization is the adjuvant effect of bacterial DNA *per se* (Sato *et al.*, 1996; Roman *et al.*, 1997). Short sequences of bacterial DNA containing CpG motifs are highly immunostimulatory. These sequences, which are 20 times more common in bacterial than in mammalian DNA, are capable of stimulating innate immune responses in NK and APC cells to release IFN-α, -β, and -γ. The cytokines produced in response to such sequences act as natural adjuvants and serve to augment the T_{H1} bias of naked DNA vaccines. However, IFN production may also adversely affect antibody responses by inhibiting antigen synthesis of transfected cells. As a consequence, DNA immunization typically results in strong T_{H1} and cytotoxic T-cell responses, with weak or variable antibody responses. Overall, DNA vaccines hold great promise as an immunization mode for intracellular pathogens such as *C. trachomatis*, because of their ability to induce strong T_{H1} immune responses, and thus we embarked on a detailed study of its effectiveness in a murine model of *C. trachomatis* mouse pneumonitis infection (Yang *et al.*, 1996).

We initially evaluated two genes, the gene for the MOMP and a gene for the cytoplasmic enzyme, CTP synthetase. The chlamydial genes were inserted into a eukaryotic expression plasmid, pcDNA3, and used to intramuscularly immunize groups of Balb/c mice (Zhang *et al.*, 1997).

The CTP synthetase construct was immunogenic, as it elicited antibodies to the recombinant protein, but it failed to induce DTH reactions to whole EBs and did not produce protective immunity to challenge infection. The MOMP gene produced distinctly different results. This construct was also immunogenic, producing antibodies that reacted to the MOMP by Western blot and, more importantly, that bound to the surface of EBs in an ELISA assay. The vaccine also elicited a DTH response and produced protective immunity to challenge infection (Figure 15.4).

Compared to blank vector immunization, MOMP DNA immunization reduced the peak growth of MoPn by more than 100-fold following lung challenge infection. These results show that DNA vaccination is immunogenic and protective, and that protective immunity is better elicited by a surface than a cytoplasmic protein gene from *C. trachomatis*.

We next characterized the cellular immune responses elicited by MOMP DNA vaccination and noted that both antigen-specific lymphoproliferative and IFN-γ cytokine secretion responses are reliably produced (Zhang *et al.*, 1998) (Figure 15.5).

Analysis of individual serum antibody responses was markedly different, with only one-half to two-thirds of the animals developing serum antibodies to the chlamydial MOMP (Figure 15.6). The finding that DNA vaccination preferentially induced IgG2a antibodies was consistent with the polarized induction of CD4 Th1 cells, as these cells secrete IFN-γ, which is known to act as a switch factor for IgG2a antibodies.

The reason for the discrepancy between the consistent induction of cellular immune responses and the variable antibody responses may relate to interference in antigen synthesis by transfected cells secondary to the adjuvant DNA sequences found both within the plasmid vector and the MOMP gene itself, as described previously.

Although these responses to MOMP DNA vaccination are promising, it seems likely that both antibody and cellular immune responses will be necessary for chlamydial immunity, and additional vaccine design features will be needed. A second-generation DNA vaccine may require a protein DNA conjugate exploiting the T-cell stimulation of MOMP DNA, the adjuvant property of the non-coding DNA, and the B-cell stimulation of MOMP protein. Lastly, induction of mucosal immunity will be essential to a successful chlamydial vaccine. How to deliver an optimal construct to the mucosal structure in a fashion that stimulates IgA class switching and T_{H1} cell induction at the mucosal level also requires further research.

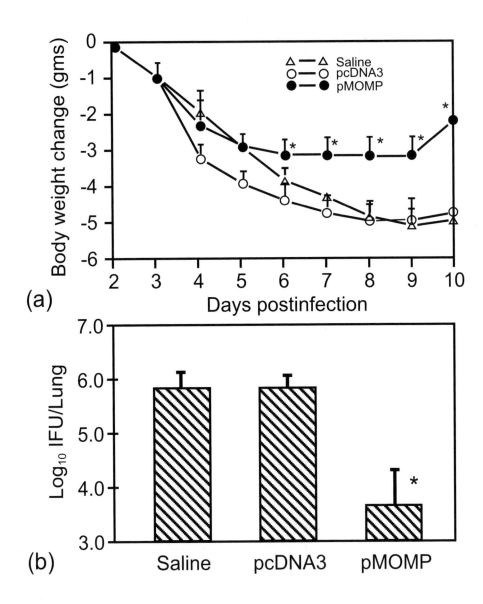

FIGURE 15.4 DNA vaccination with the MOMP gene enhanced lung clearance of *Chlamydia trachomatis* mouse pneumonitis (MoPn) infection. (a) The body weight of mice measured daily after challenge infection. (b) MoPn growth in lung analyzed by quantitative tissue culture at postinfection day 10. * means $P < 0.05$ (from Zhang et al., 1997).

Chlamydial Genomics and Its Vaccine Implications

Clearly, immunization with nucleic acids offers conceptual and practical advantages over whole killed, live attenuated, peptide and protein-based vaccines. The nucleic acid sequence of the plasmid insert can be manipulated to present all or

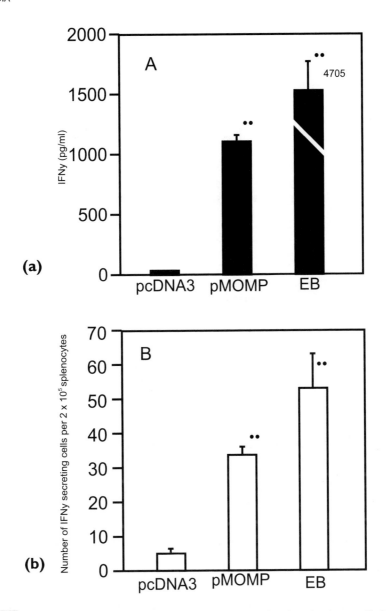

FIGURE 15.5 IFN-γ production by antigen-stimulated spleen cells isolated at day 60 following MOMP DNA vaccination. (a) The amount of IFN-γ secreted into the supernatant of spleen cell cultures as measured by ELISA. (b) The number of IFN-γ-producing cells per 2 × 10⁵ spleen cells, tested by ELISPOT.

part of the genome of interest, thereby tailoring the immune response to the pathogen. Also, genes that lead to undesired immunological inhibition or cross-reactivity, such as the chlamydial hsp60, can be deleted altogether. To take full advantage of the power of DNA vaccination, it is helpful to know the complete array of virulence genes and potential immunogens encoded by a particular pathogen (Moxon, 1997). Identification and characterization of chlamydia

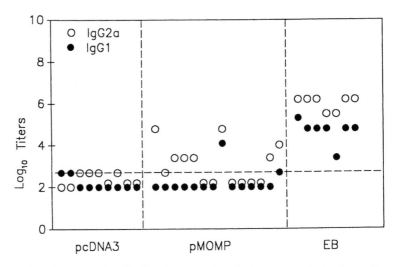

FIGURE 15.6 The profile of IgG subclass antibody to *Chlamydia trachomatis* mouse pneumonitis elementary bodies in the day 60 serum of MOMP DNA-immunized mice.

virulence factors has been limited by the fact that, to date, there is no system for genetic transformation of this organism. Furthermore, chlamydia's obligate intracellular mode of growth has made it difficult to prepare sufficient quantities of organism to allow for the purification and subsequent *in vitro* testing of putative virulence factors. The quantitative dominance of MOMP in the outer membrane of chlamydia has diverted the attention of investigators from searching for alternative, potentially protective cell surface antigens.

The recently completed nucleotide sequencing of the entire *C. trachomatis* serovar D genome provides a complete catalogue of the genes for every virulence factor and potential immunogen from which to select potential vaccine candidates ('Chlamydia Genome Project', R.S. Stephens, S. Kalman, C. Fenner, R. Davis). Genomic analysis indicates that the chlamydial genome contains at least 920 putative open reading frames (ORFs). In keeping with observations from other bacterial genome sequencing projects, approximately 60% of the ORFs can be assigned a specific biological function, whereas the remaining 40% remain unassigned. Although a significant percentage of the unassigned ORFs show homology to hypothetical open reading frames in the public databases, many of them probably encode proteins unique to chlamydia. As additional sequences are deposited in the databases, especially from other facultatively intracellular bacteria (*Mycobacteria*, *Francisella*, *Legionella* and *Salmonella*) and other chlamydiae (*C. trachomatis* serovar L2, biovar mouse pneumonitis and *C. pneumoniae*), comparative analysis will reveal which ORFs are specific to chlamydia. Many of the chlamydial-specific ORFs may prove useful as vaccine candidates.

By comparing the chlamydial genome sequence against sequences deposited in the public databases it is possible to identify homologues for putative virulence factors and potential cell surface immunogens. The chlamydial homologues so identified can be readily tested as potential vaccine candidates using DNA-based immunization in the murine model of *C. trachomatis* mouse pneumonitis infection. Promising candidates can subsequently be tested in primate models of *C. trachomatis* infection, which more closely mimics human disease pathogenesis. Through comparative genomic analysis and subsequent experimental testing, it may be possible to identify previously unrecognized cell surface antigens that are protective and conserved throughout the *C. trachomatis* species, and thus able to engender species- rather than serovar-specific immunity.

A survey of the annotated chlamydial genome sequences suggests that this approach is feasible. In addition to MOMP, there are approximately 20 ORFs which have been putatively assigned to the outer membrane, and which therefore may well be surface exposed (Table 15.5). Most intriguing is a series of nine putative outer membrane proteins, pmpA to pmpl. These are clustered in three separate physical locations on the *C. trachomatis* chromosome, and in total encompass approximately 30 kb of sequence, about 3% of the total genome. The pmps share sequence homology with previously identified *C. psittaci* immunogens. Another potentially significant locus, *omp*B, encodes a putative outer membrane protein that shows 33% homology to the chlamydial MOMP.

TABLE 15.5 Putative outer membrane protein genes based on the *Chlamydia trachomatis* serovar D genome sequencing project (from 'Chlamydia Genome Project' R.S. Stephens, S. Kalman, C. Lammel et al. (1998). 282: 754–759).

Contig.	Gene	Protein function
3.5.053	?	OMP85 homologue
10.4.656	pmpD	Putative membrane protein D
10.4.647	pmpD'	Putative membrane protein D
11.1.706	pmpE	Putative membrane protein E
11.2.707	pmpF	Putative membrane protein F
11.2.708	pmpG	Putative membrane protein G
11.2.709	pmpH	Putative membrane protein H
11.3.712	ompl	Putative membrane protein I
6.0.260	omcC	CHLTR 15 kDa cysteine-rich outer membrane protein
6.0.261	omcB	CHLTR 60 kDa cysteine-rich outer membrane protein
6.0.262	omcA	CHLTR 9 kDa lipoprotein
6.3.292	?	Possible membrane protein homologue
5.6.230	pmpA	Putative membrane protein A
5.6.231	pmpB	Putative membrane protein B
5.7.232	pmpC	Putative membrane protein C
5.7.234	?	Possible adhesin protein homologue
9.1.535	ompB	Outer membrane protein B
8.6.501	ompA	Major outer membrane protein (MOMP)

As ompB homologues from other chlamydia strains are sequenced, it will be interesting to see whether this MOMP homologue also contains antigenic variable regions. Although much work needs to be done, including confirmation of outer membrane localization/cell surface exposure and isolation, and sequencing of homologous sequences from other *C. trachomatis* serovars, so that amino acid sequence variation can be assessed, this example highlights how rapidly one can move from genomic sequence analysis to testing of a vaccine by genetic immunization.

Genomic sequence analysis has also shown that chlamydia contain homologues of several of the proteins encoded by the *Yersinia* YOP virulon (Mecsas and Strauss, 1996). This is an integrated system allowing extracellular bacteria to communicate with the host cell's cytosol by injection of effector proteins. It is composed of four components: a contact or type III secretion system, a delivery system, a control element, and finally a set of effector YOP proteins. The first three essentially form a physical apparatus which allows for the regulated passage of the effector molecules from *Yersinia* to the host cell cytoplasm. It seems likely that with chlamydia a similar system would allow for communication between intracellular RBs and the host cell cytoplasm. It is possible that the type III secretion apparatus represents the ultrastructurally observed spike-like surface projections that have been speculated to form a channel between RBs, which are typically in tight contact with the luminal surface of the inclusion, and the host cell cytoplasm. Of significance to vaccine development, chlamydial effector proteins which are injected into the host cell cytoplasm would be available to enter the MHC class I antigen processing and presentation pathway. In the case of *Yersinia*, of the four effector proteins that have been identified, one (YpkA) encodes a serine/threonine kinase that shows noticeable sequence similarity to eukaryotic counterparts. Interestingly, chlamydia genome analysis reveals three open reading frames that show sequence homology to the catalytic domain of established serine/threonine kinases. Although it is speculative, it is possible that the serine/threonine kinase orthologues are chlamydial effector proteins that interfere with a signal transduction pathway of the eukaryotic cell. Following conclusive identification of such chlamydial effector proteins it may also prove possible to use mutated inactive versions in a DNA-based vaccine to enhance the host's CTL response to chlamydia-infected cells.

In the longer term, genomics could lead to the establishment of a live attenuated chlamydial vaccine strain. In the past, the isolation of an attenuated chlamydial strain has been limited by two factors: the lack of a system for genetic transformation, and the limited knowledge regarding chlamydial virulence factors. As mentioned above, comparative genomic analysis can be used to identify putative virulence factors. Clearly, to firmly establish the role of potential virulence genes it will be necessary to have a system enabling genetic manipulation of chlamydia. Genomics has also aided this cause. Sequence analysis indicates that chlamydia encode the DNA recombination/repair enzymes required for homologous recombination/genetic exchange. In addition, genome analysis indicates that there are several genes that are duplicated in the genome, such as heat shock protein genes. Even if the products encoded by these genes are essential, it may be possible to disrupt one copy and maintain viability. The development of a genetic system for chlamydia is aided by the establishment of genes that can potentially be targeted for disruption. With the development of a genetic system it will be possible to approach attenuation at the genome level, not only by identifying and deleting the minimal set of genes required for virulence, but also by deleting undesirable genes that lead to immunopathology.

CONCLUSION

The goal of developing a chlamydial vaccine is closer than it has ever been. A key discovery was the need to base a vaccine on a subunit design. The identification of the MOMP as the principal target for protective immunity and the delivery of MOMP as naked DNA provided solutions to long-standing problems of how best to induce protective immunity with a MOMP subunit design. Further work is needed to optimize the vaccine for the induction of both antibody and cellular immunity at the mucosal sites of infection, and to overcome the problem of antigenic diversity of the MOMP. Information just be coming available from the chlamydia genome projects will surely advance vaccine development and open new avenues of investigation.

REFERENCES

Baehr, W., Zhang, Y-X., Joseph, T. et al. (1988). Mapping antigenic domains expressed by Chlamydia trachomatis major outer membrane protein genes. Proc. Natl. Acad. Sci. USA 85: 4000–4004

Bailey, R.L., Holland, M.J., Whittle, H.C. and Mabey, D.C.W. (1995). Subjects recovering from human ocular chlamydial infection have enhanced lymphoproliferative responses to chlamydial antigens compared with those of persistently diseased controls. Infect. Immun. 63: 389–392

Barry III, C.E., Hayes, S.F. and Hackstadt, T. (1992). Nucleoid condensation in Escherichia coli that express a chlamydial histone homolog. Science 256: 377–379

Batteiger, B.E., Rank, R.G., Bavoil, P.M. and Soderberg, L.S.F. (1993). Partial protection against genital reinfection by immunization of guinea-pigs with isolated outer-membrane proteins of the chlamydial agent of guinea-pig inclusion conjunctivitis. J. Gen. Microbiol. 139: 2965–2972

Beatty, P.R. and Stephens, R.S. (1994). CD8+ T lymphocyte-mediated lysis of Chlamydia-infected L cells using an endogenous antigen pathway. J. Immunol. 153: 4588–4595

Beem, M.O. and Saxon, E.M. (1977). Respiratory-tract colonization and a distinctive pneumonia syndrome in infants infected with Chlamydia trachomatis. N. Engl. J. Med. 296: 306–310

Berger, R.E., Alexander, E.R., Monda, G.D., Ansell, J., McCormick, G. and Holmes, K.K. (1978). Chlamydia trachomatis as a cause of acute 'idiopathic' epididymitis. N. Engl. J. Med. 298: 301–304

Bernstein, D.I., Hubbard, T., Wenman, W.M. et al. (1984). Mediastinal and supraclavicular lymphadenitis and pneumonitis due to Chlamydia trachomatis serovars L1 and L2. N. Engl. J. Med. 311: 1543–1546

Birkelund, S. and Stephens, R.S. (1992). Construction of physical and genetic maps of Chlamydia trachomatis serovar L2 by pulsed-field gel electrophoresis. J. Bacteriol. 174: 2742–2747

Brunham, R.C., Kimani, J., Bwayo, J. et al. (1996). The epidemiology of Chlamydia trachomatis within a sexually transmitted diseases core group. J. Infect. Dis. 173: 950–956

Brunham, R.C., Kuo, C-C., Cles, L. and Holmes, K.K. (1983). Correlation of host immune responses with quantitative recovery of Chlamydia trachomatis from the human endocervix. Infect. Immun. 39: 1491–1494

Brunham, R.C., Martin, D.H., Kuo, C-C. et al. (1981). Cellular immune response during uncomplicated genital infection with Chlamydia trachomatis in humans. Infect. Immun. 34: 98–104

Brunham, R.C., Paavonen, J., Stevens, C.E. *et al.* (1984). Mucopurulent cervicitis – The ignored counterpart in women of urethritis in men. *N. Engl. J. Med.* **311**: 1–6

Brunham, R.C. and Peeling, R.W. (1994). *Chlamydia trachomatis* antigens: Role in immunity and pathogenesis. *Infect. Agents Dis.* **3**: 218–233

Brunham, R.C., Plummer, F.A. and Stephens, R.S. (1993). Bacterial antigenic variation, host immune response, and pathogen-host coevolution. *Infect. Immun.* **61**: 2273–2276

Caldwell, H.D., Kromhout, J. and Schachter, J. (1981). Purification and partial characterization of the major outer membrane protein of *Chlamydia trachomatis*. *Infect. Immun.* **31**: 1161–1176

Campos, M., Pal, S., O'Brien, T.P. *et al.* (1995). A chlamydial major outer membrane protein extract as a trachoma vaccine candidate. *Invest. Ophthalmol. Vis. Sci.* **36**: 1477–1491

Collier, L.H. (1961). Experiments with trachoma vaccines. *Lancet.* April, 795–800

de la Maza, M.A. and de la Maza, L.M. (1995). A new computer model for estimating the impact of vaccination protocols and its application to the study of *Chlamydia trachomatis* genital infections. *Vaccine* **13**: 119–127

Donnelly, J.J., Ulmer, J.B., Shiver, J.W. and Liu, M.A. (1997). DNA vaccines. *Ann. Rev. Immunol.* **15**: 617–648

El-Asrar, A.M.A., van den Oord, J.J., Geboes, K., Missotten, L., Emarah, M.H. and Desmet, V. (1989). Immunopathology of trachomatous conjunctivitis. *Br. J. Ophthalmol.* **73**: 276–282

Fan, T., Lu, H., Hu, H. *et al.* (1998). Inhibition of apoptosis in chlamydia-infected cells: Blockade of mitochondrial cytochrome c release and caspase activation. *J. Exp. Med.* **187**: 487–496

Fu, T-M., Ulmer, J.B., Caulfield, M.J. *et al.* (1997). Priming of cytotoxic T lymphocytes by DNA vaccines: Requirement for professional antigen presenting cells and evidence for antigen transfer from myocytes. *Molec. Med.* **3**: 362–371

Grayston, J.T. and Wang, S-P. (1978). The potential for vaccine against infection of the genital tract with *Chlamydia trachomatis*. *Sex. Transm. Dis.* **5**: 73–77

Grayston, J.T., Wang, S.-P., Yeh, L.-J. and Kuo, C.-C. (1985). Importance of reinfection in the pathogenesis of trachoma. *Rev. Infect. Dis.* **7**: 717–725

Grayston, J.T. and Wang, W.-P. (1975). New knowledge of chlamydiae and the diseases they cause. *J. Infect. Dis.* **132**: 87–105

Hackstadt, T., Fischer, E.R., Scidmore, M.A., Rockey, D.D. and Heinzen, R.A. (1997). Origins and functions of the chlamydial inclusion. *Trends Microbiol.* **5**: 288–293

Harrison, H.R., English, M.G., Lee, C.K. and Alexander, E.R. (1978). *Chlamydia trachomatis* infant pneumonitis: Comparison with matched controls and other infant pneumonitis. *N. Engl. J. Med.* **298**: 702–708

Hatch, T.P., Al-Hossainy, E. and Silverman, J.A. (1982). Adenine nucleotide and lysine transport in *Chlamydia psittaci*. *J. Bacteriol.* **150**: 662–670

Hatch, T.P., Allan, I. and Pearce, J.H. (1984). Structural and polypeptide differences between envelopes of infective and reproductive life cycle forms of *Chlamydia spp*. *J. Bacteriol.* **157**: 13–20

Hayes, L.J., Conlan, J.W., Everson, J.S., Ward, M.E. and Clark, I.N. (1991). *Chlamydia trachomatis* major outer membrane protein epitopes expressed as fusions with LamB in an attenuated *aroA* strain of *Salmonella typhimurium*; their application as potential immunogens. *J. Gen. Microbiol.* **137**: 1557–1564

Hillis, S.D. and Wasserheit, J.N. (1996). Screening for chlamydia – A key to the prevention of pelvic inflammatory disease. *N. Engl. J. Med.* **334**: 1399–1401

Holland, M.J., Bailey, R.L., Conway, D.J. *et al.* (1996). T helper type-1 (Th1)/Th2 profiles of peripheral blood mononuclear cells (PBMC); responses to antigens of *Chlamydia trachomatis* in subjects with severe trachomatous scarring. *Clin. Exp. Immunol.* **105**: 429–435

Holland, M.J., Bailey, R.L., Hayes, L.J., Whittle, H.C. and Mabey, D.C.W. (1993). Conjunctival scarring in trachoma is associated with depressed cell-mediated immune responses to chlamydial antigens. *J. Infect. Dis.* **168**: 1528–1531

Holmes, K.K., Handsfield, H.H., Wang, S-P. *et al.* (1975). Etiology of nongonococcal urethritis. *N. Engl. J. Med.* **292**: 1199–1205

Ingallis, R.R., Rice, P.A., Qureshi, N., Takayama, K., Lin, J.S. and Golenbock, D.T. (1995). The inflammatory cytokine response to *Chlamydia trachomatis* infection is endotoxin mediated. *Infect. Immun.* **63**: 3125–3130

Jones, B.R. (1975). The prevention of blindness from trachoma. *Trans. Ophthalmic Soc.* UK 95: 16–33

Jones, G.E., Jones, K.A., Machell, J., Brebner, J., Anderson, I.E. and How, S. (1995). Efficacy trials with tissue-culture grown, inactivated vaccines against chlamydial abortion in sheep. *Vaccine* 13: 715–723

Kimani, J., Maclean, I.W., Bwayo, J.J. *et al.* (1996). Risk factors for *Chlamydia trachomatis* pelvic inflammatory disease among sex workers in Nairobi, Kenya. *J. Infect. Dis.* **173**: 1437–1444

Kiviat, N.B., Paavonen, J.A., Brockway, J. *et al.* (1985). Cytologic manifestations of cervical and vaginal infections. *JAMA* **253**: 989–996

Kunimoto, D. and Brunham. R.C. (1985). Human immune response and *Chlamydia trachomatis* infection. *Rev. Infect. Dis.* **7**: 665–673

Laga, M., Plummer, F.A., Piot, P. *et al.* (1988). Prophylaxis of gonogoccal and chlamydial ophthalmia neonatorum: A comparison of silver nitrate and tetracycline. *N. Engl. J. Med.* **318**: 653–657

Mårdh, P-A., Ripa, K.T., Svensson, L. and Weström, L. (1976). *Chlamydia trachomatis* infection in patients with acute salpingitis. *N. Engl. J. Med.* **296**: 1377–1379

Mabey, D. and Bailey, R. (1996). Immunity to *Chlamydia trachomatis*: Lessons from a Gambian village. *J. Med. Microbiol.* **45**: 1–2

Maclean, I.W., Peeling, R.W. and Brunham, R.C. (1988). Characterization of *Chlamydia trachomatis* antigens with monoclonal and polyclonal antibodies. *Can. J. Microbiol.* **34**: 141–147

Martin, D.H., Pollock, S., Kuo, C-C., Wang, S-P., Brunham, R.C. and Holmes, K.K. (1984). *Chlamydia trachomatis* infections in men with Reiter's syndrome. *Ann. Intern. Med.* **100**: 207–213

McClarty, G. (1994). Chlamydiae and the biochemistry of intracellular parasitism. *Trends Microbiol.* **2**: 157–164

Mecsas, J. and Strauss, E.J. (1996). Molecular mechanisms of bacterial virulence: Type III secretion and pathogenicity islands. *Emerg. Infect. Dis.* **2**: 271–288

Morrison, R.P., Manning, D.S. and Caldwell, H.D. (1992). In: 'Sexually Transmitted Diseases' (ed. T.C. Quinn), Immunology of *Chlamydia trachomatis* infection: Immunoprotective and immunopathogenetic responses, pp.57–84. Raven Press, New York

Moulder, J.W. (1991). Interaction of chlamydiae and host cells in vitro. *Microbiol. Rev.* **55**: 143–190

Moxon, E.R. (1997). Applications of molecular microbiology to vaccinology. *Lancet* 350: 1240–1244

Murdin, A.D., Su, H., Klein, M.H. and Caldwell, H.D. (1995) Poliovirus hybrids expressing neutralization epitopes from variable domains I and IV of the major outer membrane protein of *Chlamydia trachomatis* elicit broadly cross-reactive *C. trachomatis*-neutralizing antibodies. *Infect. Immun.* **63**: 1116–1121

Murdin, A.D., Su, H., Manning, D.S., Klein, M.H., Parnell, M.J. and Caldwell, H.D. (1993). A poliovirus hybrid expressing a neutralization epitope from the major outer membrane protein of *Chlamydia trachomatis* is highly immunogenic. *Infect. Immun.* **61**: 4406–4414

Pal, S., Theodor, I., Peterson, E.M. and de la Maza, L.M. (1997). (Immunization with an acellular vaccine consisting of the outer membrane complex of *Chlamydia trachomatis* induces protection against a genital challenge. *Infect. Immun.* **65**: 3361–3369

Pearce, J.H. (1983). Immune responses and chlamydial infections. *Br. Med. Bull.* **39**: 187–193

Plummer, F.A., Simonsen, J.N., Cameron, D.W. *et al.* (1991). Cofactors in male-female sexual transmission of human immunodeficiency virus type 1. *J. Infect. Dis.* **163**: 233–239

Quinn, T.C., Goodell, S.E., Mkrtichian, E. *et al.* (1981). *Chlamydia trachomatis* proctitis. *N. Engl. J. Med.* **305**: 195–200

Reacher, M.H., Pe'er, J., Rapoza, P.A., Whittum-Hudson, J.A. and Taylor, H.R. (1991). *Ophthalmology* 98: 334–341

Reynolds, D.J. and Pearce, J.H. (1991). Endocytic mechanisms utilized by chlamydiae and their influence on induction of productive infection. *Infect. Immun.* **59**: 3033–3039

Roman, M., Martin-Orozco, E., Goodman, J.S. *et al.* (1997). Immunostimulatory DNA sequences function as T helper-1-promoting adjuvants. *Nature Med.* **3**: 849–854

Sato, Y., Roman, M., Tighe, H. *et al.* (1996). Immunostimulatory DNA sequences necessary for effective intradermal gene immunization. *Science* 273: 352–354

Schachter, J. (1978). Chlamydial infections. *N. Engl. J. Med.* **298**: 428–435

Schachter, J. and Caldwell, H.D. (1980). Chlamydiae. *Annu. Rev. Microbiol.* **34**: 285–309

Scholes, D., Stergachis, A., Heidrich, F.E., Andrilla, H., Holmes, K.K. and Stamm, W.E. (1996). Prevention of pelvic inflammatory disease by screening for cervical chlamydial infection. *N. Engl. J. Med.* **334**: 1362–1366

Sowa, S., Sowa, J., Collier, L.H. and Blyth W.A. (1969). Trachoma vaccine field trials in The Gambia. *J. Hyg. Camb.* **67**: 699–717

Starnbach, M.N., Bevan, M.J. and Lampe, M.F. (1994). Protective cytotoxic T lymphocytes are induced during murine infection with *Chlamydia trachomatis*. *J. Immunol.* **153**: 5183–5189

Starnbach, M.N., Bevan, M.J. and Lampe, M.F. (1995). Murine cytotoxic T lymphocytes induced following *Chlamydia trachomatis* intraperitoneal or genital tract infection respond to cells infected with multiple serovars. *Infect. Immun.* **63**: 3527–3530

Stephens, R.S., Mullenbach, G., Sanchez-Pescador, R. and Agabian, N. (1986). Sequence analysis of the major outer membrane protein gene from *Chlamydia trachomatis* serovar L$_2$. *J. Bacteriol.* **168**: 1277–1282

Stephens, R.S., Sanchez-Pescador, R., Wagar, E.A., Inouye, C. and Urdea, M.S. (1987). Diversity of *Chlamydia trachomatis* major outer membrane protein genes. *J. Bacteriol.* **169**: 3879–3885

Su, H., Parnell, M. and Caldwell, H.D. (1995). Protective efficacy of a parenterally administered MOMP-derived synthetic oligopeptide vaccine in a murine model of *Chlamydia trachomatis* genital tract infection: Serum neutralizing IgG antibodies do not protect against chlamydial genital tract infection. *Vaccine* **13**: 1023–1032

Tan, T-W., Herring, A.J., Anderson, I.E. and Jones, G.E. (1990). Protection of sheep against *Chlamydia psittaci* infection with a subcellular vaccine containing the major outer membrane protein. *Infect. Immun.* **58**: 3101–3108

Taraska, T., Ward, D.M., Ajioka, R.S. *et al.* (1996). The late chlamydial inclusion membrane is not derived from the endocytic pathway and is relatively deficient in host proteins. *Infect. Immun.* **64**: 3713–3727

Taylor, H.R., Whittum-Hudson, J., Schachter, J., Caldwell. H.D. and Prendergast, R.A. (1988). Oral immunization with chlamydial major outer membrane protein (MOMP). *Invest. Ophthalmol. Vis. Sci.* **29**: 1847–1853

Tipples, G. and McClarty, G. (1993). The obligate intracellular bacterium *Chlamydia trachomatis* is auxotrophic for three of the four ribonucleoside triphosphates. *Molec. Microbiol.* **8**: 1105–1114

Tuffrey, M., Alexander, F., Conlan, W., Woods, C. and Ward, M. (1992). Heterotypic protection of mice against chlamydial salpingitis and colonization of the lower genital tract with a human serovar F isolate of *Chlamydia trachomatis* by prior immunization with recombinant serovar L1 major outer-membrane protein. *J. Gen. Microbiol.* **138**: 1707–1715

Ulmer, J.B., Donnelly, J.J., Parker, S.E. *et al.* (1993). Heterologous protection against influenza by injection of DNA encoding a viral protein. *Science* **259**: 1745–1749

Ward, M.E. (1995). The immunobiology and immunopathology of chlamydial infections. *APMIS* **103**: 769–796

West, S.K., Muñoz, B., Lynch, M., Kayongoya, A., Mmbaga, B.B.O. and Taylor, H.R. (1996). Risk factors for constant, severe trachoma among preschool children in Kongwa, Tanzania. *Am. J. Epidemiol.* **143**: 73–78

Wolff, J.A., Malone, R.W., Williams, P. *et al.* (1990). Direct gene transfer into mouse muscle in vivo. *Science* **247**: 1465–1468

Wølner-Hanssen, P., Weström, L. and Mårdh, P-A. (1980). Perihepatitis and chlamydial salpingitis. *Lancet* April, 901–903

Woolhouse, M.E.J., Dye, C., Etard, J-F. *et al.* (1997). Heterogeneities in the transmission of infectious agents: Implications for the design control programs. *Proc. Natl. Acad. Sci. USA* **94**, 338–342

Yang, X., HayGlass, K.T. and Brunham, R.C. (1996). Genetically determined differences in IL-10 and IFN-γ responses correlate with clearance of *Chlamydia trachomatis* mouse pneumonitis infection. *J. Immunol.* **156**: 4338–4344

Zhang, D-J., Yang, X., Berry, J., Shen, C., McClarty, G. and Brunham, R.C. (1997). DNA vaccination with the major outer-membrane protein gene induces acquired immunity to *Chlamydia trachomatis* (mouse pneumonitis) infection. *J. Infect. Dis.* **176**: 1035–1040

Zhang, D.J., Yang, X., Shen, C., Brunham, R.C. (1998). Characterization of immune responses following intramuscular DNA immunization with the MOMP gene of *Chlamydia trachomatis* mouse pneumonitis strain. *Immunology* **96**: 314–321

Zhang, J.P. and Stephens, R.S. (1992). Mechanism of C. *trachomatis* attachment to eukaryotic host cells. *Cell* **69**: 861–869

Zhang, Y-X., Fox, J.G., Ho, Y., Zhang, L., Stills Jr, H.F. and Smith, T.F. (1993). Comparison of the major outer-membrane protein (MOMP) gene of mouse pneumonitis (MoPn) and hamster SFDP strains of *Chlamydia trachomatis* with other *Chlamydia* strains. *Mol. Biol. Evol.* **10**: 1327–1342

Zhong, D-J. (1998). Intracellular DNA immunization with the MOMP gene of *Chlamydia trachomatis* evokes a dominant Th1 but variable humoral immune response. *Immunology* (in press)

16

GONORRHEA: EPIDEMIOLOGY, CONTROL AND PREVENTION

JONATHAN M. ZENILMAN*
AND
CAROLYN D. DEAL†

** Johns Hopkins University School of Medicine,
Baltimore, Maryland, USA*

*† Center for Biologics Evaluation and Research,
Food and Drug Administration,
Bethesda, Maryland, USA*

MICROBIOLOGY OF *NEISSERIA GONORRHOEAE*

Neisseria gonorrhoeae is an aerobic Gram-negative coccus. Members of the genus *Neisseria* inhabit the mucous membranes of humans and other animals. Some species, such as *N. lactamica*, *N. sicca* and *N. subflava*, can be found in the normal flora in the nasopharynx and are considered non-pathogenic. *N. gonorrhoeae* and *N. meningitidis*, which cause gonorrhea and meningitis, are the primary pathogenic species and infect only humans. Normally an aerobe, the organism can grow under anaerobic conditions when supplied with a terminal electron acceptor such as nitrite (Knapp and Clark, 1984).

Morphologically, the bacteria appear as Gram-negative kidney bean-shaped diplococci. In direct smears of urethral or endocervical exudates the organisms are usually located within polymorphonuclear leukocytes (PMNs), though they can also appear extracellularly. When they are cultured on solid agar, different colony types are observed. Freshly isolated strains form small raised glistening colonies that reflect incident light. During non-selective transfer in the laboratory, the strains revert to a colony type that is larger, slightly convex, and does not reflect incident light. Different colony types were found to vary in virulence in a male intraurethral challenge model (Kellogg *et al.*, 1963). Antibody to the pilin protein can be found in the sera and vaginal fluid of convalescent patients (Lammel *et al.*, 1985). These observations formed the basis for selecting pilin as one of the first vaccine candidates.

Structurally, the gonococci are similar to other Gram-negative bacteria. The cell envelope is composed of a cytoplasmic and an outer membrane, separated by the periplasmic space. The outer membrane consists of proteins, lipo-oligosaccharide (LOS) and phospholipids. Several of the outer membrane proteins have been studied in great detail. Subunits of pilin, plus minor pilus-associated proteins, form the pili, a long filamentous structure extending from the cell surface that mediates attachment to eukaryotic cells. The Por protein is an integral outer membrane protein present on all strains of gonococci. It aggregates in the outer membrane with Rmp, a reduction-modifiable protein, to form a channel which functions as a porin. The opacity proteins, Opa, which are variably expressed on gonococcal strains, account for different colony phenotypes. The cell surface location of these proteins has made them attractive vaccine candidates from a microbiologist's point of view.

Although antibodies to gonococcal antigens have been found in convalescing patient sera (Lammel *et al.*, 1985), to date no antigen has been clearly linked to a protective immune response. Clinically, it is not unusual to see patients presenting with multiple repeat infections. This clinical picture has led many investigators to conclude that there is no, or only a short-lived, natural immunity to gonococcal infection. Only recently have studies in a cohort of commercial sex workers by Plummer *et al.* (1994) suggested that limited immunity can develop after repeated exposure to infection. Because of this limited understanding of the human immune response to gonococcal infection, the major criteria for selection of vaccine candidates has been the cellular architecture of the bacteria.

CLINICAL DISEASE CAUSED BY *NEISSERIA GONORRHOEAE*

Gonococci preferentially invade columnar epithelial surfaces. These correlate with the characteristic infection sites: the urethra, endocervix, rectal mucosa and conjunctiva. At these sites the gonococcus elicits an inflammatory response characterized by its rapidity and a predominance of polymorphonuclear leukocytes (PMNs). Mucosal infection is often characterized by a purulent discharge, with potential direct invasion and damage caused to surrounding areas by the inflammatory infiltrate and the associated cytokine response. Gonorrhea has therefore been classified as one of the exudative sexually transmitted diseases. In unusual circumstances local infection can disseminate systemically and cause septicemia and its attendant complications.

Gonococcal Infection in Women

In women, gonococcal cervicitis is characterized by a purulent exudate and cervical edema, which can usually be detected by vaginal speculum examination (McCormack *et al.*, 1977a; Curran *et al.*, 1975). Non-specific symptoms and signs, such as increased vaginal discharge, abnormal menses (increased flow or dysmenorrhea, dyspareunia) or dysuria, are common. Dysuria may represent coexisting urethral infection. Over half of women with gonococcal cervicitis are entirely asymptomatic. Local complications, which are occasionally observed, include infection of Bartholins' or Skenes' glands, and urethritis, which can often be confused with the lower urinary tract infections typically seen in women. Although clinical data are sparse, most experts believe women with an asymptomatic infection can be culture positive as long as 3–6 months after the initial infection.

The differential diagnosis of cervicitis includes chlamydial infection, herpes simplex virus infection, vaginal infections such as trichomoniasis, and muco-purulent cervicitis with uncharacterized etiology. Diagnosis of cervical infection is made by direct culture of cervical secretions using selective media. Newer DNA amplifications methods are also useful and can be used.

Women with untreated gonococcal cervicitis can develop upper tract infection, pelvic inflammatory disease (PID) (Platt *et al.*, 1983; Wasserheit *et al.*, 1986; McCormack *et al.*, 1977b; Bowie and Jones, 1981). Semantically, PID represents infection of the soft tissue upper genital tract structures, and includes endometritis, salpingitis, oophoritis and pelvic peritonitis. Fatalities from PID are rare. However, acute severe PID may lead to tubo-ovarian abscess and the necessity, at times, for hysterectomy. The chronic complications are more common. Women with a previous history of PID are more likely to have tubal factor infertility and are at higher risk for ectopic pregnancy (Washington *et al.*, 1991; Hillis *et al.*, 1993; Westrom, 1975). The overall direct economic impact of PID alone was estimated to be $4.2 billion in 1990 (Washington and Katz, 1991).

Estimates of PID incidence vary, and are made difficult by the lack of prospective studies. Early studies of penicillinase-producing *Neisseria gonorrhoeae* (PPNG), in which treatment with penicillin was largely ineffective, suggest that PID occurs in approximately 30% of women with untreated gonorrhea (Jaffe *et al.*, 1981). In the US, approximately 200 000 cases were estimated to have required hospitalization in 1988, with a total burden of cases estimated to be more than 500 000 (Rolfs *et al.*, 1992).

Women with ascending gonococcal infection or PID often have symptoms during their menses, when cervical defenses effective in preventing the ascent of *Neisseria gonorrhoeae* are compromised. For example, the cervical mucus plug has disintegrated and the vaginal pH rises above 4.5. In addition there is necrotic tissue present in the cervical canal, a perfect setting for bacterial growth. With the breakdown in local defenses, STD organisms (*Neisseria gonorrhoeae* and *Chlamydia trachomatis*) and commensal organisms, including Gram-negative rods, coccobacilli, Gram-positive cocci and anerobes, gain access to the upper tract and can establish an inflammatory infectious process.

Gonococcal Infection in Men

In men, urethritis is the most common syndrome (Sherrard and Barlow, 1996). Discharge or dysuria usually appears within 1 week of exposure, although as many as 5–10% of patients never have any signs or symptoms (McNagny *et al.*, 1992). The discharge is characteristically purulent, and Gram-negative intracellular diplococci can be easily seen in Gram stain of the exudate (Figure 16.1). Inguinal lymphadenopathy is often present, and occasionally a frank lymphangitis can develop on the penile shaft and corona. Asymptomatic disease can exist in men up to several weeks after infection. There is some controversy as to whether asymptomatic disease represents presymptomatic disease or true asymptomatic infection. Nevertheless, the organism is potentially transmissible to sexual partners. The differential diagnosis of gonococcal urethritis is chlamydial urethritis or non-gonococcal urethritis (NGU) due to non-chlamydial

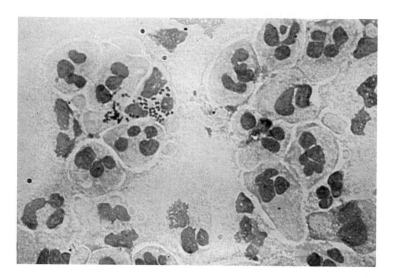

FIGURE 16.1 Gonococcal urethritis: Gram-stained slide of an urethral exudate from a 33-year-old male with gonococcal urethritis. The slide consists of a cellular component of polymorphonuclear leukocytes with intracellular Gram-negative diplococci characteristic of *Neisseria gonorrhoeae*.

etiologies (Stamm *et al.*, 1995). NGU can be often differentiated from gono-coccal urethritis by Gram stain, which has extremely high sensitivity and specificity (>95%) (Jacobs and Kraus, 1975). However, rapid Gram stain microscopy is not available in most acute clinical settings, and so syndromic treatment is typically offered.

Symptomatic anorectal gonococcal disease occurs in men with a history of receptive rectal intercourse (Lebedeff and Hochman, 1980). Approximately 50% have symptoms, which include rectal pain, discharge, constipation and tenesmus. The differential diagnosis includes other acute rectal sexually trans-mitted infections, such as herpes simplex, chlamydia proctitis and syphilitic proctitis. Anorectal diseases can also occur in women with endocervical gonor-rhea and who have not necessarily had receptive rectal intercourse. In these cases infection is presumed to have occurred via tracking of secretions across the perineum. Indeed, up to 30% of such women often have coexistent rectal infec-tion, but it is usually asymptomatic.

Because rectal gonorrhea in men implies a history of unprotected rectal intercourse, surveillance of rectal gonorrhea has been useful as a surrogate marker for HIV risk in homosexual men. For example, early in the AIDS epidemic the rate of rectal gonorrhea declined (CDC, 1984). Similarly, troubling trends were reported in 1997 of an increase of gonorrhea in homo-sexual men, which correlated with observed increases in HIV-risk behavior among homosexual adolescents (CDC, 1997a).

Gonococcal pharyngitis (Wiesner *et al.*, 1973; Hutt and Judson, 1986) occurs in men or women after oral sexual exposure. The disease is clinically indistinguishable from any other bacterial pharyngitis, and is asymptomatic in as many as 60% of cases. Natural history studies have demonstrated that gonococcal pharyngitis is a self-limited syndrome and uninfected patients become culture negative after 4–6 weeks. Throat swabs should be routinely obtained in persons at risk.

Disseminated Gonococcal Infection

Disseminated gonococcal infection (DGI, gonococcal septicemia) occurs in approximately 0.1–0.5% of total gonococcal cases. DGI is more commonly seen in women, in particular pregnant women, and is also more common in persons with deficiencies of the terminal complement components (Peterson et al., 1979).

DGI (Wise et al., 1994; O'Brien et al., 1983) is characterized by constitutional symptoms, fevers, chills, and a distinct set of syndromes which consist of oligoarticular septic arthritis, tenosynovitis and a rash. The rash consists of typically sparse (< 20), pustular lesions on the extensor surfaces of the upper and lower extremities. There are two clinical typical patterns. Patients with tenosynovitis and rash are more likely to have positive blood cultures. In patients with septic arthritis, positive blood and/or synovial fluid cultures are present in only 50% of cases. The diagnosis is made clinically in concert with the results of genital and mucosal site cultures. The diagnosis of DGI should be suspected in any young sexually active patient who presents with fever, dermatitis and rheumatological symptoms. The differential diagnosis includes septic etiologies (Rompalo et al., 1987) such as *Staphylococcus aureus*, *Neisseria menigitidis* and systemic rheumatological disorders. Occasionally, disseminated gonococcal infection can lead to endocarditis (Jackman and Glamann, 1991) or meningitis, both of which can be particularly devastating.

Perinatal Disease

Perinatal infection is relatively rare in the United States, but is still often found in the developing world (Chaisilwattana et al., 1997). Perinatal gonococcal disease is transmitted by direct exposure to an infant passing through the birth canal of an infected cervix. Gonoccal ophthalmia is a severe public health problem in the developing world, and is effectively prevented by inexpensive prophylaxis (Laga et al., 1988). The incidence of gonococcal ophthalmia is estimated to be 42% of infants exposed to an infected cervical canal (Laga et al., 1986). These infants develop a purulent conjunctivitis, which can then rapidly progress to keratitis and subsequent corneal blindness. Because of the severe sequelae of this, public health authorities instituted postnatal prophylaxis with ophthalmic silver nitrate as early as 1910, effectively preventing corneal infection in 95% of exposed infants. Prevention of perinatal infection is the rationale for aggressive screening programs in pregnant women. Ophthalmia is also occasionally seen in adults, usually as a result of self-inoculation.

Aspiration of secretions by infants during parturition can lead to gonococcal pneumonia. This can have severe sequelae as the infant, by definition, is an immunosuppressed host. One of the sequelae can be disseminated gonococcal infection, therefore women who present for delivery without the benefit of prenatal care should be evaluated for gonococcal (and chlamydial and syphilitic) infection.

DIAGNOSIS AND TREATMENT OF GONORRHEA

Traditional approaches to the diagnosis of gonorrhea have used direct smear or culture. Culture for *Neisseria gonorrhoeae* is typically performed on media such as chocolate (cooked sheep's blood) agar, Thayer-Martin, or GC medium incubated in a 5% CO_2 atmosphere. Selective medium which incorporates antibiotics such as vancomycin, colistin and amphotericin is used to inhibit the growth of commensal organisms and other potential pathogens in the genitourinary tract which could overgrow the gonococci. Identification of *Neisseria* is confirmed by the oxidase reaction, and species-specific confirmation is performed utilizing either carbohydrate fermentation or monoclonal antibody techniques.

Because of the specific transport and environmental requirements for gonococcal culture, newer non-culture techniques have evolved in the past 10 years. Genetic probe methods include techniques which are as sensitive as culture but use genetic hybridization technology for diagnosis (van Ulsen, 1986). In the past 3 years major advances in DNA and RNA amplification have allowed the development of nucleic acid amplification techniques. The most commonly used and recently approved for use in the United States include polymerase chain reaction and ligase chain reaction (Ching *et al.*, 1995). Both techniques are at least as sensitive as culture, and may be more sensitive in certain situations. Furthermore, the high sensitivity of these techniques have allowed them to be used in urine (Smith *et al.*, 1995). The use of urine as a diagnostic technique has facilitated the expansion of gonorrhea screening efforts (Zenilman, 1997), especially in populations where clinical service provision is difficult and/or inadequate.

Treatment for mucosal gonorrhea infections is based on providing single-dose regimens, preferably oral, that are effective against most or all of the known resistant determinants (CDC, 1998). Periodic antimicrobial susceptibility testing of a sample of isolates is recommended, preferably as part of an ongoing surveillance program (see below). Current single-dose oral regimens include: ciprofloxacin 500 mg, ofloxacin 400 mg and cefixime 400 mg. Ceftriaxone 125 mg i.m. is the recommended parenteral regimen for uncomplicated gonorrhea. For patients with pelvic inflammatory disease or epididymitis, 250 mg ceftriaxone is the recommended dose. All patients treated for gonorrhea should also be treated for chlamydia, with either azithromycin 1 g (single dose) or doxycycline 100 mg twice daily for 1 week. These represent the general recommendations. Advice on the treatment of more complex disease, or in pregnant patients or children, is provided in the CDC STD treatment guidelines (CDC, 1998).

EPIDEMIOLOGY AND EXPECTED IMPACT OF AN EFFECTIVE VACCINE

In 1996, 325 883 cases of gonorrhea were reported to the Centers for Disease Control by local and state health departments (CDC, 1997b), which represented the lowest rate in over 25 years (Figure 16.2). An additional 1–2 cases are thought to occur for every reported case. The highest rates are seen in adolescent women (age 15–19) and young adult men (age 20–24), representing in part sexual behavior patterns (Figure 16.3). Although the incidence of gonorrhea has decreased by nearly 70% since the peak of the epidemic in the mid-1970s, the infection has become increasingly prevalent in inner-city areas strongly affected by other social problems, especially drug use and pregnancy.

STD rates in minority groups, especially African-Americans, remain substantially higher than those in other ethnic groups. For example, in 1995 overall gonorrhea rates for African-Americans were 1087 per 100 000, compared to 29 per 100 000 for whites and 91 per 100 000 for Hispanics. This phenomenon is not limited to the US. For example, in Britain similar racial disparities are observed, despite major differences in the ethnic make-up of the population, as well as access to clinical services. For example, in Leeds, Lacey and colleagues (1997) evaluated 1664 gonorrhea reports from 1416 patients between 1989 and 1993. The population-based rate in blacks was 558 per 100 000, compared to 56 per 100 000 in whites. Although black ethnicity was associated with lower socioeconomic status (SES), in multivariate regression models that controlled for SES blacks still had a tenfold risk compared to whites. Studies in London by Daker-White and Barlow (1997) and Low *et al.* (1997) supported these findings. Even when controlling for socioeconomic status, the odds (based on odds ratios) of blacks being diagnosed with gonorrhea was 10 times that of whites.

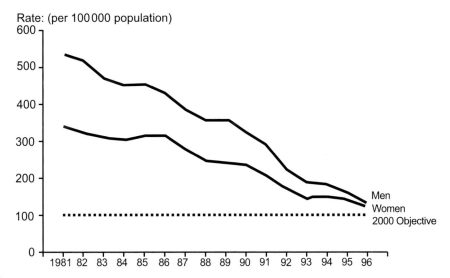

FIGURE 16.2 Gonorrhea rates in the United States by gender from 1981 to 1996. Since 1972 rates have steadily declined. Male cases consistently outnumber female cases, largely because of the increased proportion of symptoms and more specific diagnosis in males. Rates are not equally distributed across all population subgroups (see text). The dotted line indicates the Year 2000 Objective for gonorrhea. Figure courtesy of Division of STD Prevention, Centers for Disease Control (CDC, 1997b).

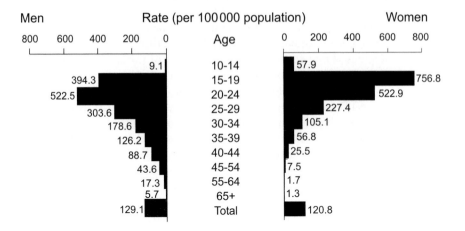

FIGURE 16.3 Gonorrhea rates in the United States in 1996 by age and gender. The highest rates are seen in women aged 15–19 and men aged 20–24 (see text). Figure courtesy of the Division of STD Prevention, Centers for Disease Control (CDC, 1997b).

The aforementioned studies from the UK strongly suggest that socioeconomic and health-care access factors do not alone explain the sharp disparity between STD rates. In the US, controlling for health-care access is difficult, one factor being the inconsistencies in current insurance plans. In a San Francisco study, Ellen *et al.* (1995) attempted to control for these variables by using census-based data, which had valid residential addresses, in analyzing gonorrhea and chlamydia cases. He found that poverty, or working class status, was not associated with STDs, but that ethnicity was a very strong predictor. For adolescents, the gonorrhea rates were 1873/100 000 for blacks and 80/100 000 for whites; for chlamydia, the respective rates were 1673/100 000 for blacks and 180/100 000 for whites. The relative risk of blacks compared to whites was 28.7 for gonorrhea, and for chlamydia it was 8.8.

Historically, gonococcal disease has also been a major factor in areas where access to health-care services has been poor. This is reflected by high gonorrhea rates in the developing world, especially sub-Saharan Africa and Asia, and rates that are currently increasing in the former Eastern Bloc/Soviet Union.

The health and economic impact of gonococcal disease is difficult to measure. The impact of pelvic inflammatory disease in the United States has been estimated to be over 500 000 cases annually, with a total cost of $6 billion. Although this estimate is combined with chlamydia infection, gonococcal-induced PID is a substantial part. In the developing world untreated gonococcal disease is especially associated with high rates of perinatal morbidity, and occasionally perinatal mortality. Studies in developing countries have indicated that the prevalence of gonorrhea may be as high as 10–15% in prenatal population highly affected areas.

VACCINE RESEARCH AND OTHER STRATEGIES FOR CONTROL

Appreciating the clinical and epidemiological aspects of gonococcal disease is important in developing control strategies. Traditional intervention strategies have been oriented towards four goals: reducing the potential for exposure through modifying risky behavior patterns and primary STD prevention; reducing the infected pool of individuals through screening programs; providing treatment based on regimens which are known to be effective; and treatment of infected partners through contact notification.

Partner notification has been extraordinarily ineffective in controlling gonococcal infection because of the short incubation period (Cowan *et al.*, 1996), estimated to be between 1 and 2 days. In this context, it is therefore impossible to 'surround' an epidemic focus of gonococcal disease and capture and treat the infected partners. In contrast, screening strategies, traditionally oriented towards women, have had a significant impact in lowering disease prevalence. With the availability of new non-invasive urine diagnostic tests, screening strategies for men in high-risk settings have become a practical option.

Treatment strategies for gonorrhea have become more complex because of the development of antimicrobial resistance. For 30 years after World War II, penicillin was the antimicrobial therapy of choice (Jaffe *et al.*, 1976). In 1976, the first case of penicillinase-producing *Neisseria gonorrhoeae* (PPNG) was reported. PPNG rapidly developed into a major public health problem and made the penicillin class of antibiotics obsolete by the mid-1980s (CDC, 1987). Chromosomally mediated resistance to penicillin (CMRNG) was described in the early 1980s (Faruki *et al.*, 1985), high-level plasmid-mediated tetracycline resistance (TRNG) developed in 1985 (Knapp *et al.*, 1987), and more recently quinolone antibiotic resistance (QRNG) has been described in a number of cities in the United States, as well as in Southeast Asia (Gordon *et al.*, 1996). The development of standardized effective gonococcal treatment strategies (Moran and Levine, 1995) has required accurate surveillance, which has occurred since the mid-1980s with the implementation of the national Gonococcal Isolate Surveillance Program (GISP; Schwarcz *et al.*, 1990), which has provided standardized mechanisms for determining antimicrobial resistance. Baseline behavioral and treatment data are also collected. As a byproduct of that system, the GISP recently described an increased incidence of anorectal infection in the GISP cohort, suggesting that risky practices were beginning to develop within the homosexual community served by the GISP program. This presents a very troubling public health issue.

With the intersection of high-risk behaviors, poverty and poor public health services, the importance of health-care access is critical for an effective control strategy (Eng and Butler, 1997). Furthermore, many patients with gonorrhea have recurrent disease which is not usually the result of treatment failure, as with the implementation of adequate single-dose effective regimens, treatment failures are rare. Recurrence often occurs through re-exposure, and with the exception of exposure to a similar serotype to gonorrhea (an extremely rare event), there is little evidence of the effective development of protective immunity.

An effective vaccine strategy would result in preventing substantial morbidity, especially pelvic inflammatory disease (PID) and infertility. It would also be expected to have subsidiary effects, such as health promotion effects and the prevention of HIV infection. Within the past 5 years gonococcal infection has also been associated with HIV transmission. Studies performed in Africa on a cross-sectional and prospective basis have demonstrated that gonococcal infection is associated with an increased risk of HIV infection even when behavioral factors are adequately controlled. The odds ratio and/or risk ratios are between 3 and 5 (Laga *et al.*, 1993; Kassler *et al.*, 1994). A potential pathophysiologic basis has been demonstrated in HIV-infected persons who are coinfected with gonorrhea. In these patients the HIV viral load in urethral secretions is increased by 1–2 logs, compared to those without gonorrhea (Moss *et al.*, 1995). Looking at HIV transmission from the standpoint of susceptibility, gonorrhea causes increased tissue edema, friability and inflammation, settings that could increase the efficiency of biological transmission to the vasculature.

Vaccine strategies are therefore attractive options for adolescents and young adults at high risk for gonorrhea, commercial sex workers at high risk for gonorrhea, military deployments where exposures to commercial sex workers would be expected, and developing countries and other settings where access to primary preventive health-care services is limited (Adimora *et al.*, 1994).

The history of attempts to control gonococcal infection through vaccination dates back to the early 1900s (Lees, 1919). In the 1970s, Greenberg tested a whole-cell autolyzed vaccine in Inuvik in the Canadian Northern Territories. The vaccine was immunogenic, but non-protective in this limited field trial (Greenberg *et al.*, 1974). Based on observations by Kellogg and colleagues (1963) that there was an association between pili and gonococcal pathogenesis, a number of investigators focused on pili as a potential subunit vaccine. Brinton and colleagues (1978) immunized volunteers with pilus fibers isolated from a hyperpiliated gonococcal strain and evaluated protection against intraurethral challenge in a human male volunteer model. This study demonstrated that a parenteral vaccine could alter the number of gonococci necessary to establish infection. In expanded studies, Tramont (1981) demonstrated that vaccine recipients had antibodies in sera and genital secretions that blocked attachment to human buccal cells and cross-reacted with heterologous gonococcal strains. In 1983, the US Army conducted a double-blind randomized placebo-controlled trial in approximately 3000 military personnel in Korea (Boslego *et al.*, 1991; Johnson *et al.*, 1991). Unfortunately, the vaccine failed to demonstrate efficacy: gonorrhea occurred in 7.2% of vaccine recipients compared to 6.7% of controls. In retrospect this failure is not unexpected, based on our current understanding of the enormous potential for antigenic variation of the pilin molecule during natural and experimental infection (Zak *et al.*, 1984; Swanson *et al.*, 1987; Seifert *et al.*, 1994).

In 1985, Buchanan and Hook investigated the ability of a vaccine based on the gonococcal porin protein Por to protect against infection. These investigators demonstrated no protection against infection when volunteers where challenged intraurethrally with the homologous gonococcal strain. A recent retrospective analysis of sera from this trial showed that vaccine recipients had an increase in antibodies against the reduction modifiable protein Rmp (Rice *et al.*, 1994). These antibodies have been shown to block the bactericidal effect of antibodies directed to the Por protein (Wetzler *et al.*, 1992a), thus suggesting that future vaccines without Rmp may have a more protective effect. Formulations of vaccines containing Por isolated from gonococcal mutants lacking Rmp are currently being evaluated (Parmar *et al.*, 1997).

Although the trials discussed above focused on a limited number of vaccine candidates, evaluation of them has provided insight into potentially protective, as well as potentially deleterious, human immune responses to these gonococcal antigens. Currently, efforts in various laboratories are focused on the identification and evaluation of new vaccine candidate antigens, as well as the improved purification and delivery of previously studied antigens.

One of the major obstacles to the development of a gonococcal vaccine is the current lack of a practical animal model that represents an accurate model of human gonococcal disease. Urethral infection of chimpanzees is similar to human gonorrhea (Kraus *et al.*, 1975), but this model is not practical for routine laboratory investigation. To overcome this obstacle, various *in vitro* assays and models have been used to evaluate the potential protective effect *in vivo*. These models include, but are not limited to, primary cell adherence assays (Apicella *et al.*, 1996), tissue culture cell adherence and/or invasion (Virji *et al.*, 1991; de la Paz *et al.*, 1995; Mertz and So, 1997), Fallophian tube organ culture (McGee *et al.*, 1981, 1988), bactericidal assays (Wetzler *et al.*, 1992a; McQuillen *et al.*, 1994; Robbins *et al.*, 1995; Bos *et al.*, 1997), and opsonic phagocystosis assays (Kim *et al.*, 1992; Rest and Frangipane, 1992). Although each model has individual merit, to date there is a lack of clear consensus regarding which surrogates may best predict a protective effect in clinical trials.

CLINICAL STUDIES FOR VACCINE

Clinical studies have been carried out in the evaluation of gonococcal vaccines since the 1960s. As pointed out earlier, one of the challenges has been to identify antigenic determinants that are able to invoke a protective immune response. Generally, the multiple antigenic serotypes and the rapid phase variation in the outer membrane proteins has thwarted attempts to develop a stable antigenic determinant. Nevertheless, findings from studies in commercial sex workers (Plummer *et al.*, 1994) suggest that a limited degree of natural immunity may develop after repeated infections, and that there may be a stable antigenic determinant on gonococcal strains.

The evaluation of potential vaccine candidates in human volunteers is a detailed and stepwise process. The initial stages include the identification of antigenic determinants and the induction of an immune response, either humoral or mucosal, by vaccination. In the ideal setting these responses would react with a variety of gonococcal serotypes. However, the demonstration of immunity does not necessarily translate into a demonstration of protection. Efficacy trials are required to demonstrate the effectiveness of a vaccine, but moving a gonococcal vaccine into efficacy trials is complicated by several factors. First, because no animal model currently represents an accurate model for human gonococcal infection, a vaccine candidate cannot be evaluated in an animal model for a potential protective effect prior to human trials. An alternative is the human urethral challenge model (Swanson *et al.*, 1987; Cohen *et al.*, 1994; Schneider *et al.*, 1995), which also has limitations. Because gonorrhea is associated with development of PID and other systemic complications in women, challenge studies have been limited to male volunteers. Therefore, any data obtained in this model must be extrapolated to women. This may or may not lead to viable conclusions, as the disease process is different in the two genders. However, to date this model has proved useful in providing information on the role of gonococcal antigens, such as pili, opacity proteins, lipo-oligosaccharide (LOS) and iron-regulated/iron-binding proteins, in the initial stages of the disease process and human immune response (Jerse *et al.*, 1994; Seifert *et al.*, 1994; Ramsey *et al.*, 1995; Schneider *et al.*, 1996; Cornelissen *et al.*, 1996).

Initial human trials can be used to evaluate the potential of vaccine candidates to induce a humoral and/or mucosal immune response. The role of mucosal immunity in preventing gonococcal infection is not well characterized, but is presumed to be important because of the non-systemic characteristics of most gonococcal infections. These studies would be performed in a small number of subjects to assess initial safety. In addition, the route of vaccine administration, dose ranging and scheduling needs to be evaluated.

Once immunogenicity and dose selection have been established, efficacy trials would be required to demonstrate that a vaccine can be clinically useful. Initial evaluations may be possible in the intraurethral challenge model discussed above. In this model, volunteers could be immunized and then challenged with an experimental inoculation of gonococci. As previously noted, these experiments would be limited to males and could have limited generalizability for a number of reasons. First, the inoculum size in natural infection may not be comparable to that used in experimental infection. Second, the mucosal exposures are different. Third, there may be characteristics of organisms passed in the laboratory setting that are different from those seen in the wild state. Nevertheless, these experiments could provide insight into the potential protective capability of a vaccine candidate.

After these evaluations, a large-scale trial would be required to demonstrate the efficacy of a vaccine in the field. Vaccine field trials require an enormous logistical and methodological effort. In addition to the usual problems encountered in such efforts, trials to evaluate the prevention of an STD such as gonorrhea present additional challenges. Some of the difficulties are based on the following issues, which have also been encountered in trials of vaccines for other STDs such as herpes simplex virus (Langenberg *et al.*, 1995; Mertz *et al.*, 1990).

A randomized double-blind placebo-controlled clinical trial would require the recruitment of a large number of individuals into a study. The population size is determined by the infection rate in the study population. Because of these sample size considerations (which are already large), it would be necessary to recruit a population at high risk for gonococcal disease exposure. For example, in the Baltimore STD clinics, during a 6-month period, repeated gonococcal infections are seen in approximately 20% of patients who have been previously infected. The study procedures would involve an informed consent process, recruitment into the study, immunization with placebo or control, and volunteer follow-up on a frequent basis to determine whether infection occurred or not. However, populations at high risk for STDs have also traditionally been at high risk for non-compliance. This would then require a substantial investment in resources for patient follow-up and incentives. From an ethical prospective all subjects would be required to have STD prevention counseling, which would reduce infection rates overall in both the vaccine and the placebo cohorts. Condom use would also be encouraged.

The technical issues related to follow-up and ascertainment of disease, the ethical issues involved in counseling patients to abstain from sexual behavior, and using measurements that identify high-risk sexual behavior result in highly complex studies from a methodological and ethical standpoint.

REFERENCE

Adimora, A.A., Sparling, P.F. and Cohen, M.S. (1994). Vaccines for classic sexually transmitted diseases. *Infect. Dis. Clin. N. Am.* **8**: 859–878

Adu-Sarkodie, Y.A., Brook, M.G., Clark, S. *et al.* (1995). The potential effect on *Neisseria gonorrhoeae* of the use of clindamycin vaginal cream in the empirical treatment of vaginal discharge. *J. Antimicrob. Chemother.* **36**: 557–560

Apicella, M.A., Ketterer, M., Lee, F.K., Zhou, D., Rice, P.A. and and Blake, M.S. (1996). The pathogenesis of gonococcal urethritis in men: confocal and immunoelectron microscopic analysis of urethral exudates from men infected with *Neisseria gonorrhoeae*. *J. Infect. Dis.* **173**: 636–646

Bos, M.P., Hogan, D. and Belland, R.J. (1997). Selection of Opa+ *Neisseria gonorrhoeae* by limited availability of normal human serum. *Infect. Immun.* **65**: 645–650

Boslego, J.W., Tramont, E.C., Chung, R.C. *et al.* (1991). Efficacy trial of a parenteral gonococcal pilus vaccine in men. *Vaccine* **9**: 15–62

Bowie, W.R. and Jones, H. (1981). Acute pelvic inflammatory disease in outpatients: association with *Chlamydia trachomatis* and *Neisseria gonorrhoeae*. *Ann. Intern. Med.* **95**: 685–688

Brinton, C.C., Bryan, J. and Dillon, J. (1978). Uses of pili in gonorrhea control: role of bacterial pili in disease, purification, and propoerties of gonococcal pili, and progress in the development of a gonococcal pilus vaccine for gonorrhea. In: Immunobiology of *Neisseria gonorrhoeae*. G.F. Brooks, E.C. Gotschlich and K.K. Holmes, eds. American Society for Microbiology, Washington, D.C., pp. 1242–1245

Centers for Disease Control. (1984). Declining rates of rectal and pharyngeal gonorrhea among males-New York City. *MMWR* **33**: 295–297

Centers for Disease Control. (1987). Antibiotic-resistant strains of *Neissena gonorrhoeae* – Policy guidelines for detection, management and control. *MMWR*, 36(Sup:5S):

Centers for Disease Control. (1997a). Gonorrhea among men who have sex with men–selected sexually transmitted disease clinics: 1993–1996. *MMWR* **46**: 889–892

Centers for Disease Control, Division of HIV/STD Prevention. (1997b). Sexually transmitted disease surveillance report, 1996. Atlanta, GA

Centers for Disease Control. (1998). 1998 guidelines for treament of sexually transmitted diseases. *MMWR*, 47(RR-1)

Chaisilwattana, P., Chuachoowong, R., Siriwasin, W. *et al.* (1997). Chlamydial and gonococcal cervicitis in HIV-seropositive and HIV-seronegative pregnant women in Bangkok: prevalence, risk factors, and relation to perinatal HIV transmission. *Sex. Transm. Dis.* **24**: 495–502

Ching, S., Lee, H., Hook, E.W., Jacobs, M.R. and Zenilman, J.M. (1995). Ligase chain reaction for detection of *Neisseria gonorrhoeae* in urogenital swabs. *J. Clin. Microbiol.* **33**: 3111–3114

Cohen, M.S., Cannon, J.G., Jerse, A.E., Charniga, L.M., Isbey, S.F. and Whicker, L.G. (1994). Human experimentation with *Neisseria gonorrhoeae*: Rationale, methods and implications for the biology of infection and vaccine development. *J. Infect. Dis.* **169**: 532–537

Cornelissen, C., Anderson, J., Kashkari, M. *et al.* (1996). Function and virulence studies of the gonococcal transferrin receptor. In: Proceedings of the Tenth International Pathogenic Neisseria Conference. W. Zollinger, R. French and C. Deal, eds. Baltimore, MD, pp. 552–553

Cowan, F.M., French, R. and Johnson, A.M. (1996). The role and effectiveness of partner notification in STD control: a review. *Genitourinary Med.* **72**: 247–252

Curran, J.W., Rendtorff, R.C., Chandler, R.W., Wiser, W.L. and Robinson, H. (1975). Female gonorrhea: its relation to abnormal uterine bleeding, urinary tract symptoms, and cervicitis. *Obstet. Gynecol.* **45**: 195–198

Daker-White, G. and Barlow, D. (1997). Heterosexual gonorrhoea at St Thomas – I: Patient characteristics and implications for targeted STD and HIV prevention strategies. *Int. J. STD. AIDS,* **8**: 32–35

de la Paz, H., Cooke, S.J. and Heckels, J.E. (1995). Effect of sialylation of lipopolysaccharide of *Neisseria gonorrhoeae* on recognition and complement-mediated killing by monoclonal antibodies directed against different outer-membrane antigens. *Microbiology* **141**: 913–920

Ellen, J.M., Kohn, R.P., Bolan, G.A., Shiboski, S. and Krieger, N. (1995). Socioeconomic differences in sexually transmitted disease rates among black and white adolescents, San Francisco, 1990 to 1992. *Am. J. Public Health* **85**: 1546–1548

Eng, T.R. and Butler, W.T. (1997). The hidden epidemic-confronting sexually transmitted diseases. National Academy Press

Faruki, H., Kohmescher, R.N., McKinney, W.P. and Sparling, P.F. (1985). A community-based outbreak of infection with penicillin-resistant *Neisseria gonorrhoeae* not producing penicillinase (chromosomally-mediated resistance). *N. Engl. J. Med.* **313**: 607–611

Gordon, S.M., Carlyn, C.J., Doyle, L.J. *et al.* (1996). The emergence of *Neisseria gonorrhoeae* with decreased susceptibility to ciprofloxacin in Cleveland, Ohio: epidemiology and risk factors [see comments]. *Ann. Intern. Med.* **125**: 465–470

Greenberg, L., Diena, B.B., Ashton, F.A. *et al.*, (1974). Gonococcal vaccine studies in Inuvik. *Can. J. Public Health* **65**: 29–33

Hillis, S.D., Joseoef, R., Marchbanks, P.A., Wasserheit, J.N., Cates, W. and Westrom, L. (1993). Delayed care for pelvic inflammatory disease as a risk factor for impaired ferility. *Am. J. Obstet. Gynecol* **168**: 1503–1509

Hutt, D.M. and Judson, F.N. (1986). Epidemiology and treatment of oropharyngeal gonorrhea. *Ann. Intern. Med.* **104**: 655–658

Jackman, J.D. and Glamann, D.B. (1991). Southwestern internal medicine conference: Gonococcal endocarditis: twenty five years experience. *Am. J. Med. Sci.* **301**: 221–230

Jacobs, N.F. and Kraus, S.J. (1975). Gonococcal and nongonococcal urethritis in men. Clinical and laboratory differentiation. *Ann. Intern. Med.* **82**: 7–12

Jaffe, H.W., Biddle, J.W., Thornsberry, C. *et al.* (1976). National gonorrhea therapy monitoring study: in vitro antibiotic susceptibility and its correlation with treatment results. *N. Engl. J. Med.* **294**: 5–9

Jaffe, H.W., Biddle, J.W., Johnson, S.R. and Wiesner, P.J. (1981). Infections due to penicillinase-producing *Neisseria gonorrhoeae* in the United States: 1976–1980. *J. Infect. Dis.* **144**: 191–197

Jerse, A.E., Cohen, Drown, P.M. *et al.* (1994). Multiple gonococcal opacity proteins are expressed during experimental urethral infection in the male. *J. Exp. Med.* **179**: 911–920

Johnson, S.C., Chung, R.C., Deal, C.D. *et al.* (1991). Human immunization with Pgh 3–2 gonococcal pilus results in cross-reactive antibody to the cyanogen bromide fragment-2 of pilin. *J. Infect. Dis.* **163**: 128–134

Kassler, W.J., Zenilman, J.M., Erickson, B., Fox, R., Peterman, T. and Hook, E.W. (1994). HIV seroconversion in STD clinic patients. *AIDS* **8**: 351–355

Kellogg, D.S., Peacock, W.L. and Deacon, W.E. (1963). *Neisseria gonorrhoeae* I: Virulence genetically linked to clonal variation. *J. Bacteriol.* **85**: 1274–1279

Kim, J.J., Zhou, D., Mandrell, R.E. and Griffiss, J.M. (1992). Effect of exogenous sialylation of the lipooligosaccharide of *Neisseria gonorrhoeae* on opsonophagocytosis. *Infect. Immun.* **60**: 4439–4442

Knapp, J.S. and Clark, V.L. (1984). Anaerobic growth of *Neisseria gonorrhoeae* coupled to nitrite reduction. *Infect. Immun.* **46**: 176–181

Knapp, J.S., Zenilman, J.M., Biddle, J.W. *et al.* (1987). Frequency and distribution in the United States of strains of *Neisseria gonorrhoeae* with plasmid-mediated, high-level resistance to tetracycline. *J. Infect. Dis.* **155**: 819–822

Kraus, S.J., Brown, W.J. and Arko, R.J. (1975). Acquired and natural immunity to gonococcal infection in chimpanzees. *J. Clin. Invest.* **55**: 1349–1356

Lacey, C.J.N., Merrick, D.W., Bensley, D.C. and Fairley, I. (1997). Analysis of the sociodemography of gonorrhea in Leeds, 1989–93. *Br. Med. J.* **314**: 1715–1718

Laga, M., Plummer, F.A., Nzanze, H. *et al.* (1986). Epidemiology of ophthalmia neonatorum in Kenya. *Lancet* **2**: 1145–1149

Laga, M., Plummer, F.A., Piot, P. *et al.* (1988). Prophylaxis of gonococcal and chlamydial ophthalmia neonatorum. A comparison of silver nitrate and tetracycline. *N. Engl. J. Med.* **318**: 653–657

Laga, M., Manoka, A. and Kivuvu, M. (1993). Non-ulcerative sexually transmitted diseases as risk factors for HIV-1 transmission in women-results from a cohort study. *AIDS* **7**: 95–102

Lammel, C.J., Sweet, R.L., Rice, P.A., *et al.* (1985). Antibody-antigen specificity in the immune response to infection with *Neisseria gonorrheae*. *J. Infect. Dis.* **152**: 990–1001

Langenberg, A.G., Burke, R.L., Adair, S.F. *et al.* (1995). A recombinant glycoprotein vaccine for herpes simplex virus type 2: safety and immunogenicity. *Ann. Intern. Med.* **122**: 889–898

Lebedeff, D.A. and Hochman, E.B. (1980). Rectal gonorrhea in men: diagnosis and treatment. *Ann. Intern. Med.* **92**: 463–466

Lees, D. (1919). Detoxicated vaccines in the treatment of gonorrhoea. *Lancet* **1**: 1107–1112

Low, N., Daker-White, G., Barlow, D. and Pozniak, A.L. (1997). Gonorrhoea in inner London: results of a cross sectional study [see comments]. *Br. Med. J.* **314**: 1719–1723

McCormack, W.M., Stumacher, R.J., Johnson, K. and Donner, A. (1977a). Clinical spectrum of gonococcal infection in women. *Lancet* **1**: 1182–1185

McCormack, W.M., Nowroozi, K., Alpert, S. *et al.* (1977b). Acute pelvic inflammatory disease: characteristics of patients with gonococcal and nongonococcal infection and evaluation of their response to treatment with aqueous procaine penicillin G and spectinomycin hydrochloride. *Sex. Transm. Dis.* **4**: 125–131

McGee, Z.A., Johnson, A.P. and Taylor-Robinson, D. (1981). Pathogenic mechanisms of *Neisseria gonorrhoeae*: Observations on damage to human fallophian tubes in organ culture by gonococci of type 1 and type 4. *J. Infect. Dis.* **143**: 413–422

McGee, Z.A., Gorby, G.L., Wyrick, P.B., Hodinka, R. and Hoffman, L.H. (1988). Parasite-directed endocytosis. *Rev. Infect. Dis.*, *10 Suppl* **2**: S311–S316

McNagny, S.E., Parker, R.M., Zenilman, J.M. and Lewis, J.S. (1992). Evaluation of urinary leukocyte esterase as a screening test for the detection of asymptomatic *Chlamydia trachomatis* and *Neissera gonorrhoeae* in men. *J. Infect. Dis.* **165**: 573–576

McQuillen, D.P., Hagblom, P. and Rice, P.A. (1994). Complement-mediated bacterial killing assays. *Meth. Enzymol.* **236**: 137–147

Mertz, G.J., Ashley, R., Burke, R.L., Benedetti, J., Critchlow, C. and Jones, C.C. (1990). Double-blind, placebo-controlled trial of a herpes simplex virus type 2 glycoprotein vaccine in persons at high risk for genital herpes infection. *J. Infect. Dis.* **161**: 653–660

Merz, A.J. and So, M. (1997). Attachment of piliated, Opa- and Opc-gonococci and meningococci to epithelial cells elicits cortical actin rearrangements and clustering of tyrosine-phosphorylated proteins. *Infect. Immun.* **65**: 4341–4349

Moran, J.S. and Levine, W.C. (1995). Drugs of choice for the treatment of uncomplicated gonococcal infection. *Clin. Infect. Dis.* **20**(*Suppl*):S47–65

Moss, G.B., Overbaugh, J., Welch, M. *et al.* (1995). Human immunodeficiency virus DNA in urethral secretions in men: association with gonococcal urethritis and CD4 cell depletion. *J. Infect. Dis.* **172**: 1469–1474

O'Brien, J.P., Goldenberg, D.L. and Rice, P.A. (1983). Disseminated gonococcal infection: a prospective analysis of 49 patients and a review of pathophysiology and immune mechanisms. *Medicine* **62**: 395–406

Parmar, M.M., Blake, M.S. and Madden, T.D. (1997). Biophysical and antigenic characterization of gonococcal protein I incorporated into liposomes. *Vaccine* **15**: 1641–1651

Peterson, B.H., Lee, T.J., Snyderman, R. and Brooks, G.F. (1979). *Neisseria meningitidis* and *Neisseria gonorrhoeae* bacteremia associated with C6, C7, or C8 deficiency. *Ann. Intern. Med.* **90**: 917–920

Platt, R., Rice, P.A. and McCormack, W.M. (1983). Risk of acquiring gonorrhea and prevalence of abnormal adnexal findings among women recently exposed to gonorrhea. *JAMA* **250**: 3205–3209

Plummer, F.A., Simonsen, J.N. and Chubb, I. (1994). Epidemiologic evidence for the development of serovar-specific immunity after gonococcal infection. *J. Clin. Invest.* **93**: 2744–2749

Ramsey, K.H., Schneider, H., Cross, A.S. *et al.* (1995). Inflammatory cytokines produced in response to experimental human gonorrhea. *J. Infect. Dis.* **172**: 186–191

Rest, R.F. and Frangipane, J.V. (1992). Growth of *Neisseria gonorrhoeae* in CMP-N-acetylneuraminic acid inhibits nonopsonic (opacity-associated outer membrane protein-mediated) interactions with human neutrophils. *Infect. Immun.* **60**: 989–997

Rice, P.A., Hook, E.W., Blake, M.S. *et al.* (1994). A possible influence of vaccine-induced Por, LOS, and Rmp antibodies on the outcome of intraurethral challenge with *Neisseria gonorrhoeae*. In: Neisseria 94: Proceedings of the Ninth International Pathogenic Neisseria Conference. J.S. Evans, S.E. Yost, M.C. Maiden and I.M. Feavers, eds. National Institute for Biological Standards and Control, Potters Bar, UK, pp. 483–484

Robbins, J.B., Schneerson, R. and Szu, S.C. (1995). Perspective: hypothesis: serum IgG antibody is sufficient to confer protection against infectious diseases by inactivating the inoculum. *J. Infect. Dis.* **171**: 1387–1398

Rolfs, R.T., Galaid, E.I. and Zaidi, A.A. (1992). Pelvic inflammatory disease: trends in hospitalization and office visits: 1979 through 1988. *Am. J. Obstet. Gynecol.* **166**: 983–990

Rompalo, A.M., Hook, E.W., Robinson, H., Ramsey, P.G., Handsfield, H.H. and Holmes, K.K. (1987). The acute arthritis-dermatitis syndrome. *Arch. Intern. Med.* **147**: 281–283

Schneider, H., Cross, A.S., Kuschner, R.A. *et al.* (1995). Experimental human gonococcal urethritis: 250 *Neisseria gonorrhoeae* MS11mkC are infective. *J. Infect. Dis.* **172**: 180–185

Schneider, H., Schmidt, K.A., Skillman, D.R. *et al.* (1996). Sialylation lessens the infectivity of *Neisseria gonorrhoeae* MS11mkC. *J. Infect. Dis.* **173**: 1422–1427

Schwarcz, S.K., Zenilman, J.M., Schnell, D. *et al.* (1990). National surveillance of antimicrobial resistance in *Neisseria gonorrhoeae*. The Gonococcal Isolate Surveillance Project [see comments]. *JAMA* **264**: 1413–1417

Seifert, H.S., Wright, C.J., Jerse, A.E., Cohen, M.S. and Cannon, J.G. (1994). Multiple gonococcal pilin antigenic variants are produced during experimental human infection. *J. Clin. Invest.* **93**: 2744–2749

Sherrard, J. and Barlow, D. (1996). Gonorrhoea in men: clinical and diagnostic aspects. *Genitourinary Med.* **72**: 422–426

Smith, K.R., Ching, S., Lee, H. *et al.* (1995). Evaluation of ligase chain reaction for use with urine for identification of *Neisseria gonorrhoeae* in females attending a sexually transmitted disease clinic. *J. Clin. Microbiol.* **33**: 455–457

Stamm, W.E., Hicks, C.B., Martin, D.H. *et al.* (1995). Azithromycin for empirical treatment of the nongonococcal urethritis syndrome in men. A randomized double-blind study [see comments]. *JAMA* **274**: 545–549

Swanson, J., Robbins, K., Barrera, O. *et al.* (1987). Gonococcal pilin variants in experimental gonorrhea. *J. Exp. Med.* **165**: 1344–1357

Tramont, E.C., Sadoff, J.C., Boslego, J.W. *et al.* (1981). Gonococcal pilus vaccine. Studies of antigenicity and inhibition of attachment. *J. Clin. Invest.* **68**: 881–888

van Ulsen, J., Michel, M.F., vanStrik R., vanEijk R.V., vanJoost T. and Stolz, E. (1986). Experience with a modified solid-phase enzyme immunoassay for detection of gonorrhea in prostitutes. *Sex. Transm. Dis.* **13**: 1–4

Virji, M., Kayhty, H., Ferguson, D.J. *et al.* (1991). The role of pili in the interactions of pathogenic Neisseria with cultured human endothelial cells. *Molec. Microbiol.* **5**: 1831–1841

Washington, A.E., Aral, S.O., Wolner-Hanssen, P., Grimes, D.A. and Holmes, K.K. (1991). Assessing risk for pelvic inflammatory disease and its sequelae [see comments]. *JAMA* **266**: 2581–2586

Washington, A.E. and Katz, P. (1991). Cost of and payment source for pelvic inflammatory disease. Trends and projections, 1983 through 2000 [see comments]. *JAMA* **266**: 2565–2569

Wasserheit, J.N., Bell, T.A., Kiviat, N.B. *et al.* (1986). Microbial causes of proven pelvic inflammatory disease and efficacy of clindamycin and tobramycin. *Ann. Intern. Med.* **104**: 187–193

Westrom, L. (1975). Effect of acute pelvic inflammatory disease on fertility. *Am. J. Obstet. Gynecol,* **121**: 707–713

Wetzler, L.M., Blake, M.S., Barry, K. and Gotschlich, E.C. (1992a). Gonococcal porin vaccine evaluation: Comparison of Por proteosomes, lipsomes, and blebs isolated from rmp deletion mutants. *J. Infect. Dis.* **166**: 551–555

Wetzler, L.M., Barry, K., Blake, M.S. and Gotschlich, E.C. (1992b). Gonococcal lipooligosaccharide sialylation prevents complement-dependent killing by immune sera. *Infect. Immun.* **60**: 39–43

Wiesner, P.J., Tronca, E., Bonin, P., Pedersen, A.H.B. and Holmes, K.K. (1973). Clinical spectrum of pharyngeal gonococcal infection. *N. Engl. J. Med.* **288**: 181–185

Wise, C.M., Morris, C.R., Wasilauska, B.L. and Salzer, W.L. (1994). Gonococcal arthritis in an era of increasing penicillin resistance. *Arch. Intern. Med.* **154**: 2690–2695

Zak, K., Diaz, J., Heckels, J.E. (1984). Antigenic variation during infection with *Neisseria gonorrhoeae:* Detection of antibodies to surface proteins in sera of patients with gonorrhea. *J. Infect. Dis.* **149**: 166–174

Zenilman, J.M. (1997). Surveys and Urine STD Diagnostic tests: A new paradigm for survey design and validity measurement (editorial). *Sex. Transm. Dis.* **24**: 310–311

17
SYPHILIS

ADRIAN MINDEL
AND
CLAUDIA ESTCOURT

Academic Unit of Sexual Health Medicine,
Sydney Hospital, Sydney, Australia

CLINICAL DISEASE: SPECTRUM OF ILLNESS, INCLUDING PERINATAL TRANSMISSION AND RESULTING DISEASE

Introduction and Historical Background

Syphilis has coloured history for centuries. It has been present in Europe since the Middle Ages, and until relatively recently was an extremely common disease. In the pre-penicillin era in the first half of this century, a prevalence of 5–10% at autopsy was described in some studies in the US (Rosahn, 1947).

The condition has been known by a variety of names in the past, among them 'the great pox', 'the French disease and 'the Italian disease'. Its current name comes from a poem by the 16th-century poet Frascotoro, who wrote of an afflicted shepherd named 'Syphilus'.

The origin of syphilis as a disease entity is unclear. Two main hypotheses exist. The Columbian theory suggests that the advent of syphilis in Europe in the late 15th century was attributable to the return of Christopher Columbus from America in 1493. It was assumed that European sailors contracted the disease from the indigenous peoples of the West Indies, and then introduced it into a naïve European population. The continent was a virtual war zone at the time, which would have greatly facilitated the sexual spread of the disease. Such a disease has not been documented in American peoples, however, but descriptions of possible cutaneous syphilis do occur in ancient writings from China. It is possible that syphilis was already endemic in Europe at that time, and that Columbus' return merely coincided with enhanced transmission at a time of war.

The Environmental or Unitarian theory suggests that syphilis originated in the tropics as a disease similar to yaws, and was spread by casual contact. In colder climates, where clothing was worn, sexual contact became the major method of transmission. It is believed that *T. pallidum* is really the same species as *T. pertenue* and *T. carateum*, and that the disease processes are merely variants of the same condition whose expression has been modified by the environment (Hudson, 1968). The DNA homology noted between the sub-species, as described later, gives weight to this hypothesis.

It was not until the 18th century that syphilis was understood to be caused by a sexually transmitted agent.

In 1890 Professor Boeck of Oslo, Norway, started to enrol patients into a prospective study of the natural history of the disease. He believed that existing therapies were of little worth, and that their toxicities may have been worse than syphilis itself. Patients were recruited over a 20-year period and not treated until 1951, when penicillin became widely available (Gjestland, 1955). Extensive follow-up of these patients advanced our understanding of the disease considerably. These findings were largely confirmed by a more recent study by the US Public Health Service in Tuskegee, USA. From 1932, African-American men were recruited to a 30-year prospective study during which they were carefully followed up but not treated (Rockwell *et al.*, 1964). The ethics of the study have received much scrutiny and the findings were never published. The American government recently made a formal apology to the participants and their families.

A large autopsy study has also been historically important. In 1947, Rosahn published a series of autopsy cases at Yale University School of Medicine between 1917 and 1941. Overall, 9.7% showed evidence of syphilis. In general, all the studies showed that syphilis was associated with an increased mortality rate of up to 20% loss of life expectancy at 12 years of follow-up in the Tuskegee study (Heller and Bruyere, 1946), and a probability of death as a result of syphilis of 17.1% for women and 8% for men after 40 years of infection in the Oslo study. They also reported that between 15 and 40% of untreated individuals progressed to recognizable late complications, but that around 60% of untreated patients did not demonstrate any evidence of complications.

The early 1900s were a time of great excitement in treponemal medicine. In 1905 Schaudinn and Hoffmann discovered the causative agent of syphilis, and a year later Wassermann described the serological test, the Wassermann reaction. Salvarsan, the organic arsenical compound with antitreponemal action, was discovered by Erlich in 1909, and was the forerunner of more effective arsenical treatments. The discovery of penicillin in the 1940s meant that an effective treatment was at last available.

Laboratory tests for syphilis came into prominence at around the same time. Initially the tests centered on the detection of non-specific lipophilic antibodies. In 1946 the Venereal Disease Reference Laboratory (VDRL) test was described by Harris, Rosenberg and Riedel, which has since been modified to the Rapid plasma reagin test (RPR). In 1949 Nelson and Mayer developed the *Treponema pallidum* immobilization assay (TPI). This test has been largely superseded by the *Treponema pallidum* haemagglutination assay (TPHA). More recently, the polymerase chain reaction (PCR) has been applied to syphilis testing. These tests will be discussed in a later section.

Definitions

As with other age-old diseases, some of the terminology used in syphilis medicine can be confusing. Sexually transmitted syphilis is known as *acquired* syphilis, whereas transmission from mother to fetus is known as *congenital* syphilis. Both can be subdivided into *early* and *late* stages. In acquired syphilis 'early' is arbitrarily taken to mean within the first 2 years of primary infection in the UK, but 1 year in the US, and corresponds with the period of highest risk of transmission. Early acquired syphilis can be further categorized into an *incubation* period of 9–90 days, with an average of 21 days; a *primary* stage, with the appearance of the primary lesion or *chancre*, which heals within 3–10 weeks; a *secondary* stage, which occurs 4–10 weeks after the appearance of the primary lesion and takes the form of generalized skin and mucous membrane lesions, possibly with non-specific constitutional symptoms; and a *latent* phase, which occurs after the resolution of any secondary-stage symptoms. In this context 'latent' implies that there are neither symptoms nor signs of disease, and that the patient has not been treated. Diagnosis is by serology.

It should be noted that latency often occurs in individuals who have never had recognizable primary or secondary syphilis. Latency is referred to as early (within 2 years) and late (present over 2 years). Late syphilis may therefore be latent or, in approximately 35% of patients, accompanied by complications of the cardiovascular and neurological systems and/or gummatous lesions. The term *late benign syphilis* has been used historically to describe proliferative or destructive (gummatous) lesions in body structures not essential to the maintenance of vital functions. However, this is not a particularly useful definition, as gumma can form in a variety of organs. *Gummatous disease* of the relevant organ is a better term. Gumma may occur in both congenital and acquired syphilis (figure 17.1).

Primary Syphilis

It has been estimated that the risk of transmission following intercourse with an infected individual is around 30%. The primary lesion or chancre appears between 9 and 90 days after infection, at the site of entry of the organism, and usually coincides with areas of genital trauma. The common sites in men are the frenulum, prepuce, glans, penile shaft and coronal sulcus, but the peri and intra-anal areas may also be affected if anal intercourse has taken place (Mindel *et al.*, 1989). In women, chancres may be found on the vulva (most commonly around the fourchette), in the vagina and on the cervix. Extragenital sites may be involved, as infection may occur at any skin surface or mucous membrane, and primary lesions have been described in a variety of sites, including the lip, tongue, mouth, pharynx, breast, eyelid and finger.

The chancre is usually solitary and painless and may be asymptomatic if, for example, it is on the cervix or in the rectum. Classically it begins as a reddish macule, rapidly becomes papular, and then ulcerates to leave a round, clean ulcer with an indurated base and rubbery edge (Stokes *et al.*, 1934). With time the base becomes grey and sloughy. There is often a degree of mild to moderate, non-tender non-suppurative rubbery lymphadenopathy that develops within 1

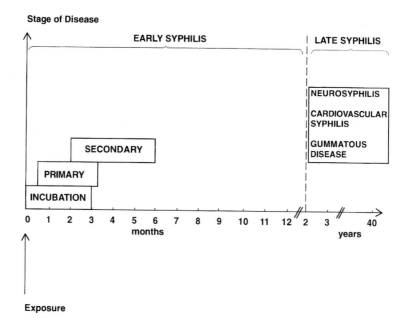

Stage of Disease

FIGURE 17.1 Stages of syphilis, with time of onset

week of the chancre's appearance. The lymphadenopathy is usually bilateral with genital lesions, but unilateral with extragenital lesions. Appearances may not always be classic, especially if the patient presents very early or with an extragenital lesion. Secondary infection of the chancre sometimes occurs, which may make the lesion and nodes painful.

Without treatment the chancre will heal within 3–6 weeks, but may persist in the secondary stage. Rarely, a healed chancre may reappear during the secondary stage. Occasionally a gumma may form at the site of the initial lesion.

A primary syphilitic lesion may not appear at all like the classically described chancre, and it is important to exclude treponemal disease in the assessment of any potentially sexually acquired genital ulcer.

Secondary Syphilis

Four to eight weeks after the appearance of the primary chancre, the features of secondary syphilis supervene. This reflects the bacteraemic phase of infection, with dissemination of the organism throughout the body. Systemic symptoms, including malaise, headache and myalgia, occur and may be severe (Mindel *et al.*, 1989). Some patients present with a sore throat or a rash. There is often palpable widespread non-tender rubbery lymphadenopathy, which may involve the epitrochlear nodes. It is important to distinguish this presentation from other conditions such as HIV seroconversion illness, toxoplasmosis, acute hepatitis B and infectious mononucleosis, which may present in a similar way.

The dermatological manifestations are varied: 75–100% of patients will develop a rash. This starts as round, pale pink, 0.5–1 cm lesions on the trunk and arms, then becomes darker and spreads over the rest of the body. A characteristic rash with trunk and sole involvement is shown in Figure 17.2

The rash may be follicular, annular, rarely pustular or rupial, in which central necrosis occurs as a result of endarteritis; corymbose, in which one central papule is surrounded by a circle of smaller lesions; or papulosquamous, which may appear psoriasiform on the palms and soles. A dense crop of papules may occur on the forehead, and this is known as corona veneris. The anatomical site can influence the morphology of the lesions: split papules may occur at intertriginous areas as a result of fissuring, and in warm moist opposed areas – usually the genital and perianal area – the lesions may coalesce to form large fleshy warty-looking masses teeming with treponemes, known as condylomata lata. The base of these lesions is characteristically flat, which helps to distinguish them from viral warts (condylomata accuminata) (Figure 17.3).

Mucosal lesions occur in less than 25% of patients (Mindel *et al.*, 1989). These are superficial and painless erosions with a dull red or grey sloughy base. They may occur on any mucosal surface, but are commonly seen on the buccal mucosa, the pharynx and the genitals. In the mouth they are often known as snail-track ulcers.

Alopecia may occur, but this tends to be late in the secondary infection. The term 'moth-eaten' is used to describe the patchy nature of hair loss, though diffuse thinning may also occur.

As previously stated, in secondary syphilis the treponemes are disseminated throughout the body. Any organ may be affected, and a plethora of clinical presentations have been described. In practice these features are rare and only occur in a very small percentage of patients (Mindel *et al.*, 1989). Neurological involvement includes meningitis, cranial nerve palsies and transverse myelitis. Iridocyclitis and choroidoretinitis may be present. Arthritis, periostitis and bursitis are also seen. Hepatitis with or without hepatomegaly has been described, and may be associated with a disproportionately high alkaline phosphatase. However, in many of these cases other causes for hepatitis, in particular hepatitis A, B or C, could not be excluded. Splenomegaly, gastritis and gastric ulceration have also been reported. The renal tract may be involved in the form of glomerulonephritis and nephrotic syndrome.

The symptoms and signs of secondary syphilis eventually resolve spontaneously. However, the rash may heal with depigmentation, especially around the back of the neck. This distribution of hypopigmented macules is known as the collar of Venus.

The severity of the clinical features of primary or secondary syphilis may vary greatly between individuals. It is unclear whether severe symptoms are more likely to be associated with an adverse outcome.

Latency

The resolution of the manifestations of secondary syphilis occurs as a direct result of the generation of an adequate immune response to control treponemal replication. As previously stated, in two-thirds of patients the infection will not progress to late complications.

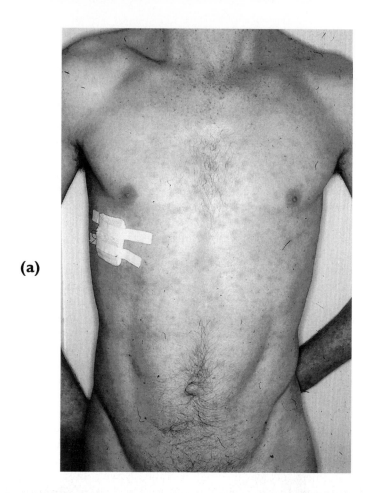

(a)

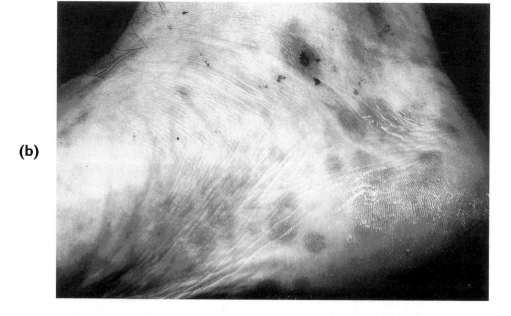

(b)

FIGURE 17.2 Rash of secondary syphilis showing (a) trunk and (b) sole of foot involvement.

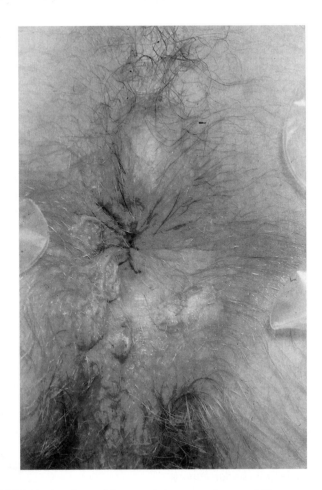

FIGURE 17.3 Perianal condylomata lata: secondary syphilis

Late Complications

Cardiovascular Syphilis

Around 10% of patients will develop late cardiovascular complications, which typically present after a latent period of 15–30 years. Large vessel disease predominates, but smaller vessels can also be involved. The treponemes induce a chronic inflammatory process in the vasa vasorum of the artery walls which results in endarteritis. The media, elastic tissue, adventitia and intima are all involved in the disease process. The aorta is most commonly affected, especially the ascending section, but lesions of the coronary ostia, the valves and myocardium may occur.

Clinically, the common manifestations include aortitis with or without coronary ostial stenosis; aneurysm of the ascending aorta; and aortic valve incompetence. The aneurysms may remain asymptomatic for years, but as they enlarge pressure is put on adjacent structures. This can result in persistent chest pain, or even hoarseness due to pressure on the laryngeal nerve; Horner's syndrome due to compression of the sympathetic chain; stridor and cough caused by tracheal compression; or superior vena caval compression. Aortic incompetence can present as angina, acute left ventricular failure and paroxysmal nocturnal dyspnoea.

Examination findings are similar to those of aortic incompetence and aneurysm from other causes. Chest X-ray may show a mediastinal mass with calcification surrounding the aneurysm, or may be normal.

Neurological Syphilis

Dissemination of treponemes into the central nervous system occurs early in the infection, but only 10% of individuals experience late neurological sequelae. Numerous manifestations have been described, many of which are now only rarely seen. In most developed nations, where there is widespread use of antibiotics, it is possible that some of the classic presentations have been modified by unrelated prescribing of an antibiotic with antitreponemal activity. Late neurosyphilis can be classified into three subgroups: asymptomatic early or late; meningovascular, cerebral and/or spinal; and parenchymatous, general paresis and tabes dorsalis.

Asymptomatic Neurosyphilis

This describes patients with positive cerebrospinal fluid treponemal tests in whom there are no symptoms or signs. 'Early' refers to patients within the first 5 years of initial infection. The CSF contains fewer than 100 lymphocytes per mm^3, a normal or slightly raised protein, and a non-treponemal test is positive in over 90% of cases (Merritt *et al.*, 1946). Serology is almost always positive, but occasionally negative results occur, especially in HIV infection.

Meningovascular Disease

As previously described, secondary syphilis may be complicated by a number of neurological symptoms. In late disease the onset of symptoms is often gradual. Cerebral involvement is due to infarction resulting from endarteritis. It may include cranial nerve palsies – usually the third, sixth, seventh and eighth cranial nerves; pupillary changes – the pupils become small and irregular and do not react to light, but retain the accommodation reflex, known as 'Argyll Robertson' pupils; and syndromes resulting from vessel occlusion. Occlusive disease can present suddenly, in a similar way to cerebrovascular accidents due to other causes, or be preceded by headache, dizziness or even psychiatric symptoms (Merritt *et al.*, 1946). Spinal involvement occurs only rarely and is almost always associated with cerebral pathology.

Parenchymatous Disease

General Paresis (General Paralysis of the Insane, Dementia Paralytica, Paretic Neurosyphilis)   Direct treponemal invasion of the cerebral cortex leads to a slowly evolving meningoencephalitis, which usually presents in middle age. The condition is progressive and will result in death if untreated. The onset of symptoms is usually insidious, beginning with impairment of executive functions, personality changes and memory loss, and progressing to include emotional lability, confusion and disorientation. In some, florid delusions occur which were once thought to be characteristic, but which probably occur in only 10–20% (Dawson-Butterworth and Heathcote, 1970). Seizures may also occur. Examination reveals a lack of facial expression, known as paralytic facies; pupillary changes and optic atrophy; tremors of the facial muscles, tongue and lips; changes in handwriting; dysarthria, and occasionally aphasia; and exaggerated tendon reflexes and extensor plantar reflexes. Untreated, the process progresses with features of dementia, incontinence, progressive paralysis and seizures.

Tabes Dorsalis This is an uncommon manifestation of late syphilis. Features of tabes occur after a long latent period of 20–25 years, and correspond to dorsal column degeneration. Initially, lightning pains, which are sudden, shooting pains in the legs lasting a few minutes at a time, occur with parasthesiae. Tendon reflexes are reduced, vibration and joint position sensation is diminished and Romberg's sign is positive. Ataxia with a broad-based stamping gait, incontinence, poor vision due to optic atrophy, and tabetic crises occur later. Tabetic crises are related to lightning pains and most commonly affect the stomach. Intense pain is experienced which may mimic an acute surgical abdomen. A classic, though relatively rare, manifestation is the Charcot joint. This is a grossly distorted joint with osteophyte formation around it and with loose bodies inside the joint. The joint itself is hypermobile and the destruction is painless, giving rise to great disability and perpetuating further joint destruction. Even without treatment tabes can burn out with time, but treatment cannot reverse the late changes.

Gummatous Disease

Gummas are granulomatous lesions which occur between 3 and 12 years after the primary infection. They appear to represent a hypersensitivity reaction to a small number of organisms (Magnuson *et al.*, 1956). Gummas appear as brownish-red nodules of varying size in any tissue or organ, but are most commonly seen in skin and bone. They are composed of coagulative necrosis surrounded by a mononuclear infiltrate and encapsulated by proliferating connective tissue. If mucous membranes are involved the gumma may break down to form well circumscribed, punched-out ulcers. Healing of ulcerative lesions occurs with extensive scarring, known as tissue-paper scarring. Gummas can produce a great deal of tissue damage and considerable disfigurement.

Congenital Syphilis

In contrast to most other congenitally acquired sexually transmitted infections, syphilis is transmitted *in utero* by transplacental passage of the treponemes. The risk of transmission is related to the maternal treponemal load, hence early maternal infection carries a much greater risk of transmission than late infection. Indeed, almost all babies born to mothers with untreated primary or secondary infection will be infected, but the likelihood of transmission and the severity of infection will fall with subsequent pregnancies, as the mothers' treponemal load will fall with time.

In a similar way to acquired syphilis, the underlying pathological process in congenital syphilis is a vasculitis, typically around small vessels, with resulting necrosis and fibrosis.

Syphilis in pregnancy has a number of potential outcomes: spontaneous abortion, which may occur during the second and early third trimesters; early congenital syphilis, which is a syndrome similar to adult secondary disease and which occurs during the first 2 years of life; late congenital syphilis, a syndrome which usually presents around puberty and is similar to late adult disease; and latent congenital syphilis, in which no signs nor symptoms are present and the diagnosis is made on incidental positive syphilis serology.

Early Congenital Syphilis

There is often a delay in diagnosis of several weeks, as most neonates do not display signs at birth (Nabarro, 1954). The early clinical manifestations usually become apparent in the third to eighth weeks of life. Most babies present with failure to thrive, rhinitis or pneumonia. Classically the infant is snuffly, with a clear then purulent then bloody nasal discharge which is a rich source of treponemes. Chondritis with destruction of the nasal septum may ensue, and laryngeal involvement may lead to an aphonic cry. Skin lesions occur in 30–60% and are generally similar to the adult lesions of secondary syphilis. In addition, vesiculobullous rash may occur, with involvement of the palms and soles with desquamation. Mucous membrane lesions also occur. Many other clinical manifestations have been reported, and include osteochondritis, lymphadenopathy, anaemia, nephrotic syndrome (rare), choroidoretinitis, glaucoma and uveitis.

Late Congenital Syphilis

Late congenital syphilis is similar to late adult disease except for the relative sparing of the cardiovascular system. The manifestations can be thought of as malformations (stigmata), which occur because of infection at a crucial stage of development or growth; and inflammatory lesions, which are presumed to arise once the fetus has an adequate immune capacity (Thomas, 1946).

Stigmata include frontal bossing; saddle-shaped nose; anterior bowing of the tibia ('sabre tibia'); incisors with central notching, known as Hutchinson's incisors; and Moon's molars, which are poorly enamelized, oddly shaped molars.

Of the inflammatory lesions, interstitial keratitis is the most common (Stokes *et al.*, 1934). It usually presents in childhood as misty or blurred vision, photophobia and excess lacrimation. Cochlear degeneration may occur, resulting in deafness. When interstitial keratitis occurs together with Hutchinson's incisors and deafness, the collection of signs is known as Hutchinson's triad and is strongly predictive of congenital syphilis. Musculoskeletal involvement takes the form of osteoperiostitis or symmetrical, painless swelling of the knees, known as Cluttton's joints. Gummas of the hard and soft palate begin in late childhood. Juvenile paresis occurs in 1–5% of cases of congenital syphilis, and tends to be more rapidly progressive than the adult form.

Diagnosis

Immediate diagnosis is sometimes possible if dark-field microscopy is performed on nasal secretions or skin lesions. Serology at birth can be difficult to interpret as maternal antibody can cross the placenta. After 6 weeks, positive infant serology is consistent with infection. The fluorescent treponemal antibody-absorbed assay (FTD-abs) can be used on the baby's blood for IgM, which is more likely to represent neonatal infection, but sensitivity and specificity can be a problem (Kaufman *et al.*, 1974; Reimer *et al.*, 1975).

Prevention of Congenital Syphilis

Strategies for prevention have been classified into three groups (Schulz *et al.*, 1989): (1) case detection: treatment of all cases of primary and secondary syphilis in the community with contact tracing, together with rescreening of high-risk pregnant women after their initial serology; this would reduce the incidence of early syphilis in the childbearing population; (2) follow-up of positive serologies by investigating positive reports from the laboratory, and selective screening of high-risk groups; this would reduce the prevalence of latent syphilis; and (3) routine prenatal screening.

Antenatal testing is also a major line of prevention in many developed countries. In the majority of cases this takes the form of serology testing on one occasion in pregnancy, but this approach has inherent problems: a woman may have very early syphilis and therefore negative serology, or she may acquire syphilis after screening; positive serology in a woman with a previous history of treated syphilis may be attributed to the old infection and a reinfection missed; a positive result may not be given sufficient priority and treatment may be given late or inadequately. Despite these potential pitfalls, antenatal screening is extremely useful and has helped to dramatically reduce the incidence of congenital syphilis in most developed countries.

Diagnostic Tests for Syphilis (Table 17.1)

The diagnosis of syphilis has a number of components: the presence of compatible symptoms and signs; consistent laboratory test results; and, in the case of complicated syphilis, the results of chest X-rays and CSF examination.

Tests for syphilis fall into two main categories: direct visualization of the organism, which may be possible in primary and secondary syphilis; and serological testing using both treponemal and non-treponemal tests, which should be performed whenever a diagnosis of syphilis is considered. Newer tests include a treponemal enzyme-linked immunosorbent assay (ELISA), though this is not in common usage and, in an experimental context, polymerase chain reactions for treponemal DNA have been used.

Direct Visualization

Dark-field Microscopy

As previously described, *T. pallidum* has such small diameter that it is not visible under light microscopy. However, dark-field microscopy can be used (Figure 17.4) and is an important diagnostic tool in primary, secondary and

TABLE 17.1 Diagnostic tests for syphilis: percentage of cases at different stages of disease in which each test is reactive

Stage of Disease	Tests: Percentage Reactive		
	VDRL (RPR)	TPHA	FTA-Abs
Primary	50–75	50–70	90
Secondary	> 99	> 99	> 99
Latent	75	97	97
Late	75	> 99	> 99
Treated	50	50–80	50–80
Biological false positive	100	0	0

In patients treated in early syphilis, serology usually becomes unreactive after the first year following treatment.
Serology may remain positive for life in patients with late syphilis, despite adequate treatment. The VDRL is usually only present at low titre.
VDRL, venereal Disease reference laboratory test;
RPR, rapid plasma reagin test;
TPHA, *Treponema pallidum* haemagglutination assay;
FTA-Abs, fluorescent treponemal antigen-absorbed test.

early congenital syphilis, when chancres, skin lesions, nasal secretions (congenital syphilis) and lymph nodes can be a rich source of the organism (Manual of Tests for Syphilis, 1969).

The technique approaches 80% sensitivity (Daniels and Ferneyhough, 1977), but relies on the careful obtaining of specimens. Wet lesions should be irrigated with normal saline, lightly abraded with gauze, then firmly squeezed to obtain a serous exudate – not blood – which can then be applied to the slide. The skin over lymph nodes should be cleansed with saline, and a small amount of saline injected and then aspirated back. Nasal secretions may be examined directly. Diagnosis depends on the demonstration of spiral motile treponemes, which are characteristically said to display corkscrew, watchspring and angular motion. In the absence of demonstrable treponemes, sampling should be performed on 3 successive days before the test is said to be negative. An antibiotic history is essential, as prior treatment with an antibiotic with antitreponemal activity may reduce the number of organisms in a lesion such that microscopy is negative in the presence of active disease. In practice, any undiagnosed genital ulcer should be sampled for dark-field microscopy. This method has the advantage of instantaneous diagnosis.

Direct Fluorescence Antibody Tests

Specimens collected under the same conditions as above may also be examined with fluorescent antibody stains. This method has the advantage of use in biopsy and autopsy specimens.

Serological Tests

There are two types of serological test used for the diagnosis of syphilis: non-treponemal and treponemal. In practice, both are used. At present, none of

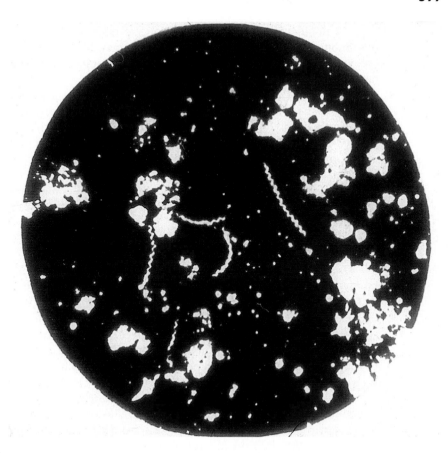

FIGURE 17.4 Dark-field microscopic preparation of *Treponema pallidum*.

the routinely used tests permit differentiation of one treponemal infection from another, and this may cause diagnostic difficulty, especially when trying to disentangle yaws from latent syphilis.

Non-treponemal Tests

In response to treponemal infection, non-specific antibodies directed towards anticardiolipin, cholesterol and lecithin are generated which can be detected in the serum at titres that relate to disease activity. This forms the basis of the Venereal Disease Reference Laboratory (VDRL) and rapid plasma reagin (RPR) tests, which have largely superseded the Wassermann reaction (Larsen *et al.*, 1995). It is important to note that these tests are not positive in very early syphilis, and only become positive 3–5 weeks after contracting the infection. The VDRL is a flocculation assay and positive results are expressed as a titre, indicating the lowest dilution at which the test remains positive. This may be helpful in assessing the stage and activity of disease, as active and early disease and reinfection are usually associated with high titres. In treated patients, a fourfold increase in titre is taken to be an indicator of reinfection or reactivation. The VDRL may stay positive for life, even after effective treatment.

If antibody is present in very high concentration, serum agglutination may be inhibited and, paradoxically, the VDRL may be negative. This effect is known as the prozone reaction, and the sample must be diluted prior to assay to determine the true titre.

The antibodies detected by the VDRL test are not specific and can be detected in the serum in a small percentage of the population in the absence of treponemal disease. This is known as a biological false positive reaction. The antibodies may occur transiently in response to an acute viral or mycoplasma infection or vaccination, or be a feature of a healthy pregnancy. The chronic forms may be associated with connective tissue disorders, most commonly systemic lupus erythematosus. The specific treponemal tests must be negative for this diagnosis to be made, and it is usual to investigate such cases to search for underlying pathology.

Treponemal Tests

These tests establish past or present treponemal disease. Two are commonly used: the fluorescent treponemal antibody absorbed test (FTA-Abs) (Scotti and Logan, 1968) and a *Treponema pallidum* haemagglutination assay (TPHA), or microhaemagglutination assay for *Treponema pallidum* (MHA-TP) (Jaffe *et al.*, 1978). The FTA-Abs is an indirect fluorescent antibody test and is the very first test to become positive, at around 3–4 weeks after infection. It may be the only test to be positive in primary syphilis, and is positive in 85–90% of such cases. The FTA-Abs test normally remains positive for life, except in some cases of primary syphilis if treatment is given very early. The TPHA/MHA-TP is a qualitative haemagglutination test which relies on passive agglutination of sheep erythrocytes which have been sensitized with material from *T. pallidum* (Kennedy, 1990). It is the last test to become positive. Like the FTA-Abs, it usually remains positive for life even if the infection has been fully treated.

The serological tests can also be applied to cerebrospinal fluid, in the search for evidence of neurological involvement. All three are used, as the VDRL can be negative in patients with active neurosyphilis. In general, cases of neurosyphilis are accompanied by positive serum and CSF tests, but it is possible to have positive CSF serology in the absence of positive serum serology.

Molecular Biology-based Tests – PCR

Although still largely experimental, PCR has a potentially useful role in certain clinical scenarios in which existing tests may be unable to confirm or exclude a diagnosis of syphilis. These include neurosyphilis, where serology only achieves 50% sensitivity; early primary syphilis; differentiation of new from old infection; and finally congenital syphilis, in which passive transfer of maternal antibody makes serological tests difficult to interpret (Larsen *et al.*, 1995). The test is based on recognition of treponemal DNA which codes for outer membrane lipoproteins (Grimprel *et al.*, 1991; Noordhoek *et al.*, 1991).

Syphilis and HIV Infection

The advent of HIV has been partly responsible for the increased worldwide interest in syphilis, provoked initially by the observation that positive syphilis serology was common in HIV-positive individuals (Rogers *et al.*, 1983). The interplay between HIV and syphilis is complex and may affect the ease of transmission of both diseases, the clinical presentation, the natural history, the interpretation of serological tests and the treatment of syphilis in HIV-infected individuals.

The presence of genital ulceration has been shown to enhance the transmission of HIV (Cameron *et al.*, 1989). In the case of syphilis, ulcerative lesions provide a break in the skin's normal defenses through which virus could more easily pass. It is also possible that cells of the monocyte/macrophage lineage, which would be present in increased numbers at ulcerative sites and which act as natural reservoirs for HIV, could also function as transporters of virus both within the infected individual and between individuals (Pantaleo *et al.*, 1993).

The natural history of syphilis in HIV-infected individuals may be different from classically described features in the immunocompetent patient. There have been a number of case reports of very rapidly progressive syphilis or florid clinical presentations (Musher *et al.*, 1990; Hook 1989), but larger studies have failed to demonstrate conclusive differences in the clinical presentation or rate of progression of syphilis in HIV-positives (Hutchinson *et al.*, 1991). The possibility of neurosyphilis should always be borne in mind in the investigation of neurological symptoms in an HIV-positive individual (Centres for Disease Control and Prevention, 1993).

For the majority of patients coinfected with syphilis and HIV, the principles of interpretation of the routine serological tests for syphilis apply (Centres for Disease Control and Prevention, 1993). However, abnormal serological responses are possible: delayed or absent seroconversion has been reported (Hicks *et al.*, 1987); extremely high titres in non-treponemal tests giving rise to false negative results (Musher, 1991), perhaps due to polyclonal B-cell activation, have been described; and failure of non-treponemal tests to decline with time following adequate treatment is well recognized (Larsen *et al.*, 1995). A high index of suspicion should be maintained and increased use made of alternative diagnostic tests, such as biopsy of lesions, aspiration of lymph nodes and dark-field examination (Centres for Disease Control and Prevention, 1993).

Current guidelines for the treatment of syphilis in an HIV-infected patient emphasize the need for aggressive, prolonged treatment with close follow-up (Centres for Disease Control and Prevention, 1993). They are summarized in a later section.

Management of Syphilis

The aims of treatment are twofold: to eliminate the bacteria and prevent the development of complications, and to prevent spread of the infection either vertically or horizontally. Penicillin, which so radically changed the face of syphilis in the 1940s, is still the mainstay of treatment. The duration of the antibiotic regimen is largely determined by how long the infection has been present, and by the presence or absence of complications.

Treponemal replication in late infection is believed to be slow, and high levels of penicillin are necessary to penetrate the central nervous system. The half-life of penicillin in the blood may be prolonged and tissue penetrance enhanced by the concomitant use of oral probenecid (Goh *et al.*, 1984), which is favoured by some clinicians. A number of different regimens are in use for the treatment of syphilis, though few have been tested in randomized controlled trials. Treatment choice is often guided principally by the local availability of drugs and, to some extent, patient acceptability. Treatment regimens have been comprehensively detailed in the following publications: Centres for Disease Control and Prevention, 1993; Holmes *et al.*, 1989; Thin *et al.*, 1995. A number of commonly used treatment options are shown in Table 17.2. The relative convenience of benzathine penicillin has made it a popular choice in many parts of the world. Of concern are the reports of poor penetrance into the CSF (Rein, 1981), suboptimal CSF concentrations with conventional dosages (Dunlop, 1985), and treatment failures following its use in HIV infection (Berry *et al.*, 1987). Some experts recommend avoiding benzathine penicillin if suitable alternatives are available, particularly for the treatment of syphilis in HIV-infected individuals.

A flu-like reaction with headache, myalgia and fever may occur within the first 24 hours of starting therapy; this is known as the Jarisch–Herxheimer reaction, and is believed to be due to a rapid reduction in treponemal load. It occurs in 50% of cases of primary syphilis and 70% of cases of secondary syphilis (Catterall, 1974). The symptoms are transient and can be relieved by simple antipyretics. In contrast, the Jarisch–Herxheimer reaction associated with the treatment of late syphilis can be severe if, for example, the aorta or cerebral vasculature is involved in the disease process. Systemic steroids are often prescribed for 2–3 days prior to treatment of late syphilis, to good effect. In pregnancy, the Jarisch–Herxheimer reaction that arises from the treatment of primary and secondary disease in the second and third trimesters carries a risk of fetal loss and should be closely monitored.

As previously described, syphilis may have different characteristics in the HIV-infected patient. It has been suggested that intramuscular treatment should be continued for at least 17 days regardless of the stage of syphilis, though the rationale for prolonged treatment has not been tested in clinical trials.

The prognosis following adequate treatment of primary, secondary and latent syphilis is excellent. In advanced disease, tissue destruction may be so advanced in critical sites such as the aorta that treatment has little influence on morbidity and mortality. Monitoring of the patient following therapy is essential, as organisms may persist at low levels in sanctuary sites such as the CSF, despite clinical cure. The VDRL test is useful to monitor disease activity, and a persistent fourfold increase in titre is usually indicative of reinfection, reactivation or treatment failure.

EPIDEMIOLOGY AND EXPECTED IMPACT OF AN EFFECT OF VACCINE

In the early part of the 20th century syphilis was endemic throughout Europe, North America and probably most of Africa and Asia, and the problem was exacerbated by social breakdown and huge population movements during the First World War (Adler, 1980). Following that war, the availability of clinical services and social stabilization led to a reduction in the incidence of syphilis.

■ **TABLE 17.2 Management of syphilis**

Type of syphilis	Treatment	Alternative
Primary Secondary Early or late latent (CSF negative)	Procaine penicillin 600 mg i.m. daily for 10 days	1. Doxycycline 100 mg (O) three times a day for 21 days 2. Benzathine penicillin G 2.4 million units i.m. single dose
Late syphilis: cardiovascular gummatous disease, or neurosyphilis	Procaine penicillin 1.8 g i.m. daily for 17 days and probenecid 500 mg (O) four times a day for 17 days with prednisolone 30 mg (O) daily for 3 days, starting the day before treatment	1. Doxycycline 100 mg (O) three times a day for 28 days with prednisolone 30 mg (O) daily for 3 days, starting the day before treatment 2. Benzathine penicillin G 7.2 million units total, as 3 doses of 2.4 million units i.m. at weekly intervals
Syphilis of any stage in an HIV-positive patient	As for late syphilis	
Pregnancy: primary secondary early or late latent	Procaine penicillin regimen as in non-pregnant above	1. Erythromycin 500 mg (O) four times a day for 21 days* 2. Benzathine penicillin G 2.4 million units i.m. single dose
Pregnancy: late syphilis cardiovascular gummatous disease, or neurosyphilis	Procaine penicillin regime, as in non-pregnant, above	1. Erythromycin 500 mg (O) four times a day for 28 day* 2. Benzathine penicillin G 7.2 million units total, as 3 doses of 2.4 million units i.m. at weekly intervals
Congenital syphilis: aged under 2 years	Procaine penicillin 50 mg/kg i.m. daily for 14 days	
Congenital syphilis: aged over 2 years	As for latent/late acquired syphilis	

i.m, = intramuscular; O, = orally; * = treat baby at birth with procaine penicillin.

However, this reduction was short-lived and the incidence of syphilis increased rapidly during and immediately following the Second World War (Aral and Holmes, 1990). In the late 1940s and early 1950s syphilis again declined rapidly as a consequence of the availability of clinical services, and in particular the introduction of penicillin (Adler, 1991).

In the 1960s and 1970s there was an increase in syphilis in many parts of the developed world, particularly in males, largely as a consequence of homosexual sex (Mindel *et al.*, 1987). Most of this increase was seen in primary, secondary and early latent syphilis, whereas late and congenital syphilis have continued to decrease in most parts of the developed world.

The 1980s saw the widespread promotion of safer sex messages as a consequence of the HIV epidemic, and this led to a reduction in the incidence of infectious syphilis in many parts of Europe, North America and Australia (Adler, 1991). In Western Europe, Australia and some parts of North America the incidence of syphilis has continued to decrease during the 1990s (Adler, 1991), and in many European and Australian cities, early infectious syphilis is only rarely diagnosed.

Unfortunately, Eastern Europe and some inner city areas in North America have seen an alarming rise in syphilis in recent years (Tichonova *et al.*, 1997). In several of the newly independent states of the former Soviet Union, syphilis has increased 15–30 times, from 5–50/100 000 in 1990 to 120–170/100 000 in 1994 (WHO, 1992). A considerable part of this is probably due to social breakdown in the former USSR, resulting in poverty, unemployment and large-scale migration. In addition, the availability of quality of health care has declined and the process of partner notification has all but evaporated. Among other factors that may be important are the rise in prostitution, the availability of sexually oriented materials, including pornography, and rapid and profound changes in sexual mores and behaviours (WHO, 1992).

The situation in North America has been quite different. The incidence of syphilis began to increase in the United States amongst heterosexuals in the mid-1980s, from 13.7 cases/100 000 in 1981 to 184/100 000 in 1989 (CDC, 1993). Most of this was in minority groups, particularly in inner cities of the East and the South. Much was associated with crack cocaine use, particularly when the drug was smoked (Dunn and Rolfs, 1991). A consequence of the epidemic was a dramatic increase in neonatal syphilis, from 158 cases in 1983 to 7219 cases in 1990. Fortunately, over the last 5 years the incidence has again decreased. However, an estimated 130 000 new infections still occur each year in the USA (CDC, 1996).

The situation in the developing world is different. Syphilis remains a major burden in many developing countries, for both adults and children (Meheus and de Schryver, 1989). For example, in Ethiopia in 1995 syphilis seroprevalence was 12.8% among blood donors (Rahlenbeck *et al.*, 1997), and a study in South Africa among antenatal patients showed a seroprevalence of 6.5–9.0% (Wilkinson *et al.*, 1997). WHO estimates in 1995 (in adults aged 15–49) for sub-Saharan Africa were 3.14% in males and 3.89% in females (Gerbase, 1998). In Asia, syphilis remains a common sexually transmitted disease, particularly in some of the larger cities, where commercial sex work appears to be a major source of infections (Kunawarak *et al.*, 1995). In South and Southeast Asia in 1995, WHO estimates in adults aged 15–49 put the prevalence at 1.42% for males and 1.77% for females (Gerbase, 1998). Worldwide, the World Health Organization (WHO) has estimated that up to 12 million new cases of syphilis occur each year in adults (Gerbase, 1998).

Where they have been introduced, public health measures to control syphilis have been very successful. Such measures include health promotion, including easy and cheap availability of condoms, early detection of cases, effective diagnosis and treatment and contact tracing (Adler, 1996). The importance of a breakdown of public health services to control syphilis is highlighted in the re-emergence of syphilis in Russia and other Eastern European countries. In these nations, public health measures were very successful in controlling syphilis until 1992, when social, political and economic problems led to a major reduction in public health services and a dramatic rise in infectious and congenital syphilis (see Chapter 2).

The likely impact of an effective vaccine is difficult to determine, as many authorities believe that most cases of syphilis are transmitted by so-called core transmitters, and whether an effective vaccine campaign could effectively target such individuals has yet to be determined. In addition, given that public health measures, including diagnosis and treatment, are relatively cheap and proven effective, the likely impact of vaccines is further undermined. None-the-less, an effective vaccine would enhance current public health programs and would be particularly helpful in some 'high-risk' communities.

Vaccine Research and Other Strategies for Control: Background

As discussed above, natural history studies suggest that approximately one-third of individuals infected with *Treponema pallidum* are able to clear their infection without long-term consequences. This protection is at best partial, as is evidenced by the fact that some individuals may acquire the infection for a second time (Mindel *et al.*, 1987), and studies in the 1950s in prisoners in Singapore showed that subjects with previously treated primary or secondary syphilis (11 individuals) developed skin lesions when challenged with *T. pallidum*, as evidenced by the presence of spirochetes on dark-ground microscopy, whereas only half of 26 individuals with previously treated late latent syphilis and none of five with untreated late latent syphilis developed lesions (Magnuson *et al.*, 1956).

Vaccine Strategies

Vaccine strategies may be divided into three broad categories: eradication, elimination and containment (Barbosa-Cesnik *et al.*, 1997). Eradication is the complete removal of the organism, and consequently all disease processes associated with it. This has been achieved with smallpox. Elimination is the disappearance of the disease, whereas the organism may remain in humans and host animals. Containment is the control of the disease without elimination or removal of the organism, so that it is no longer a public health problem. Recently, consideration has been given to a fourth strategy, namely disease modification, where the vaccine decreases the severity of subsequent infection but does not necessarily prevent it. In relation to syphilis, a successful vaccine would need to prevent the acquisition of disease and thereby the long-term consequences, prevent reinfection, prevent onward transmission, and prevent the acquisition of congenital syphilis. An additional benefit would be to reduce the transmission of HIV.

In order for eradication to be successful, the disease has to be confined to humans, requires identifiable clinical manifestations, there should be no subclinical or latent infection, the disease should be of low infectivity, and immunity should be lifelong (Adimora *et al.*, 1994; Begg and Miller, 1990; Noah, 1983). Unfortunately, syphilis does cause subclinical infection, and the infectivity is moderate to high.

The implementation of an effective vaccine program requires the identification and accessibility of the target population. In practice this has never been tested as a public health measure for the control of an STD (with the possible exception of hepatitis B). It is likely that most STD vaccine strategies, including that for the control of syphilis, would involve containment rather than eradication or elimination. The target population would be sexually active young adults. Ideally the vaccine would be given to preadolescent children prior to the commencement of sexual activity. Such children are readily accessible through schools, and although this strategy poses a number of ethical and moral dilemmas, it would probably be the most effective and cost-efficient method for controlling syphilis. Other target groups could include men and women attending STD clinics, and women attending antenatal clinics (Barbosa-Cesnik *et al.*, 1997).

One strategy that has been postulated to achieve maximum coverage would focus on two goals: first, containment by targeting sexually active young adults, mostly in the 15–49-year age group, with particular priority given to those aged between 15 and 25; and second, aiming at the possibility of elimination by including syphilis vaccination as part of an adolescent program. All of this would need to be linked with effective education, counselling, availability of condoms, and early disease detection and management (WHO, 1992).

The development of an effective vaccine against syphilis has been hampered by the inability to continuously cultivate *T. pallidum* subspecies *pallidum* (Lukehart, 1992). Consequently, much of the earlier research was concentrated on the use of inactivated organisms, vaccines produced from different species of spirochetes, the use of passive immunity, including immune serum, and the passive transfer of specific immune cells.

Inactivated *T. pallidum* Vaccines and Vaccines Produced with Other Treponemes

Numerous animal experiments, mainly in rabbits, have demonstrated partial protection with inactivated *T. pallidum* (Miller, 1967, 1973; Jones *et al.*, 1976; Metzger and Smogor, 1975) (Table 17.3) and, as mentioned above, the single human experiment conducted in Singapore supports this effect (Magnuson *et al.*, 1956). Other experiments with different species of treponemes, including *Treponema minutum*, *T. ambigua*, *T. refringens*, *T. microdentium* and *T. aurantia*, showed partial protection in rabbits when rechallenged with *T. pallidum* (Al-Samarrai and Henderson, 1976; Graves *et al.*, 1984) (Table 17.4). Further experiments in hamsters suggested that prior exposure to *Treponemas* could prevent reinfection with the same species. This was demonstrated with *T. pallidum* Nichols strain (the cause of syphilis), *T. pallidum* Bosnia A (the cause of endemic syphilis), and *T. pallidum* pertenue (the cause of yaws), and cross-resistance was demonstrated with *T. pallidum* Bosnia A to both of the other types (Schell *et al.*, 1982).

TABLE 17.3 Animal experiments with inactivated T. Pallidum vaccines

Animal	No of research animals	No of controls	Antigen	Adjuvant	Outcome measure	Outcome	Author
Dutch male rabbits	15 for 12 weeks 17 for 24 weeks	10	γ-irradiated *T. pallidum* suspension	Nil	ID challenge with 500 virulent T pallidum	No protection after 12 weeks Partial protection after 17 weeks	Miller, 1967
Dutch male rabbits	24 experiment 1	22 experiment 1	γ-irridiated *T. pallidum* suspension	Nil	Clinical, dark-ground and serological response to intradermal infection with *T. pallidum* and	Complete protection for the 94-day observation period	Miller, 1973
	10 experiment 2	10 experiment 2			Clinical, dark-ground and serological response to intradermal infection with *T. pertenue* Haiti strain	Partial protection	
New Zealand White rabbits	18	5	Glutaraldehyde-fixed *T. pallidum* Nichols strain	Six different adjuvants were used: Incomplete Freunds (IFA) Complete Freunds (CFA) Zymosan (ZYM), Polyadenylic–polyuridylic acid (PUA) Phytohaemagglutinin P (PHA) Alumina C gel (ALC)	Clinical and immunological response to intratesticular and intradermal challenge to *T. pallidum*	Partial protection with vaccines containing PHA or ALC	Jones *et al.*, 1976
Danish White rabbits	8 pre-treated i.v. with non viable *T. pallidum* suspension 4 pre-treated i.v. with virulent *T. pallidum*	6 normal rabbits 3 infected with *T. pallidum*	Non-viable *T. pallidum* suspension	Nil	Clinical response to ID challenge	Lymphocytes from immunized donors and from those with syphilis conferred partial immunity	Metzger and Smogor, 1975

TABLE 17.4 Animal experiments with other spirochetes

Animal	No of research animals	No of controls	Antigen	Adjuvant	Outcome measure	Outcome	Author
Adult male New Zealand White rabbits	8	1	Suspension of *T. pallidum* Nichols (non pathogenic) *T. pallidum* Kazan 2,4,5 & 8, *T. refringens*, *T. minutum*, *T. ambigua*, *T. microdentium*	Freunds	Response to Intratesticular injection with *T. pallidum* Nichols	Partial protection	Al Samarrai & Henderson, 1976
Male rabbits	15	8	*Spirochaeta aurantia*	Complete and incomplete Freunds	Clinical response to intradermal infection with *T. pallidum*	No effect	Graves *et al*, 1984

Cell-derived Vaccines

The importance of cell-mediated immunity was first demonstrated by Metzger and Smogor (1975), who showed that lymphocytes from rabbits infected with *T. pallidum* gave some partial protection to uninfected rabbits. These results were duplicated by other workers in rabbits, hamsters and guinea pigs (Metzger *et al.*, 1980; Schell *et al.*, 1980; Schell *et al.*, 1981; Pavia, 1986). Further studies showed that both T-lymphocyte helper and T-lymphocyte cytotoxic suppressor cells derived from infected hamsters were able to confer some protection in uninfected hamsters (Liu *et al.*, 1991). All of these experiments using intra-dermal challenge with *T. pallidum* documented clinical disease at the inoculation site. Additional endpoints in some experiments included serological response and documenting the number of treponemes in draining lymph nodes.

Serum-derived Vaccines (Table 17.5)

In addition to cell-mediated immunity, it is evident that passive transfer of immune serum is also protective (Perine *et al.*, 1973; Sepetjian *et al.*, 1973; Turner *et al.*, 1973; Bishop and Miller, 1976; Weiser *et al.*, 1976; Graves and Alden, 1979; Smogor and Metzger, 1984; Pavia *et al.*, 1985). Studies by Wicher *et al* (1992) demonstrated that purified immunoglobulin antibody to *T. pallidum* was less effective than whole immune serum, possibly supporting the contention of Engel (1985) that immune complexes were important in conferring some protection.

Specific and Cloned Treponemal Antigen Vaccines

Recent experiments have concentrated on specific antigens, including endoflagella antigen, TmpA, TmpB and TmpC (Champion *et al.*, 1990; Wicher *et al.*, 1992) (Table 17.6). TmpB, also known as TpN36, appears to offer some protection, as does another antigen, TpN19 (Adimora *et al.*, 1994). Endoflagella antigen TmpA and TmpC do not appear to be immunogenic.

More recently, cloned *T. pallidum* genes have opened up a number of new approaches, including the possibility of purified recombinant antigens, synthetic peptide antigens or vector antigens. However, as yet there are no reports of the safety, immunogenicity and efficacy of such vaccines in either animals or humans. The possibility of using BCG as a vector for antigens to a number of infective agents, including *T. pallidum*, has been postulated, and this may prove to be a highly efficient method for producing a vaccine (Barbosa-Cesnik *et al.*, 1977; Sell and Hsu, 1993).

Human Experiments

As discussed above, the only human experiment was conducted in prisoners in Singapore in the early 1950s, using live *T. pallidum* (Magnuson *et al.*, 1956). This experiment suggested that this approach offered at best only partial protection. Concerns about safety have prevented further human experimentation. However, the availability of cloned *T. pallidum* antigens may reawaken interest.

TABLE 17.5 Animal experiments with serum-derived vaccines

Animal	No of Research animals	No of controls	Antigen	Adjuvant	Outcome measure	Outcome	Author
New Zealand White male rabbits	Experiment 1 8	8	Hyperimmune serum from rabbits infected with *T. pallidum*	Nil	Clinical and immunological response 24 h after intradermal infection with *T. pallidum*	Recession of chancres in experimental animals after day 10	Perine *et al.*, 1973
	Experiment 2 4	9			Clinical and immunological response to intradermal infection with *T. pallidum* 24 h after infection	Delay in the onset of chancres	
Fawn male Burgundy rabbits	2	3	Pooled sera from rabbits infected with syphilis	Nil	Clinical and immunological response to intradermal challenge with *T. pallidum*	Partial protection	Sepetjian *et al.*, 1973
New Zealand White male rabbits	4–6 in each experiment (8 experiments)	4–6 in each experiment (8 experiments)	Hyperimmune serum (HIS) from rabbits infected with *T. pallidum*	Nil	Clinical response to intradermal infection with *T. pallidum*	Partial protection against local and disseminated infection	Turner *et al.*, 1973
NZ Albino rabbits	5	5 with non immune rabbit serum 5 with saline daily injections for 37 days	Immune rabbit serum (IRS) daily injections for 37 days	Nil	Time to lesion development and lesion characteristics	IRS rabbits had a longer incubation period and developed atypical lesions	Bishop and Miller, 1976
Zealand White male rabbits	6	6	Hyperimmune serum (HIS) from rabbits infected with *T. pallidum*	Nil	Clinical and immunological response to intradermal infection with *T. pallidum*. HIS was given for 2 days before and 6 days after challenge	Partial protection with both experiments. However only experiment 1 resulted in a delay in the development of lesions	Weiser *et al.*, 1976
					Clinical and immunological response to intradermal infection with *T. pallidum* HIS was given 2 h before and 8 days after challenge		

(continues)

TABLE 17.5 (continued)

Host	Number	Immune serum	Adjuvant	Assessment	Results	Reference
Adult rabbits	15	Immune serum	Nil	Number of lesions and latent period following ID injection with *T. pallidum*	Serum from 3, 4 and 6 months decreased the number of lesions and increased the latent period	Graves and Alden, 1979
Danish White rabbits	41	Rabbits infected with *T. pallidum* and then treated with horse antirabbit thymocyte serum (ATS), or normal horse serum (NHS)	Nil	Clinical and immunological response to ID challenge	ATS treatment caused a marked reduction in T lymphocytes and CMI and lack of resistance to reinfection	Smogor and Metzger, 1984
Strain 2 guinea pigs	12	Immune serum from previously infected guinea pigs	NA	Clinical, and serological response to intradermal infection with *T. pallidum*	Fewer lesions containing fewer organisms. Slower and less intense immune response	Pavia et al., 1985
Female outbred rabbits	12	Immune complexes derived from patients with syphilis	Freunds	Clinical and serological response to intradermal infection with *T. pallidum*	Partial protection and delayed lesion development	Engel et al., 1985
Inbred strain-2 Guinea pigs	24 injected with *T. pallidum* Nichols strain; 18 injected with *T. pallidum* Nichols strain	Immune serum from pathogenic treponemes (TPI IgG), non-pathogenic treponemes (NTPI IgG) or from normal guinea pigs (NGPS IgG)	Complete and incomplete Freund's	Resistance to ID challenge	TPI IgG offered partial protection, less protection was given by NTPI IgG and no protection by NGPS IgG	Wicher et al., 1992

TABLE 17.6 Specific treponemal antigen vaccines

Animal	No of Research animals	No of controls	Antigen	Adjuvant	Outcome measure	Outcome	Author
Male C4D guinea pigs	30	10	Recombinant *T. pallidum* antigens TmpA, TmpB, TmpC, TmpA and TmpB and TmpC, and *E. coli* membranes	Monophospholipid (MPL)	Clinical and immunological response to intradermal challenge with *T. pallidum*	Partial protection with TmpB antigen	Wicher *et al.*, 1991
New Zealand White male rabbits	4	Number not mentioned	*T. pallidum* endoflagella	Freund's complete and incomplete	Clinical, dark ground and serological response to intradermal infection with *T. pallidum*	No protection from infection and an altered course of infection	Champion *et al.*, 1990

MICROBIOLOGY

Classification

Treponema pallidum, subspecies pallidum, the causative organism of syphilis, is a spirochaetal organism belonging to the genus Treponema, which includes three other human pathogens and a number of non-pathogenic treponemes. Of the pathogenic organisms, T. pallidum subspecies pertenue causes yaws; T. pallidum subspecies endemicum is the organism responsible for bejel, and T. carateum is the causative organism for pinta. T. carateum is currently ascribed a separate subspecies owing to a lack of knowledge of its genetic profile. The human pathogenic treponemes are very closely related. They are morphologically indistinguishable and share 95% DNA sequence homology by hybridization (Fieldsteel and Maio, 1982). They also exhibit a close correlation in their respective protein profiles and reactivity to monoclonal antibodies.

Structure

T. pallidum assumes a spiral shape between 6 and 20 μm long and around 0.18 μm in diameter. The organism has a tight helical structure, with between 6 and 14 helices per cell. It appears wave-like on fixation, with a wavelength of 1.1 μm and an amplitude of 0.2–0.3 μm. Dark-field microscopy of fresh specimens reveals corkscrew motility, which is a rapid rotation of the organism about its own axis caused by the action of flagellae inserted into each end and extending down within the periplasmic space. Flexion and reversal of rotation also occur, which are said to be hallmarks of the pathogenic treponemes (Clarkson, 1956). The flagellae themselves are complex in structure. They are composed of three core proteins and a single unrelated sheath protein. The organism has an outer membrane, which contains only a small number of poorly immunogenic transmembrane proteins (Radolf, 1995). Beneath the outer membrane lie a peptidoglycan layer and a cytoplasmic membrane. Embedded in the periplasmic leaflet of the cytoplasmic membrane are highly immunogenic lipoproteins of uncertain function, which are thought to play a crucial role in immune responses during infection (Figure 17.5) (Radolf, 1995).

Physiology, Culture and Metabolism

T. pallidum is an obligate human parasite. Continuous culture of T. pallidum in vitro is currently impossible, although limited replication has been described in the presence of mammalian cells (Cumberland and Turner, 1949; Fieldsteel and Maio, 1982), with a slow rate of replication of about 30 hours. It is now known to be a microaerophilic organism (Norris et al., 1987) which metabolizes glucose and pyruvate aerobically, is capable of nucleic acid synthesis, and possesses a cytochrome system (Austin et al., 1981).

Infection/Invasion

The treponemes probably enter the body via microabrasions in the epithelium, caused by the trauma of intercourse, and then migrate down into the dermis.

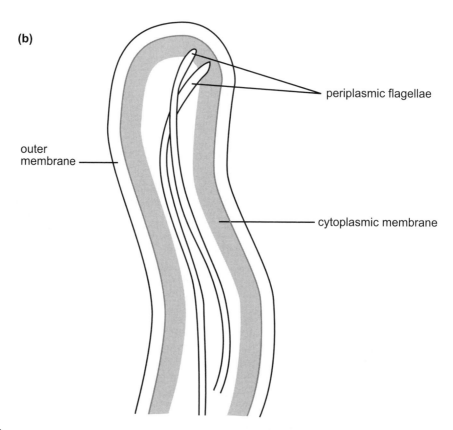

FIGURE 17.5 Schematic representations of the morphology of *Treponema pallidum*. (a) Whole organism to show helical structure. (b) Detail of ultra-structure to show periplasmic flagellae and cytoplasmic membrane.

They attach rapidly to the surface of the host cells, although *in vitro* they do not appear to enter the cell (Fitzgerald *et al.*, 1975) but penetrate endothelial junctions and tissue layers (Thomas *et al.*, 1989).

IMMUNOLOGY/IMMUNOBIOLOGY OF INFECTION

Localized infection at the site of invasion (chancre) is followed by wide dissemination of the organism throughout the body in the secondary stage of infection. Late complications are largely the result of a chronic inflammatory reaction (Sell and Norris, 1983), which continues over many years, even when the treponemal load is very low. In gummatous disease it is difficult to isolate the organism from within the lesions, but heavy infiltrates of lymphocytes and mononuclear cells are seen around the area of central necrosis. Two suggestions have been put forward to explain gumma formation (Grin, 1993): they may arise as a result of a hypersensitivity reaction to superinfection in an already sensitized individual, or they may result from reactivation of treponemes in a sensitized individual with untreated or partially treated disease.

In cardiovascular syphilis an obliterative endarteritis occurs which is characterized by lymphocytic and plasma cell perivascular infiltrates. As with gumma formation, this process continues many years into the infection, again supporting an immune-based pathogenesis rather than a direct infective process.

The pathological changes of late neurosyphilis also reflect an ongoing inflammatory reaction: lymphocytes and plasma cells infiltrate the meninges and perivascular areas, there is iron deposition in vessel walls and microglia, and degeneration of nerve cells is seen (Swartz, 1989). In this case, treponemes may be detected in brain tissue.

T. pallidum's immune-evasiveness is the key to its success as a pathogen. Despite an active immune response in early infection, eradication of the organism only occurs in one-third of cases. In the majority of infected individuals it persists at low levels throughout the clinically latent phase, although it is unclear whether the organism itself is truly latent. Quite how this is achieved is uncertain, but may be due to its sequestration in immune sanctuary sites; masking of the organism's surface by host proteins; or the paucity of outer membrane proteins to act as antigenic targets (Norris, 1993).

Role of Outer Membrane Proteins and Cytoplasmic Lipoproteins in Immunopathogenesis

Recent work has overturned previous dogma on the nature of immunogenicity of *T. pallidum*. Early studies suggested that an outer coat made up of serum proteins and mucopolysaccharides surrounded the organism and effectively provided it with a poorly immunogenic shield. However, reports that this structure could not be seen by electron microscopy questioned the hypothesis (Cox *et al.*, 1992). This led to the suggestion that the immunogenic proteins may be hidden away from immune effectors, under the cell surface. They were subsequently found to be associated with the cytoplasmic membrane (Radolf *et al.*, 1988), and then further localized to the periplasmic leaflet (Cox *et al.*, 1992). The finding that these proteins are only present in very small numbers (Blanco *et al.*, 1994) further restricts their potential as useful immune targets. Further studies revealed that these molecules are in fact lipoprotein, and that it is the lipid component that is so intensely immunogenic (Radolf *et al.*, 1994). Research continues to characterize these proteins, and it will be interesting to see whether they will play a role in future vaccine development.

Role of Humoral Immunity

IgM and IgG to certain polypeptides can be detected before the appearance of clinical disease (Baker-Zander *et al.*, 1985). In primary and secondary infection, both IgM and IgG to treponemal polypeptides are present in the serum. IgM titres decline in later disease and with treatment (Norris, 1993). By secondary and early latent infection, antibodies to most of the major polypeptides are present. As the infection persists antibody titres fall, but some IgG remains in the serum and accounts for the persistently positive VDRL tests in some patients. In patients who receive adequate treatment very early in the infection, antibody may decline so low that it becomes undetectable by conventional diagnostic serology

Role of Cell-mediated Immunity

Previous studies have implicated a role for cell-mediated immunity in treponemal infection (Sell and Norris, 1983). At present, the precise character of these T-cell responses is unclear.

CONCLUSIONS

Animal experiments suggest that in rabbits, hamsters and guinea pigs, and perhaps in humans, partial immunity can be achieved by using either live or inactivated *T. pallidum*, and more recent studies suggest that specific antigens can be used to produce an effective immune response. There is a growing body of evidence to suggest that delayed-type hypersensitivity is the major mechanism in controlling syphilis infection. This is believed to be mediated by T cells secreting lymphokines, which in turn activate macrophages, which destroy the organisms. In future, the production of vaccines will need to take into consideration the importance of delayed-type hypersensitivity at the serum and mucosal level.

Whether recombinant antigens alone will be sufficient to produce protective immunity remains unknown, and the importance of adjuvants to boost the response will need to be evaluated. The results of further studies are awaited with interest.

REFERENCES

Adimora, A.A., Sparling, P.F. and Cohen, M.S. (1994). Vaccines for classic sexually transmitted diseases (review). *Infect. Dis. Clin. N. Am.* 8: 859–876

Adler, M.W. (1980) Medical history. The terrible peril: a historical perspective on the venereal diseases. *Br. Med. J.* 281: 206–211.

Adler, M.W. (1991). In: 'Oxford Textbook of Public Health. Sexually Transmitted Diseases'. (eds: Holland, W.W., Detels, R. and Knox, G.), Sexually Transmitted Diseases. Oxford pp. 345–355.

Adler, M.W. (1996). Sexually transmitted diseases control in developing countries. *Genitourinary Med.* 72: 83–88

Al-Samarrai, H.T. and Henderson, W.G. (1976). Immunity in syphilis. Studies in active immunity. *Br. J. Vener. Dis.* 52(5): 300–308

Aral, S.O. and Holmes, K.K. (1990). Epidemiology of sexual behavior and sexually transmitted diseases. In: Sexually Transmitted Diseases (eds. Holmes, K.K. Bardls, P.A., Sparling, P.E., Wiesner, P.J.). McGraw Hill, New York.

Austin, F.E., Barbieri, J.T., Corin, R.E., Grigas, K.E. and Cox. C.D. (1981). Distribution of superoxide dismutase, catalase and peroxidase activities among *Treponema pallidum* and other spirochaetes. *Infect. Immun.* **33** (2): 372–379

Baker-Zander, S.A., Hook, E.W. III, Bonin, P., Handsfield, H.H. and Lukehart, S.A. (1985). Antigens of *Treponema pallidum* recognised by IgG and IgM antibodies during syphilis in humans. *J. Infect. Dis.* **151**: 264–272

Barbosa-Cesnik, C.T., Gerbase, A. and Heymann, D. (1997). STD vaccines – an overview. *Genitourinary Med.* **73**: 336–342

Begg, N. and Miller, E. (1990). Role of epidemiology in vaccine policy. *Vaccine* **8**: 180–189

Berry, C.D., Hooton, T.M., Collier, A.C. and Lukehart, S.A. (1987). Neurological relapse after benzathine penicillin therapy for secondary syphilis in a patient with HIV infection. *N. Engl. J. Med.* **316** (25): 1587–1589

Bishop, N.H. and Miller, J.N. (1976). Humoral immunity in experimental syphilis. I. The demonstration of resistance conferred by passive immunization. *J. Immunol.* **117** (1): 191–196

Blanco, D.R., Reimann, K., Skare, J. *et al.* (1994). Isolation of the outer membrane from Treponema pallidum and Treponema vincentii. *J. Bacteriol.* **176** (19): 6088–6099

Cameron, D.W., Simonsen, J.N., D'Costa, L.J. *et al.* (1989). Female to male transmission of human immunodeficiency virus type 1: risk factors for seroconversion in men. *Lancet* **2**: 403–407

Catterall, R.D. (1974). In: 'A Short Textbook of Venereology' 2nd edition. (ed. Catterall, R.D.), The treatment of syphilis and the treponematoses, pp.152–153. English Universities Press Limited, London

Centers for Disease Control and Prevention. (1993). Sexually transmitted disease guidelines. MMWR 42 (RR-14: 1–102)

Champion, C.I., Miller, J.N., Borenstein, L.A., Lovett, M.A. and Blanco, D.R. (1990). Immunization with *Treponema pallidum* endoflagella alters the course of experimental rabbit syphilis. *Infect. Immun.* **58** (9): 158–161

Clarkson, K.A. (1956). Technique of dark field examination. *Med. Tech. Bull.* **7**: 199

Cox, D.L., Chang, P., McDowall, A. and Radolf, J.D. (1992). The outer membrane, not a coat of host proteins, limits antigenicity of virulent *Treponema pallidum*. *Infect. Immun.* **60** (3): 1076–1083

Cumberland, M.C. and Turner, T.B. (1949). The rate of multiplication of *Treponema pallidum* in normal and immune rabbits. *Am. J. Syph. Gon. Vener. Dis.* **33**: 201–211

Daniels, K.C. and Ferneyhough, H.S. (1977). Specific direct fluorescence antibody detection of treponema pallidum. *Health Lab. Sci.* **14**: 164–171

Dawson-Butterworth, K. and Heathcote, P.E.M. (1970). Review of hospitalised cases of general paralysis of the insane. *Br. J. Vener. Dis.* **46** (4): 295–302

Dunlop, E.M.C. (1985). Survival of treponemes after treatment: Comments, clinical conclusions and recommendations. *Genitourinary Med.* **61** (5): 293–301

Dunn, R.A. and Rolfs, R.T. (1991). The resurgence of syphilis in the United States. *Curr. Opin. Infect. Dis.* **4**: 3–11

Engel, S., Diezel, W. and Sonnichsen, N. (1985). Protective immunization against *Treponema pallidum* using specific immune complexes – an attempt. *Acta Dermatol. Venereol.* **65** (6): 484–8

Fieldsteel, A.H. and Maio, R.M. (1982). In: 'Pathogenesis and Immunology of Treponemal Infection' (eds. Schell, R.F. and Musher, D.M.), *Genetics of Treponema*. Marcel Dekker, New York, pp. 39–54

Fitzgerald, T.J., Miller, J.N. and Sykes, J.A. (1975). Treponema pallidum (Nichols strain) in tissue cultures: cellular attachment, entry, and survival. *Infect. Immun.* **11** (5): 1141–1146

Gerbase, A.C., Rowley, J.T., Heymann, D.H.L., Berkley, S.F.B. and Piot P. (1998). Global prevalence and incidence estimates of selected curable STDs. *Sex. Transm. Infect.*, **74** (Suppl 1): S12–S16.

Gjestland, T. (1955). The Oslo study of untreated syphilis: An epidemiologic investigation of the natural course of syphilitic infection based on a restudy of the Boeck-Bruusgaard material. *Acta Dermatol. Venereol.* **35** (Suppl) (Stockh) 34: 1

Goh, B.T., Smith, G.W., Samarasinghe, L., Singh, V. and Lim, K.S. (1984). Penicillin concentrations in serum and cerebrospinal fluid after intramuscular injection of aqueous procaine penicillin 0.6 MU with and without probenecid. *Br. J. Vener. Dis.* **60**: 371–373

Graves, S. and Alden J. (1979). Limited protection of rabbits against infection with *Treponema pallidum* by immune rabbit sera. *Br. J. Vener. Dis.* **55** (6): 399–403

Graves., S., Drummond, L. and Strugnell, R. (1984). Immunization of rabbits with *Spirochaeta aurantia* does not induce resistance to *Treponema pallidum*. *Sex. Transm. Dis.* **11** (1): 1–5

Grimprel, E.P.J., Sanchez, G.D., Weudel, J.M. *et al.* (1991). Use of polymerase chain reaction and rabbit infectivity testing to detect *Treponema pallidum* in amniotic fluid. *J. Clin. Microbiol.* **29**: 1711–1718

Grin, E.L. (1993). Epidemiology and control of endemic syphilis: Report on a mass treatment campaign in Bosnia. WHO Monograph Series, Geneva, WHO

Harris, A., Rosenberg, A.A. and Riedel, L.M. (1946). A microflocculation test for syphilis using cardiolipin antigen: preliminary report. *J. Vener. Dis. Inform.* **27**: 159–72

Heller, J.R. Jr and Bruyere, P.T. (1946). Untreated syphilis in the male negro. II: mortality during 12 years of observation. *J. Vener. Dis. Inform.* **27**: 34

Hicks, C.B., Benson, P., Lupton, G.P. and Tramont, E.C. (1987). Seronegative secondary syphilis in a patient infected with HIV with Kaposi's sarcoma: a diagnostic dilemma. *Ann. Intern. Med.* **107**: 492–495

Holmes, K.K., Mardh, P-A., Sparling, P.F. *et al.* (1989). In: 'Sexually Transmitted Diseases', 2nd edn. Appendix: Sexually Transmitted Diseases Treatment Guidelines, pp.A7-A11 McGraw-Hill

Hook, E.W. III. (1989). Syphilis and HIV infection. *J. Infect. Dis.* **160**: 530–534

Hudson, E.H. (1968). Christopher Columbus and the history of syphilis. *Acta Tropica* **25** (1): 1–16

Hutchinson, C.M., Rompalo, A.M., Reichart, C.A., and Hook, E.W. III. (1991). Characteristics of patients with syphilis attending Baltimore STD clinics: multiple, high-risk subgroups and interaction with human immunodeficiency virus infection. *Arch. Intern. Med.* **151**: 511–516

Jaffe, H.W., Larsen, S.A., Jones, O.G. and Dans, P.E. (1978). Haemagglutination tests for syphilis antibody. *Am. J. Clin. Pathol.* **70**: 230–233

Jones, A.M., Ziegler, J.A. and Jones, R.H. (1976). Experimental syphilis vaccines in rabbits. I. Differential protection with an adjuvant spectrum. *Br. J. Vener. Dis.* **52** (1): 9–17

Kaufman, R.E., Olansky, D.C. and Wiesner, P.J. (1974). The FTA-ABS (IgM) test for neonatal congenital syphilis: a critical review. *J. Am. Vener. Dis. Assoc.* **1**(2): 79–84

Kennedy, E.J. Jr. (1990). In: 'A manual of tests for syphilis', 8th edition. (eds. Larsen, S.A., Hunter, E.F. and Kraus, S.J.), Microhaemagglutination assay for antibodies to *Treponema pallidum* (MHA-TP), pp.153–166. American Public Health Association, Washington DC

Kunawararak, P., Beyrer, C., Natpratan, C. *et al.* (1995). The epidemiology of HIV and syphilis among male commercial sex workers in northern Thailand, **9** (5): 517–521

Larsen, S.A., Steiner, B.M. and Rudolph, A.H. (1995). Laboratory diagnosis and interpretation of tests for syphilis. *Clin. Microbiol. Rev.* **8**(1): 1–21

Liu, H., Alder, J.D., Steiner, B.M., Stein-Streilein, J., Lim, L. and Schell, R.F. (1991). Role of L3T4+ and 38+ T-cell subsets in resistance against infection with *Treponema pallidum* subsp. pertenue in hamsters. *Infect. Immun.* **59** (2): 529–536

Lukehart, S.A. (1985). Prospects for development of a treponemal vaccine. *Rev. Infect. Dis.* **7** Suppl 2: S305–S313

Lukehart, S.A. (1992). In: 'Sexually Transmitted Diseases' (ed. Quinn, T.), Immunology and pathogenesis of syphilis, pp.141–163. Raven Press, New York

Magnuson, H., Thomas, E. and Olansky, S. (1956). Inoculation syphilis in human volunteers. *Medicine* **35**: 33–82

Manual of Tests for Syphilis. Atlanta Venereal Disease Program, US Communicable Disease Centre. 1969 (USPHS Publication 411)

Matej, H., Metzger, M. and Smogor, W. (1973). Artificial immunization of rabbits against experimental syphilis. II. Imunological side effects of the treponemal vaccine. *Arch. Immunol. Ther. Exp.* **21** (2): 263–20

Meheus, A. and De Schryver, A. (1989). Sexually transmitted diseases in developing countries. *Curr. Opin. Infect. Dis.* **2**: 25–30

Merritt, H.H., Adams, R.D. and Solomon, H.C. (1946). In 'Neurosyphilis', p.21. New York, Oxford

Metzger, M. and Smogor, W. (1975). Passive transfer of immunity to experimental syphilis in rabbits by immune lymphocytes. *Arch. Immunol. Ther. Exp.* **23** (5): 625–630

Metzger, M., Podwinska, J. and Smogor, W. (1980). Development of immunological responsiveness and resistance to infection with *Treponema pallidum* in rabbits given immune lymphocyte preparations. *Arch. Immunol. Ther. Exp.* **28** (2): 329–336

Miller, J.N. (1967). Immunity in experimental syphilis. V. The immunogenicity of *Treponema pallidum* attenuated by gamma-irradiation. *J. Immunol.* **99** (5): 1012–1016

Miller, J.N. (1973). Immunity in experimental syphilis. VI. Successful vaccination of rabbits with *Treponema pallidum*, Nichols strain, attenuated by gamma-irradiation. *J. Immunol.* **119** (5): 1206–1215

Mindel, A., Tovey, S.J. and Williams, P. (1987). Primary and secondary syphilis, 20 years' experience. 1 Epidemiology. *Genitourinary Med.* **63**: 361–364

Mindel, A., Tovey, S.J., Timmins, D.J. and Williams, P. (1989). Primary and secondary syphilis, 20 years' experience. 2. Clinical features. *Genitourinary Med.* **65**: 1–3

Musher, D.M., Hamil, R.J. and Baughn, R.E. (1990). Effect of human immunodeficiency virus infection on the course of syphilis and on the response to treatment. *Ann. Intern. Med.* **113**: 872–881

Musher, D.M. (1991). Syphilis, neurosyphilis, penicillin and AIDS. *J. Infect. Dis.* **163**: 1201–1206

Nelson, R.A. and Mayer, M.M. (1949). Immobilisation of *Treponema pallidum in vitro* by antibody produced in syphilitic lesions. *J. Exp. Med.* **89**: 369–383

Noah, N. (1983). The strategy of immunization. *Commun. Med.* **5**: 140–147

Noordhoek, G.T., Wolters, E.C., De Jonge, M.E.J. and van Embden, J.D. (1991). Detection by polymerase chain reaction of *Treponema pallidum* DNA in cerebrospinal fluid from neurosyphilis patients before and after antibiotic treatment. *J. Clin. Microbiol.* **29**: 1976–1984

Norris, S.G., Alderete, J.F., Axelsen, N.H. *et al.* (1987). Identity of *Treponema pallidum* subsp. pallidum polypeptides: correlation of sodium dodecyl sulfate-polyacrylamide gel electrophoresis results from different laboratories. *Electrophoresis* **8**: 77–92

Norris, S.J. and the Treponema Pallidum Polypeptide Research Groups. (1993). Polypeptides of *Treponema pallidum*: progress toward understanding their structural, functional and immunologic roles. *Microbiol. Rev.* **57** (3): 750–779

Pantaleo, G., Graziosi, C. and Fauci, A.S. (1993). The immunopathogenesis of human immunodeficiency virus infection. *N. Engl. J. Med.* **328**: 327–335

Pavia, C.S., Niederbuhl, C.J. and Saunders. J. (1985). Antibody-mediated protection of guinea-pigs against infection with *Treponema pallidum*. *Immunology*, **56** (2): 195–202

Pavia, C.S. (1986). Transfer of resistance to syphilitic infection from maternal to newborn guinea pigs. *Infect. Immun.* **51** (1): 365–368

Perine, P.L., Weiser, R.S. and Klebanoff, S.J. (1973). Immunity to syphilis. I. Passive transfer in rabbits with hyperimmune serum. *Infect. Immun.* **8**(5): 787–790

Radolf, J.D., Chamberlain, N.R., Clausell, A. and Norgard, M.V. (1988). Identification and localisation of integral membrane proteins of virulent *Treponema pallidum* subsp pallidum by phase partitioning with the nonionic detergent triton X-114. *Infect. Immun.* **56** (2): 490–498

Radolf, J.D., Arndt, L.L., Atkins, D.R. *et al.* (1994). Treponema pallidum and Borrelia burgdorferi lipoproteins and synthetic lipopeptides activate monocytes/macrophages. *J. Immunol.* **154** (6): 2866–2877

Radolf, J.D. (1995). Treponema pallidum and the quest for the outer membrane proteins. *Molec. Microbiol.* **16** (6): 1067–1073

Rahlenbeck, S.I., Yohannes, G., Molla, K., Reifen, R. and Assefa, A. (1997). Infections with HIV, syphilis and hepatitis B in Ethiopia: a survey in blood donors. *Int. J. STD AIDS* **8** (4): 261–264

Reimer, C.B., Black, C.M., Phillips, D.J. *et al.* (1975). The specificity of fetal IgM: Antibody or antiantibody? *Ann. NY. Acad. Sci.* **254**: 77–93

Rein, M.F. (1981). Treatment of neurosyphilis. *JAMA* **246** (22): 2613–2614

Rockwell, D.H., Yobs, A.R. and Moore, M.B. (1964). Tuskegee study of untreated syphilis. *Arch. Intern. Med.* **114**: 792–798

Rogers, M.F., Morens, D.M., Stewart, J.A. *et al.* and the Task Force on Acquired Immune Deficiency Syndrome. (1983). National case control study of Kaposi's sarcoma and *Pneumocystis carinii* pneumonia in homosexual men. Part 2. Laboratory results. *Ann. Intern. Med.* **99**: 151–158

Rosahn, P.D. (1947). Autopsy studies in syphilis. *J. Vener. Dis. Inf.* Suppl 21, US Public Health Service, Venereal Disease Division

Schaudinn, F.R. and Hoffmann, E. (1905). Vorläufiger Bericht über das Vorkomen von Spirochaeten in syphilitischen Krakheitsproducten und bei Papillomen. Arbeiten aus dem K Gesundheitsamte, **22**: 527

Schell, R.F., Chan, J.K., LeFrock, J.L. and Bagasra, O. (1980). Endemic syphilis: transfer of resistance to *Treponema pallidum* strain Bosnia A in hamsters with a cell suspension enriched in thymus-derived cells. *J. Infect. Dis.* **141** (6): 752–758

Schell, R.F., LeFrock, J.L., Chan, J.K. and Bagasra, O. (1981). SLH hamster model of syphilitic infection and transfer of resistance with immune T cells. *Adv. Exp. Med. Biol.* **134**: 291–300

Schell, R.F., Azadegan, A.A., Nitskansky, S.G. and LeFrock, J.L. (1982). Acquired resistance of hamsters to challenge with homologous and heterologous virulent treponemes. *Infect. Immun.* **37** (2): 617–621

Schulz, K.F., Murphy, F.K., Patamasucon, P. and Meheus, A.Z. (1989). Congenital Syphilis. In: 'Sexually Transmitted Diseases', 2nd ed, (eds. Holmes, K.K., Mardh, P.-A., Sparling, P.F. *et al.* pp. 821–842, McGraw-Hill

Scotti, A.T. and Logan, L. (1968). A specific IgM antibody test in neonatal congenital syphilis. *J. Paediatr.* **73**: 242–243

Sell, S. and Norris, S.J. (1983). The biology, pathology and immunology of syphilis. *Int. Rev. Exp. Pathol.* **24**: 204–276

Sell, S. and Hsu, P.L. (1993). Delayed hypersensitivity, immune deviation, antigen processing and T-cell subset selection in syphilis pathogenesis and vaccine design. *Immunol. Today* **14**: 576–582

Sepetjian, M., Salussola, D. and Thivolet, J. (1973). Attempt to protect rabbits against experimental syphilis by passive immunization. *Br. J. Vener. Dis.* **49** (4): 335–337

Smogor, W. and Metzger, M. (1984). Administration of antithymocyte serum to syphilitic rabbits inhibits development of resistance to reinfection with *Treponema pallidum*. *Arch. Immunol. Ther. Exp.* **32** (1): 37–42

Stokes, J.H., Beerman, H. and Ingraham, N.R. (1934). In: 'Modern Clinical Syphiliology'. Philadelphia, Saunders

Swartz, M.N. (1989). Neurosyphilis. In: 'Sexually Transmitted Diseases', 2nd edn, (eds. Holmes, K.K., Mardh, P-A., Sparling, P.F. *et al.* pp. 231–250, McGraw-Hill

Thin, R.N., Barlow, D., Bingham, J.S. and Bradbeer, C. (1995). Investigation and management guide for sexually transmitted disease (excluding HIV). *Int. J. STD. AIDS.* **6**: 130–136

Thomas, D.D., Fogelman, A.M., Miller, J.N. and Lovett, M.A. (1989). Interactions of treponema pallidum with endothelial cell monolayers. *Eur. J. Epidemiol.* **5**: 15–21

Thomas, E.N. (1946). Syphilis: its course and management. Macmillan, New York

Tichonova, L., Borisenko, K., Ward, H., Meheus, A., Gromyko, A. and Renton, A. (1997). Epidemics of syphilis in the Russian Federation: trends, origins and priorities for control. *Lancet* **350**: 210–213

Turner, T.B., Hardy, P.H. Jr, Newman, B. and Nell, E.E. (1973). Effects of passive immunization on experimental syphilis in the rabbit. *Johns Hopkins Med. J.* **133** (5): 251–251

Weiser R.S., Erickson D., Perine P.L., Pearsall N.N. (1976). Immunity to syphilis: passive transfer in rabbits using serial doses of immune serum. *Infect. Immun.* **13** (5): 1402–1407

Wicher, K., Schouls, L.M., Wicher, V., Van Embden, J.D. and Nakeeb, S.S. (1991). Immunization of guinea pigs with recombinant TmpB antigen induces protection against challenge infection with *Treponema pallidum* Nichols. *Infect. Immun.* **59** (12): 4343–4348

Wicher, K., Zabek, J. and Wicher, V. (1992). Effect of passive immunization with purified specific or cross-reacting immunoglobulin G antibodies against *Treponema pallidum* on the courses of infections in guinea pigs. *Infect. Immun.* **60** (8): 3217–3223

Wilkinson, D., Sach, M. and Connolly, C. (1997). Epidemiology of syphilis in pregnancy in rural South Africa: opportunities for control. *Trop. Med. Int. Health* **2**(1): 57–62

World Health Organization. (1992). Global Program on AIDS. Second meeting on AIDS drug and vaccine supply. Potential vaccination strategies using HIV vaccines in developing countries. Geneva: WHO, 30 July 1992

18
HIV VACCINES: MILESTONES AND NEW INSIGHTS

BONNIE J. MATHIESON

Office of AIDS Research,
National Institutes of Allergy and Infectious Diseases,
Bethesda, Maryland, USA

INTRODUCTION

Despite the fact that extensive research has gone into the pathogenesis and etiology of the human immunodeficiency virus (HIV), a vaccine to prevent infection is still elusive more than 15 years after the first milestone in HIV/AIDS vaccine research was achieved – the identification of HIV as the causative agent of acquired immunodeficiency syndrome (AIDS) (Gallo *et al.*, 1983; Barre-Sinnoussi *et al.*, 1983; Levy *et al.*, 1984). There have been several reviews on HIV and AIDS vaccine research published in recent years (Letvin, 1998; Baltimore and Heilman, 1998; Schultz, 1998; Frey, 1999). The reader is referred to these and other more specific reviews for further insight into a particular topic. Unfortunately, the need for an effective HIV vaccine is even greater now that it was in 1983–84. The UNAIDS estimated in November 1998 that 16 000 new cases of HIV infection occur annually worldwide, and over 30 million adults and children are living with AIDS. This chapter will focus on why there is a growing conviction, with many obstacles and difficulties still unresolved at the time of writing, that an HIV vaccine is possible.

WHAT IS NEEDED IN AN HIV VACCINE?

From a public health standpoint the goal for preventive HIV vaccines is to slow and eventually halt the HIV pandemic and to protect individuals from infection.

Complete or 'Sterilizing' Protection

Under ideal circumstances an HIV vaccine would provide complete protection from infection. This is often referred to as sterilizing protection, and almost certainly requires the induction and maintenance of high-titer antibody that can neutralize a broad range of potential incoming virus isolates, interfere with viral spread from the site of infection, and also possibly target the initial infected cells through antibody-dependent cell-mediated cytotoxicity (ADCC). This type of vaccine would benefit an individual who was vaccinated and prevent overt infection or detectable viremia by the various routes by which HIV might be acquired, e.g. sexual transmission at mucosal surfaces, either through vaginal or penile exposure, anal receptive intercourse or oral exposure to HIV; bloodborne transmission through shared needles and/or paraphernalia involved with injected drug use or untested blood products (which is still occurring in many parts of the developing world); or mother-to-infant oral transmission which occurs through breastfeeding. An effective vaccine would stop the epidemic once adequate vaccine coverage was attained in groups of individuals with risk of exposure. In any of these modes, mother-to-infant transmission could be prevented by effective vaccine-induced protection of women.

Control Viral Replication and Course of Disease

An HIV vaccine might not provide complete protection from infection, but could induce cellular immune effectors and memory cells (either MHC class I restricted cytotoxic T lymphocytes (CTL) or CD8+ T cells which release effector chemokines that can control viral replication) against cells that harbor the pathogenic virus. Alternatively, effective cellular immunity might be able to reduce the first rounds of viral replication, contributing either to a decrease in the the peak of the viremia or to a suppression of the total level of virus in the body. Early and sustained immunological control of HIV replication could by itself reduce set-point viral load and/or viral shedding, thereby preventing secondary transmission of HIV to sexual partners and from mother to infant. Many antiviral vaccines may work in the latter fashion. Because HIV is an integrating virus, this type of vaccine might not effectively prevent eventual progression to AIDS in an individual, but it could halt the epidemic if the viral load were sufficiently reduced to prevent viral shedding.

Vaccine-induced diminution in viral load also might be correlated with prevention or a delay in disease progression in a vaccinated individual. A correlation between low viral load and long-term non-progression has been observed in both epidemiologic investigations of HIV in adults (Mellors *et al.*, 1995) and children (Dickover *et al.*, 1998), and in SIV intervention studies in macaques, e.g. vaccine studies (Hirsch *et al.*, 1996), and studies with passive administration of immunoglobulin (Haigwood *et al.*, 1996) or immune serum (Van Rompay *et al.*, 1998). The efficacy of vaccines in preventing disseminated HIV infection can be determined relatively easily in the conduct of a vaccine trial by quantitative assessment of viral RNA in plasma, culture of viruses from peripheral blood mononuclear cells (PBMC), or PCR of plasma or cells. However, assessment of long-term control of viral replication, immunosuppression and disease progression, which may be critical for assessing efficacy, may be unfeasible in populations with access to highly effective, multiple-drug anti-retroviral regimens for early HIV therapy. At the very least, the conduct of efficacy trials will be substantially more complex.

Combination Vaccine Approaches

A combined approach that engages both cellular effectors and the humoral arm of the immune system for effective systemic antibodies, as well as mucosal barriers that might be needed for a highly effective HIV vaccine, is currently favored by most investigators. Previous studies with a wide range of viral vaccines and immune studies would support the notion that prestimulated humoral immunity can be effective in reducing the initial viral peak, even if it is not effective at preventing the initial focus of infection. MHC class I-restricted CTLs are unlikely to directly eliminate virus. However, they are uniquely designed to eliminate infected cells which display processed HIV antigen epitopes on the cell surface.

This series of goals serially parallels the concepts that have been tested for HIV/AIDS vaccines and the evolving understanding of the complex pathogenesis of the HIV. At the same time HIV continues to present a series of challenges to vaccine design. It has been established that HIV that can vary its genes, particularly in the envelope protein, to evade both the humoral and the cellular immune response (McKeating et al., 1993; Lewis et al., 1998; Borrow et al., 1997); can integrate into the host genome and remain present in immune memory T-cell reservoirs for very long periods of time (Finzi et al., 1997; Wong et al., 1997; Chun et al., 1997a,b; Finzi et al., 1999); and can infect and kill or deregulate the various cells, CD4+ T cells, macrophages and dendritic cells required to generate an effective immune response against this pathogen (reviewed in Levy, 1998).

WHAT HAVE WE LEARNED FROM VACCINE APPROACHES AND CANDIDATES THAT HAVE BEEN TRIED?

Whole inactivated virus vaccines

One of the first 'standard' approaches used to vaccinate non-human primate models was chemically inactivated HIV-1 and SIV, or whole inactivated virus (WIV). Early experiments in macaques looked very promising (Desrosiers et al., 1989; Murphey-Corb et al., 1990; Carlson et al., 1990). Inactivated SIV from various stocks provided high levels of protective immunity and a second milestone was reached. However, when studies to define the mechanism of protection were undertaken, it was determined that experimental control macaques immunized with uninfected human cells in which the challenge virus was grown were also partially protected (Stott, 1991). Subsequent studies (Arthur et al., 1992, 1995) demonstrated that MHC class II proteins, incorporated into the SIV envelope as it budded from human cells, was inducing potent neutralizing activity and in vivo protection. At the same time, chimpanzee experiments with chemically inactivated HIV had failed to demonstrate sterilizing protection against challenge virus (Gibbs *et al.*, 1991; Niedrig *et al.*, 1993). Other factors contributed to the lack of interest in whole inactivated HIV vaccines. The viruses that had been tried were highly susceptible to loss of envelope protein, particularly gp120, during production and purification. Also, as the immune responses to MHC antigens were being defined in the macaques, it also was recognized that non-syncytium inducing (NSI) isolates, and not the syncytia-inducing (SI) isolates of the virus, which grew well in transformed cell lines, were the dominant forms of the virus early in the disease course (Schuitemaker *et al.*, 1992; van't Wout *et al.*, 1994). Added to these factors was the low level of production that was encountered with even the best T-cell line-adapted SI isolates in transformed cell lines. These combined data led to the widespread abandonment of whole inactivated vaccine approaches.

Several investigators have been looking afresh at how some of these problems might be reassessed. Viral strains have been identified that may retain envelope much better than the isolates that were previously tried. Also, new agents for inactivation that do not disrupt the envelope components, but which affect internal proteins such as the nucleocapsid protein, through Zn finger disruption, are being explored (Arthur *et al.*, 1998). In addition, a new pseudo-virion vaccine candidate, incorporating the *env* of a non-syncytia inducing (NSI) R5 HIV-1 isolate (BX08) is approaching phase I testing in the clinic (Rovinski *et al.*, 1999). This 'genetically killed' candidate vaccine contains multiple genetic deletions for improved safety, is non-infectious, and is somewhat comparable to a chemically killed virus. Currently it is designed for incorporation as a boost in 'prime–boost' strategies that first vaccinate with avipox-HIV-1 candidate vaccines.

One important insight that was provided by the early studies should not be disregarded: HIV and SIV isolates that were relatively resistant to neutralization with specific antibodies to the virus, were not resistant to neutralization by the anticellular antibodies. Thus, the viruses themselves were not inherently resistant to neutralization; however, a better understanding of the weak links in their armor was needed. In addition, the underlying principle of using essentially an intact form of the virus to present viral proteins accurately in their native configuration to the immune system may still have an important role. However, the immunogen may need to be carefully selected and shown to induce neutralizing antibodies. Any inactivated virion approach is unlikely to efficiently induce CTLs that are class I MHC-restricted immune effectors, and so this approach is likely to be optimal as a boost for vaccine candidates that prime the cellular arm immunity by some other mechanism.

Recombinant Envelope Vaccines

The initial isolations of HIV closely followed the successful development of recombinant protein vaccines for hepatitis, and there was optimism that HIV could be conquered with similar recombinant protein vaccines. Several groups independently cloned and expressed the envelope protein(s) of early T-cell line-adapted isolates, HIV-1 IIIB, MN and SF2, in different protein expression systems (Rusche *et al.*, 1987; Barr *et al.*, 1987; Berman *et al.*, 1989). In 1987, another milestone was reached when the first of the recombinant envelope proteins entered trials in human subjects: an HIV-1 LAI gp160 produced in insect cells with a baculovirus expression system (Kovacs *et al.*, 1993; Dolin *et al.*, 1991). In parallel, trials of a recombinant HIV-1 LAI gp160 envelope in a vaccinia vector were conducted in human volunteers (Cooney *et al.*, 1991). Initially a number of safety concerns were raised about all envelope vaccine approaches, because *in vitro* data suggested that HIV-1 gp120 protein would cause extensive apoptosis or anergy by cross-linking of the CD4 T cells that were needed for T-cell help for effective vaccination (reviewed in Chirmule and Pahwa, 1996). This proved to be unfounded in human vaccine trials (Keefer *et al.*, 1997). Individuals vaccinated with recombinant gp120 were able to mount immune responses to HIV gp120 or gp160, although many of the products, produced in various cell lines, induced only low levels of antibody (reviewed in Walker and Fast, 1994). The gp120 products were generally more immunogenic than the gp160 products and the responses were longer lasting (> 2 years) than those seen in non-human primate models. Two of these vaccine candidates, HIV-1 MN gp120 and HIV-1 SF2 gp120, produced in mammalian (Chinese hamster ovary) cell lines required only two to four vaccinations to achieve what appeared to be maximal response for the products. Neutralization of laboratory-adapted SI strains of HIV-1 was consistently induced by these vaccine candidates, and some level of cross-reactivity to heterologous HIV-1 was observed (Graham, 1994) These vaccines were advanced into the first phase II trial in 1992, achieving another milestone for HIV vaccines.

A number of intercurrent infections were observed in the vaccine arms of the phase II trial with recombinant gp120 vaccine candidates, and serious concerns about their potential effectiveness were raised (Berman, 1997; Graham *et al.*, 1998; Connor *et al.*, 1998). In addition, all of monovalent gp120 (or gp160) candidate vaccines failed to induce broadly reactive antibodies that were capable of neutralizing primary NSI isolates, which were subsequently shown to be utilizing the CCR5 coreceptor (Mascola *et al.*, 1996a; Hanson, 1994; Graham, 1994). Furthermore, attempts to induce CTL with the gp120 proteins failed even when novel adjuvants, which were useful in small animals, were employed in human vaccine candidates (McElrath *et al.*, 1997). There were a number of proposals to resolve these issues, which are reviewed elsewhere (Schultz, 1998).

One of the hypotheses for the inadequacy of the gp120 vaccine candidates to induce broadly reactive antibodies was that the antigenic form, monomeric gp120 proteins, could not present the relevant conformational epitopes to the immune system. Being monomeric and not oligomeric, as one would expect for the native conformational form on the surface of the virion, monomeric vaccine candidates would be missing key conformational structures where the subunits abutted each other. Data from subsequent experiments that generated monoclonal antibodies to oligomeric proteins in mice suggested that oligomers of gp160 recognized an additional series of determinants that might be important for successful vaccines (Broder *et al.*, 1994). Current attempts to generate better antibodies to conformational determinants with gp160 or truncated gp140 vaccine products have not dramatically improved the neutralizing antibody responses, but additional epitopes have been detected in murine immunizations (Earl *et al.*, 1997; Sugiura, 1999).

A second hypothesis proposed that the failure of the original gp120 vaccine candidates to induce broadly reactive antibodies to primary NSI isolates of HIV was because the wrong type of isolates had been used as the source material for the recombinant gp120 vaccines. After the seminal demonstrations that HIV replication was inhibited by several chemokines (Cocchi *et al.*, 1995), and that the virus was using CD4 plus a seven-transmembrane coreceptor for viral fusion and entry (Feng *et al.*, 1996), a flurry of papers showed that HIV-1 isolates broadly known as SI and NSI or T-cell tropic and macrophage–tropic were using two different types of chemokine coreceptors, CCX4 (X4) and CCR5 (R5), respectively, to enter the different kinds of cells (D'Souza and Harden, 1996; Berger, 1998; Berger *et al.*, 1999). The resulting corollary hypothesis for HIV vaccine development is that if one chooses or adds envelope(s) from NSI R5-using isolates to the vaccine, then protective immunity to NSI isolates will be generated. An attempt to stimulate antibodies to gp120 from a dual-tropic isolate failed to improve the immune responses of human volunteers to the R5-using isolates (Beddows *et al.*, 1999), indicating that cross-typing neutralizing antibodies were generated by a clade E R5 gp120 immunogen, but this still appears to be largely limited to neutralization of SI isolates (VanCott *et al.*, 1999).

A third hypothesis for the inadequacy of monovalent HIV-1 rgp120 vaccine candidates proposed that because there are many variants of HIV, immunizing with only one envelope would not be adequate. It follows that some cocktail will need to be generated for an effective vaccine candidate. Incorporated in this rationale is the assumption that there will be some finite number of different HIV isolates whose envelope proteins will be able to display the range of correct epitopes to the immune system and generate a broadly protective immune response. One rationale to resolve the inadequacy of monomeric gp120 vaccines has been termed a sieve hypothesis. Under this scheme, the new variants to be added to a multiple component vaccine would be selected according to whether or not they are blocked by monovalent vaccine immunity. This approach has been followed in the generation of the bivalent B/B and B/E vaccines (Berman, 1998; Francis *et al.*, 1998). These vaccine candidates entered phase III trials in the United States and in Thailand in 1998 and 1999, respectively, marking another milestone for HIV/AIDS vaccine research. This is the same underlying concept that generated a trivalent polio vaccine and a pneumcoccal vaccine with 23 valencies. However, in this case the criteria for selection of the next gp120s to be included are not very precise and require an inefficient system of identification of vaccine failures in large-scale trials to identify new proteins to be included. An alternative to this selection process to increase relevant variants incorporates envelopes from sequential HIV-infected patient isolates (Rencher *et al.*, 1995) or some combination of the above to broaden the immune response (Caver *et al.*, 1999; Richmond *et al.*, 1997).

Live Attenuated Virus Vaccines

In non-human primate experiments attenuated lentiviruses have been one of the most effective vaccine approaches against pathogenic challenge (Marthas *et al.*, 1990; Daniel *et al.*, 1992; Clements *et al.*, 1995; Putkonen *et al.*, 1995; Almond *et al.*, 1995; Stahl-Hennig *et al.*, 1996; Wyand *et al.*, 1996; Cranage *et al.*, 1997; Joag *et al.*, 1998; Johnson *et al.*, 1999). In some cases protection has been observed against both intravenous and mucosal challenge. Together, these data demonstrate that protection from superinfection or disease can be achieved, and provide the basis for another milestone in HIV/AIDS vaccine development. Because several of the most effective vaccines now in public health use are attenuated viral vaccines, this was not unexpected. However, serious concerns about the safety of any attenuated vaccine will need to be systematically addressed in HIV because of the high rate of mutation, the integration into the genome and the ability of the virus to be reactivated out of reservoirs.

Attenuation of viral replication *in vivo* can be generated by a number of different mechanisms, not all of which will necessarily attenuate the viral pathogenicity. Studies of specific genetic deletions have indicated that deletions in *nef* have provided substantial decreases in viral replication *in vivo* (Daniel *et al.*, 1992; Almond *et al.*, 1995). This seems to parallel observations in human subjects that have been identified with *nef* deletions in HIV (Kirchhoff *et al.*, 1995; Learmont *et al.*, 1992; Mills *et al.*, 1998). However, further data indicate that singular deletion of the *nef* gene, which has a profound effect on the ability of the virus to replicate *in vivo*, is not adequate to attenuate the pathogenicity of SIV or HIV-1 (Connor *et al.*, 1998; Lewis *et al.*, 1999; Greenough *et al.*, 1999; Learmont *et al.*, 1999), and loss of CD4 cells and the development of AIDS was observed in some macaques and in some humans. These and additional safety studies in animals have indicated that the permissible level for continued viral replication in animals infected with attenuated virus is vanishingly low. However, variants produced by genetic changes in other regions of the viral genome have revealed potential additional sites in genes of other regulatory proteins, *env*, *gag*, and the integrase region of the *pol* gene, which may provide additional or alternative targets for the attenuation of AIDS viruses (von Gegerfelt and Felber, 1997; Luciw *et al.*, 1998).

Vectors, DNA Vaccines and Prime–Boost Strategies

Vectors can be engineered to deliver genetic information in a live, replicating virus or bacterium; as purified DNA, plasmids or replicons, i.e. single round packages of genetic material that have little or no likelihood of replicating; as mRNA to 'infected' cell cytoplasm for the synthesis of protein; or in some cases as the protein presentation system of an inserted epitope. Live replicating viral or intracellular bacterial vaccine candidates that carry copies of the HIV genome in their genetic backbone offer, in principle, the advantage of an active infection and multiple rounds of replication to amplify the input material, without the safety risks of an attenuated HIV. The chief benefit is the intracellular expression of proteins that allows processing into epitopes in intracellular pathways for MHC class I antigen presentation, i.e. short peptides of 9–11 amino acids that can be processed and bound into the groove of MHC class I molecules for presentation to T-cell receptors (TcR) on CD8+ T lymphocytes. The concern for any replicating vector is the stability of the insert, because if there is strong selective pressure against the insert, because of either replication efficiency or immune selective pressure, an 'empty' vector provides only the benefit of the vector itself. DNA vaccines, although not capable of replicating, if either delivered or formulated to provide sufficient copies of the input material to 'professional' antigen-presenting cells (dendritic cells or activated macrophages) or designed to produce high levels of transcription and/or translation, could accomplish the same goals. Replicon vectors that deliver genetic material or RNA vectors that produce a single cycle of infection without replication and spread of the viral vector may also need to be optimized for expression or cell targeting of the HIV genetic material. The virus vectors reviewed here and elsewhere (Cairns and Sarver, 1998) are being explored as recombinant HIV vaccines because of the advantages they offer.

Selected replicating recombinant vectors can be designed to provide a large supply of antigen over a few days or weeks that can be used to optimize the immune response (compared to an intramuscular injection of a bolus of protein that is dispersed, even in the presence of common adjuvants, over a few days.) However, because most vectors are themselves immunogenic, they may not be suitable for repeated immunizations. Therefore, the use of any specific recombinant vector must be balanced against existing immunity in any population that must benefit from the vaccine. A corollary, but reciprocal, principle to be considered is whether a vaccine vector can provide immunity as a 'two for one' vaccine to benefit a particular population, i.e. immunity to HIV plus immunity to the vector.

Poxviruses

Poxviruses have a large genome that can accommodate extensive insertions of foreign genes which can be expressed either under native poxvirus promoters or under synthetic promoters. Diverse sources of pathogen DNA have been inserted into various poxviruses (Moss and Flexner, 1987). Another milestone for HIV vaccines was the use of a recombinant vaccinia vector for the expression of HIV envelope (Hu *et al.*, 1987), and testing of this vaccine approach in human volunteers who had either received a prior vaccinia vaccination or who were vaccinia naive (Cooney *et al.*, 1991; Graham *et al.*, 1992). The HIV-1 specific immune responses that were generated were substantially improved in the vaccinia-naive individuals. The clinical trial with a more complex vaccinia vector containing *env*, *gag* and *pol* (Panicali *et al.*, 1992) marked another milestone for the incorporation of all three of the major genes for HIV structural proteins. The clinical trial has compared intradermal injection to scarification or subcutaneous injection (Keefer *et al.*, 1998).

Because smallpox was eradicated and vaccination halted in the 1970s, in 1999 most persons under the age of 25 have not been exposed to vaccinia. Thus candidate HIV vaccines developed from vaccinia vectors would be appropriate for most of the world's young people who are at the highest risk of acquiring HIV. However, concern for the use of standard vaccinia as a vector is its safety profile and the risk associated with disseminated vaccinia in immune-compromised hosts who might acquire vaccinia by contact in high prevalence areas. Nearly all of the vaccinia vectors under study have at least minimal attenuating deletions that are used for the insertion of genetic material.

Several more highly attenuated vaccinia vectors have been explored for AIDS vaccines, for example modified vaccinia Ankara (MVA) and NYVAC have been studied in animal models (Hirsch *et al.*, 1996; Hanke *et al.*, 1998a; Tartaglia *et al.*, 1992; Myagkikh *et al.*, 1996; Andersson *et al.*, 1996). These experiments have indicated that, to some extent, CTL and protection effects against a high viral load and delay of disease can be induced by these vaccines, even when sterilizing protection cannot be achieved. The vaccinia vector experiments, like the envelope protein experiments, indicated that the ability to protect against the challenge virus was often confined to specific protection against viruses that were genetically matched to the envelope in the vaccine (Polacino *et al.*, 1999a,b).

More highly attenuated poxvirus vectors, such as fowlpox (Jenkins *et al.*, 1991) and canarypox (ALVAC) vectors (Tartaglia *et al.*, 1997), have been explored as candidate vaccines, both in animals and in human volunteers. As the complexity increased in the number of HIV genes included in the vector, the immunogenicity of these candidate HIV vaccines has improved (Weinhold *et al.*, 1997). One of these vaccine candidates (vCP 205), containing *env*, *gag* and protease genes, entered clinical trials in Uganda to test the level of cross-clade CTL (8 February, 1999), marking another important milestone with the first trials in Africa.

From early testing it became apparent that the various poxvirus candidate vaccines did not induce high levels of antibody, particularly in humans. However, a series of studies, both in animals (Hu *et al.*, 1993; Giavedoni *et al.*, 1993; Israel *et al.*, 1994; Daniel *et al.*, 1994; Ahmad *et al.*, 1994) and in human volunteers (Cooney *et al.*, 1993; Graham *et al.*, 1993; Corey *et al.*, 1998; Belshe *et al.*, 1998; Clements-Mann *et al.*, 1998), demonstrated that priming with the poxvirus vectors induced a strong memory response in the vaccinees that could be demonstrated when they were subsequently boosted with a recombinant protein or particles. The titers of antibody after a single protein boost often exceeded that observed with three to four immunizations with the protein itself. These were the experimental bases for further exploration of combination vaccine approaches using a vector-based candidate vaccine to prime the immune system, particularly the cellular components (both T-cell help and CTL) and a second protein-containing vaccine to boost the humoral immune response.

Adenovirus

Adenovirus vectors containing segments of the HIV genome have been studied in animal models that were able to assess the immunity that could be induced by the vector and the insert (Prevec *et al.*, 1991; Natuk *et al.*, 1992; Lubeck *et al.*, 1994, 1997; Robert-Guroff *et al.*, 1998). The primary interest in this family of vectors is the potential opportunity to induce mucosal immunity where transmission is occurring. Recent surveys of the seroprevalence of different adenovirus serotypes in young military recruits indicate that the susceptibility is 66% and 73%, respectively, for serotypes 4 and 7 (Ludwig *et al.*, 1998). This indicates that with multivalent adenovirus vectors that could be derived from the many serotypes of adenovirus, one might be able to achieve over 90% coverage of this vaccine in the age group which we know is at highest risk in the United States. The issues that have been raised about this approach are the limited amount of genetic material that can be inserted; the concerns about safety of delivery to the nasal passages instead of via oral/gastric delivery, which has been used in the past; and the optimal selection of serotypes for different populations. However, the long-lived protection observed in a few chimpanzees against a primary NSI isolate (5016) (Robert-Guroff et al., 1998) warrants further phase I, safety and immunogenicity studies in human volunteers.

Poliovirus

Poliovirus vectors for HIV or SIV have been devised in several ways and demonstrate several advantages and disadvantages of the different approaches to viral vectors. Poliovirus is a picornavirus with a relatively small, highly effici- ent genome which permits only small changes to its genetic makeup. The first polio–HIV chimeric vaccine approach tested was the development of a Sabin strain of polio with a gp41 HIV epitope (<20 amino acids) inserted into the S1 protein of polio (Vella et al., 1993). This generated a 'display' vector which permitted the immune response to recognize the linear sequence that was inserted and to neutralize both laboratory-adapted HIV and the poliovirus constructs.

A second approach has developed polio replicons that contain HIV or SIV protein genes substituted for the polio capsid protein gene (Porter et al., 1996). When these replicons are produced in *trans* with a vaccinia virus that produces the polio capsid, the HIV or SIV recombinant genomes are incorporated into the replicon, creating a virus-like particle which cannot replicate without a coin- fection that will produce the polio capsid. Further studies have demonstrated that this type of vaccine candidate can produce replicons that enter cells and produce protein that is physically and immunologically equivalent to proteins produced by the AIDS virus from which it was derived (Anderson et al., 1997). This replicon approach incorporates a safety feature that neither the polio nor the recombinant with the HIV or SIV genome insert is capable of replication on its own. However, because of this design the amount of genetic material delivered to the target tissue, particularly mucosal sites, is limited by the input replicons, and the production of viral proteins must be high to obtain effective immunization.

The third poliovirus vector approach for AIDS vaccines relies on engineer- ing the poliovirus with sequences inserted into the polyprotein sequence for new proteolytic cleavage sites to yield mature SIV or HIV proteins (Andino et al., 1994; Tang et al., 1997). The advantage of this approach is that the recombi- nant virus can replicate up to wild-type efficiency and amplify the vaccine material delivered. There is a limit to the amount of genetic material in the insert, but it is is possible to insert the full length of *gag* and a large segment of *env* protein, which is glycosylated in the host cells (Tang et al., 1997).

Rhinovirus

So far, rhinovirus vectors have been limited to the production of a recom- binant virus that creates a protein epitope display at the surface of the viral capsid (Smith et al., 1998). Although this is a genetically stable approach, the rhinovirus genome also has restrictions on the amount of genetic information that can be inserted, i.e. only small epitopes have been successfully inserted. Larger insertions disrupt the structure and infectivity of the rhinovirus chimeras.

Influenza Virus

Chimeric influenza viruses have been developed which insert a conserved epitope of HIV recognized by a mAb (2F5) that is broadly neutralizing (Ferko et al., 1998) or a segment of the V3 loop of HIV (Li et al., 1993; Gonzalo et al., 1999). These recombinant vectors have been able to induce immune responses in mice, even when delivered intranasally (Ferko et al., 1998), to epitope inserts that are capable of neutralizing HIV. The major advantages of this vector system include the potential development of non-needle delivery through an intranasal mist that might also induce effective immune responses at the genital mucosa (Palese et al., 1997).

Alphaviruses: Semiliki Forest Virus (SFV) and Venezuelan Equine Encephalitis Virus (VEE)

Several groups of investigators have developed SFV vaccines for HIV or SIV that have been evaluated in macaques (Mossman et al., 1996; Berglund et al., 1997; Notka et al., 1999). At least partial protection was demonstrated by a reduction in viral load and protection from early lethal disease in the SIV-PBj14 challenge (Mossman et al., 1996), or decreased viral load in the non-pathogenic SHIV challenge (Berglund et al., 1997; Notka et al., 1999). In addition, the development of stably transformed alphavirus packaging cell lines for Sindbis and Semliki Forest virus-derived vectors (Polo et al., 1999) resolves one of the concerns with these types of vectors, and presents an improved safety profile.

The vaccine currently being used for Venezuelan equine encephalitis virus (VEE) is an attenuated form of the virus. Further attenuation and development of a replicon-helper system to permit the expression of genetic inserts (Davis et al., 1996; Pushko et al., 1997) indicates that there are some very intriguing aspects of this vector system that might be exploited in HIV vaccine development. Experimental application as vectors suggests that inserted gene are expressed strongly in the draining lymph nodes, and that the viral promoters are capable of driving high levels of gene expression and immunity to mucosal challenge (Davis et al., 1996). Additional experiments have shown that SIV genes for gag and env can be inserted into this vector and delay disease development in vaccinated animals challenged with a pathogenic isolate of SIVsm (E660) (Johnston RE, Davis, NL, Johnson, P. personal communication).

Other Virus Vectors

Attenuated vesicular stomatitis virus (VSV) vectors are being developed as vaccine constructs (Roberts et al., 1999). Studies have shown that HIV envelope protein was incorporated when the cytoplasmic tail was replaced with that of the VSV glycoprotein (Gibbs et al., 1991; Johnson et al., 1998). There are substantial concerns about the safety of vectors from this virus that will have to be resolved in non-human primates, but the possibility of presenting viral envelope in a proper trimeric configuration at the surface of a particle that could be produced in large quantities for boosting a response primed by other vectors or DNA presents an interesting possibility.

Adeno-associated virus (AAV) vectors are being developed for gene transfer to human cells. At present the widespread usage of AAV which have been adapted to incorporate vector genes of interest is limited by the ability to generate recombinant virions on an adequate scale for preclinical and clinical studies. The development of a stable cell line for adeno-inducible *rep* and *cap* genes that permits assessment of AAV infectivity is an important step for studies of this vector system (Clark *et al.*, 1996). If immunogenicity is adequate with these vectors, issues surrounding the random integration of viral vector genetic material will still need to be resolved.

Because the immune response may need to be repeatedly triggered to optimize immune responses for HIV, several groups have considered the development of vectors with latent infection profiles. In particular, two viruses that have the ability to maintain long-term stable integration in human cells have been considered for HIV vector development: varicella zoster virus (VZV) and herpes simplex virus type 1 (HSV-1). Both of these are viral vectors with large genomes that may have a number of sites that could be targeted for gene insertion. In HSV-1, thymidine kinase (TK) has been used to insert SIV envelope to evaluate expression in monkey cell lines (Murphy *et al.*, 1998). In a proof of concept, the glycoprotein D of HSV-2 (gD2) has been inserted into an intragenic site on the Oka vaccine strain of VZV, which is a vaccine strain now in use in children for chickenpox (Heineman *et al.*, 1995). Concerns about the long-term latency and reactivation of vaccine vectors such as Oka and HSV-1 will require long-term follow-up for use in populations at risk for HIV infection.

Bacterial Vectors

Several bacterial vectors have been pursued as approaches for HIV vaccines. Most of those selected have a strong propensity for intracellular expression of antigen, and also a strong bias toward Th1 helper-cell induction, as well as CTL. Among the first bacterial vectors that were tested for AIDS vaccines were *Mycobacteruim bovis* Bacille Calmette–Guérin (BCG) vectors (Stover *et al.*, 1991; Fuerst *et al.*, 1992). These offer a number of important advantages. Although current BCG vaccines are highly variable in their efficacy, the advantage is that they have been administered to many infants in endemic areas of HIV infection and millions of infants receive BCG vaccines without any apparent adverse health effects. If a recombinant BCG–HIV vaccine could be developed for children in the developing world, where breastfeeding is a necessity, this would provide an enormous public health benefit. Monkeys were shown to develop CTL responses to an MHC-defined epitope with vaccination, particularly those boosted with a lipopeptide (Yasutomi *et al.*, 1993); however, they failed to show protection against SIV challenge. At the time when the animals were challenged, virus infection was the only endpoint measured and the ability of an SIV-specific immune response to control viral load was not determined. More recently, others have evaluated a somewhat different approach with HIV V3 (or *gag*) antigen expressed as a fusion protein, either inserted in the bacterial genome or in plasmids in the BCG vector (Honda *et al.*, 1995). Alternatively, plasmids have been used to express T-cell epitopes or minigenes of HIV (Berggren *et al.*, 1995). An improved vaccine for *Mycobacterium tuberculosis* that also could be used as a basis for these vaccine approaches is most desirable.

Different *Salmonella* species have been investigated as potential HIV vaccines because these vectors appear to be able to selectively target the mucosal inductive immune system. It is widely expected that strong mucosal immunity will provide improved protection against sexual exposure to HIV. Recombinant *Salmonella* – HIV vectors may provide a dual benefit to many individuals in countries where GI tract infections increase morbidity. Further, the goal of a stable, easy-to-deliver vaccine might be met with an orally administered bacterial vaccine.

AIDS vaccine candidates using *Salmonella typhimurium* in mice and macaques have been investigated (Tijhaar *et al.*, 1994; Franchini *et al.*, 1995; Wu *et al.*, 1997). Localization of the site for antigen expression was a strategy used to improve antigen stability and availability to the immune system (Hone *et al.*, 1996; Pascual *et al.*, 1997). *Salmonella typhimurium* vectors for human use with alternative forms of attenuation have been developed and tested in human subjects (Hohmann *et al.*, 1996). These vectors may be particularly useful for inducing mucosal immunity to inserted genes, and may also be more efficacious because of the parental vector design.

Listeria monocytogenes also has been explored as a vector to investigate whether a strong bias in the immune system introduced early in the course of immunization is effective in skewing the balance after multiple immunizations. In one case, an attenuated *Listeria monocytogenes* vector has been developed (Frankel *et al.*, 1995) which is being evaluated for SIV immunogenicity in non-human primates. These kinds of vectors often have serious pathogenicity concerns, but these can be addressed by a number of avenues during the early stages of development in non-human primates, when warranted, by early preclinical studies in small animals.

DNA Vaccine Candidates and for Prime and/or Boost Approaches

The above discussion focused on the ability of recombinant viral or bacterial vectors engineered to carry genetic information into a host and for the production of proteins for presentation to the immune system through actual viral or bacterial infection or some mechanism of viral entry or fusion, to deliver genetic material to the cytoplasm where proteins can be produced. Almost immediately after the direct injection of DNA was shown to produce proteins (Wolff *et al.*, 1990), several groups demonstrated the ability of injected DNA to generate immune responses to the encoded proteins (Tang *et al.*, 1992; Ulmer *et al.*, 1993; Fynan *et al.*, 1993; Wang *et al.*, 1993; Cox *et al.*, 1993; Davis and Whalen, 1995). Several of the potential advantages of DNA as a vaccine for HIV were immediately obvious: DNA vaccines could be adapted quickly to a stream of emerging variants; could avoid the requirement for live viral or bacterial vectors, with their incumbent concerns for safety, population immunity and insert stability; would be stable under a number of conditions needed for worldwide use, including heat; appeared to require only small amounts of material for immunization of animals; could be purified for human use with minimal manipulation from bacterial plasmids; and might therefore be inexpensive and transferrable to economically disadvantaged settings.

Exploration of DNA vaccines for use in AIDS vaccines quickly followed the first reports of DNA immunogenicity. Reviews have covered many of the aspects of DNA vaccines for HIV/AIDS (Robinson, 1997; Kim and Weiner, 1997; Liu, 1998), which include issues of vector construction, promoters, immunogenicity and mechanisms of antigen presentation, as well as some of the limits of this approach. Substantial information now has been gained about both the advantages and limitations of DNA vaccines in non-human primates and from initial tests in human volunteers. In general, DNA vaccines induce low levels of antibodies, in contrast to the relative efficiency of antibody induction with proteins, and the effect is magnified by comparison between experiments in mice or guinea pigs and primates. This is not true for all antigens, but it is apparent from several studies with HIV antigens, including DNA for gp120 or gp160 (Wang *et al.*, 1993, 1995a; Barnett *et al.*, 1997; Fuller *et al.*, 1997a; Fomsgaard *et al.*, 1998). It is not clear whether this represents species differences in the ability to respond to the specific antigen or viral pathogen (Krieg *et al.*, 1998), or whether the DNA preparations used in small animals may not be prepared with the same rigor and purity as that tested in human subjects. By a reciprocal interpretation, it may be relatively easy to adjuvant DNA vaccines with formulations that stimulate the specific immune response via signals to the inflammatory response. In this light, utilization of the CpG stimulatory dinucleotides appropriate for the antigen and species may enable improved responses (Krieg *et al.*, 1995, 1998; Sato *et al.*, 1996; Jakob *et al.*, 1999).

DNA vaccines by themselves have been shown to provide partial protection from relatively apathogenic challenges or the ability to reduce viral load (Wang *et al.*, 1995b; Boyer *et al.*, 1996; Verschoor *et al.*, 1999). Codelivery and expression of cytokines from DNA have been utilized to improve immune responses in DNA vaccination (Kim *et al.*, 1999), which has provided some improvement in the protective effect (Weiner, D.B., personal communication). Immune responses against HIV-1 *rev*, *tat* and *nef* gene immunizations have been generated in mice (Hinkula *et al.*, 1997), monkeys (Putkonen *et al.*, 1998), and HIV-infected subjects with no prior response to these antigens (Calarota *et al.*, 1998). However, no protection from infection or reduction in viral load was observed in this monkey model.

One of the most intriguing observations is the relatively consistent ability of DNA vaccines to prime for subsequent boosts of immunity to a wide variety of immunogens in several host species. The optimal strategies for generating both T-cell responses and antibodies to HIV-1 antigens appear to be priming with DNA and boosting with protein (Barnett *et al.*, 1997) or vectors such as the poxvirus vectors (Richmond *et al.*, 1997; Fuller *et al.*, 1997b; Ramsay *et al.*, 1997; Hanke *et al.*, 1998a,b, 1999; Kent, 1998; Hanke and McMichael, 1999; Robinson *et al.*, 1999; Haigwood *et al.*, 1999; Caver *et al.*, 1999). Additional boosting of antibody responses has been observed with sequential immunizations of DNA, poxvirus vector and protein (Caver *et al.*, 1999).

It is not clear what mechanism(s) are responsible for this pattern of optimal induction of immunity, particularly of the humoral immune response, which is not seen with equivalent repeated boosting of either the DNA or the vector. One possibility is that the vectors are able to deliver relatively large segments of HIV (or SIV) genomic material that is transcribed and translated in the context of inflammatory signals generated by active infection after optimal DNA priming of Th1 plus CTL via dendritic cells. However, DNA vaccine priming appears to work equally for the fowlpox vectors, which have relatively little evidence of an inflammatory response, and with the more robust, replicating vaccinia vectors.

Because protection in monkey experiments has been observed in several experiments with DNA followed by recombinant vector immunization (Fuller et al., 1997b; Kent et al., 1998; Robinson et al., 1999; Haigwood et al., 1999), there is strong reason to pursue and optimize this approach to achieve both strong CTL and antibody responses.

HIV Peptide Vaccines

The initial enthusiasm for using specific peptides for an HIV-1 vaccine arose out of the observations that neutralization of the laboratory-adapted HIV-1 isolates was primarily dependent on antibodies to the third variable loop (V3-loop) of the HIV-1 envelope. A series of studies demonstrated that T-cell proliferation and neutralizing antibodies could be generated by several different strategies; the first of these involved V3 peptides, either by themselves (Sastry and Arlinghaus, 1991; Nehetete et al., 1995) or linked to a helper epitope from HIV (Ahlers et al., 1993, 1996; Haynes et al., 1993). Two of these candidate vaccines have been used to immunize HIV-infected volunteers (Bartlett et al., 1998; Berzofsky, J., personal communication) with limited immunogenicity.

Several other V3 loop-based candidate immunogens have been generated with non-HIV carrier proteins. The first of these was a segment of the V3 loop linked to tuberculin-purified protein derivative (PPD), which was tested in a phase I trial of seronegative volunteers (Rubinstein et al., 1995). With this immunogen, and also the approach above, an HIV-specific CTL response was only detected in HLA-B7 individuals. Other groups have linked V3 or conserved peptides from other regions of HIV, e.g. HGP30p17, to keyhole limpet hemocyanin (KLH) (Okuda et al., 1993; Sarin et al., 1999) or to hepatitis B surface antigen (HBsAg) (Eckhart et al., 1996), or through a DNA construct with HBsAg (Borgne et al., 1998; Bryder et al., 1999). Although immunogenicity can usually be shown in mice with these various peptides and constructs, the immunity in human volunteers usually does not exceed 50% of responders, even in HLA-selected individuals. This reflects the genetic immune response limitations that are the major drawback for any individual peptide approach.

At least three other approaches have been tested for their ability to improve the response to the V3 peptide region. The first involves multiply branched peptides that present octamers of the V3 peptide from one or more HIV-1 isolates in various delivery or adjuvant vehicles as the immunogen. These peptide-based immunogens have been tested in several international sites as well as in the United States (Gorse et al., 1996; Kelleher et al., 1997; Phanuphak et al., 1997; Li et al., 1997a). Because of the low levels of immunogenicity seen in these trials, and the evidence that there was minimal neutralizing antibody, even against the laboratory-adapted strains of HIV-1, these products have been abandoned. An alternative approach fused multiple V3 epitopes from six HIV-1 laboratory strains of HIV-1 with short five amino acid spacers, and demonstrated responses in rabbits to multiple V3 determinants (Gomez et al., 1999). A third approach attempted to deal with the variability seen in the V3 region by synthesizing hypervariable epitope constructs (HEC) that would reflect the known HIV variation (Meyer *et al.*, 1998).

There are several other types of HIV-1 peptide immunogens which have raised immune responses to conformational epitopes in small animals that appear to be worth pursuing for possible 'boost' products. The first two are variants of repeating units of the CD4-binding region of HIV-1 gp120 (Kelker *et al.*, 1994; Robey *et al.*, 1996; Frey *et al.*, 1997). These immunogens share features which suggest they may present conformational regions that are conserved across HIV-1 strains. Neither of these has been tested in an optimized regimen in non-human primates to determine whether this immunogenicity will be observed in species other than mice.

Finally, several groups have attempted to detect mimetopes that would mimic conserved neutralizing sites present on HIV-1 envelope proteins. Although epitopes can be detected (Boots *et al.*, 1997), there was little or no evidence until recently that mimetopes could be useful immunogens. Now, several regions of the HIV envelope capable of generating neutralizing antibodies have been detected on phage display libraries that permit the evaluation of immunogenicity (Scala *et al.*, 1999).

Passive Immunity to HIV, SIV or SHIV Chimeric Viruses

The administration of a complex regimen of zidovudine (ZDV) treatment of mother and infant to prevent perinatal transmission of HIV was highly effective (ACTG trial 076) (Connor *et al.*, 1994), despite a minimal effect on the maternal plasma viral load (Sperling *et al.*, 1996). Subsequent studies have suggested that drug therapy, even given only to the HIV-infected mother or to non-breastfeeding infants, if started immediately after birth, can be effective (Wade *et al.*, 1998; Fiscus *et al.*, 1999; Shaffer *et al.*, 1999). In the developing world, where the majority of HIV infection of women and children is occurring, a passive immunoglobulin (Ig) plus active vaccine regimen that could be given once or for a limited time, easily administered, non-toxic to both mother and child, inexpensive, and with the potential to prevent transmission via colostrum or breast milk, would be ideal.

In the past passive Ig studies have been used to guide the design of vaccines as well as providing direct benefit in high-risk exposure settings. As in the model of hepatitis B perinatal intervention (Beasley *et al.*, 1983), passive administration of HIV-specific Ig, if delivered in the interpartum period and to the infant to prevent breast-milk transmission, might be equally effective. Passive immunity in mother-to-infant transmission or other post-exposure prophylaxis settings should be considered in the context of available antiretrovirals, but it must be recognized that, in general, passive Ig has a prolonged half-life compared to antiretoviral drugs, might have prolonged benefit, and is likely to have a different mode of action – neutralizing virus before cell entry and/or targeting of HIV-infected cells through ADCC. Passive immunity might be provided by purified Ig from asymptomatic HIV-infected subjects with or without added monoclonal antibodies (mAbs), or as a cocktail of mAbs and supported by active vaccination to prevent infection over a longer period of time.

Several passive immunity studies in macaque models have shown that transmission of SIV, HIV-2 or SHIV chimeric viruses can be blocked (Putkonen et al., 1991; Van Rompay et al., 1998; Mascola et al., 1999; Joag et al., 1999; Shibata et al., 1999). Alternatively, the viral load and/or clinical progression can be greatly altered by virus-specific antibody (Haigwood et al., 1996; Mascola et al., 1999; Igarashi et al., 1999). Of particular importance for intervention in a breastfeeding population is the report that immune serum was able to protect against oral SIV transmission (Van Rompay et al., 1998). Other experiments with passive administration of HIV Ig or specific mAbs to HIV-1 that were tested in chimpanzees (Prince et al., 1991; Emini et al., 1992; Conley et al., 1996), a model in which disease outcomes could not be assessed, had indicated that it was possible to block infection and/or affect viral load even of primary isolates, if adequate levels of Ig were achieved.

In addition, experiments in the Hu-PBL-SCID mouse model have shown that the administration of high doses of individual mAbs can block HIV infection of human PBL (Safrit et al., 1994; Gauduin et al., 1995, 1997; Parren et al., 1995; Andrus et al., 1998). A particular advantage of the hu-PBL-SCID model for these analyses is the ability to use primary isolates of HIV with different tropisms and coreceptor use (Picchio et al., 1997), which can be reflected in differential kinetics of growth and human cell depletion (Picchio et al., 1998). These studies have revealed that individual mAbs are unlikely to be effective against a range of HIV isolates, and that viral escape can be generated relatively quickly in vivo, even without an intact immune system.

The clinical study ACTG 185 provided pooled Ig from asymptomatic HIV-infected subjects (HIVIG) (Cummins et al., 1991) or a standard preparation of intravenous Ig (IVIG) to pregnant women and their infants, who also received the 076 regimen of ZDV. The trial was terminated early, with only about half of the projected sample size, because the maternal transmission of HIV had continued to decrease with ZDV treatment. Analyses of the ACTG 185 passive immunity study (Stiehm et al., 1999) indicate that infection clearly was not prevented in all infants; although there were somewhat fewer infections in the HIVIG arm of the trial, the effect was not statistically significant. However, none of the infants who became infected in the HIVIG arm of the study had detectable viruses by culture at birth, and only one of seven infants that were eventually shown to be HIV-infected was DNA PCR positive but plasma RNA negative at birth. In contrast, about one-third of the infected infants in the IVIG arm of the study had a culturable virus from cells at the time of birth (P = 0.05). Analysis of the transmission, by maternal CD4 count and/or prior ZDV use, the original criteria for trial entry, indicated a strong trend (P = 0.59) for a reduction of HIV-1 transmission with HIVIG in women who were more immunocompromised. These results reflect the range of outcomes that has been observed in animal studies. A few infections may have been averted; others appear to have been delayed in virus replication and detection. Improved products for passive immunity continue to be developed. With evidence of antibody synergy for virus neutralization (Li et al., 1997b; Mascola et al., 1999), the future development of effective cocktails to prevent viral escape from a monovalent product is highly desirable.

WHAT OTHER FACTORS ARE CONTRIBUTING TO THE OPTIMISM FOR AN HIV VACCINE?

In 1999, in addition to the partial success in the animal models for HIV/AIDS vaccines, we are better positioned at nearly every level to design, construct and test new HIV vaccine candidates. New tools and technology to measure vaccine-induced immune responses; new insights from structural studies; data from multiple geographic sites about the immune relatedness of different HIV-1 isolates despite genetic variation; better information about mucosal transmission events, particularly in animal models, are all available to us now.

New tools and animal models

Insights from new tools, reagents and methodologies for the assessment of human immune responses are now available, and are even being driven by HIV research. For example, the technology has been developed to detect individual T cells with specific T-cell receptors (TCR) by dimeric (O'Herrin et al., 1997) or tetrameric (Davis et al., 1997) MHC molecules associated with specific peptide epitopes. It is now possible to study, in selected human subjects as well as in animals, the magnitude and duration of immune T-cell responses to viral infections and vaccines. The analyses of T-cell binding of MHC tetramers to the study of responses in HIV-infected persons (Ogg et al., 1998) and SIV-infected animals (Kuroda et al., 1998) has been rapidly developed and applied to the analysis of vaccine-induced immune responses. Flow-cytometric and ELI-SPOT technologies for the assessment of immune cells producing interferon-δ and other cytokines that have been correlated with non-HIV viral control in other species, are being adapted for use in HIV vaccine studies. Studies to crosswalk flow cytometry and TCR technologies are being conducted in trials of HIV candidate vaccines (McElrath, J. and Evans, T, personal communication).

With the development of methods to measure viral load for lentiviruses in animals as well as HIV human subjects, it has been possible to compare the *in vivo* replicative capacity of stocks of different virus isolates and the impact of prior vaccination. Key developments for the various vaccine approaches noted above have resulted from studies using pathogenic SIV in macaque models (Schultz and Stott, 1994; Schultz, 1998; Letvin, 1998; Nathanson et al., 1999). With plasma viral loads comparable to HIV in human subjects, an AIDS-like syndrome that develops in several species of macaques, disease endpoints, and the different degrees of pathogenicity that reflect the spectrum of HIV infections that occur in humans, SIV has become the primary model for AIDS vaccine studies. The SIV/macaque models have provided a framework to evaluate challenges via different routes not only sterilizing immunity capable of preventing any infection, but also vaccine-induced modulation of viral replication and protection against disease.

Chimeric recombinant viruses SIV/HIV (SHIVs) that contain HIV *env* and other selected regulatory genes (e.g. *tat*) of HIV inserted into an SIV backbone have been developed and provide important models to evaluate vaccines (Shibata et al., 1991; Li et al., 1992; Luciw et al., 1995; Uberla et al., 1995; Shibata et al., 1997; Bogers et al., 1997; Heeney et al., 1998). SHIV isolates infect and persist in macaques, and several animal-adapted isolates have induced rapid CD4 T-cell depletion, AIDS and death (Reimann *et al.*, 1996; Lu, 1997; Joag *et al.*, 1997; Lu *et al.*, 1998a,b; Agy *et al.*, 1998). Thus, chimeric SHIV isolates with inserted HIV-1 *env* genes can permit the direct evaluation of HIV envelope-based vaccines in a disease model. In addition, SHIV pathogenic constructs now available for vaccine testing have incorporated macrophage-tropic, R5-using HIV-1 *env* genes (Harouse *et al.*, 1999). SHIV isolates with non-clade B HIV *env* are being developed to evaluate vaccines against divergent isolates. However, there is clearly a preference for replication and growth in macaques of the X4 or dual-tropic HIV-1 variants, and it appears that SIV macrophage-tropic isolates may be using a different constellation of chemokine receptors for infection (Edinger *et al.*, 1998).

The ability of HIV-1 to infect chimpanzees was established soon after HIV-1 was first isolated, and the chimpanzee provided an early model for HIV-1 vaccines (Gibbs *et al.*, 1991) and passive immunity studies (Prince *et al.*, 1991). However, HIV-1 replication is limited and apathogenic in most chimpanzees, and nearly all of the isolates that have infected chimpanzees have been of the X4 type. The description of AIDS in chimpanzees which were injected more than 10 years earlier with several HIV-1 isolates, demonstrates that HIV-1 can recombine and adapt to cause slow but profound pathogenesis in these animals (Novembre *et al.*, 1997; Mwaengo and Novembre, 1998; Fultz *et al.*, 1998; Wei and Fultz, 1998). However, with all of these considerations in mind, it is unlikely that large numbers of chimpanzees will ever be available or used for vaccine studies.

Other lentivirus animal models continue to be explored because of the unique characteristics that make them relevant to AIDS vaccines. Baboons (Locher *et al.*, 1998) and pigtailed macaques (Looney *et al.*, 1998) can be infected with HIV-2 for studies of vaccines that utilize a human pathogen and not a surrogate, such as SIV or SHIV. Feline immunodeficiency virus (FIV) presents opportunities to evaluate vaccine concepts with diverse variants and pathogenic clones, where cross-strain and antibody-independent protection has been observed (Tellier *et al.*, 1998; Hosie *et al.*, 1998). At the same time, FIV vaccine studies provide opportunities to test vaccines relevant for household pets. Equine infectious anemia virus (EIAV) in horses permits the *in vivo* evaluation of antibody-dependent enhancement of a highly macrophage-tropic viral infection and pathogenesis that is induced with EIAV envelope protein vaccines (Raabe *et al.*, 1998).

Structural HIV Studies

In the last few years, reports about the structure and glycosylation of HIV (and SIV) envelope proteins have revealed critical information about both the surface (SU) component gp120 and the transmembrane (TM) component gp41, and their interaction. These new insights have come after many years of attempts, and only after several modifications of the viral envelope proteins to obtain stable structures for analysis. The findings on the core components of gp41 came first, with the demonstration that there was a trimeric coiled coil component that was basically resistant to proteolytic digestion and that it resembled the spring region of the influenza hemagglutinin protein (Chan *et al.*,

1997; Weissenhorn *et al.*, 1997). Further data indicate that this region of gp41 undergoes dynamic changes with viral interaction with the cell membrane it is infecting, and that critical regions on the gp41 molecule are not available until CD4 binds to the virus gp120 envelope (Furuta *et al.*, 1998). A number of other findings also indicated that this was an important segment of the molecule for neutralization of the virus or for blocking viral entry. Peptides of this region were initially shown to block viral fusion events (Wild *et al.*, 1992), and have had a remarkable ability to decrease viral load in HIV-infected patients (Kilby *et al.*, 1998). Monoclonal antibodies to this region have had variable effects on neutralization (Muster *et al.*, 1994; Cotropia *et al.*, 1996; Hioe *et al.*, 1997), but several mAbs recognize highly conserved regions in HIV-1 gp41. In addition, they may be particularly effective if present during the transition stages of viral to cell entry, and not just at the initial stages of viral binding, or if they are used in combination with other antibodies.

Partial structural analysis of gp120 was permitted with the systematic development of a core structure that was stabilized with the binding unit of CD4 and an mAb, 17b, which recognizes a relatively conserved internal domain which is exposed upon CD4 receptor binding (Kwong *et al.*, 1998; Wyatt *et al.*, 1998). The original recombinant gp120 products were carefully prepared and analyzed for the structure and glycoslyation, which was thought to be critical for their 'native' structure (Leonard *et al.*, 1990). The more recent studies have confirmed that a large fraction of the predicted accessible surface of the gp120 molecule in its native trimeric complex was heavily glycosylated or covered with variable loops that protected the receptor and coreceptor sites. The use of these data to expose sites of the virus envelope to immune responses and to design new immunogens is now under way in several laboratories.

As noted above in the section on recombinant HIV envelope vaccines, the identification of the chemokine receptors, particularly CXCR4 (X4) and CCR5 (R5), as major coreceptors for HIV, has also led to studies to develop new vaccines and drugs that might attack this new target of virus and host cell interaction. Focus on the virus–cell membrane fusion events led to a series of experiments in mice that have revealed a conserved fusion intermediate that is immunogenic if the virus and host cells are fixed by chemical treatment after some period of time to permit virus attachment and formation of the fusion intermediate (LaCasse *et al.*, 1999). Cell lines designed to express specific co-receptors are being used as tools to evaluate the ability of immune sera from vaccinees to neutralize primary HIV isolates with specific coreceptor use, and genetically engineered small animal models with human coreceptors (Garber *et al.*, 1998) are being explored for future vaccine studies.

Issues Related to Genetic Variation and Immune Responses

HIV vaccine design efforts continue to be complicated by the many unique aspects of HIV infection, including an unprecedented degree of genetic variation seen in epidemiological samples (McCutchan *et al.*, 1996). This is is not easily resolved into simple serological subtypes. However, the obstacle of almost infinite genetic variation in HIV, and the concern that this genetic variation would be insurmountable, has been dissipating under the weight of new data. As mentioned above, genetic variation has been shown to affect recognition by and escape from both humoral and cellular immune responses, particularly where strong immune responses are generated to a single dominant HIV-1 variant.

At least two major groups of HIV-1 (M and O) exist, with major genotypic clades within the M group, which is dispersed worldwide. Virus from one clade (B) still predominates in the United States. However, recent isolates from geographic sites with evolving or newly emerging epidemics have revealed epidemiologically important recombinant HIV-1 variants between A and G (McCutchan *et al.*, 1999), as well as A/E and B/F recombinants that have been previously identified. Epidemiological studies of viral variants in women have indicated that vaginal transmission, possibly enhanced by endogenous or exogenous hormonal factors, permits the entry of multiple variants from their infected partners (Overbaugh *et al.*, 1999), providing a plausible scenario for the generation of diverse recombinants. Intravenous injection of several strains of HIV-1 into individual chimpanzees has resulted in at least one incident of recombination between strains, and the generation of a more pathogenic variant, as noted above. Thus, multiple modes of transmission could permit the generation of recombinant strains and add further to the identification of the types of virus that must be incorporated in any vaccine.

Despite the high degree of variation in HIV, sera derived from some infected individuals are able to effect broad, cross-clade neutralization of HIV-1 subtypes (Weiss *et al.*, 1986; Moore *et al.*, 1996; Nyambi *et al.*, 1996; Weber *et al.*, 1996). However, this is not the case for all sera, with higher-titered samples from clade E-infected individuals demonstrating limited cross-reactivity to other clades (Mascola *et al.*, 1996b). This suggests the existence of common or conserved sites on the HIV-1 envelope, possibly related to the requirement of these sites, expressed on divergent HIV isolates, to interact with the cellular receptors and coreceptors. Vaccine strategies that have been tested in clinical trials have been unable thus far to induce this type of broad cross-reactivity. Attempts to identify and define virus immunotypes for serological typing (Nyambi *et al.*, 1998; Zolla-Pazner *et al.*, 1999) will require additional characterization of both the viral isolates and the serological reagents.

With regard to the characterization of the MHC class I-restricted T-cell responses, there have been a number of concerns about the level of conservation of sequence that would be required for immune cross-reactivity to occur. Substantial data now exist which indicate that CTL, both from HIV vaccine recipients (Ferrari *et al.*, 1997) and HIV-1-infected individuals (Betts *et al.*, 1997; Cao *et al.*, 1997; Lynch *et al.*, 1998), can recognize epitopes from widely divergent HIV subtypes. It is not yet clear how major histocompatibility (MHC) differences in persons from vastly different ethnic and racial backgrounds will affect their ability to make cytolytic T-lymphocyte (CTL) responses to candidate vaccines. A phase I vaccine trial in Uganda will be the first to address this important basic vaccine research issue.

Issues Related to Mucosal Transmission

Transmission rates via mucosal routes are relatively low compared to those for hepatitis or other sexually transmitted diseases. A number of concerns were raised in the mid-1980s that it might not be possible to produce a vaccine for HIV that protected against sexual transmission routes. In experiments completed over several years, it actually appears to be easier to protect against rectal or vaginal challenge than has been previously observed against intravenous challenge (Lohman *et al.*, 1994; Cranage *et al.*, 1997; Miller *et al.*, 1997; Benson *et al.*, 1998).

It had also been claimed that cell-associated virus would be difficult to block with a vaccine strategy. Linked to this, sexual exposure to HIV-1 through abrasions in the mucosal surface would make it possible to permit cellular Trojan horses filled with HIV to enter through the genital mucosa. There are several arguments against this method of HIV transmission. First, any incoming cell would be recognized as foreign because of its MHC molecules; these might in turn provide recognition signals that would generate strong rejection responses to eliminate incoming cells. Even if the incoming cells were able to get across both the physical and the 'autologous' barriers, the virus released from those cells becomes a 'cell-free' component that must be dealt with by the host. The second issue is the actual difficulty of demonstrating cell-associated SIV transmission by the intravaginal route, despite multiple attempts (Miller *et al.*, 1994).

Vaccines also may serve as immunomodulators to diminish HIV disease progression and transmission by improving immune function or by stimulating newly generated T cells. Combining HIV vaccines with early and highly active antiretroviral therapy (HAART) may be particularly advantageous in recently infected infants, adolescents and young adults, where long-term adherence to complex drug regimens might be the most difficult to achieve, and successful reconstitution of immune cell subsets may be highly feasible. One of the most vulnerable populations for HIV infection worldwide, women of childbearing age, would also benefit greatly from immunotherapeutic vaccine regimens, both for themselves and for the interruption of HIV transmission to their infants.

SUMMARY

Various new strategies to stimulate a protective immune response against HIV infection are being explored through basic research, evaluation in animal models, and clinical trials. There are many inherent uncertainties, not only in the identification of the immune responses required for protection against HIV infection or disease, but also in the vaccine designs and strategies needed to induce such protective responses. More intensive use of the SIV and SHIV chimeric viruses in macaque models for concept testing and comparative studies of different vaccines is warranted. This will permit multiple conceptual and practical approaches for the development of candidate vaccines to be pursued at the preclinical level. We have learned a great deal about responses in human volunteers, and the parallel advancement of vaccine candidates to small phase I trials should be encouraged as soon as products and appropriate protocols to evaluate safety and immunogenicity in human volunteers are developed. Conducting small clinical trials early in product development affords the opportunity to rapidly return to the laboratory and incorporate improvements or refine promising vaccine candidates that induce less than optimal immune responses. Because of the complexity of HIV, it is unlikely that only a single immune correlate will be observed. Therefore, intensive analysis of immune parameters in early studies will permit better opportunities to define correlates of immune protection for second and third-generation HIV vaccines.

The information, opinions, data and statements contained herein are not necessarily those of the US. Government, the National Institutes of Health (NIH) or the Office of AIDS Research (OAR), and should not be interpreted, acted upon or represented as such.

The author wishes to thank the many colleagues who have contributed to the continuing efforts for an HIV vaccine, particularly those who have provided us with early copies of manuscripts and work in progress. A special thank you to Richard Billingsley, who guided me through the perils of referencing and editing this review.

REFERENCES

Agy, M.B., Thompson, J., Grant, R.F. *et al.* (1998). Enhanced pathogenicity of SHIV HXBc2 following whole blood passage in macaca nemestrina. *Int. Conf. AIDS* 25–26

Ahlers, J.D., Pendleton, C.D., Dunlop, N., Minassian, A., Nara, P.L. and Berzofsky, J.A. (1993). Construction of an HIV-1 peptide vaccine containing a multideterminant helper peptide linked to a V3 loop peptide 18 inducing strong neutralizing antibody responses in mice of multiple MHC haplotypes after two immunizations. *J. Immunol.* **150:** 5647–5665

Ahlers, J.D., Dunlop, N., Pendleton, C.D., Newman, M., Nara, P.L. and Berzofsky, J.A. (1996). Candidate HIV type 1 multideterminant cluster peptide-P18MN vaccine constructs elicit type 1 helper T cells, cytotoxic T cells, and neutralizing antibody, all using the same adjuvant immunization. *AIDS Res. Hum. Retroviruses* **12:** 259–272

Ahmad, S., Lohman, B., Marthas, M. *et al.* (1994). Reduced virus load in rhesus macaques immunized with recombinant gp 160 and challenged with simian immunodeficiency virus. *AIDS Res. Hum. Retroviruses* **10:** 195–204

Almond, N., Kent, K., Crange, M., Rud, E., Clarke, B. and Stott, E.J. (1995). Protection by attenuated simian immunodeficiency virus in macaques against challenge with virus-infected cells. *Lancet* **345:** 1342–1344

Anderson, M.J., Porter, D.C., Moldoveanu, Z., Fletcher, T.M., McPherson, S. and Morrow, C.D. (1997). Characterization of the expression and immunogenicity of poliovirus replicons that encode simian immunodeficiency virus SIVmac239 Gag or envelope SU proteins. *AIDS Res. Hum. Retroviruses* **13:** 53–62

Andersson, S., Makitalo, B., Thorstensson, R. *et al.* (1996). Immunogenicity and protective efficacy of a human immunodeficiency virus type 2 recombinant canarypox (ALVAC) vaccine candidate in cynomolgus monkeys. *J. Infect. Dis.* **174:** 977–985

Andino, R., Silvera, D., Suggett, S.D. *et al.* (1994). Engineering poliovirus as a vaccine vector for the expression of diverse antigens. *Science* **265:** 1448–1451

Andrus, L., Prince, A.M., Bernal, I. *et al.* (1998). Passive immunization with a human immuno-deficiency virus type 1- neutralizing monoclonal antibody in Hu-PBL-SCID mice: isolation of a neutralization escape variant. *J. Infect. Dis.* **177:** 889–897

Arthur, L.O., Bess, J.W., Sowder, R.C. *et al.* (1992). Cellular proteins bound to immunodeficiency viruses: implications for pathogenesis and vaccines. *Science* **258:** 1935–1938

Arthur, L.O., Bess, J.W., Urban, R.G. *et al.* (1995). Macaques immunized with HLA-DR are protected from challenge with simian immunodeficiency virus. *J. Virol.* **69:** 3117–3124

Arthur, L.O., Bess, J.W.J., Chertova, E.N. *et al.* (1998). Chemical inactivation of retroviral infectivity by targeting nucleocapsid protein zinc fingers: a candidate SIV vaccine. *AIDS Res. Hum. Retroviruses* **14 Suppl 3, S311–9:** S311–S319

Barnett, S.W., Rajasekar, S., Legg, H. *et al.* (1997). Vaccination with HIV-1 gp120 DNA induces immune responses that are boosted by a recombinant gp120 protein subunit. *Vaccine* **15**: 869–873

Barr, P.J., Steimer, K.S., Sabin, E.A. *et al.* (1987). Antigenicity and immunogenicity of domains of the human immunodeficiency virus (HIV) envelope polypeptide expressed in the yeast *Saccharomyces cerevisiae*. *Vaccine* **5**: 90–101

Barre-Sinnoussi, F., Chermann, J. and Rey, F. (1983). Isolation of a T-lymphotropic retrovirus from a patient at risk for acquired immune deficiency syndrome (AIDS). *Science* **220**: 868–871

Bartlett, J.A., Wasserman, S.S., Hicks, C.B. *et al.* (1998). Safety and immunogenicity of an HLA-based HIV envelope polyvalent synthetic peptide immunogen. DATRI 010 Study Group. Division of AIDS Treatment Research Initiative. *AIDS* **12**: 1291–1300

Beasley, R.P., Hwang, L.Y., Stevens, C.E. *et al.* (1983). Efficacy of hepatitis B immune globulin for prevention of perinatal transmission of the hepatitis B virus carrier state: final report of a randomized double-blind, placebo-controlled trial. *Hepatology* **3**: 135–141

Beddows, S., Lister, S., Cheingsong, R., Bruck, C. and Weber, J. (1999). Comparison of the antibody repertoire generated in healthy volunteers following immunization with a monomeric recombinant gp120 construct derived from a CCR5/CXCR4-using human immunodeficiency virus type 1 isolate with sera from naturally infected individuals. *J. Virol.* **73**: 1740–1745

Belshe, R.B., Gorse, G.J., Mulligan, M.J. *et al.* (1998). Induction of immune responses to HIV-1 by canarypox virus (ALVAC) HIV-1 and gp120 SF-2 recombinant vaccines in uninfected volunteers. NIAID AIDS Vaccine Evaluation Group. *AIDS* **12**: 2407–2415

Benson, J., Chougnet, C., Robert-Guroff, M. *et al.* (1998). Recombinant vaccine-induced protection against the highly pathogenic simian immunodeficiency virus SIVmac 251: dependence on route of challenge exposure. *J. Virol.* **72**: 4170–4182

Berger, E.A. (1998). Introduction: HIV co-receptors solve old questions and raise many new ones. *Semin. Immunol.* **10**: 165–168

Berger, E.A., Murphy, P.M. and Farber, J.M. (1999). Chemokine receptors as HIV-1 coreceptors: roles in viral entry, tropism, and disease. *Annu. Rev. Immunol* **17**: 657–700, 657–700

Berggren, R.E., Wunderlich, A., Ziegler, E. *et al.* (1995). HIV gp120-specific cell-mediated immune responses in mice after oral immunization with recombinant Salmonella. *J. AIDS Hum. Retrovirol.* **10**: 489–495

Berglund, P., Quesada-Rolander, M., Putkonen, P., Biberfeld, G., Thorstensson, R. and Liljestrom, P. (1997). Outcome of immunization of cynomolgus monkeys with recombinant Semliki Forest virus encoding human immunodeficiency virus type 1 envelope protein and challenge with a high dose of SHIV-4 virus. *AIDS Res. Hum. Retroviruses* **13**: 1487–1495

Berman, P.W., Riddle, L., Nakamura, G. *et al.* Matthews, T. and Gregory, T. (1989). Expression and immunogenicity of the extracellular domain of the human immunodeficiency virus type 1 envelope glycoprotein, gp160. *J. Virol.* **63**: 3489–3498

Berman, P. (1997). Genetic and immunologic characterization of viruses infecting MN-rgp 120-vaccinated volunteers. *J. Infect. Dis.* **176**: 384–397

Berman, P.W. (1998). Development of bivalent rgp 120 vaccines to prevent HIV type 1 infection. *AIDS Res. Hum. Retroviruses* **14 Suppl 3**, S277–89, S277–S289

Betts, M.R., Krowka, J., Santamaria, C. *et al.* (1997). Cross-clade human immunodeficiency virus (HIV)-specific cytotoxic T-lymphocyte responses in HIV-infected Zambians. *J. Virol.* **71**: 8908–8911

Bogers, W.M., Dubbes, R., Ten Haaft, P. *et al.* (1997). Comparison of in vitro and in vivo infectivity of different clade B HIV-1 envelope chimeric simian/human immunodeficiency viruses in *Macaca mulatta*. *Virology* **236**: 110–117

Boots, L.J., McKenna, P.M., Arnold, B.A. *et al.* (1997). Anti-human immunodeficiency virus type 1 human monoclonal antibodies that bind discontinuous epitopes in the viral glycoproteins can identify mimotopes from recombinant phage peptide display libraries. *AIDS Res. Hum. Retroviruses* **13**: 1549–1559

Borgne, S.L., Mancini, M., Grand, R.L. *et al.* (1998). In vivo induction of specific cytotoxic T lymphocytes in mice and rhesus macaques immunized with DNA vector encoding an HIV epitope fused with hepatitis B surface antigen. *Virology* **240**: 304–315

Borrow, P., Lewicki, H., Wei, X. *et al.* (1997). Antiviral pressure exerted by HIV-1-specific cytotoxic T lymphocytes (CTLs) during primary infection demonstrated by rapid selection of CTL escape virus. *Nature Med.* **3**: 205–211

Boyer, J.D., Wang, B., Ugen, K.E. *et al.* (1996). In vivo protective anti-HIV immune responses in non-human primates through DNA immunization. *J. Med. Primatol.*, **25**: 242–250

Broder, C.C., Earl, P.L., Long, D., Abedon, S.T., Moss, B. and Doms, R.W. (1994). Antigenic implications of human immunodeficiency virus type 1 envelope quaternary structure: oligomer-specific and -sensitive monoclonal antibodies. *Proc. Natl. Acad. Sci. USA* **91**: 11699–11703

Bryder, K., Sbai, H., Nielsen, H.V. *et al.*, (1999). Improved immunogenicity of HIV-1 epitopes in HBsAg chimeric DNA vaccine plasmids by structural mutations of HBsAg. *DNA Cell Biol.* **18**: 219–225

Cairns, J.S. and Sarver, N. (1998). New viral vectors for HIV vaccine delivery. *AIDS Res. Hum. Retroviruses* **14**: 1501–1508

Calarota, S., Bratt, G., Nordlund, S. *et al.* (1998). Cellular cytotoxic response induced by DNA vaccination in HIV-1- infected patients. *Lancet* **351**: 1320–1325

Cao, H., Kanki, P., Sankale, J.L. *et al.* (1997). Cytotoxic T-lymphocyte cross-reactivity among different human immunodeficiency virus type 1 clades: implications for vaccine development. *J. Virol.* **71**: 8615–8623

Carlson, J.R., McGraw, T.P., Keddie, E. *et al.* (1990). Vaccine protection of rhesus macaques against simian immunodeficiency virus infection. *AIDS Res. Hum. Retroviruses* **6**: 1239–1246

Caver, T.E., Lockey, T.D., Srinivas, R.V., Webster, R.G. and Hurwitz, J.L. (1999). A novel vaccine regimen utilizing DNA, vaccinia virus and protein immunizations for HIV-1 envelope presentation. *Vaccine* **17**: 1567–1572

Chan, D.C., Fass, D., Berger, J.M. and Kim, P.S. (1997). Core structure of gp41 from the HIV envelope glycoprotein. *Cell* **89**: 263–273

Chirmule, N. and Pahwa, S. (1996). Envelope glycoproteins of human immunodeficiency virus type 1: profound influences on immune functions. *Microbiol. Rev.* **60**: 386–406

Chun, T.W., Carruth, L., Finzi, D. *et al.* (1997a). Quantification of latent tissue reservoirs and total body viral load in HIV-1 infection. *Nature* **387**: 183–188

Chun, T.W., Stuyver, L., Mizell, S.B. *et al.* (1997b). Presence of an inducible HIV-1 latent reservoir during highly active antiretroviral therapy. *Proc. Natl. Acad. Sci. USA* **94**: 13193–13197

Clark, K.R., Voulgaropoulou, F. and Johnson, P.R. (1996). A stable cell line carrying adenovirus-inducible rep and cap genes allows for infectivity titration of adeno-associated virus vectors. *Gene Ther.* **3**: 1124–1132

Clements, J., Montelaro, R., Zink, M.C. *et al.* (1995). Cross-protective immune responses induced in rhesus macaques by immunization with attenuated macrophage-tropic simian immunodeficiency virus. *J. Virol.* **69**: 2737–2744

Clements-Mann, M., Weinhold, K., Matthews, T.J. *et al.* and NIAID AIDS Vaccine Evaluation Group (1998). Immune responses to human immunodeficiency virus (HIV) type 1 induced by canarypox expressing HIV-1mn gp120, HIV-1sf2 recombinant gp120, or both vaccines in seronegative adults. *J. Infect. Dis.* **177**: 1230–1246

Cocchi, F., DeVico, A.L., Garzino-Demo, A., Arya, S.K., Gallo, R.C. and Lusso, P. (1995). Identification of RANTES, MIP-1 alpha, and MIP-1 beta as the major HIV-suppressive factors produced by CD8+ T cells. *Science* **270**: 1811–1815

Conley, A.J., Kessler, J.A., II, Boots, L.J. *et al.* (1996). The consequence of passive administration of an anti-human immunodeficiency virus type 1 neutralizing monoclonal antibody before challenge of chimpanzees with a primary virus isolate. *J. Virol.* **70**: 6751–6758

Connor, E.M., Sperling, R.S., Gelber, R. *et al.* (1994). Reduction of maternal-infant transmission of human immunodeficiency virus type 1 with zidovudine treatment. Pediatric AIDS Clinical Trials Group Protocol 076 Study Group. *N. Engl. J. Med* **331**: 1173–1180

Connor, R., Korber, B., Graham, B. *et al.* (1998). Immunological and virological analyses of persons infected by human immunodeficiency virus type 1 while participating in trials of recombinant gp 120 subunit vaccines. *J. Virol.* **72**: 1552–1576

Cooney, E.L., Collier, A.C., Greenberg, P.D. *et al.* (1991). Safety of and immunological response to a recombinant vaccinia virus vaccine expressing HIV envelope glycoprotein. *Lancet* **337**: 567–572

Cooney, E.L., McElrath, M.J., Corey, L. *et al.* (1993). Enhanced immunity to human immuno-deficiency virus (HIV) envelope elicited by a combined vaccine regimen consisting of priming with a vaccinia recombinant expressing HIV envelope and boosting with gp160 protein. *Proc. Natl. Acad. Sci. USA* **90**: 1882–1886

Corey, L., McElrath, M.J., Weinhold, K. *et al.* (1998). Cytotoxic T cell and neutralizing antibody responses to human immunodeficiency virus type 1 envelope with a combination vaccine regimen. AIDS Vaccine Evaluation Group. *J. Infect. Dis.* **177**: 301–309

Cotropia, J., Ugen, K.E., Kliks, S. *et al.* (1996). A human monoclonal antibody to HIV-1 gp41 with neutralizing activity against diverse laboratory isolates. *J. AIDS Hum. Retrovirol.* **12**: 221–232

Cox, G.J., Zamb, T.J. and Babiuk, L.A. (1993). Bovine herpesvirus 1: immune responses in mice and cattle injected with plasmid DNA. *J. Virol.* **67**: 5664–5667

Cranage, M., Whatmore, A., Sharpe, S. *et al.* (1997). Macaques infected with live attenuated SIVmac are protected against superinfection via the rectal mucosa. *Virology* **229**: 143–154

Cummins, L.M., Weinhold, K.J., Matthews, T.J. *et al.* (1991). Preparation and characterization of an intravenous solution of IgG from human immunodeficiency virus-seropositive donors. *Blood* **77**: 1111–1117

D'Souza, M.P. and Harden, V.A. (1996). Chemokines and HIV-1 second receptors. Confluence of two fields generates optimism in AIDS research. *Nature Med.* **2**: 1293–1300

Daniel, M., Kirchhoff, F., Czajak, S., Sehgal, P. and Desrosiers, R.C. (1992). Protective Effects of a Live Attenuated SIV Vaccine with a Deletion in the nef Gene. *Science* **258**: 1938–1941

Daniel, M., Mazzara, G., Simon, M. *et al.* (1994). High-titer immune responses elicited by recombinant vaccinia virus priming and particle boosting are ineffective in preventing virulent SIV infection. *AIDS Res. Hum. Retroviruses* **10**: 839–851

Davis, H.L. and Whalen, R.G. (1995). DNA-based immunization. *Mol. Cell Biol. Hum. Dis. Ser.* **5**: 368–387

Davis, M.M., Lyons, D.S., Altman, J.D. *et al.* (1997). T cell receptor biochemistry, repertoire selection and general features of TCR and Ig structure. *Ciba Found. Symp.* **204**: 94–100

Davis, N.L., Brown, K.W. and Johnston, R.E. (1996). A viral vaccine vector that expresses foreign genes in lymph nodes and protects against mucosal challenge. *J. Virol.* **70**: 3781–3787

Desrosiers, R.C., Wyand, M.S., Kodama, T. *et al.* (1989). Vaccine protection against simian immunodeficiency virus infection. *Proc. Natl. Acad. Sci. USA* **86**: 6353–6357

Dickover, R.E., Dillon, M., Leung, K.M. *et al.* (1998). Early prognostic indicators in primary perinatal human immunodeficiency virus type 1 infection: importance of viral RNA and the timing of transmission on long-term outcome. *J. Infect. Dis.* **178**: 375–387

Dolin, R., Graham, B.S., Greenberg, S.B. *et al.* (1991). The safety and immunogenicity of a human immunodeficiency virus type 1 (HIV-1) recombinant gp160 candidate vaccine in humans. NIAID AIDS Vaccine Clinical Trials Network. *Ann. Intern. Med.* **114**: 119–127

Earl, P.L., Broder, C.C., Doms, R.W. and Moss, B. (1997). Epitope map of human immunodeficiency virus type 1 gp41 derived from 47 monoclonal antibodies produced by immunization with oligomeric envelope protein. *J. Virol.* **71**: 2674–2684

Eckhart, L., Raffelsberger, W., Ferko, B. *et al.* (1996). Immunogenic presentation of a conserved gp41 epitope of human immunodeficiency virus type 1 on recombinant surface antigen of hepatitis B virus. *J. Gen. Virol.* **77**: 2001–2008

Edinger, A.L., Hoffman, T.L., Sharron, M., Lee, B., O'Dowd, B. and Doms, R.W. (1998). Use of GPR1, GPR15, and STRL33 as coreceptors by diverse human immunodeficiency virus type 1 and simian immunodeficiency virus envelope proteins. *Virology* **249**: 367–378

Emini, E.A., Schleif, W.A., Nunberg, J.H. *et al.* (1992). Prevention of HIV-1 infection in chimpanzees by gp120 V3 domain-specific monoclonal antibody. *Nature* **355**: 728–730

Feng, Y., Broder, C.C., Kennedy, P.E. and Berger, E.A. (1996). HIV-1 entry cofactor: functional cDNA cloning of a seven-transmembrane, G protein-coupled receptor. *Science* **272**: 872–877

Ferko, B., Katinger, D., Grassauer, A. *et al.* (1998). Chimeric influenza virus replicating predomi-nantly in the murine upper respiratory tract induces local immune responses against human immunodeficiency virus type 1 in the genital tract. *J. Infect. Dis.* **178**: 1359–1368

Ferrari, G., Berend, C., Ottinger, J. *et al.* (1997). Replication-defective canarypox (ALVAC) vectors effectively activate anti-human immunodeficiency virus-1 cytotoxic T lymphocytes present in infected patients: implications for antigen-specific immunotherapy. *Blood* **90**: 2406–2416

Finzi, D., Hermankova, M., Pierson, T. et al. (1997). Identification of a reservoir for HIV-1 in patients on highly active antiretroviral therapy. Science 278: 1295–1300

Finzi, D., Blankson, J., Siliciano, J.D. et al. (1999). Latent infection of CD4+ T cells provides a mechanism for lifelong persistence of HIV-1, even in patients on effective combination therapy. Nature Med. 5: 512–517

Fiscus, S.A., Schoenbach, V.J. and Wilfert, C. (1999). Short courses of zidovudine and perinatal transmission of HIV. N. Engl. J. Med. 340: 1040–1041

Fomsgaard, A., Nielsen, H.V., Nielsen, C. et al. (1998). Comparisons of DNA-mediated immunization procedures directed against surface glycoproteins of human immunodeficiency virus type-1 and hepatitis B virus. APMIS 106: 636–646

Franchini, G., Robert-Guroff, M., Tartaglia, J. et al. (1995). Highly attenuated HIV type 2 recombinant poxviruses, but not HIV-2 recombinant Salmonella vaccines, induce long-lasting protection in rhesus macaques. AIDS Res. Hum. Retroviruses 11: 909–920

Francis, D.P., Gregory, T., McElrath, M.J. et al. (1998). Advancing AIDSVAX to phase 3. Safety, immunogenicity, and plans for phase 3. AIDS Res. Hum. Retroviruses 14 Suppl 3:S325–31, S325–S331

Frankel, F.R., Hegde, S., Lieberman, J. and Paterson, Y. (1995). Induction of cell-mediated immune responses to human immunodeficiency virus type 1 Gag protein by using Listeria monocytogenes as a live vaccine vector. J. Immunol. 155: 4775–4782

Frey, A., Neutra, M.R. and Robey, F.A. (1997). Peptomer aluminum oxide nanoparticle conjugates as systemic and mucosal vaccine candidates: synthesis and characterization of a conjugate derived from the C4 domain of HIV-1MN gp120. Bioconjug. Chem. 8: 424–433

Frey, S.E. (1999). HIV vaccines. Infect. Dis. Clin. N. Am. 13: 95–112

Fuerst, T.R., de, 1.C., V, Bansal, G.P. and Stover, C.K. (1992). Development and analysis of recombinant BCG vector systems. AIDS Res. Hum. Retroviruses 8: 1451–1455

Fuller, D.H., Corb, M.M., Barnett, S., Steimer, K. and Haynes, J.R. (1997a). Enhancement of immunodeficiency virus-specific immune responses in DNA- immunized rhesus macaques. Vaccine 15: 924–926

Fuller, D., Simpson, L., Cole, K. et al. (1997b). Gene gun-based nucleic acid immunization alone or in combination with recombinant vaccinia vectors suppresses virus burden in rhesus macaques challenged with a heterologous SIV. Immuno. Cell Biol. 75: 389–396

Fultz, P., Wei, Q., Davis, I. and Girard, M. (1998). Retroviruses of human AIDS and related animal diseases (Girard, M. and Dodet, B., Eds.) Elsevier, Paris. 261–265

Furuta, R.A., Wild, C.T., Weng, Y. and Weiss, C.D. (1998). Capture of an early fusion-active conformation of HIV-1 gp41. Nat. Struct. Biol. 5: 276–279

Fynan, E.F., Webster, R.G., Fuller, D.H., Haynes, J.R., Santoro, J.C. and Robinson, H.L. (1993). DNA vaccines: protective immunizations by parenteral, mucosal, and gene- gun inoculations. Proc. Natl. Acad. Sci. USA 90: 11478–11482

Gallo, R., Sarin, P. and Gelmann, E. (1983). Isolation of human T-cell leukemia virus in acquired immunodeficiency syndrome (AIDS). Science 220: 865–867

Garber, M.E., Wei, P., KewalRamani, V.N. et al. (1998). The interaction between HIV-1 Tat and human cyclin T1 requires zinc and a critical cysteine residue that is not conserved in the murine CycT1 protein. Genes Dev. 12: 3512–3527

Gauduin, M.C., Safrit, J.T., Weir, R., Fung, M.S. and Koup, R.A. (1995). Pre- and postexposure protection against human immunodeficiency virus type 1 infection mediated by a monoclonal antibody. J. Infect. Dis. 171: 1203–1209

Gauduin, M.C., Parren, P.W., Weir, R., Barbas, C.F., Burton, D.R. and Koup, R.A. (1997). Passive immunization with a human monoclonal antibody protects hu-PBL- SCID mice against challenge by primary isolates of HIV-1. Nature Med. 3: 1389–1393

Giavedoni, L., Planelles, V., Haigwood, N.L. et al. (1993). Immune response of rhesus macaques to recombinant simian immunodeficiency virus gp130 does not protect from challenge infection. J. Virol. 67: 577–583

Gibbs, C.J.J., Peters, R., Gravell, M. et al. (1991). Observations after human immunodeficiency virus immunization and challenge of human immunodeficiency virus seropositive and seronegative chimpanzees. Proc. Natl. Acad. Sci. USA 88: 3348–3352

Gomez, C.E., Navea, L., Lobaina, L. et al. (1999). The V3 loop based multi-epitope polypeptide TAB9 adjuvated with montanide ISA720 is highly immunogenic in nonhuman primates and induces neutralizing antibodies against five HIV-1 isolates. *Vaccine* **17**: 2311–2319

Gonzalo, R.M., Rodriguez, D., Garcia-Sastre, A., Rodriguez, J.R., Palese, P. and Esteban, M. (1999). Enhanced CD8+ T cell response to HIV-1 env by combined immunization with influenza and vaccinia virus recombinants [in process citation]. *Vaccine* **17**: 887–892

Gorse, G.J., Keefer, M.C., Belshe, R.B. *et al.* (1996). A dose-ranging study of a prototype synthetic HIV-1MN V3 branched peptide vaccine. The National Institute of Allergy and Infectious Diseases AIDS Vaccine Evaluation Group. *J. Infect. Dis.* **173**: 330–339

Graham, B.S. (1994). Serological responses to candidate AIDS vaccines. *AIDS Res. Hum. Retroviruses* **10** Suppl 2: S145–8, S145–S148

Graham, B.S., Belshe, R.B., Clements, M.L. *et al.* (1992). Vaccination of vaccinia-naive adults with human immunodeficiency virus type 1 gp160 recombinant vaccinia virus in a blinded, controlled, randomized clinical trial. The AIDS Vaccine Clinical Trials Network. *J. Infect. Dis.* **166**: 244–252

Graham, B.S., Matthews, T.J., Belshe, R.B. *et al.* (1993). Augmentation of human immunodeficiency virus type 1 neutralizing antibody by priming with gp 160 recombinant vaccinia and boosting with rgp160 in vaccinia-naive adults. The NIAID AIDS Vaccine Clinical Trials Network. *J. Infect. Dis.* **167**: 533–537

Graham, B., McElrath, J.M., Connor, R. *et al.* (1998). Analysis of intercurrent human immundeficiency virus type 1 infections in phase I and II trials of candidate AIDS vaccines. *J. Infect. Dis.* **177**: 310–319

Greenough, T.C., Sullivan, J.L. and Desrosiers, R.C. (1999). Declining CD4 T-cell counts in a person infected with nef-deleted HIV-1. *N.Engl.J Med.* **340**: 236–237

Haigwood, N.L., Watson, A. and Sutton, W. (1996). Passive immune globulin therapy in SIV/macaque model: early intervention can alter disease profile. *Immunol. Lett.* **51**: 107–114

Haigwood, N.L., Pierce, C.C., Robertson, M.N. *et al.* (1999). Protection from pathogenic SIV challenge using multigenic DNA vaccines. *Immunol. Lett.* **66**: 183–188

Hanke, T., Blanchard, T.J., Schneider, J. *et al.* (1998a). Immunogenicities of intravenous and intramuscular administrations of modified vaccinia virus Ankara-based multi-CTL epitope vaccine for human immunodeficiency virus type 1 in mice. *J. Gen. Virol.* **79**: 83–90

Hanke, T., Blanchard, T.J., Schneider, J. *et al.* (1998b). Enhancement of MHC class I-restricted peptide-specific T cell induction by a DNA prime-MVA boost vaccination regime. *Vaccine* **16**: 439–445

Hanke, T. and McMichael, A. (1999). Pre-clinical development of a multi-CTL epitope-based DNA prime MVA boost vaccine for AIDS. *Immunol. Lett.* **66**: 177–181

Hanke, T., Neumann, V.C., Blanchard, T.J. *et al.* (1999). Effective induction of HIV-specific CTL by multi-epitope using gene gun in a combined vaccination regime. *Vaccine* **17**: 589–596

Hanson, C.V. (1994). Measuring vaccine-induced HIV neutralization: report of a workshop. *AIDS Res. Hum. Retroviruses* **10**: 645–648

Harouse, J.M., Gettie, A., Tan, R.C., Blanchard, J. and Cheng-Mayer, C. (1999). Distinct pathogenic sequela in rhesus macaques infected with CCR5 or CXCR4 utilizing SHIVs. *Science* **284**: 816–819

Haynes, B.F., Torres, J.V., Langlois, A.J. *et al.* (1993). Induction of HIVMN neutralizing antibodies in primates using a prime- boost regimen of hybrid synthetic gp 120 envelope peptides. *J. Immunol.* **151**: 1646–1653

Heeney, J., Mooij, P., Bogers, W. *et al.* (1998). *Retroviruses of human AIDS and related Animal diseases* (Girard, M. and Dodet, B., Eds.) Colloque des Cent Gardes edition. Elsevier, Paris. 281–285

Heilman, C. and Baltimore, D. (1998). HIV Vaccines – where are we going? *Nature Med. Vaccine Supplement* **4**: 532–534

Heineman, T.C., Connelly, B.L., Bourne, N., Stanberry, L.R. and Cohen, J. (1995). Immunization with recombinant varicella-zoster virus expressing herpes simplex virus type 2 glycoprotein D reduces the severity of genital herpes in guinea pigs. *J. Virol.* **69**: 8109–8113

Hinkula, J., Svanholm, C., Schwartz, S. *et al.* (1997). Recognition of prominent viral epitopes induced by immunization with human immunodeficiency virus type 1 regulatory genes. *J. Virol.* **71**: 5528–5539

Hioe, C.E., Xu, S., Chigurupati, P. *et al.* (1997). Neutralization of HIV-1 primary isolates by polyclonal and monoclonal human antibodies. *Int. Immunol.* **9**: 1281–1290

Hirsch, V., Fuerst, T., Sutter, G. *et al.* (1996). Patterns of viral replication correlate with outcome in simian immunodeficiency virus (SIV)-infected macaques: effect of prior immunization with a trivalent SIV vaccine in modified vaccinia virus Ankara. *J. Virol.* **70**: 3741–3752

Hohmann, E.L., Oletta, C.A., Killeen, K.P. and Miller, S.I. (1996). phoP/phoQ-deleted *Salmonella typhi* (Ty800) is a safe and immunogenic single-dose typhoid fever vaccine in volunteers. *J. Infect. Dis.* **173**: 1408–1414

Honda, M., Matsuo, K., Nakasone, T. *et al.* (1995). Protective immune responses induced by section of a chimeric soluble protein from a recombinant *Mycobacterium bovis* Bacillus Calmette–Guérin vector candidate vaccine for human immunodeficiency virus type 1 in small animals. *Proc. Natl. Acad. Sci. USA* **92**: 10693–10697

Hone, D.M., Wu, S., Powell, R.J. *et al.* (1996). Optimization of live oral *Salmonella*-HIV-1 vaccine vectors for the induction of HIV-specific mucosal and systemic immune responses. *J. Biotechnol.* **44**: 203–207

Hosie, M.J., Flynn, J.N., Rigby, M.A. *et al.* (1998). DNA vaccination affords significant protection against feline immunodeficiency virus infection without inducing detectable antiviral antibodies [published erratum appears in J Virol 1998 Oct;72(10):8460]. *J. Virol.* **72**: 7310–7319

Hu, S.L., Fultz, P., McClure, H.M. *et al.* (1987). Effect of immunization with a vaccinia-HIV env recombinant of HIV infection of chimpanzees. *Nature* **328**: 721723

Hu, S.L., Stallard, V., Abrams, K. *et al.* (1993). Protection of vaccinia-primed macaques against SIVmne infection by combination immunization with recombinant vaccinia virus and SIVmne gp160. *J. Med. Primatol.* **22**: 92–99

Igarashi, T., Brown, C., Azadegan, A. *et al.* (1999). Human immunodeficiency virus type 1 neutralizing antibodies accelerate clearance of cell-free virions from blood plasma. *Nature Med.* **5**: 211–216

Israel, Z., Edmonson, P., Maul, D. *et al.* (1994). Incomplete protection, but suppression of virus burden, elicited by subunit simian immunodeficiency virus vaccines. *J. Virol.* **68**: 1843–1853

Jakob, T., Walker, P.S., Krieg, A.M., von Stebut, E., Udey, M.C. and Vogel, J.C. (1999). Bacterial DNA and CpG-containing oligodeoxynucleotides activate cutaneous dendritic cells and induce IL-12 production: implications for the augmentation of Th1 responses. *Int. Arch. Allergy Immunol.* **118**: 457–461

Jenkins, S., Gritz, L., Fedor, C.H., O'Neill, E.M., Cohen, L.K. and Panicali, D.L. (1991). Formation of lentivirus particles by mammalian cells infected with recombinant fowlpox virus. *AIDS Res. Hum. Retroviruses* **7**: 991–998

Joag, S.J., Adany, I., Li, Z. *et al.* (1997). Animal model of mucosally transmitted human immuno-deficiency virus type 1 disease: intravaginal and oral deposition of simian/human immunode-ficiency virus in macaques results in systemic infection, elimination of CD4+T cells and AIDS. *J. Virol.* **71**: 4016–4023

Joag, S.V., Liu, Z.Q., Stephens, E.B. *et al.* (1998). Oral immunization of macaques with attenuated vaccine virus induces protection against vaginally transmitted AIDS. *J. Virol.* **72**: 9069–9078

Joag, S.V., Li, Z., Wang, C. *et al.* (1999). Passively administered neutralizing serum that protected macaques against infection with parenterally inoculated pathogenic simian-human immuno-deficiency virus failed to protect against mucosally inoculated virus. *AIDS Res. Hum. Retro-viruses* **15**: 391–394

Johnson, J.E., Rodgers, W. and Rose, J.K. (1998). A plasma membrane localization signal in the HIV-1 envelope cytoplasmic domain prevents localization at sites of vesicular stomatitis virus budding and incorporation into VSV virions. *Virology* **251**: 244–252

Johnson, R.P., Lifson, J.D., Czajak, S.C. *et al.* (1999). Highly attenuated vaccine strains of simian immunodeficiency virus protect against vaginal challenge: inverse relationship of degree of protection with level of attenuation. *J. Virol.* **73**: 4952–4961

Keefer, M.C., McElrath, M.J., Weinhold, K. *et al.* (1998). A phase I trial of vaccinia-env/gag/pol (TBC-3B) given by alternative routes, boosted with rgp120MN (AVEG 014C). *12th World AIDS Conference, Geneva*, Abstract 21199

Keefer, M.C., Wolff, M., Gorse, G.J. *et al.* (1997). Safety profile of phase I and II preventive HIV type 1 envelope vaccination: experience of the NIAID AIDS Vaccine Evaluation Group. *AIDS Res. Hum. Retroviruses* **13**: 1163–1177

Kelker, H.C., Schlesinger, D. and Valentine, F.T. (1994). Immunogenic and antigenic properties of an HIV-1 gp1 20-derived multiple chain peptide. *J. Immunol.* **152**: 4139–4148

Kelleher, A.D., Emery, S., Cunningham, P. *et al.* (1997). Safety and immunogenicity of UBI HIV-1MN octameric V3 peptide vaccine administered by subcutaneous injection. *AIDS Res. Hum. Retroviruses* 13: 29–32

Kent, S.J., Zhao, A., Best, S.J., Chandler, J.D., Boyle, D.B. and Ramshaw, I.A. (1998). Enhanced T-cell immunogenicity and protective efficacy of a human immunodeficiency virus type 1 vaccine regimen consisting of consecutive priming with DNA and boosting with recombinant fowlpox virus. *J. Virol.* 72: 10180–10188

Kilby, J.M., Hopkins, S., Venetta, T.M. *et al.* (1998). Potent suppression of HIV-1 replication in humans by T-20, a peptide inhibitor of gp41-mediated virus entry. *Nature Med.* 4: 1302–1307

Kim, J.J. and Weiner, D.B. (1997). DNA gene vaccination for HIV. *Springer Semin. Immunopathol.* 19: 175–194

Kim, J.J., Simbiri, K.A., Sin, J.I. *et al.* (1999). Cytokine molecular adjuvants modulate immune responses induced by DNA vaccine constructs for HIV-1 and SIV. *J. Interferon. Cytokine Res.* 19: 77–84

Kirchhoff, F., Greenough, T.C., Brettler, D.B., Sullivan, J.L. and Desrosiers, R.C. (1995). Brief report: absence of intact nef sequences in a long-term survivor with nonprogressive HIV-1 infection. *N. Engl. J. Med.* 332: 228–232

Kovacs, J.A., Vasudevachari, M.B., Easter, M. *et al.* (1993). Induction of humoral and cell-mediated anti-human immunodeficiency virus (HIV) responses in HIV sero-negative volunteers by immunization with recombinant gp 160. *J. Clin. Invest.* 92: 919–928

Krieg, A.M., Yi, A.K., Matson, S. *et al.* (1995). CpG motifs in bacterial DNA trigger direct B-cell activation. *Nature* 374: 546–549

Krieg, A.M., Wu, T., Weeratna, R. *et al.* (1998). Sequence motifs in adenoviral DNA block immune activation by stimulatory CpG motifs. *Proc. Natl. Acad. Sci. USA* 95: 12631–12636

Kuroda, M.J., Schmitz, J.E., Barouch, D.H. *et al.* (1998). Analysis of Gag-specific cytotoxic T lymphocytes in simian immunodeficiency virus-infected rhesus monkeys by cell staining with a tetrameric major histocompatibility complex class I-peptide complex. *J. Exp. Med.* 187: 1373–1381

Kwong, P.D., Wyatt, R., Robinson, J., Sweet, R.W., Sodroski, J. and Hendrickson, W.A. (1998). Structure of an HIV gp120 envelope glycoprotein in complex with the CD4 receptor and a neutralizing human antibody. *Nature* 393: 648–659

LaCasse, R.A., Follis, K.E., Trahey, M., Scarborough, J.D., Littman, D.R. and Nunberg, J.H. (1999). Fusion-competent vaccines: broad neutralization of primary isolates of HIV. *Science* 283: 357–362

Learmont, J., Tindall, B. and Evans, L. (1992). Long-term symptomless HIV-1 infection in recipients of blood products from a single donor. *Lancet* 340: 863–867

Learmont, J.C., Geczy, A.F., Mills, J. *et al.* (1999). Immunologic and virologic status after 14 to 18 years of infection with an attenuated strain of HIV-1 – a report from the Sydney Blood Bank cohort. *N. Engl. J. Med.* 340: 1715–1722

Leonard, C.K., Spellman, M.W., Riddle, L., Harris, R.J., Thomas, J.N. and Gregory, T.J. (1990). Assignment of intrachain disulfide bonds and characterization of potential glycosylation sites of the type 1 recombinant human immunodeficiency virus envelope glycoprotein (go 120) expressed in chinese hamster ovary cells. *J. Biol. Chem.* 265: 10373–10382

Letvin, N.L. (1998). Progress in the development of an HIV-1 vaccine. *Science* 280: 1875–1880

Levy, J., Hoffman, A.D., Kramer, S.M., Landis, J.A., Shimabukuro, J.M. and Oshiro, L.S. (1984). Isolation of lymphocytopathic retroviruses from San Francisco patients with AIDS. *Science* 2: 840–842

Levy, J. (1998). *HIV and the pathogenesis of AIDS*, 2nd edn. ASM Press

Lewis, J., Balfe, P., Arnold, C., Kaye, S., Tedder, R.S. and McKeating, J.A. (1998). Development of a neutralizing antibody response during acute primary human immunodeficiency virus type 1 infection and the emergence of antigenic variants [in process citation]. *J. Virol.* 72: 8943–8951

Lewis, M.G., Yalley-Ogunro, J., Greenhouse, J.J. *et al.* (1999). Limited protection from a pathogenic chimeric simian-human immunodeficiency virus challenge following immunization with attenuated simian immunodeficiency virus. *J. Virol.* 73: 1262–1270

Li, J., Lord, C., Hastetine, W., Letvin, N. and Sodroski, J. (1992). Infection of cynomolgus monkeys with a chimeric HIV-1/SIV mac virus that expresses the HIV-1 envelope glycoproteins. *J. AIDS* 5: 639–646

Li, S., Polonis, V., Isobe, H. *et al.* (1993). Chimeric influenza virus induces neutralizing antibodies and cytotoxic T cells against human immunodeficiency virus type 1. *J. Virol.* **67**: 6659–6666

Li, D., Forrest, B.D., Li, Z. *et al.* (1997a). International clinical trials of HIV vaccines: II. phase I trial of an HIV-1 synthetic peptide vaccine evaluating an accelerated immunization schedule in Yunnan, China. *Asia. Pac. J. Allergy Immunol.* **15**: 105–113

Li, A., Baba, T.W., Sodroski, J. *et al.* (1997b). Synergistic neutralization of a chimeric SIV/HIV type 1 virus with combinations of human anti-HIV type 1 envelope monoclonal antibodies or hyperimmune globulins. *AIDS Res. Hum. Retroviruses* **13**: 647–656

Liu, M.A., Fu, T.M., Donnelly, J.J., Caulfield, M.J. and Ulmer, J.B. (1998). DNA vaccines. Mechanisms for generation of immune responses. *Adv. Exp. Med. Biol.* **452**: 187–191

Locher, C.P., Barnett, S.W., Herndier, B.G. *et al.* (1998). Human immunodeficiency virus-2 infection in baboons is an animal model for human immunodeficiency virus pathogenesis in humans. *Arch. Pathol. Lab. Med.* **122**: 523–533

Lohman, B.L., McChesney, M.B., Miller, C.J. *et al.* (1994). A partially attenuated simian immunodeficiency virus induces host immunity that correlates with resistance to pathogenic virus challenge. *J. Virol.* **68**: 7021–7029

Looney, D.J., McClure, J., Kent, S.J. *et al.* (1998). A minimally replicative HIV-2 live-virus vaccine protects M. nemestrina from disease after HIV-2(287) challenge. *Virology* **242**: 150–160

Lu, Y. (1997). HIV-1 Vaccine Candidate Evaluation in Non-Human Primates. *Crit. Rev. Oncogenesis* **8**: 273–291

Lu, Y., Pauza, C.D., Montefiori, D. and Miller, C. (1998a). Rhesus macaques that become systemically infected with pathogenic SHIV 89.6-PD after intravenous, rectal, or vaginal inoculation and fail to make an antiviral antibody response rapidly develop AIDS. *J AIDS Hum. Retrovirol.* **19**: 6–18

Lu, Y., Salvato, M.S., Pauza, C.D. *et al.* (1998b). Utility of SHIV for testing HIV-1 vaccine candidates in macaques. *J AIDS Hum. Retrovirol.* **12**: 99–106

Lubeck, M.D., Natuk, R.J., Chengalvala, M. *et al.* (1994). Immunogenicity of recombinant adenovirus-human immunodeficiency virus vaccines in chimpanzees following intranasal administration [published erratum appears in *AIDS Res. Hum. Retroviruses* 1995 Jan;11(1):189]. *AIDS Res. Hum. Retroviruses* **10**: 1443–1449

Lubeck, M.D., Natuk, R., Myagkikh, M. *et al.* (1997). Long-term protection of chimpanzees against high-dose HIV-1 challenge induced by immunization. *Nature Med.* **3**: 651–658

Luciw, P.A., Pratt-Lowe, E., Shaw, K.E., Levy, L.A. and Cheng-Mayer, C. (1995). Persistent infection of rhesus macaques with T-cell-line-tropic and macrophage-tropic clones of simian/human immunodeficiency viruses (SHIV). *Proc. Nat. Acad. Sci. USA* **92**: 7490–7494

Luciw, P.A., Shaw, K.E., Shacklett, B.L. and Marthas, M.L. (1998). Importance of the intracytoplasmic domain of the simian immunodeficiency virus (SIV) envelope glycoprotein for pathogenesis. *Virology* **252**: 9–16

Ludwig, S.L., Brundage, J.F., Kelley, P.W. *et al.* (1998). Prevalence of antibodies to adenovirus serotypes 4 and 7 among unimmunized US army trainees: results of a retrospective nationwide seroprevalence survey. *J. Infect. Dis.* **178**: 1776–1778

Lynch, J.A., deSouza, M., Robb, M.D. *et al.* (1998). Cross-clade cytotoxic T cell response to human immunodeficiency virus type 1 proteins among HLA disparate North Americans and Thais. *J. Infect. Dis.* **178**: 1040–1046

Marthas, M., Sutjipto, S. and Higgins, J. (1990). Immunization with a live, attenuated simian immunodeficiency virus (SIV) prevents early disease but not infection in rhesus macaques challenged with pathogenic SIV. *J. Virol.* **64**: 3694–3700

Mascola, J., Snyder, S., Weislow, O. *et al.* and NIAID AIDS Vaccine Evaluation Group (1996a). Immunization with envelope subunit vaccine products elicits neutralizing antibodies against laboratory-adapted but not primary isolates of human immunodeficiency virus type 1. *J. Infect. Dis.* **173**: 340–348

Mascola, J.R., Louder, M.K., Surman, S.R. *et al.* (1996b). Human immunodeficiency virus type 1 neutralizing antibody serotyping using serum pools and an infectivity reduction assay. *AIDS Res. Hum. Retroviruses* **12**: 1319–1328

Mascola, J.R., Lewis, M.G., Stiegler, G. *et al.* (1999). Protection of macaques against pathogenic simian/human immunodeficiency virus 89.6PD by passive transfer of neutralizing antibodies. *J. Virol.* **73**: 4009–4018

McCutchan, F.E., Salminen, M.O., Carr, J.K. and Burke, D.S. (1996). HIV-1 genetic diversity. *AIDS* **10** Suppl 3: S13–20, S13–S20

McCutchan, F.E., Carr, J.K., Bajani, M. *et al*. (1999). Subtype G and multiple forms of A/G intersubtype recombinant human immunodeficiency virus type 1 in Nigeria. *Virology* **254**: 226–234

McElrath, J.M., Siliciano, R. and Weinhold, K.J. (1997). HIV type 1 vaccine-induced cytotoxic T cell responses in phase I clinical trials: Detection, characterization, and quantitation. *AIDS Res. Hum. Retroviruses* **13**: 211–216

McKeating, J.A., Bennett, J., Zolla-Pazner, S. *et al*. (1993). Resistance of a human serum-selected human immunodeficiency virus type 1 escape mutant to neutralization by CD4 binding site monoclonal antibodies is conferred by a single amino acid change in gp120. *J. Virol*. **67**: 5216–5225

Mellors, J.W., Rinaldo, C.R., Gupta, P., White, R.M., Todd, J.A. and Kingsley, L.A. (1995). Prognosis in HIV-1 infection predicted by the quantity of virus in plasma. *Science* **272**: 1167–1170

Meyer, D., Anderson, D.E., Gardner, M.B. and Torres, J.V. (1998). Hypervariable epitope constructs representing variability in envelope glycoprotein of SIV induce a broad humoral immune response in rabbits and rhesus macaques. *AIDS Res. Hum. Retroviruses* **14**: 751–760

Miller, C.J., Marthas, M., Torten, J. *et al*. A.G. (1994). Intravaginal inoculation of rhesus macaques with cell-free simian immunodeficiency virus results in persistent or transient viremia. *J. Virol*. **68**: 6391–400

Miller, C.J., McChesney, M.B., Lu, X. *et al*. (1997). Rhesus macaques previously infected with simian/human immunodeficiency virus are protected from vaginal challenge with pathogenic SIVmac239. *J. Virol*. **71**: 1911–1921

Mills, J., Deacon, N., McPhee, D. *et al*. (1998). *Retroviruses of human AIDS and related animal diseases* (Girard, M. and Dodet, B., Eds.) Elsevier, Paris. 293–298

Moore, J.P., Cao, Y., Leu, J., Qin, L., Korber, B. and Ho, D.D. (1996). Inter- and intraclade neutralization of human immunodeficiency virus type 1: genetic clades do not correspond to neutralization serotypes but partially correspond to gp 120 antigenic serotypes. *J. Virol*. **70**: 427–444

Moss, B. and Flexner, C. (1987). Vaccinia virus expression vectors. *Annu. Rev. Immunol*. **5**: 305–24, 305–324

Mossman, S.P., Bex, F., Berglund, P. *et al*. (1996). Protection against lethal simian immunodeficiency virus SIV smmPBj14 disease by a recombinant Semliki Forest virus gp 160 vaccine and by a gp 120 subunit vaccine. *J. Virol*. **370**: 1953–1960

Murphey-Corb, M., Martin, L.N., Davison-Fairburn, B. *et al*. (1990). A formalin inactivated whole SIV vaccine and a glycoprotein-enriched subunit vaccine confers protection against experimental challenge with pathogenic live SIV in rhesus monkeys. *Dev. Biol. Standards* **72**, 273–85, 273–285

Murphy, C.G., Lucas, W.T. and Knipe, D.M. (1998). Herpes simplex virus as a vaccine vector for simian immunodeficiency virus. *Int. Conf. Emerg. Infect. Dis*.: 77:

Muster, T., Guinea, R., Trkola, A. *et al*. (1994). Cross-neutralizing activity against divergent human immunodeficiency virus type 1 isolates induced by the gp41 sequence ELDKWAS. *J. Virol*. **68**: 4031–4034

Mwaengo, D.M. and Novembre, F.J. (1998). Molecular cloning and characterization of viruses isolated from chimpanzees with pathogenic human immunodeficiency virus type 1 infections. *J. Virol*. **72**: 8976–8987

Myagkikh, M., Alipanah, S., Markham, P.D. *et al*. (1996). Multiple immunizations with attenuated poxvirus HIV type 2 recombinants and subunit boosts required for protection of rhesus macaques. *AIDS Res. Hum. Retroviruses* **12**: 985–992

Nathanson, N., Hirsch, V.M. and Mathieson, B.J. (1999). The role of nonhuman primates in the development of an AIDS vaccine. *AIDS* (in press)

Natuk, R.J., Chanda, P.K., Lubeck, M.D. *et al*. (1992). Adenovirus-human immunodeficiency virus (HIV) envelope recombinant vaccines elicit high-titered HIV-neutralizing antibodies in the dog model. *Proc. Natl. Acad. Sci. USA* **89**: 7777–7781

Nehete, P.N., Casement, K.S., Arlinghaus, R.B. and Sastry, K.J. (1995). Studies on in vivo induction of HIV-1 envelope-specific cytotoxic T lymphocytes by synthetic peptides from the V3 loop region of HIV-1 IIIB gp 120. *Cell Immunol*. **160**: 217–223

Niedrig, M., Gregersen, J.P., Fultz, P.N., Broker, M., Mehdi, S. and Hilfenhaus, J. (1993). Immune response of chimpanzees after immunization with the inactivated whole immunodeficiency virus (HIV-1), three different adjuvants and challenge. *Vaccine* 11: 67–74

Notka, F., Stahl-Hennig, C., Dittmer, U., Wolf, H. and Wagner, R. (1999). Construction and characterization of recombinant VLPs and Semliki-Forest virus live vectors for comparative evaluation in the SHIV monkey model. *Biol. Chem.* 380: 341–352

Novembre, F.J., Saucier, M., Anderson, D.C. *et al.* (1997). Development of AIDS in a chimpanzee infected with human immunodeficiency virus type 1. *J. Virol.* 71: 4086–4091

Nyambi, P.N., Nkengasong, J., Lewi, P. *et al.* (1996). Multivariate analysis of human immunodeficiency virus type 1 neutralization data. *J. Virol.* 70: 6235–6243

Nyambi, P.N., Gorny, M.K., Bastiani, L., Groen, G.V., Williams, C. and Zolla-Pasner, S. (1998). Mapping of epitopes exposed on intact human immunodeficiency virus type 1 (HIV-1) virons: a new strategy for studying the immunologic relatedness of HIV-1. *J. Virol.* 72: 9384–9391

O'Herrin, S.M., Lebowitz, M.S., Bieler, J.G. *et al.* (1997). Analysis of the expression of peptide-major histocompatibility complexes using high affinity soluble divalent T cell receptors. *J. Exp. Med.* 186: 1333–1345

Ogg, G.S., Jin, X., Bonhoeffer, S. *et al.* (1998). Quantitation of HIV-1-specific cytotoxic T lymphocytes and plasma load of viral RNA. *Science* 279: 2103–2106

Okuda, K., Kaneko, T., Yamakawa, T. *et al.* (1993). Strong synergistic effects of multicomponent vaccine for human immunodeficiency virus infection. *J. Clin. Lab Immunol.* 40: 97–113

Overbaugh, J., Kreiss, J., Poss, M. *et al.* (1999). Studies of human immunodeficiency virus type 1 mucosal viral shedding and transmission in kenya. *J. Infect. Dis.* 179 **Suppl 3:S401–4**, S401–S404

Palese, P., Zavala, F., Muster, T., Nussenzweig, R.S. and Garcia-Sastre, A. (1997). Development of novel influenza virus vaccines and vectors. *J. Infect. Dis.* 176 Suppl 1:S45–9: S45–S49

Panicali, D.L., Mazzara, G., Sullivan, J. *et al.* (1992). Use of Lentivirus-like Particles Alone and In Combination with Live Vaccinia-virus-based Vaccines. *AIDS Res. Hum. Retroviruses* 8: 1449

Parren, P.W., Ditzel, H.J., Gulizia, R.J. *et al.* (1995). Protection against HIV-1 infection in hu-PBL-SCID mice by passive immunization with a neutralizing human monoclonal antibody against the gp120 CD4-binding site. *AIDS* 9: F1–F6

Pascual, D.W., Powell, R.J., Lewis, G.K. and Hone, D.M. (1997). Oral bacterial vaccine vectors for the delivery of subunit and nucleic acid vaccines to the organized lymphoid tissue of the intestine. *Behring Inst. Mitt.* 143–152

Phanuphak, P., Teeratakulpixarn, S., Sarangbin, S. *et al.* (1997). International clinical trials of HIV vaccines: I. Phase I trial of an HIV-1 synthetic peptide vaccine in Bangkok, Thailand. *Asia. Pac. J. Allergy Immunol.* 15: 41–48

Picchio, G.R., Gulizia, R.J. and Mosier, D.E. (1997). Chemokine receptor CCR5 genotype influences the kinetics of human immunodeficiency virus type 1 infection in human PBL-SCID mice. *J. Virol.* 71: 7124–7127

Picchio, G.R., Gulizia, R.J., Wehrly, K., Chesebro, B. and Mosier, D.E. (1998). The cell tropism of human immunodeficiency virus type 1 determines the kinetics of plasma viremia in SCID mice reconstituted with human peripheral blood leukocytes. *J. Virol.* 72: 2002–2009

Polacino, P., Stallard, V., Klaniecki, J.E. *et al.* (1999a). Limited breadth of the protective immunity elicited by simian immunodeficiency virus SIVmme gp160 vaccines in a combination immunization regimen. *J. Virol.* 73: 618–630

Polacino, P., Stallard, V., Montefiori, D.C. *et al.* (1999b). Protection of macaques against intrarectal infection by a combination immunization regimen with recombinant simian immunodeficiency virus SIVmne gp160 vaccines. *J. Virol.* 73: 3134–3146

Polo, J.M., Belli, B.A., Driver, D.A. *et al.* (1999). Stable alphavirus packaging cell lines for Sindbis virus and Semliki Forest virus-derived vectors. *Proc. Natl. Acad. Sci. USA* 96: 4598–4603

Porter, D.C., Melsen, L.R., Compans, R.W. and Morrow, C.D. (1996). Release of virus-like particles from cells infected with poliovirus replicons which express human immunodeficiency virus type 1 Gag. *J. Virol.* 70: 2643–2649

Prevec, L., Christie, B.S., Laurie, K.E. *et al.* (1991). Immune response to HIV-1 gag antigens induced by recombinant adenovirus vectors in mice and rhesus macaque monkeys. *J. A. D. S.* 4: 568–576

Prince, A.M., Reesink, H., Pascual, D. *et al.* (1991). Prevention of HIV infection by passive immunization with HIV immunoglobulin. *AIDS Res. Hum. Retroviruses* 7: 971–973

Pushko, P., Parker, M., Ludwig, G.V., Davis, N.L., Johnston, R.E. and Smith, J.F. (1997). Replicon-helper systems from attenuated Venezuelan equine encephalitis virus: expression of heterologous genes in vitro and immunization against heterologous pathogens in vivo. *Virology* 239: 389–401

Putkonen, P., Thorstensson, R., Ghavamzadeh, L. *et al.* (1991). Prevention of HIV-2 and SIVsm infection by passive immunization in cynomolgus monkeys. *Nature* 352: 436–438

Putkonen, P., Walther, L., Zhang, Y.J. *et al.*, (1995). Long-term protection against SIV-induced disease in macaques vaccinated with a live attenuated HIV-2 vaccine. *Nature Med*: 1: 914–918

Putkonen, P., Quesada-Rolander, M., Leandersson, A.C. *et al.* (1998). Immune responses but no protection against SHIV by gene-gum delivery of HIV-1 DNA followed by recombinant subunit protein boosts. *Virology* 250: 293–301

Raabe, M.L., Issel, C.J., Cook, S.J., Cook, R.F., Woodson, B. and Montelaro, R.C. (1998). Immunization with a recombinant envelope protein (rgp90) of EIAV produces a spectrum of vaccine efficacy ranging from lack of clinical disease to severe enhancement. *Virology* 245: 51–162

Ramsay, A.J., Leong, K.H. and Ramshaw, I.A. (1997). DNA vaccination against virus infection and enhancement of antiviral immunity following consecutive immunization with DNA and viral vectors. *Immunol. Cell Biol.* 75: 382–388

Reimann, K.A., Li, J.T., Veazey, R. *et al.* (1996). A chimeric simian/human immunodeficiency virus expressing a primary patient human immunodeficiency virus type 1 isolate env causes an AIDS-like disease after *in vivo* passage in rhesus monkeys. *J. Virol.* 70: 6922–6928

Rencher, S.D., Slobod, K.S., Dawson, D.H., Lockey, T.D. and Hurwitz, J.L. (1995). Does the key to a successful HIV type 1 vaccine lie among the envelope sequences of infected individuals? *AIDS Res. Hum. Retroviruses* 11: 1131–1133

Richmond, J.F., Mustafa, F., Lu, S. *et al.* (1997). Screening of HIV-1 Env glycoproteins for the ability to raise neutralizing antibody using DNA immunization and recombinant vaccinia virus boosting. *Virology* 230: 265–274

Robert-Guroff, M., Kaur, H., Patterson, L.J. *et al.* (1998). Vaccine protection against a heterologous, non-syncytium-inducing, primary human immunodeficiency virus. *J. Virol.* 72: 10275–10280

Roberts, A., Buonocore, L., Price, R., Forman, J. and Rose, J.K. (1999). Attenuated vesicular stomatitis viruses as vaccine vectors. *J. Virol.* 73: 3723–3732

Robey, F.A., Harris-Kelson, T., Robert-Guroff, M. *et al.* (1996). A synthetic conformational epitope from the C4 domain of HIV Gp120 that binds CD4. *J. Biol. Chem.* 271: 17990–17995

Robinson, H.L. (1997). DNA vaccines for immunodeficiency viruses. *AIDS* 11 Suppl A, S109–S119

Robinson, H.L., Montefiori, D.C., Johnson, R.P. *et al.* (1999). Neutralizing antibody-independent containment of immunodeficiency virus challenges by DNA priming and recombinant pox virus booster immunizations. *Native Med.* 5: 526–534

Rovinski, B., Dekaban, G.A., Cao, S.X. *et al.* (1999). Engineering of noninfectious HIV-1-like particles containing mutant gp41 glycoproteins as vaccine candidates that allow vaccinees to Be distinguished from HIV-1 infectees. *Virology* 257: 438–448

Rubinstein, A., Goldstein, H., Pettoello-Mantovani, M. *et al.* (1995). Safety and immunogenicity of a V3 loop synthetic peptide conjugated to purified protein derivative in HIV-seronegative volunteers. *AIDS* 9: 243–251

Rusche, J.R., Lynn, D.L., Robert-Guroff, M. *et al.* (1987). Humoral immune response to the entire human immunodeficiency virus envelope glycoprotein made in insect cells. *Proc. Natl. Acad. Sci. USA* 84: 6924–6928

Safrit, J.T., Andrews, C.A., Zhu, T., Ho, D.D. and Koup, R.A. (1994). Characterization of human immunodeficiency virus type 1-specific cytotoxic T lymphocyte clones isolated during acute seroconversion: recognition of autologous virus sequences within a conserved immunodominant epitope. *J. Exp. Med.* 179: 463–72

Sarin, P.S., Talmadge, J.E., Heseltine, P. *et al.* (1999). Booster immunization of HIV-1 negative volunteers with HGP-30 vaccine induces protection against HIV-1 virus challenge in SCID mice. *Vaccine* 17: 64–71

Sastry, K.J. and Arlinghaus, R.B. (1991). Identification of T-cell epitopes without B-cell activity in the first and second conserved regions of the HIV Env protein. *AIDS* 5: 699–707

Sato, Y., Roman, M., Tighe, H. *et al.* (1996). Immunostimulatory DNA sequences necessary for effective intradermal gene immunization. *Science* 273: 352–354

Scala, G., Chen, X., Liu, W. *et al.* (1999). Selection of HIV-specific immunogenic epitopes by screening random peptide libraries with HIV-1-positive sera. *J. Immunol.* (in press)

Schuitemaker, H., Koot, M., Kootstra, N.A. *et al.* (1992). Biological phenotype of human immunodeficiency virus type 1 clones at different stages of infection: progression of disease is associated with a shift from monocytotropic to T-cell-tropic virus population. *J. Virol.* 66: 1354–1360

Schultz, A. and Stott, E.J. (1994). Primate Models for AIDS Vaccines. *AIDS* 8: S203–S212

Schultz, A. (1998). *Vaccines* (Bergstrom, T., Perlmann, P. and Wigzell, H., Eds.) Springer Verlag, 357–396

Shaffer, N., Bulterys, M. and Simonds, R.J. (1999). Short courses of zidovudine and perinatal transmission of HIV. *N. Engl. J. Med.* 340: 1042–1043

Shibata, R., Kawamura, M., Sakai, H. *et al.* (1991). Generation of chimeric human and simian immunodeficiency virus infections to monkey peripheral blood mononuclear cells. *J. Virol.* 65: 3514–3520

Shibata, R., Maldarelli, F., Siemon, C. *et al.* (1997). Infection and pathogenicity of chimeric simian-human immunodeficiency viruses in macaques; determinants of high virus loads and CD4 cell killing. *J. Infect. Dis.* 176: 362–373

Shibata, R., Igarashi, T., Haigwood, N. *et al.* (1999). Neutralizing antibody directed against the HIV-1 envelope glycoprotein can completely block HIV-1/SIV chimeric virus infections of macaque monkeys. *Nature Med.* 5: 204–210

Smith, A.D., Geisler, S.C., Chen, A.A. *et al.* (1998). Human rhinovirus type 14:human immunodeficiency virus type 1 (HIV-1) V3 loop chimeras from a combinatorial library induce potent neutralizing antibody responses against HIV-1. *J. Virol.* 72: 651–659

Sperling, R.S., Shapiro, D.E., Coombs, R.W. *et al.* (1996). Maternal viral load, zidovudine treatment, and the risk of transmission of human immunodeficiency virus type 1 from mother to infant. Pediatric AIDS Clinical Trials Group Protocol 076 Study Group. *N. Engl. J. Med.* 335: 1621–1629

Stahl-Hennig, C., Dittmer, U., NiBlein, T. *et al.* (1996). Rapid Development of Vaccine Protection in Macaques by Live-attenuated Simian Immunodeficiency Virus. *J. Gen. Virol.* 77: 2969–2981

Stiehm, E.R., Lambert, J.S., Mofenson, L.M. *et al.* (1999). Efficacy of zidovudine and human immunodeficiency virus (HIV) hyperimmune immunoglobulin for reducing perinatal HIV transmission from HIV-infected women with advanced disease: results of Pediatric AIDS Clinical Trials Group protocol 185. *J. Infect. Dis.* 179: 567–575

Stott, E.J. (1991). Anti-cell antibody in macaques. *Nature* 353: 393–393

Stover, C.K., de, I.C., V, Fuerst, T.R. *et al.* (1991). New use of BCG for recombinant vaccines. *Nature* 351: 456–460

Sugiura, W., Broder, C.C., Moss, B. and Earl, P.L. (1999). Characterization of conformation-dependent anti-gp120 murine monoclonal antibodies produced by immunization with monomeric and oligomeric human immunodeficiency virus type 1 envelope proteins. *Virology* 254: 257–267

Tang, D.C., DeVit, M. and Johnston, S.A. (1992). Genetic immunization is a simple method for eliciting an immune response. *Nature* 356: 152–154

Tang, S., van Rij, R., Silvera, D. and Andino, R. (1997). Toward a poliovirus-based simian immunodeficiency virus vaccine: correlation between genetic stability and immunogenicity. *J. Virol.* 71: 7841–7850

Tartaglia, J., Perkus, M.E., Taylor, J. *et al.* (1992). NYVAC: a highly attenuated strain of vaccinia virus. *Virology* 188: 217–232

Tartaglia, J., Benson, J., Cornet, B. *et al.* (1997). Potential improvements for poxvirus-based immunization vehicles. *Onzieme Colloque Des Cent Gardes*

Tellier, M.C., Pu, R., Pollock, D. *et al.* (1998). Efficacy evaluation of prime-boost protocol: canarypoxvirus-based feline immunodeficiency virus (FIV) vaccine and inactivated FIV-infected cell vaccine against heterologous FIV challenge in cats. *AIDS* 12: 11–18

Tijhaar, E.J., Zheng-Xin, Y., Karlas, J.A. *et al.* (1994). Construction and evaluation of an expression vector allowing the stable expression of foreign antigens in a Salmonella typhimurium vaccine strain. *Vaccine* 12: 1004–1011

Uberla, K., Stahl-Hennig, C., Bottiger, D. *et al.* (1995). Animal model for the therapy of acquired immunodeficiency syndrome with reverse transcriptase inhibitors. *Proc. Natl. Acad. Sci. USA* **92**: 8210–8214

Ulmer, J.B., Donnelly, J.J., Parker, S.E. *et al.* (1993). Heterologous protection against influenza by injection of DNA encoding a viral protein. *Science* **259**: 1745–1749

van't Wout, A.B., Kootstra, N.A., Mulder-Kampinga, G.A. *et al.* (1994). Macrophage-tropic variants initiate human immunodeficiency virus type 1 infection after sexual, parenteral, and vertical transmission. *J. Clin. Invest.* **94**: 2060–2067

Van Rompay, K., Berardi, C., Dillard-Telm, S. *et al.* (1998). Passive immunization of newborn rhesus macaques prevents oral simian immunodeficiency virus infection. *J. Infect. Dis.* **177**: 1247–1259

VanCott, T.C., Mascola, J.R., Loomis-Price, L.D. *et al.* (1999). Cross-subtype neutralizing antibodies induced in baboons by a subtype E gp120 immunogen based on an R5 primary human immunodeficiency virus type 1 envelope. *J. Virol.* **73**: 4640–4650

Vella, C., Ferguson, M., Dunn, G. *et al.* (1993). Characterization and primary structure of a human immunodeficiency virus type 1 (HIV-1) neutralization domain as presented by a poliovirus type 1/HIV-1 chimera. *J. Gen. Virol.* **74**: 2603–2607

Verschoor, E.J., Mooij, P., Oostermeijer, H. *et al.* (1999). Comparison of immunity generated by nucleic acid-, MF59-, and ISCOM-formulated human immunodeficiency virus type 1 vaccines in rhesus macaques: evidence for viral clearance. *J. Virol.* **73**: 3292–3300

von Gegerfelt, A. and Felber, B.K. (1997). Replacement of posttranscriptional regulation in SIVmac239 generated a Rev-independent infectious virus able to propagate in rhesus peripheral blood mononuclear cells. *Virology* **232**: 291–299

Wade, N.A., Birkhead, G.S., Warren, B.L. *et al.* (1998). Abbreviated regimens of zidovudine prophylaxis and perinatal transmission of the human immunodeficiency virus. *N. Engl. J. Med.* **339**: 1409–1414

Walker, M. and Fast, P. (1994). Clinical trials of candidate AIDS vaccines. *AIDS* **8**: S213–S236

Wang, B., Ugen, K.E., Srikantan, V. *et al.* (1993). Gene inoculation generates immune responses against human immunodeficiency virus type 1. *Proc. Natl. Acad. Sci. USA* **190**: 4156–4160

Wang, B., Boyer, J., Srikantan, V. *et al.* (1995a). Induction of humoral and cellular immune responses to the human immunodeficiency type 1 virus in nonhuman primates by in vivo DNA inoculation. *Virology* **211**: 102–112

Wang, B., Boyer, J.D., Ugen, K.E. *et al.* (1995b). Nucleic acid-based immunization against HIV-1: induction of protective in vivo immune responses. *AIDS* **9 Suppl A, S159–70**, S159–S170

Weber, J., Fenyo, E.M., Beddows, S., Kaleebu, P. and Bjorndal, A. (1996). Neutralization serotypes of human immunodeficiency virus type 1 field isolates are not predicted by genetic subtype. The WHO Network for HIV Isolation and Characterization. *J. Virol.* **70**: 7827–7832

Wei, Q. and Fultz, P.N. (1998). Extensive diversification of human immunodeficiency virus type 1 subtype B strains during dual infection of a chimpanzee that progressed to AIDS. *J. Virol.* **72**: 3005–3017

Weinhold, K., Humphrey, W., Corr, K. *et al.* (1997). Cellular Immune Responses to Candidate AIDS Vaccines. *Onzième Colloque Des Cent Gardes*

Weiss, R.A., Clapham, P.R., Weber, J.N., Dalgleish, A.G., Lasky, L.A. and Berman, P.W. (1986). Variable and conserved neutralization antigens of human immunodeficiency virus. *Nature* **324**: 572–575

Weissenhorn, W., Dessen, A., Harrison, S.C., Skehel, J.J. and Wiley, D.C. (1997). Atomic structure of the ectodomain from HIV-1 gp41. *Nature* **387**: 426–430

Wild, C., Oas, T., McDanal, C., Bolognesi, D. and Matthews, T. (1992). A synthetic peptide inhibitor of human immunodeficiency virus replication: correlation between solution structure and viral inhibition. *Proc. Natl. Acad. Sci. USA* **89**: 10537–10541

Wolff, J.A., Malone, R.W., Williams, P. *et al.* (1990). Direct gene transfer into mouse muscle in vivo. *Science* **247**: 1465–1468

Wong, J.K., Hezareh, M., Gunthard, H.F. *et al.* (1997). Recovery of replication-competent HIV despite prolonged suppression of plasma viremia. *Science* **278**: 1291–1295

Wu, S., Pascual, D.W., Lewis, G.K. and Hone, D.M. (1997). Induction of mucosal and systemic responses against human immunodeficiency virus type 1 glycoprotein 120 in mice after oral immunization with a single dose of a Salmonella-HIV vector, *AIDS Res. Hum. Retroviruses* **13**: 1187–1194

Wyand, M., Manson, K., Garcia-Moll, M., Montefiori, D. and Desrosiers, R.C. (1996). Vaccine protection by a triple deletion mutant of simian immunodeficiency virus. *J. Virol.* **70**: 3724–3733

Wyatt, R., Kwong, P.D., Desjardins, E. *et al.* (1998). The antigenic structure of the HIV gp120 envelope glycoprotein. *Nature* **393**: 705–711

Yasutomi, Y., Koenig, S., Haun, S.S. *et al.* (1993). Immunization with recombinant BCG-SIV elicits SIV-specific cytotoxic T lymphocytes in rhesus monkeys. *J. Immunol.* **150**: 3101–3107

Zolla-Pazner, S., Gorny, M.K., Nyambi, P.N., VanCott, T.C. and Nadas, A. (1999). Immunotyping of human immunodeficiency virus type 1 (HIV): an approach to immunologic classification of HIV. *J. Virol.* **73**: 4042–4051

■ INDEX

Numbers in *italics* refer to tables or illustrations

Chapter 2

THE LAW OF PROSPERITY

●　　●　　●　　●　　●　　●　　●　　●

One of the greatest messages given to us through the scriptures is that God is everyone's supply, and that a woman can release, through her spoken word, all that belongs to her by Divine Right. She must, however, have perfect faith in her spoken word.

Isaiah said, "My word shall not return unto me void, but shall accomplish that where it is sent." We know now that words and thoughts are a tremendous vibratory force, ever molding a woman's body and affairs.

A woman came to me in great distress and said she was to be sued on the 15th of the month for $3,000. She knew no way of getting the money and was in despair. I told her that God was her supply, and there is a supply for every demand.

So I spoke the word! I gave thanks that the woman would receive $3,000 at the right time in the right way. I told her she must have perfect faith and act her perfect faith. The 15th came, but no money had materialized.

She called me on the phone and asked what she was to do. I replied, "It is Saturday, so they won't sue you today. Your part is to act rich, thereby showing perfect faith that you will receive it by Monday." She asked me to lunch with her to keep up her courage. When I joined her at a restaurant, I said, "This is no time to economize. Order an expensive luncheon, act as if you have already received the $3,000. All things whatsoever ye ask in prayer, believing, ye shall receive … You must act as if you had already received." The next morning she called me on the phone and asked me to stay with her during the day. I said, "No, you are divinely protected, and God is never too late."

In the evening she phoned again, greatly excited, and said, "My dear, a miracle has happened! I was sitting in my room this morning when the doorbell rang. I said to the maid, 'Don't let anyone in.' The maid, however, looked out the window and said, 'It's your cousin with the long white beard.' So I said,

'Call him back. I would like to see him.' He was just turning the corner when he heard the maid's voice, and he came back. He talked for about an hour, and just as he was leaving he said, 'Oh, by the way, how are finances?' I told him I needed the money, and he said, 'Why, my dear, I will give you $3,000 the first of the month.' I didn't want to tell him I was going to be sued. What shall I do? I won't receive it till the first of the month, and I must have it tomorrow."

I told her, "I'll keep on treating. Spirit is never too late." I gave thanks that she had received the money on the invisible plane and that it manifested on time. The next morning her cousin called her up and said, "Come to my office this morning and I will give you the money." That afternoon, she had $3,000 to her credit in the bank and wrote checks as rapidly as her excitement would permit. If one asks for success and prepares for failure, she will get the situation she has prepared for.

For example, a woman came to me asking me to speak the word that a certain debt would be wiped out. I found that she spent her time planning what she would say to the woman when she did not pay her bill, thereby neutralizing my words. She should have seen herself paying the debt.

We have a wonderful illustration of this in the Bible, relating to the three kings who were in the desert without water for their men and horses. They consulted the prophet Elisha, who gave them this astonishing message: "Thus saith the Lord—Ye shall not see wind, neither shall ye see rain, yet make this valley full of ditches."

A woman must prepare for the thing she has asked for when there isn't the slightest sign of it in sight.

For example, a woman found it necessary to look for an apartment during the year when there was a great shortage of apartments in New York. It was considered almost an impossibility, and her friends were sorry for her and said, "Isn't it too bad, you'll have to store your furniture and live in a hotel." She replied, "You needn't feel sorry for me. I'm a strong woman, and I'll get an apartment."

She spoke the words: "Infinite Spirit, open the way for the right apartment." She knew there was a supply for every demand.

She had contemplated buying new blankets when the "tempter," the adverse thought or reasoning mind, suggested, "Don't buy the blankets. Perhaps, after all, you won't get an apartment and you will have no use for them." She promptly replied (to herself), "I'll dig my ditches by buying the blankets!" So she prepared for the apartment, acting as though she already had it.

She found one in a miraculous way, and it was given to her, although there were over 200 other applicants. The blankets showed active faith.

It is needless to say that the ditches dug by the three kings in the desert were filled to overflowing. (Read II Kings.)

Getting into the spiritual swing of things is no easy matter for the average person. The adverse thoughts of doubt and fear surge from the subconscious. They are the "army of the aliens" that must be put to flight. This explains why it is so often "darkest before the dawn."

A big demonstration is usually preceded by tormenting thoughts.

Having made a statement of high spiritual truth, one challenges the old beliefs in the subconscious, and "error is exposed" to be put out.

Make your affirmations of Truth repeatedly, and rejoice and give thanks that you have already received: "Before ye call I shall answer." This means that "every good and perfect gift" is already yours awaiting your recognition.

A woman can only receive what she sees herself receiving.

The children of Israel were told that they could have all the land they could see. This is true of every woman. She has only the land within her own mental vision. Every great work, every big accomplishment, has been brought into manifestation through holding to the vision; and often, just before the big achievement, comes apparent failure and discouragement.

When they reached the "Promised Land," the children of Israel were afraid to go in, for they said it was filled with giants who made them feel like grasshoppers. "And there we saw the giants, and we were in our own sight as grasshoppers." This is almost every woman's experience.

However, the one who knows Spiritual Law is undisturbed by appearance and rejoices while she is "yet in captivity." That is, she holds to her vision and gives thanks that the end is accomplished; she has received.

Jesus Christ gave a wonderful example of this. He said to his disciples, "Say not ye, there are yet four months and then cometh the harvest? Behold, I say unto you, lift up your eyes and look on the fields; for they are ripe already to harvest." His clear vision pierced the "world of matter," and he saw clearly the fourth-dimensional world, things as they really are, perfect and complete in Divine Mind. So a woman must ever hold the vision of her journey's end and demand the manifestation of that which she has already received. It may be her perfect health, love, supply, self-expression, home, or friends.

They are all finished and perfect ideas registered in Divine Mind (a woman's own superconscious mind) and must come *through* her, not *to* her.

For example, a woman came to me asking for treatments for success. It was imperative that she raise, within a certain time, $50,000 for her business. The time limit was almost up when she came to me in despair. No one wanted to invest in her enterprise, and the bank had flatly refused a loan. I replied, "I suppose you lost your temper while at the bank, therefore your power. You can control any situation if you first control yourself."

"Go back to the bank," I added, "and I will treat." My treatment was: "You are identified in love with the spirit of everyone connected with the bank. Let the Divine Idea come out of this situation."

She replied, "Woman, you are talking about an impossibility. Tomorrow is Saturday; the bank closes at noon, and my train won't get me there until 10, and the time limit is up tomorrow. And anyway, they won't do it. It's too late."

I replied, "God doesn't need any time, and it is never too late. With God, all things are possible." I added, "I don't know anything about business, but I know all about God."

She replied, "It all sounds fine when I sit here listening to you, but when I go out, it's terrible."

She lived in a distant city, and I did not hear from her for a week. Then came a letter. It read, "You were right. I raised the money, and will never again doubt the truth of all that you told me."

I saw her a few weeks later, and I asked, "What happened? You evidently had plenty of time, after all."

She replied, "My train was late, and I got there just 15 minutes before 12. I walked into the bank quietly and said, 'I have come for the loan,' and they gave it to me without a question."

It was the last 15 minutes of the time allotted to her, and Infinite Spirit was not too late. In this instance, this woman could never have demonstrated alone. She needed someone to help her hold to the vision. This is what one woman can do for another.

Jesus Christ knew the truth of this when he said, "If two of you shall agree on Earth as touching anything that they shall ask, it shall be done for them of my Father which is in heaven." One gets too close to their own affairs and becomes doubtful and fearful.

The friend or "healer" sees clearly the success, health, or prosperity, and never wavers, because she is not close to the situation. It is much easier to demonstrate for someone else than for oneself, so a person should not hesitate to ask for help if she feels herself wavering.

A keen observer of life once said, "No woman can fail, if some one person sees her successful."

Such is the power of the vision, and many a great woman owed her success to a spouse, or sister, or teacher, or a friend who believed in her and held without wavering to the perfect pattern!

Chapter 3

THE POWER OF
THE WORD

● ● ● ● ● ● ● ●

A person knowing the power of the word becomes very careful of her conversation. She has only to watch the reaction of her words to know that they do "not return void." Through her spoken word, a woman is continually making laws for herself.

I knew a woman who said, "I always miss a cab. It invariably pulls out just as I arrive." Her daughter said, "I always catch a cab. It's sure to come just as I get there." This occurred for years. Each had made a separate law for herself, one of failure, one of success. This is the psychology of superstitions.

The horseshoe or rabbit's foot contains no power, but a woman's spoken word and belief that it will bring good luck creates expectancy in the subconscious mind and attracts a lucky situation. I find, however, that this will not "work" when a woman has advanced spiritually and knows a higher law. One cannot turn back and must put away "graven images."

For example, two women in my class had had great success in business for several months, when suddenly everything "went to smash." We tried to analyze the situation, and I found that, instead of making their affirmations and looking to God for success and prosperity, they had each bought a "lucky monkey."

I said, "Oh, I see, you have been trusting in the lucky monkeys instead of God. Put away the lucky monkeys and call on the law of forgiveness," for a woman has power to forgive or neutralize her mistakes.

They decided to throw the lucky monkeys down a coalhole, and all went well again. This does not mean, however, that one should throw away every "lucky" ornament or horseshoe about the house, but one must recognize that the power in back of it is the one and only power, God, and that the object simply gives one a feeling of expectancy.

I was with a friend one day who was in deep despair. While crossing the street, she picked up a horseshoe. Immediately, she was filled with joy and hope. She said that God had sent her the horseshoe in order to keep up her courage.

It was indeed, at that moment, about the only thing that could have registered in her consciousness. Her hope became faith, and she ultimately made a wonderful demonstration. I wish to make the point clear that the women previously mentioned were depending on the monkeys alone, while this woman recognized the power in back of the horseshoe.

I know, in my own case, that it took a long while to get out of a belief that a certain thing brought disappointment. If the thing happened, disappointment invariably followed. I found that the only way I could make a change in the subconscious was by asserting, "There are not two powers; there is only one power, God; therefore, there are not disappointments, and this thing means a happy surprise." I noticed a change at once, and happy surprises commenced coming my way.

I have a friend who said that nothing could induce her to walk under a ladder. I said, "If you are afraid, you are giving in to a belief in two powers, good and evil, instead of one. Since God is absolute, there can be no opposing power, unless a woman makes the false law of evil for herself. To show you believe in only One Power, God, and that there is no power or reality in evil, walk under the next ladder you see."

Soon after, she went to her bank. She wished to open her box in the safe-deposit vault, and there stood a ladder in her pathway. It was impossible to reach the box without passing under the ladder. She recoiled in fear and turned back. She could not face the lion on her pathway. However, when she reached the street, my words rang in her ears, and she decided to return and walk under it. It was a big moment in her life, for ladders had held her in bondage for years. She retraced her steps to the vault, and the ladder was no longer there! This so often happens! If one is willing to do a thing she is afraid to do, she does not have to.

It is the law of nonresistance, which is so little understood.

Someone has said that courage contains genius and magic. Face a situation fearlessly, and there is no situation to face; it falls away of its own weight.

The explanation is that fear attracted the ladder on the woman's pathway, and fearlessness removed it.

Thus, the invisible forces are ever working for a woman who is always "pulling the strings" herself, though she does not know it. Owing to the vibratory power of words, whatever a woman voices, she begins to attract. People who continually speak of disease, invariably attract it.

After a woman knows the Truth, she cannot be too careful of her words. For example, I have a friend who often says on the phone, "Do come to see me and have an old-fashioned chat." This "old-fashioned chat" means an hour of about five hundred to a thousand destructive words, the principal topics being loss, lack, failure, and sickness.

I reply, "No, thank you. I've had enough old-fashioned chats in my life; they are too expensive. But I will be glad to have a new-fashioned chat, and talk about what we want, not what we don't want." There is an old saying that a woman only dares use her words for three purposes: to "heal, bless, or prosper." What a woman says of others will be said of her, and what she wishes for another, she is wishing for herself.

"Curses, like chickens, come home to roost."

If a woman wishes someone "bad luck," she is sure to attract bad luck herself. If she wishes to aid someone to success, she is wishing and aiding herself to success.

The body may be renewed and transformed through the spoken word and clear vision, and disease be completely wiped out of the consciousness. The metaphysician knows that all disease has a mental correspondence, and in order to heal the body, one must first "heal the soul."

The soul is the subconscious mind, and it must be "saved" from wrong thinking.

In the 23rd Psalm, we read, "He restoreth my soul." This means that the subconscious mind or soul must be restored with the right ideas, and the "mystical marriage" is the marriage of the soul and the spirit, or the subconscious and superconscious mind. They must be one. When the subconscious is flooded with the perfect ideas of the superconscious, God and a woman are one: "I and the Father are one." That is, she is one with the realm of perfect ideas; she is a woman made in God's likeness and image (imagination) and is given power and dominion over all created things, her mind, body, and affairs.

It is safe to say that all sickness and unhappiness come from the violation of the law of love. A new commandment I give unto you: "Love one another," and in the Game of Life, love or goodwill takes every trick.

For example, a woman I know had, for years, an appearance of a terrible skin disease. The doctors told her it was incurable, and she was in despair. She was on the stage, she feared she would soon have to give up her profession, and she had no other means of support. She, however, procured a good engagement and, on the opening night, was a great hit. She received flattering notices from the critics, and was joyful and elated. The next day she received a notice of dismissal. A woman in the cast had been jealous of her success and had caused her to be sent away. She felt hatred and resentment taking complete possession of her, and she cried out, "Oh God, don't let me hate that woman!" That night she worked for hours "in the silence."

She said, "I soon came into a very deep silence. I seemed to be at peace with myself, with the woman, and with the whole world. I continued this for two following nights, and on the third day I found that I was healed completely of the skin disease!"

In asking for love or goodwill, she had fulfilled the Law ("for love is the fulfilling of the Law"), and the disease (which came from subconscious resentment) was wiped out.

Continual criticism produces rheumatism, since critical, inharmonious thoughts cause unnatural deposits in the blood, which settle in the joints.

False growths such as cancer, tumors, and cysts are caused by jealousy, hatred, unforgiveness, fear, and so on. Every disease is caused by a mind not at ease. I said once, in my class, "There is no use asking anyone 'What's the matter with you?' We might just as well say, 'Who's the matter with you?' Unforgiveness is the most prolific cause of disease. It will harden arteries or the liver and affect the eyesight. In its train are endless ills."

I called on a woman one day who said she was ill from having eaten a poisoned oyster. I replied, "Oh no, the oyster was harmless; you poisoned the oyster. What's the matter with you?" She answered, "Oh, about 19 people." She had quarreled with that many people and had become so inharmonious that she attracted the wrong oyster.

Any disharmony on the external indicates that there is mental disharmony. "As the within, so the without."

A woman's only enemies are within herself. "And a woman's foes shall be they of her own household." Personality is one of the last enemies to be overcome, as this planet is taking its initiation in love. It was Christ's message: "Peace on Earth, goodwill toward all." The enlightened woman, therefore, endeavors to perfect herself upon her neighbor. Her work is with herself, to send out goodwill and blessings to everyone, and the marvelous thing is, that if one blesses her supposed enemy, the other person has no power to harm her.

For example, a woman came to me asking me to treat her for success in business. She was selling machinery, and a rival appeared on the scene with what she proclaimed was a better machine. My friend feared defeat. I said, "First of all, we must wipe out all fear and know that God protects your interest, and that the Divine Idea must come out of the situation. That is, the right machine will be sold by the right woman, to the right customer." And I added, "Don't hold one critical thought toward that woman. Bless her all day, and be willing not to sell your machine if it isn't the Divine Idea."

So she went to the meeting, fearless and nonresistant, and blessing the other woman. She said that the outcome was very remarkable. Her competitor's machine refused to work, and she sold hers without the slightest difficulty.

"But I say unto you, love your enemies, bless them who curse you, do good to them who hate you, and pray for them who spitefully use you and persecute you."

Goodwill produces a great aura of protection about the one who sends it and "no weapon that is formed against her shall prosper." In other words, love and goodwill destroy the enemies of oneself, therefore, one has no enemies on the external!

There is peace on Earth for the woman who sends goodwill to all!

Chapter 4

THE LAW OF NONRESISTANCE

●　　●　　●　　●　　●　　●　　●　　●

Nothing on Earth can resist an absolutely nonresistant person. The Chinese say that water is the most powerful element because it is perfectly nonresistant. It can wear away a rock and sweep all before it.

Jesus Christ said, "Resist not evil," for he knew in reality there is no evil, therefore nothing to resist. Evil has come from "vain imagination," or a belief in two powers, good and evil.

There is an old legend that Adam and Eve ate of "Maya, the Tree of Illusion," and saw two powers instead of one power, God. Therefore, evil is a false law people have made for themselves, through psychoma, or soul sleep. Soul sleep means that a society's soul has been hypnotized by the belief in sin, sickness, and death, and so on that is carnal or mortal thought, and their affairs have out-pictured their illusions.

We have read in a preceding chapter that society's soul is their subconscious mind, and whatever is felt deeply, good or bad, is out-pictured by that faithful servant. A woman's body and affairs show forth what she has been picturing. The sick woman has pictured sickness; the poor woman, poverty; the rich woman, wealth.

People often ask, "Why does a little child attract illness when it is too young even to know what it means?" I answer that children are sensitive and receptive to the thoughts of others about them, and often out-picture the fears of their parents.

I heard a metaphysician once say, "If you do not run your subconscious mind yourself, someone else will run it for you." Parents often, unconsciously, attract illness and disaster to their children by continually holding them in thoughts of fear and watching for symptoms.

For example, a friend asked a woman if her little girl had had the measles. She replied promptly, "Not yet!" This implied that she was expecting the illness and, therefore, preparing the way for what she did not want for herself and her child.

However, the woman who is centered and established in right thinking, the woman who sends out only goodwill to others and who is without fear, cannot be touched or influenced by the negative thoughts of others. In fact, she could then receive only good thoughts, as she herself sends forth only good thoughts. Resistance is hell, for it places people in a "state of torment."

A metaphysician once gave me a wonderful recipe for taking every trick in the Game of Life; it is the acme of nonresistance. He gave it in this way: "At one time in my life, I baptized children, and of course, they had many names. Now I no longer baptize children, but I baptize events, and I give every event the same name. If I have a failure, I baptize it 'Success,' in the name of the Father, and of the Son, and of the Holy Ghost!"

In this, we see the great law of transmutation, founded on nonresistance. Through his spoken word, every failure was transmuted into success.

For example, a woman who required money and who knew the Spiritual Law of opulence, was thrown continually in a business way with someone who made her feel very poor. This person talked lack and limitation, and she commenced to catch those poverty thoughts, so she disliked this person and blamed this person for her failure. She knew that in order to demonstrate her supply, she must first feel that she had received—a feeling of opulence must precede its manifestation.

It dawned on her one day that she was resisting the situation, and seeing two powers instead of one. So she blessed this person and baptized the situation "Success!"

She affirmed: "As there is only one power, God, this person is here for my good and my prosperity" (just what they did not seem to be there for). Soon after that she met, through this person, a woman who gave her several thousand dollars for a service rendered. Her former associate moved to a distant city and faded harmoniously from her life.

Make the statement: "Every person is a golden link in the chain of my good, for all people are God in manifestation, awaiting the opportunities given by others, to serve the Divine Plan of their lives. Bless your enemies, and you rob them of their ammunition." Their arrows will be transmuted into blessings. This law is true of nations as well as individuals. Bless a nation, send love and goodwill to every inhabitant, and it is robbed of its power to harm.

A woman can only get the right idea of nonresistance through spiritual understanding. My students have often said, "I don't want to be a doormat."

I reply, "When you use nonresistance with wisdom, no one will ever be able to walk over you."

Another example: One day I was impatiently awaiting an important telephone call. I resisted every call that came in and made no outgoing calls myself, reasoning that it might interfere with the one I was awaiting.

Instead of saying, "Divine Ideas never conflict; the call will come at the right time," leaving it to Infinite Intelligence to arrange, I commenced to manage things myself—I made the battle mine, not God's, and remained tense and anxious. The phone did not ring for about an hour, and I glanced at it and found the receiver had been off that length of time and was disconnected. My anxiety, fear, and belief in interference had brought on a total eclipse of the telephone. Realizing what I had done, I commenced blessing the situation at once; I baptized it "Success," and affirmed: "I cannot lose any call that belongs to me by Divine Right; I am under grace, and not under law."

A neighbor contacted the phone company on my behalf and asked them to reconnect. My phone was connected at once, and two minutes later, I received a very important call, and about an hour afterward, the one I had been awaiting.

One's ships come in over a calm sea.

So long as a woman resists a situation, she will have it with her. If she runs away from it, it will run after her.

For example, I repeated this to a woman one day, and she replied, "How true that is! I was unhappy at home, I disliked my mother, who was critical and domineering; so I ran away and was married—but I married my mother, for my spouse was exactly like my mother, and I had the same situation to face again."

"Agree with thine adversary quickly."

This means, agree that the adverse situation is good, be undisturbed by it, and it falls away of its own weight. "None of these things move me" is a wonderful affirmation.

The inharmonious situation comes from some disharmony within a woman herself. When there is, in her, no emotional response to an inharmonious situation, it fades away forever from her pathway.

People have said to me, "Give treatments to change my spouse, or my friends."

I reply, "No, I will give treatments to change you; when you change, your spouse and your friends will change."

One of my students was in the habit of lying. I told her it was a failure method, and if she lied, she would be lied to.

She replied, "I don't care, I can't possibly get along without lying."

One day she was speaking on the phone to a man with whom she was very much in love. She turned to me and said, "I don't trust him, I know he's lying to me."

I replied, "Well, you lie yourself, so someone has to lie to you, and you will be sure it will be just the person you want the truth from."

Sometime after that, I saw her, and she said, "I'm cured of lying."

I questioned, "What cured you?"

She replied: "I have been living with a roommate who lied worse than I did!"

One is often cured of her faults by seeing them in others. Life is a mirror, and we find only ourselves reflected in our associates.

Living in the past is a failure method and a violation of Spiritual Law.

Jesus Christ said, "Behold, now is the accepted time ... Now is the day of Salvation."

Lot's wife looked back and was turned into a pillar of salt.

The robbers of time are the past and the future. If a woman's past keeps her in bondage, she should bless it and forget it. She should bless the future, knowing it has in store for her endless joys, and she should live fully in the now.

For example, a woman came to me, complaining that she had no money with which to buy Christmas gifts. She said, "Last year was so different; I had plenty of money and gave lovely presents, and this year I have scarcely a cent."

I replied, "You will never demonstrate money while you are pathetic and live in the past. Live fully in the now and get ready to give Christmas presents. Dig your ditches, and the money will come." She exclaimed, "I know what to do! I will buy some tinsel twine, Christmas seals, and wrapping paper." I replied, "Do that, and the presents will come and stick themselves to the Christmas seals."

This, too, was showing financial fearlessness and faith in God, since the reasoning mind said, "Keep every cent you have, since you are not sure you will get any more."

She bought the seals, paper, and twine, and a few days before Christmas, received a gift of several hundred dollars. Buying the seals and twine had impressed the subconscious with expectancy and opened the way for the manifestation of the money. She purchased all the presents in plenty of time.

A woman must live suspended in the moment.

"Look well, therefore, to this Day! Such is the salutation of the Dawn."

Be spiritually alert, ever awaiting your leads, taking advantage of every opportunity.

One day I said continually (silently): "Infinite Spirit, don't let me miss a trick," and something very important was told to me that evening. It is most necessary to begin the day with right words.

Make an affirmation immediately upon waking.

For example: "Thy will be done this day! Today is a day of completion. I give thanks for this perfect day, miracle shall follow miracle, and wonders shall never cease." Make this a daily habit, and you will see wonders and miracles come into your life.

One morning I picked up a book and read, "Look with wonder at that which is before you!" It seemed to be my message for the day, so I repeated again and again: "Look with wonder at that which is before you."

At about noon, a large sum of money was given me, which I had been desiring for a certain purpose.

In a following chapter, I will give affirmations that I have found most effective. However, one should never use an affirmation unless it is absolutely satisfying and convincing to her own consciousness, and often an affirmation is changed to suit different people.

For example, the following has brought success to many: "I have wonderful work, in a wonderful way; I give wonderful service, for wonderful pay!"

I gave the first two lines to one of my students, and she added the last two.

It made a most powerful statement, since there should always be perfect payment for perfect service, and a rhyme sinks easily into the subconscious. She went about singing it aloud, and soon did receive wonderful work in a wonderful way, and gave wonderful service for wonderful pay.

Another student, a businesswoman, took it, and changed the word *work* to *business*. She repeated: "I have a wonderful business, in a wonderful way, and I give wonderful service for wonderful pay." That afternoon she made a $41,000 deal, though there had been no activity in her affairs for months.

Every affirmation must be carefully worded to "cover the ground" completely.

For example, I knew a woman who was in great need and made a demand for work. She received a great deal of work but was never paid anything. She now knows to add, "wonderful service for wonderful pay."

It is a woman's Divine Right to have plenty! More than enough!

"Her life should be full, and her cup should flow over!" This is God's idea for all people, and when people break down the barriers of lack in their own consciousness, the Golden Age will be theirs, and every righteous desire of their hearts fulfilled!

It is a woman's Divine Right to have plenty. More than enough!

Chapter 5

THE LAW OF KARMA AND THE LAW OF FORGIVENESS

● ● ● ● ● ● ● ●

A woman receives only that which she gives. The Game of Life is a game of boomerangs. A woman's thoughts, deeds, and words return to her sooner or later, with astounding accuracy. This is the law of *karma*, which is Sanskrit for "comeback."

"Whatsoever a woman soweth, that shall she also reap."

For example, a friend told me this story of herself, illustrating this law. She said, "I make all my karma on my aunt. Whatever I say to her, someone says to me. I am often irritable at home, and one day I said to my aunt, who was talking to me during dinner, 'No more talk, I wish to eat in peace.'

"The following day, I was lunching with someone with whom I wished to make a great impression. I was talking animatedly, when they said, 'No more talk, I wish to eat in peace!'"

My friend is high in consciousness, so her karma returns much more quickly than to one on the mental plane.

The more a woman knows, the more she is responsible for, and a person with a knowledge of Spiritual Law, which she does not practice, suffers greatly in consequence. "The fear of the Lord (Law) is the beginning of wisdom." If we read the word Lord as Law, it will make many passages in the Bible much clearer.

"Vengeance is mine, I will repay, saith the Lord (Law)." It is the Law that takes vengeance, not God. God sees a woman perfect, "created in God's own image (imagination)," and given "power and dominion."

This is the perfect idea of a woman, registered in Divine Mind, awaiting a woman's recognition; for a woman can only be what she sees herself to be, and only attain what she sees herself attaining.

"Nothing ever happens without an onlooker," is an ancient saying.

A woman sees first her failure or success, her joy or sorrow, before it swings into visibility from the scenes set in her own imagination. We have observed this in the mother picturing disease for her child, or a woman seeing success for her spouse.

Jesus Christ said, "And ye shall know the truth and the truth shall make you free."

So, we see freedom (from all unhappy conditions) comes through knowledge—a knowledge of Spiritual Law.

Obedience precedes authority, and the Law obeys a woman when she obeys the Law. The law of electricity must be obeyed before it becomes a servant. When handled ignorantly, it becomes a deadly foe. So with the laws of Mind!

For example, a woman with a strong personal will wished she owned a house that belonged to an acquaintance, and she often made mental pictures of herself living in the house. In the course of time, the owner died and she moved into the house. Several years afterward, coming into the knowledge of Spiritual Law, she said to me, "Do you think I had anything to do with that woman's death?"

I replied, "Yes, your desire was so strong, everything made way for it, but you paid your karmic debt. Your spouse, whom you loved devotedly, died soon after, and the house was a white elephant on your hands for years."

The original owner, however, could not have been affected by her thoughts had she been positive in the truth, nor her spouse, but they were both under karmic law. The woman should have said (feeling the great desire for the house): "Infinite Intelligence, give me the right house, equally as charming as this, the house that is mine by Divine Right."

The Divine Selection would have given perfect satisfaction and brought good to all. The Divine Pattern is the only safe pattern to work by.

Desire is a tremendous force and must be directed in the right channels, or chaos ensues.

In demonstrating, the most important step is the first step, to "ask right."

A woman should always demand only that which is hers by Divine Right.

To go back to the illustration: Had the woman taken this attitude: "If this house I desire is mine, I cannot lose it; if it is not, give me its equivalent," the owner might have decided to move out, harmoniously (had it been the Divine Selection for her), or another house would have been substituted. Anything forced into manifestation through personal will is always "ill-got" and has "ever bad success."

A woman is admonished, "My will be done, not thine," and the curious thing is, a woman always gets just what she desires when she does relinquish personal will, thereby enabling Infinite Intelligence to work through her.

"Stand ye still and see the salvation of the Lord (Law)."

For example, a woman came to me in great distress. Her daughter had determined to take a very hazardous trip, and her mother was filled with fear. She said she had used every argument, had pointed out the dangers to be encountered, and forbidden her to go, but the daughter became more and more rebellious and determined. I said to the mother, "You are forcing your personal will upon your daughter, which you have no right to do, and your fear of the trip is only attracting it, for a woman attracts what she fears."

I added, "Let go, and take your mental hands off; put it in God's hands, and use this statement: *I put this situation in the hands of Infinite Love and Wisdom; if this trip is the Divine Plan, I bless it and no longer resist, but if it is not divinely planned, I give thanks that it is now dissolved and dissipated.*"

A day or two after that, her daughter said to her, "Mother, I have given up the trip," and the situation returned to its "native nothingness." It is learning to "stand still" that seems so difficult for a woman. I have dealt more fully with this law in the chapter on nonresistance.

I will give another example of sowing and reaping, which came in the most curious way.

A woman came to me saying that she had received a counterfeit $20 bill, given to her at the bank. She was much disturbed, for, she said, "The people at the bank will never acknowledge their mistake."

I replied, "Let us analyze the situation and find out why you attracted it."

She thought a few moments and exclaimed, "I know it! I sent a friend a lot of stage money, just for a joke." So the Law had sent her some stage money, for it doesn't know anything about jokes.

I said, "Now we will call on the law of forgiveness and neutralize the situation."

Christianity is founded upon the law of forgiveness—Christ has redeemed us from the curse of the karmic law, and the Christ within each woman is her redeemer and salvation from all inharmonious conditions.

So I said: "Infinite Spirit, we call on the law of forgiveness and give thanks that she is under grace, and not under Law, and cannot lose this $20, which is hers by Divine Right."

"Now," I said, "go back to the bank and tell them, fearlessly, that it was given you there by mistake." She obeyed and, to her surprise, they apologized and gave her another bill, treating her most courteously.

So knowledge of the Law gives a woman power to "rub out her mistakes." A woman cannot force the external to be what she is not.

If she desires riches, she must be rich first in consciousness.

For example, a woman came to me asking treatment for prosperity. She did not take much interest in her household affairs, and her home was in great disorder.

I said to her, "If you wish to be rich, you must be orderly. People with great wealth are orderly—and order is heaven's first law."

I added, "You will never become rich with a burnt match in the pin cushion."

She had a good sense of humor and commenced immediately, putting her house in order. She rearranged furniture, straightened out bureau drawers, cleaned rugs, and soon made a big financial demonstration—a gift from a relative. The woman, herself, became made over, and keeps herself keyed up financially by being ever watchful of the external and expecting prosperity, knowing that God is her supply.

Many people are in ignorance of the fact that gifts and things are investments, and that hoarding and saving invariably lead to loss.

"There is that scattereth and yet increaseth; and there is that withholdeth more than is meet, but it tendeth to poverty."

For example, I knew a woman who wanted to buy an expensive overcoat. She went to various shops, but there was none she wanted. She said they were all too cheap-looking. At last, she was shown one the salesman said was valued at over $1,000, but which the manager would sell her for $500, since it was late in the season.

Her financial possessions amounted to about $700. The reasoning mind would have said, "You can't afford to spend nearly all you have on a coat," but she was very intuitive and never reasoned.

She thought for a while and said, "If I get this coat, I'll make a ton of money!" So she agreed to the purchase.

About a month later, she received a $10,000 commission. The coat made her feel so rich that it linked her with success and prosperity; without the coat, she would not have received the commission. It was an investment paying large dividends!

If a woman ignores these leadings to spend or to give, the same amount of money will go in an uninteresting or unhappy way.

For example, on Thanksgiving Day, a woman informed her family that they could not afford a Thanksgiving dinner. She had the money but decided to save it. A few days later, someone entered her room and took from the bureau drawer the exact amount the dinner would have cost.

The Law always stands in back of a woman who spends fearlessly, with wisdom.

For example, one of my students was shopping with her little nephew. The child clamored for a toy, which she told him she could not afford to buy.

She realized suddenly that she was seeking lack and not recognizing God as her supply! So she bought the toy, and on her way home, picked up, in the street, the exact amount of money she had paid for it.

A woman's supply is inexhaustible and unfailing when fully trusted, but faith or trust must precede the demonstration. "According to your faith, be it unto you. Faith is the substance of things hoped for, the evidence of things not seen," for faith holds the vision steady, and the adverse pictures are dissolved and dissipated, and "in due season we shall reap, if we faint not."

Jesus Christ brought the good news (the gospel) that there was a higher law than the law of karma—and that that law transcends the law of karma. It is the law of grace, or forgiveness. It is the law that frees a woman from the law of cause and effect—the law of consequence. "Under grace, and not under law."

We are told that, on this plane, a woman reaps where she has sown; the gifts of God are simply poured out upon her. "All that the Kingdom affords is hers." This continued state of bliss awaits the woman who has overcome the mortal thought.

In the mortal thought there is tribulation, but Jesus Christ said, "Be of good cheer; I have overcome the world."

Mortal thought is that of sin, sickness, and death. Jesus saw their absolute unreality and said sickness and sorrow shall pass away, and death itself, the last enemy, be overcome.

We know now, from a scientific standpoint, that death can be overcome by stamping the subconscious mind with the conviction of eternal youth and eternal life. The subconscious, being simply power without direction, carries out orders without questioning.

Working under the direction of the superconscious (the Christ or God within all), the "resurrection of the body" would be accomplished.

A woman will no longer throw off her body in death. Instead, it will be transformed into the "body electric," as sung by Walt Whitman, for Christianity is founded upon the forgiveness of sins and "an empty tomb."

A woman cannot force the external to be what she is not.

Chapter 6

CASTING THE BURDEN

• • • • • • • •

When a woman knows her own powers and the workings of her mind, her great desire is to find an easy and quick way to impress the subconscious with good, for simply an intellectual knowledge of the Truth will not bring results. In my own case, I found that the easiest way is in casting the burden.

A metaphysician once explained it in this manner. He said, "The only thing that gives anything weight in nature is the law of gravitation, and if a boulder could be taken high above the planet, there would be no weight in that boulder; and that is what Jesus Christ meant when he said: 'My yoke is easy and my burden is light.'"

He had overcome the mortal thought, and functioned in the fourth-dimensional realm, where there is only perfection, completion, life, and joy.

He said, "Come to me all ye that labor and are heavy laden, and I will give you rest … Take my yoke upon you, for my yoke is easy and my burden is light."

We are also told in the 55th Psalm, to "cast thy burden upon the Lord." Many passages in the Bible state that the battle is God's and that a woman is always to stand still and see the salvation of the Lord.

This indicates that the superconscious mind (or Christ within) is the department that fights the battle and relieves the burdens.

We see, therefore, that a woman violates Law if she carries a burden. A burden is an adverse thought or condition, and this thought or condition has its root in the subconscious.

It seems almost impossible to make any headway directing the subconscious from the conscious, or reasoning mind, as the reasoning mind (the intellect) is limited in its conceptions and filled with doubts and fears.

How scientific it then is to cast the burden upon the superconscious mind (or Christ within) where it is "made light," or dissolved into its "native nothingness."

For example, a woman in urgent need of money "made light" upon the Christ within, the superconscious, with the statement: "I cast this burden of lack on the Christ (within) and I go free to have plenty!" The belief in lack was her burden, and as she cast it upon the superconscious with its belief of plenty, an avalanche of supply was the result.

We read, "The Christ in you, the hope of glory."

Another example: One of my students had been given a new piano, and there was no room in her studio for it until she had moved out the old one. She was in a state of perplexity. She wanted to keep the old piano, but knew of no place to send it. She became desperate, since the new piano was to be sent immediately; in fact, it was on its way, with no place to put it. She said it came to her to repeat: "I cast this burden on the Christ within, and I go free."

A few moments later, her phone rang, and a friend asked if she might rent her old piano. It was moved out a few minutes before the new one arrived.

I knew a woman whose burden was resentment. She said: "I cast this burden of resentment on the Christ within, and I go free, to be loving, harmonious, and happy." The Almighty superconscious flooded the subconscious with love, and her whole life was changed. For years, resentment had held her in a state of torment and imprisoned her soul (the subconscious mind).

The statement should be made over and over and over, sometimes for hours at a time, silently or audibly, with quietness but determination. I have often compared it to winding up an old clock. We must wind ourselves up with spoken words.

I have noticed, in "casting the burden," after a little while, one seems to see clearly. It is impossible to have clear vision while in the throes of carnal mind. Doubts and fear poison the mind and body, and imagination runs riot, attracting disaster and disease.

In steadily repeating the affirmation: "I cast this burden on the Christ within, and go free," the vision clears, and with it comes a feeling of relief, and sooner or later comes the manifestation of good, be it health, happiness, or supply.

One of my students once asked me to explain the "darkness before the dawn." I referred in a preceding chapter to the fact that often, before the big demonstration, "everything seems to go wrong," and deep depression clouds the consciousness. It means that out of the subconscious are rising the doubts and fears of the ages. These old derelicts of the subconscious rise to the surface, to be put out.

It is then that a woman should clap her cymbals, like Jehoshaphat, and give thanks that she is saved, even though she seems surrounded by the enemy (the situation of lack or disease). The student continued, "How long must one remain in the dark?" and I replied, "Until one can see in the dark," for "casting the burden enables one to see in the dark."

In order to impress the subconscious, active faith is always essential.

"Faith without works is dead." In these chapters I have endeavored to bring out this point. Jesus Christ showed active faith when "he commanded the multitude to sit down on the ground" before he gave thanks for the loaves and fishes.

I will give another example showing how necessary this step is. In fact, active faith is the bridge over which a woman passes to her Promised Land.

Through misunderstanding, a woman had been separated from her husband, whom she loved deeply. He refused all offers of reconciliation and would not communicate with her in any way.

Coming into the knowledge of Spiritual Law, she denied the appearance of separation. She made this statement: "There is no separation in Divine Mind; therefore, I cannot be separated from the love and companionship that are mine by Divine Right."

She showed active faith by keeping her husband in her thoughts every day, thereby impressing the subconscious with a picture of his return. Over a year passed, but she never wavered, and one day her husband walked in.

The subconscious is often impressed through music. Music has a fourth-dimensional quality and releases the soul from imprisonment. It makes wonderful things seem possible and easy of accomplishment!

I have a friend who uses music daily, for this purpose. It puts her in perfect harmony and releases the imagination. Another woman often dances while making her affirmations. The rhythm and harmony of music and motion carry her words forth with tremendous power.

The student must remember also not to despise the "day of small things."

Invariably, before a demonstration, come "signs of land." Before Columbus reached America, he saw birds and twigs that showed him that land was near. So it is with a demonstration; but often the student mistakes it for the demonstration itself and is disappointed.

For example , a woman had spoken the word for a set of dishes. Not long afterward a friend gave her a dish that was old and cracked. She came to me and said, "Well, I asked for a set of dishes, and all I got was a cracked plate."

I replied, "The plate was only signs of land. It shows your dishes are coming—look upon it as birds and seaweed." Not long afterward, the dishes came.

Continually "making believe" impresses the subconscious. If one makes believe she is rich, and makes believe she is successful, in "due time she will reap."

Children are always making believe, and "except ye be converted, and become as little children, ye shall not enter the Kingdom of Heaven."

For example, I know of a woman who was very poor, but no one could make her feel poor. She earned a small amount of money from rich friends who constantly reminded her of her poverty and to be careful and saving. Regardless of their admonitions, she would spend all her earnings on a hat, or make someone a gift, and be in a rapturous state of mind. Her thoughts were always centered on beautiful clothes and "rings and things," but without envying others.

She lived in the world of the wondrous, and only riches seemed real to her. Before long, she married a rich man, and the rings and things became visible. I do not know whether the man was her Divine Selection, but opulence had to manifest in her life, as she had imagined only opulence.

There is no peace or happiness for a woman until she has erased all fear from the subconscious. Fear is misdirected energy and must be redirected, or transmuted into faith.

Jesus Christ said, "Why are ye fearful, O ye of little faith? . . . All things are possible to him that believeth."

I am asked so often by my students, "How can I get rid of fear?" I reply, "By walking up to the thing you are afraid of."

"The lion takes its fierceness from your fear." Walk up to the lion, and he will disappear; run away and he runs after you.

I have shown in previous chapters how the lion of lack disappeared when the individual spent money fearlessly, showing faith that God was her supply and, therefore, unfailing.

Many of my students have come out of the bondage of poverty and are now bountifully supplied through losing all fear of letting money go out. The subconscious is impressed with the truth that God is the Giver and Gift; therefore, since one is one with the Giver, we are one with the Gift. A splendid statement is: "I now thank God the Giver for God the Gift."

When a woman has long separated herself from her good and her supply through thoughts of separation and lack, sometimes it takes dynamite (a big situation) to dislodge these false ideas from her subconscious. We see in the foregoing illustration how the individual was freed from her bondage by showing fearlessness.

Watch yourself hourly to detect if your motive for action is fear or faith. "Choose ye this day whom we shall serve," fear or faith.

Perhaps one's fear is of personality. Then do not avoid the people feared; be willing to meet them cheerfully, and they will either prove "golden links in the chain of one's good" or disappear harmoniously from one's pathway.

Perhaps one's fear is of disease or germs. Then one should be fearless and undisturbed in a germ-laden situation, and she would be immune.

One can only contract germs while vibrating at the same rate as the germ, and fear drags women down to the level of the germ. Of course, the disease-laden germ is the product of carnal mind, as all thought must objectify. Germs do not exist in the superconscious or Divine Mind, and therefore are the product of "vain imagination."

"In the twinkling of an eye," a woman's release will come when she realizes there is no power in evil.

The material world will fade away, and the fourth-dimensional world, the "World of the Wondrous," will swing into manifestation.

"And I saw a new heaven, and a new Earth—and there shall be no more death, neither sorrow nor crying, neither shall there be any more pain; for the former things are passed away."

Choose ye this day whom we shall serve, fear or faith.

Chapter 7

LOVE

● ● ● ● ● ● ● ●

Every woman on this planet is taking her initiation in love. "A new commandment I give unto you, that ye love one another." Ouspensky states, in *Tertium Organum*, that "love is a cosmic phenomenon" and opens the fourth-dimensional world, "the World of the Wondrous."

Real love is selfless and free from fear. It pours itself out upon the object of its affection without demanding any return. Its joy is in the joy of giving. Love is God in manifestation and the strongest magnetic force in the Universe.

Pure, unselfish love draws to itself its own; it does not need to seek or demand. Scarcely anyone has the faintest conception of real love. When people are selfish, tyrannical, or fearful in their affections, they lose the thing they love. Jealousy is the worst enemy of love, for the imagination runs riot, seeing the loved one attracted to another, and invariably these fears objectify if they are not neutralized.

For example, a woman came to me in deep distress. The one she loved—she called him "The Cap" because he was a captain—had left her for someone else and said he never intended to marry her. She was torn with jealousy and resentment and said she hoped he would suffer as he had made her suffer. She then added, "How could he leave me when I loved him so much?"

I replied, "You are not loving that man; you are hating him," and added, "You can never receive what you have never given. Give a perfect love and you will receive a perfect love. Perfect yourself on him. Give him a perfect, unselfish love, demanding nothing in return; do not criticize or condemn, and bless him wherever he is."

She replied, "No, I won't bless him unless I know where he is!"

"Well," I said, "that is not real love. When you send out real love, real love will return to you, either from this man or an equivalent, for if this person is not the Divine Selection, you will not want him. As you are one with God, you are one with the love that belongs to you by Divine Right."

Several months passed and matters remained about the same, but she was working conscientiously with herself. I said, "When you are no longer disturbed by his cruelty, he will cease to be cruel, as you are attracting it through your own emotions."

Then I told her of a brotherhood in India who never said "Good morning" to each other. They used these words: "I salute the Divinity in you." They saluted the Divinity in everyone, and in the wild animals in the jungle, and they were never harmed, for they saw only God in every living thing. I said, "Salute the Divinity in this man, and say: 'I see your Divine Self only. I see you as God sees you, perfect, made the same in image and likeness.'"

She found she was becoming more poised and gradually losing her resentment. One day, she said, suddenly, "God bless The Cap wherever he is."

I replied: "Now that is real love, and when you have become a 'complete circle' and are no longer disturbed by the situation, you will have that love or attract its equivalent."

I was moving at this time and did not have a telephone, so was out of touch with her for a few weeks, when one morning I received a letter saying, "We are married."

At the earliest opportunity, I paid her a call. My first words were, "What happened?"

"Oh," she exclaimed, "a miracle! One day I woke up and all suffering had ceased. I saw him that evening, and he asked me to marry him. We were married in about a week, and I have never seen a more devoted man."

There is an old saying: "No one is your enemy, no one is your friend; everyone is your teacher."

So become impersonal and learn what each person has to teach, and soon you will learn your lessons and be free.

This woman's lover was teaching her selfless love, which everyone, sooner or later, must learn.

Suffering is not necessary for a woman's development; it is the result of violation of Spiritual Law, but few people seem able to rouse themselves from their "soul sleep" without it. When people are happy, they usually become selfish, and automatically the law of karma is set in action. People often suffer loss through lack of appreciation.

I knew a woman who had a very nice husband, but she said often, "I don't care anything about being married, but that is nothing against my husband. I'm simply not interested in married life."

She had other interests and scarcely remembered she had a husband. She only thought of him when she saw him. One day her husband told her he was in love with another woman, and left. She came to me in distress and resentment.

I replied, "It is exactly what you spoke the word for. You said you didn't care anything about being married, so the subconscious worked to get you unmarried."

She said, "Oh yes, I see. People get what they want, and then feel very much hurt."

She soon became in perfect harmony with the situation and knew they were both much happier apart.

A woman came to me dejected, miserable, and poor. Her close friend was interested in the "Science of Numbers" and had read her numerology chart. It seems the report was not very favorable, for she said, "My friend says I'll never amount to anything because I am a 'two.'"

I replied, "I don't care what your number is; you are a perfect idea in Divine Mind, and we will demand the success and prosperity that are already planned for you by Infinite Intelligence."

Within a few weeks, she had a very fine position, and a year or two later, she achieved brilliant success as a writer.

No woman can attract money if she despises it. Many people are kept in poverty by saying, "Money means nothing to me, and I have a contempt for people who have it." This is the reason so many artists are poor. Their contempt for money separates them from it. I remember hearing one artist say of another, "She's no good as an artist; she has money in the bank."

This attitude of mind, of course, separates a woman from her supply; she must be in harmony with something in order to attract it.

Money is God in manifestation, as freedom from want and limitation, but it must be always kept in circulation and put to right uses. Hoarding and saving react with grim vengeance.

This does not mean that a woman should not have houses and lots, stocks and bonds, for "the life of the a righteous woman shall be full." It means a woman should not hoard even the principal if an occasion arises when money is necessary. In letting it go out fearlessly and cheerfully, she opens the way for more to come in, for God is the unfailing and inexhaustible supply.

This is the spiritual attitude toward money, and the great Bank of the Universal never fails!

We see an example of hoarding in the 1925 MGM film production of *Greed*. The woman won $5,000 in a lottery, but would not spend it. She hoarded and saved, let her husband suffer and starve, and eventually she scrubbed floors for a living. She loved the money itself and put it above everything, and one night she was murdered and the money taken from her.

(Editor's note: A more contemporary example might be found in the 1994 movie *It Could Happen to You*, where a police officer played by Nicolas Cage does not have a tip for his coffee-shop waitress, played by Bridget Fonda. Not wanting to leave without showing his appreciation, he offers to share his prize money if his lottery ticket is a winner. When he does win, his greedy wife takes him to court to sue for all of the winnings, and she is successful. But she is not the ultimate winner. Eventually she squanders everything, including her marriage. The police officer and the waitress find love and receive cash gifts from well-wishers that far exceed the amount they had originally won. In his efforts to share, the police officer is rewarded with abundance. The wife's greed and desire to take everything costs her everything.)

This is an example of where "love of money is the root of all evil." Money in itself is good and beneficial, but used for destructive purposes, hoarded and saved, or considered more important than love, brings disease and disaster, and the loss of the money itself.

Follow the path of love, and all things are added, for God is love and God is supply; follow the path of selfishness and greed, and the supply vanishes or you become separated from it.

For example, I knew the case of a very rich woman who hoarded her income. She rarely gave anything away, but bought and bought things for herself. She was very fond of necklaces, and a friend once asked her how many she possessed. She replied, "Sixty-seven." She bought them and put them away, carefully wrapped in tissue paper. Had she used the necklaces, it would have been quite legitimate, but she was violating the law of use. Her closets were filled with clothes she never wore and jewels that never saw the light.

The woman's arms were gradually becoming paralyzed from holding on to things. Eventually she was considered incapable of looking after her affairs, and her wealth was handed over to others to manage.

So people, in ignorance of the Law, bring about their own destruction.

All disease, all unhappiness, comes from the violation of the law of love. A woman's boomerangs of hate, resentment, and criticism come back laden with sickness and sorrow. Love seems almost a lost art, but the woman with the knowledge of Spiritual Law knows it must be regained, for without it, she has "become as sounding brass and tinkling cymbals."

For example, I had a student who came to me, month after month, to clean her consciousness of resentment. After a while, she arrived at the point where she resented only one woman, but that one woman kept her busy. Little by little she became poised and harmonious, and one day, all resentment was wiped out.

She came in radiant and exclaimed, "You can't understand how I feel! The woman said something to me, and instead of being furious, I was loving and kind, and she apologized and was perfectly lovely to me. No one can understand the marvelous lightness I feel within!"

Love and goodwill are invaluable in business.

For example, a woman came to me, complaining of her employer. She said she was cold and critical; the woman knew her employer did not want her in the position.

"Well," I replied, "salute the Divinity in the woman and send her love."

She said, "I can't; she's a marble woman."

I answered, "You remember the story of the sculptor who asked for a certain piece of marble? He was asked why he wanted it, and he replied, 'Because there is an angel in the marble,' and out of it he produced a wonderful work of art."

"Very well," she said, "I'll try it." A week later she came back and said, "I did what you told me to, and now the woman is very kind; she even took me out to lunch."

People are sometimes filled with remorse for having done someone an unkindness, perhaps years ago. If the wrong cannot be righted, its effect can be neutralized by doing someone a kindness in the present.

"This one thing I do, forgetting those things that are behind and reaching forth unto those things that are before."

Sorrow, regret, and remorse tear down the cells of the body and poison the atmosphere of the individual. A woman said to me in deep sorrow, "Treat me to be happy and joyous, for my sorrow makes me so irritable with members of my family that I keep making more karma."

I was asked to treat a woman who was mourning for her daughter. I denied all belief in loss and separation and affirmed that God was the woman's joy, love, and peace.

The woman gained her poise at once, but sent word by her son not to treat any longer because she was "so happy, it wasn't respectable."

Mortal mind loves to hang on to its griefs and regrets.

I knew a woman who went about bragging of her troubles, so, of course, she always had something to brag about.

The old idea was if a woman did not worry about her children, she was not a good mother. Now we know that mother-fear is responsible for many of the diseases and accidents that come into the lives of children. For fear pictures vividly the disease or situation feared, and these pictures objectify if not neutralized.

Happy is the mother who can say sincerely that she puts her child in God's hands and knows, therefore, that she is divinely protected.

For example, a woman awoke suddenly in the night, feeling that her brother was in great danger. Instead of giving in to her fears, she commenced making statements of Truth, saying: "We are a perfect idea in Divine Mind and are always in the right place. Therefore, my brother is in his right place and is divinely protected."

The next day she learned that her brother had been in close proximity to an explosion in a mine but had miraculously escaped.

So a woman is her brother's keeper (in thought), and every woman should know that the thing she loves dwells in "the secret place of the most high and abides under the shadow of the Almighty."

"There shall no evil befall thee, neither shall any plague come nigh thy dwelling."

"Perfect love casteth out fear. She that feareth is not made perfect in love," and "Love is the fulfilling of the Law."

Love is the fulfilling of the Law.
Follow the path of love.

Chapter 8

INTUITION OR GUIDANCE

• • • • • • • •

There is nothing too great of accomplishment for the woman who knows the power of her word and who follows her intuitive leads. By the word, she starts in action unseen forces and can rebuild her body or remold her affairs. It is, therefore, of the utmost importance to choose the right words, and the student carefully selects the affirmation she wishes to catapult into the invisible. She knows that God is her supply, that there is a supply for every demand, and that her spoken word releases this supply.

"Ask and ye shall receive."

A woman must make the first move. "Draw nigh to God and God will draw nigh to you."

I have often been asked just how to make a demonstration. I reply, "Speak the word and then do not do anything until you get a definite lead." Demand the lead, saying: "Infinite Spirit, reveal to me the way, let me know if there is anything for me to do."

The answer will come through intuition (or hunch), a chance remark from someone, or a passage in a book. The answers are sometimes quite startling in their exactness.

For example, a woman desired a large sum of money. She spoke the words: "Infinite Spirit, open the way for my immediate supply, let all that is mine by Divine Right now reach me, in great avalanches of abundance." Then she added: "Give me a definite lead; let me know if there is anything for me to do."

The thought came quickly: "Give a certain friend [who had helped her spiritually] a hundred dollars." She told her friend, who said, "Wait and get another lead, before giving it." So she waited, and that day met a woman who said to her, "I gave someone a dollar today; it was just as much for me, as it would be for you to give someone a hundred."

This was indeed an unmistakable lead, so she knew she was right in giving the hundred dollars. It was a gift that proved a great investment, for shortly after that, a large sum of money came to her in a remarkable way.

Giving opens the way for receiving. In order to create activity in finances, one should give. Tithing, or giving one-tenth of one's income, is an old Jewish custom and is sure to bring increase. Many of the richest people in this country have been tithers, and I have never known it to fail as an investment.

The tenth-part goes forth and returns blessed and multiplied. But the gift or tithe must be given with love and cheerfulness, for "God loveth a cheerful giver." Bills should be paid cheerfully; all money should be sent forth fearlessly and with a blessing.

This attitude of mind makes a woman master of money. It is hers to obey, and her spoken word then opens vast reservoirs of wealth.

A woman, herself, limits her supply by her limited vision. Sometimes the student has a great realization of wealth but is afraid to act.

The vision and action must go hand in hand, as in the case of the woman who bought the expensive overcoat.

A woman came to me asking me to speak the word for a position. So I demanded: "Infinite Spirit, open the way for this woman's right position." Never ask for just "a position"; ask for the right position, the place already planned in Divine Mind, as it is the only one that will give satisfaction.

I then gave thanks that she had already received and that it would manifest quickly. Very soon, she had three positions offered her, two in New York and one in Palm Beach, and she did not know which to choose. I said, "Ask for a definite lead."

The time was almost up, and she was still undecided, when one day, she telephoned. "When I woke up this morning, I could smell Palm Beach," she said. She had been there before and knew its balmy fragrance.

I replied, "Well, if you can smell Palm Beach from here, it is certainly your lead." She accepted the position, and it proved a great success. Often one's lead comes at an unexpected time.

One day I was walking down the street when I suddenly felt a strong urge to go to a certain bakery a block or two away. The reasoning mind resisted, arguing, "There is nothing there that you want." However, I had learned not to reason, so I went to the bakery, looked at everything, and there was certainly nothing there that I wanted. But coming out, I encountered a woman I had thought of often and who was in great need of the help that I could give her.

So often, one goes for one thing and finds another. Intuition is a spiritual faculty and does not explain, but simply points the way.

A person often receives a lead during a treatment. The idea that comes may seem quite irrelevant, but some of God's leadings are mysterious.

In the class one day, I was treating that each individual would receive a definite lead. A woman came to me afterward and said, "While you were treating, I got the hunch to take my furniture out of storage and get an apartment." The woman had come to be treated for health. I told her I knew in getting a home of her own, her health would improve, and I added, "I believe your trouble, which is a congestion, has come from having things stored away. Congestion of things causes congestion in the body. You have violated the law of use, and your body is paying the penalty."

So I gave thanks that "Divine Order was established in her mind, body, and affairs."

People little dream of how their affairs react on the body. There is a mental correspondence for every disease. A person might receive instantaneous healing through the realization of her body being a perfect idea in Divine Mind, and, therefore, whole and perfect, but if she continues her destructive thinking, hoarding, hating, fearing, condemning, the disease will return.

Jesus Christ knew that all sickness came from sin, but admonished the leper, after the healing, to go and sin no more, lest a worse thing come.

The metaphysician is always delving deep for the correspondence in disease. In order to achieve total healing, the soul (or subconscious mind) must be washed whiter than snow.

Jesus Christ said, "Condemn not lest ye also be condemned … Judge not, lest ye be judged."

Many people have attracted disease and unhappiness through condemnation of others. What a woman condemns in others, she attracts to herself.

For example, a friend came to me in anger and distress because her husband had deserted her for another woman. She condemned the other woman and said continually, "She knew he was a married man and had no right to accept his attentions."

I replied, "Stop condemning the woman, bless her, and be through with the situation; otherwise, you are attracting the same thing to yourself." She was deaf to my words, and a year or two later, became deeply interested in a married man herself.

People pick up a live wire whenever they criticize or condemn, and may expect a shock.

Indecision is a stumbling block in many a pathway. In order to overcome it, make the statement repeatedly: "I am always under direct inspiration; I make right decisions, quickly."

These words impress the subconscious, and soon one finds herself awake and alert, making her right moves without hesitation. I have found it destructive to look to the psychic plane for guidance, as it is the plane of many minds and not the "One Mind."

As a woman opens her mind to subjectivity, she becomes a target for destructive forces. The psychic plane is the result of our mortal thought and is on the "plane of opposites." She may receive either good or bad messages. The science of numbers and the reading of horoscopes keep people down on the mental (or mortal) plane, for they deal only with the karmic path.

I know of a man who should have been dead years ago, according to his horoscope, but he is alive and a leader of one of the biggest movements in this country for the uplift of humanity.

It takes a very strong mind to neutralize a prophecy of evil. The student should declare: "Every false prophecy shall come to naught; every plan God in heaven has not planned shall be dissolved and dissipated. The Divine Idea now comes to pass."

However, if any good message has ever been given one—of coming happiness or wealth—harbor and expect it, and it will manifest sooner or later through the law of expectancy.

A woman's will should be used to back the Universal will: "I will that the will of God be done."

It is God's will to give each woman every righteous desire of her heart, and her will should be used to hold the perfect vision, without wavering.

The prodigal son said: "I will arise and go to my Father."

It is, indeed, often an effort of the will to leave the husks and swine of mortal thinking. It is so much easier, for the average person, to have fear than faith; so faith is an effort of the will.

As a woman becomes spiritually awakened, she recognizes that any external disharmony is the correspondence of mental disharmony. If she stumbles or falls, she may know she is stumbling or falling in consciousness.

One day a student was walking along the street condemning someone in her thoughts. She was saying mentally, "That woman is the most disagreeable woman on Earth," when suddenly three teenagers rushed around the corner and almost knocked her over. She did not condemn the teenagers, but immediately called on the law of forgiveness, and saluted the Divinity in the woman. Wisdom's ways are ways of pleasantness, and all her paths are peace.

When one has made her demands upon the Universal, she must be ready for surprises. Everything may seem to be going wrong, when in reality, it is going right.

For example, a woman was told that there was no loss in Divine Mind; therefore, she could not lose anything that belonged to her; anything lost would be returned, or she would receive its equivalent.

Several years previously, she had lost $2,000. She had loaned the money to a relative during her lifetime, but the relative had died, leaving no mention of it in her will. The woman was resentful and angry, and as she had no written statement of the transaction, she never received the money. So she determined to deny the loss and collect the $2,000 from the Bank of the Universal. She had to begin by forgiving the woman, as resentment and unforgiveness close the doors of this wonderful bank.

She made this statement: "I deny loss, there is no loss in Divine Mind; therefore, I cannot lose the $2,000, which belong to me by Divine Right. As one door shuts, another door opens."

She was living in an apartment house that was for sale. In the lease was a clause stating that if the house was sold, the tenants would be required to move out within 90 days. Suddenly the landlord broke the lease and raised the rent. Again, injustice was on her pathway, but this time she was undisturbed. She blessed the landlord, and said: "Since the rent has been raised, it means that I'll be that much richer, for God is my supply."

New leases were made out for the advanced rent, but by some Divine mistake, the 90-day clause had been forgotten. Soon after, the landlord had an opportunity to sell the house. On account of the mistake in the new leases, the tenants held possession for another year.

The agent offered each tenant $200 to vacate. Several families moved; three remained, including the woman. A month or two passed, and the agent again appeared. This time he said to the woman, "Will you break your lease for the sum of $1,500?" It flashed upon her: "Here comes the $2,000." She remembered having said to friends in the house, "We will all act together if anything more is said about leaving." So her lead was to consult her friends.

These friends said," Well, if they have offered you $1,500, they will certainly give $2,000." So she received a check for $2,000 for giving up the apartment. It was certainly a remarkable working of the Law, and the apparent injustice was merely opening the way for her demonstration.

It proved that there is no loss, and when a woman takes her spiritual stand, she collects all that is hers from this great Reservoir of Good.

"I will restore to you the years the locusts have eaten." The locusts are the doubts, fears, resentments, and regrets of mortal thinking.

These adverse thoughts, alone, are self-defeating, for "no woman gives to herself but herself, and no woman takes away from herself but herself."

People are here to prove God and "to bear witness to the Truth," and they can only prove God by bringing plenty out of lack and justice out of injustice.

"Prove me now herewith, saith the Lord of hosts, if I will not open you the windows of heaven, and pour out a blessing, that there shall not be room enough to receive it."

Chapter 9

PERFECT SELF-EXPRESSION OR THE DIVINE DESIGN

● ● ● ● ● ● ● ●

There is, for each woman, perfect self-expression. There is a place that she is to fill and no one else can fill; something that she is to do that no one else can do. It is her destiny!

This achievement is held as a perfect idea in Divine Mind, awaiting each woman's recognition. As the imaging faculty is the creative faculty, it is necessary for a woman to see the idea before it can manifest. So a woman's highest demand is for the Divine Design of her life.

She may not have the faintest conception of what it is, for there is, possibly, some marvelous talent hidden deep within her.

Her demand should be: "Infinite Spirit, open the way for the Divine Design of my life to manifest; let the genius within me now be released; let me see clearly the perfect plan."

The perfect plan includes health, wealth, love, and perfect self-expression. This is the square of life, which brings perfect happiness. When one has made this demand, she may find great changes taking place in her life, for nearly every woman has wandered far from the Divine Design.

In one woman's case, it was as though a cyclone had struck her affairs. But readjustments came quickly, and new and wonderful conditions took the place of old ones.

Perfect self-expression will never be labor, but of such absorbing interest that it will seem almost like play. The student knows, also, since a woman comes into the world financed by God, that the supply needed for her perfect self-expression will be at hand.

Many a genius has struggled for years with the problem of supply, when her spoken word, and faith, would have quickly released the necessary funds.

For example, after class one day, a woman came to me and handed me a penny. She said, "I have just seven cents in the world, and I'm going to give you one, for I have faith in the power of your spoken word. I want you to speak the word for my perfect self-expression and prosperity."

I spoke the word and did not see her again until a year later. She came in one day, successful and happy, with a roll of money in her pocket. She said, "Immediately after you spoke the word, I had a position offered me in a distant city, and am now demonstrating health, happiness, and supply."

A woman's perfect self-expression may be in becoming a perfect mother, a perfect homemaker, or a perfect career woman. Demand definitely leads, and the way will be made easy and successful.

One should not visualize or force a mental picture. When a woman demands that the Divine Design come into her conscious mind, she will receive flashes of inspiration and begin to see herself making some great accomplishment. This is the picture, or idea, she must hold without wavering.

The thing a woman seeks is seeking her—the telephone was seeking Bell!

Parents should never force careers and professions upon their children. With a knowledge of Spiritual Truth, the Divine Plan could be spoken for early in childhood or prenatally.

A prenatal treatment should be: "Let the God in this child have perfect expression; let the Divine Design of mind, body, and affairs be made manifest throughout life, throughout eternity."

God's will be done, not ours; God's pattern, not our pattern, is the command we find running through all the scriptures, and the Bible is a book dealing with the science of the mind. It is a book telling people how to release their soul (or subconscious mind) from bondage.

The battles described are pictures of people waging war against mortal thoughts. "Foes shall be they of their own household." Everyone is Jehoshaphat, and everyone is David who slays Goliath (mortal thinking) with the little white stone (faith).

So a woman must be careful that she is not the "wicked and slothful servant" who buried her talent. There is a terrible penalty to be paid for not using one's ability.

Often fear stands between a woman and her perfect self-expression. Stage fright has hampered many a genius. This may be overcome by the spoken word or treatment. The individual then loses all self-consciousness and feels simply that she is a channel for Infinite Intelligence to express Itself through.

She is under direct inspiration, fearless, and confident; for she feels that it is the "God within" her who does the work.

A young girl came often to my class with her mother. She asked me to speak the word for her coming examinations at school. I told her to make the statement: "I am one with Infinite Intelligence. I know everything I should know on this subject."

She had an excellent knowledge of history, but was not sure of her arithmetic.

I saw her afterward, and she said, "I spoke the word for my arithmetic and passed with the highest honors, but I thought I could depend on myself for history, and instead got a very poor mark." A woman often receives a setback when she is too sure of herself, which means that she is trusting her personality and not the God within.

Another one of my students gave me an example of this. She took an extended trip abroad one summer, visiting many countries where she was ignorant of the languages. She was calling for guidance and protection every minute, and her affairs went smoothly and miraculously. Her luggage was never delayed nor lost! Accommodations were always ready for her at the best hotels; and she had perfect service wherever she went. She returned to New York. Knowing the language, she felt God was no longer necessary, so looked after her affairs in an ordinary manner.

Everything went wrong: Her trunks were delayed, amid disharmony and confusion. The student must form the habit of practicing the presence of God every minute. "In all thy ways acknowledge God," for nothing is too small or too great.

Sometimes an insignificant incident may be the turning point in someone's life. Robert Fulton, watching some boiling water simmering in a tea kettle, saw a steamboat!

I have seen a student often keep back her demonstration through resistance. She pins her faith to one channel only and dictates just the way she desires the manifestation to come, which brings things to a standstill.

"My way, not your way!" is the command of Infinite Intelligence. Like all power, be it steam or electricity, it must have a nonresistant engine or instrument to work through, and people are the engines or instruments.

Over and over again, one is told to stand still. "Oh, Judah, fear not; but tomorrow go out against them, for the Lord will be with you. You shall not need to fight this battle; set yourselves, stand ye still, and see the salvation of the Lord with you."

We see this in the incidents of the $2,000 coming to the woman through the landlord when she became nonresistant and undisturbed, and the woman who won the man's love after "all suffering had ceased."

The student's goal is poise!

Poise is power, for it gives God-Power a chance to rush through, to "will and to do Its good pleasure." Poised, she thinks clearly, and makes "right decisions quickly . . . she never misses a trick."

Anger blurs the vision, poisons the blood, is the root of many diseases, and causes wrong decision leading to failure. It has been named one of the worst "sins," since its reaction is so harmful. The student learns that, in metaphysics, sin has a much broader meaning than in the old teaching "Whatsoever is not of faith is sin."

One finds that fear and worry are deadly sins. They are inverted faith, and through distorted mental pictures, bring to pass the thing most feared. The work is to drive out these enemies (from the subconscious mind). "When we are fearless we are finished!" Maurice Maeterlinck (who was awarded the Nobel Prize for Literature in 1911) says that "man is God afraid."

So, as we read in the previous chapters, a woman can only vanquish fear by walking up to the thing she is afraid of. When Jehoshaphat and his army prepared to meet the enemy, singing "Praise the Lord, for mercy endureth forever," they found their enemies had destroyed each other, and there was nothing to fight.

For example, a woman asked a friend to deliver a message to another friend. The woman feared to give the message, since the reasoning mind said, "Don't get mixed up in this affair; don't give that message."

She was troubled in spirit, for she had given her promise. At last, she determined to "walk up to the lion" and call on the law of Divine Protection. She met the friend to whom she was to deliver the message. She opened her mouth to speak, when her friend said, "So-and-so has left town." This made it unnecessary to give the message, since the situation depended upon the person being in town. As she was willing to do it, she was not obliged to; as she did not fear, the situation vanished.

The student often delays her demonstration through a belief in incompletion. She should make this statement: "In Divine Mind there is only completion, therefore, my demonstration is completed. My perfect work, my perfect home, my perfect health." Whatever she demands are perfect ideas registered in Divine Mind and must manifest "under grace in a perfect way." She gives thanks she has already received on the invisible, and makes active preparation for receiving on the visible.

One of my students was in need of a financial demonstration. She came to me and asked why it was not completed.

I replied, "Perhaps you are in the habit of leaving things unfinished, and the subconscious has gotten into the habit of not completing (as the without, so the within)."

She said, "You're right. I often begin things, such as clothes I'm sewing, and never finish them. I'll go home and finish something I commenced weeks ago, and I know it will be symbolic of my demonstration."

She sewed assiduously, and the article was soon completed. Shortly after, the money came in a most curious manner. Her husband was paid his salary twice that month. He told his employers of their mistake, and they sent word to keep it.

When a woman asks, believing, she must receive, for God creates His own channels!

I have been sometimes asked, "Suppose one has several talents. How is she to know which one to choose?" Demand to be shown definitely. Say: "Infinite Spirit, give me a definite lead, reveal to me my perfect self-expression, show me which talent I am to make use of now."

I have known people suddenly to enter a new line of work and be fully equipped, with little or no training. So make the statement: "I am fully equipped for the Divine Plan of my life," and be fearless in grasping opportunities.

Some people are cheerful givers but bad receivers. They refuse gifts through pride or some negative reason, thereby blocking their channels, and invariably find themselves eventually with little or nothing.

For example, a woman who had given away a great deal of money had a gift offered her of several thousand dollars. She refused to take it, saying she did not need it. Shortly after that, her finances were "tied up," and she found herself in debt for that amount. A woman should receive gracefully the bread returning to her upon the water—freely ye have given, freely ye shall receive.

There is always the perfect balance of giving and receiving, and although a woman should give without thinking of returns, she violates Law if she does not accept the returns that come to her; for all gifts are from God, people being merely the channel.

A thought of lack should never be held over the giver.

For example, when the woman gave the one cent, I did not say, "Poor woman, she cannot afford to give me that." I saw her rich and prosperous, with her supply pouring in. It was this thought that brought it. If one has been a bad receiver, she must become a good one, and take even a postage stamp if it is given her, and open up her channels for receiving.

The Lord loveth a cheerful receiver, as well as a cheerful giver.

I have often been asked why one is born rich and healthy, and another poor and sick. Where there is an effect, there is always a cause; there is no such thing as chance.

This question is answered through the law of reincarnation. A woman goes through many births and deaths, until she knows the truth that sets her free. She is drawn back to the Earth plane through unsatisfied desire, to pay her karmic debts, or to "fulfill her destiny."

The woman born rich and healthy has had pictures in her subconscious mind, in her past life, of health and riches; and the poor and sick woman, of disease and poverty. A woman manifests, on any plane, the sum total of her subconscious beliefs.

However, birth and death are mortal laws, for the "wages of sin is death." The real woman, spiritual woman, is birthless and deathless! She never was born and has never died—"As she was in the beginning, she is now, and ever shall be!"

So through the Truth, people are set free from the law of karma, sin, and death, and manifest those made in "God's image and likeness." A woman's freedom comes through fulfilling her destiny, bringing into manifestation the Divine Design of her life.

The Lord will say unto her, "Well done, thou good and faithful servant. Thou has been faithful over a few things, I will make thee ruler over many things (death itself); enter thou into the joy of thy Lord (eternal life)."

A woman's freedom comes through fulfilling her destiny.

Chapter 10

DENIALS AND AFFIRMATIONS

A ll the good that is to be made manifest in a woman's life is already an accomplished fact in Divine Mind and is released through a woman's recognition or spoken word. So she must be careful to decree that only the Divine Idea be made manifest, for often, she decrees, through her "idle words," failure or misfortune.

It is, therefore, of the utmost importance to word one's demands correctly, as stated in a previous chapter. If one desires a home, friend, position, or any other good thing, make the demand for the Divine Selection.

For example: "Infinite Spirit, open the way for my right home, my right friend, my right position. I give thanks that it now manifests under grace in a perfect way."

The latter part of the statement is most important.

For example, I knew a woman who demanded a thousand dollars. Her daughter was injured and they received a thousand-dollar indemnity, so it did not come in a "perfect way." The demand should have been worded in this way: "Infinite Spirit, I give thanks that the $1,000, which is mine by Divine Right, is now released, and reaches me under grace in a perfect way."

As one grows in financial consciousness, she should demand that the enormous sums of money, which are hers by Divine Right, reach her under grace, in perfect ways.

It is impossible for a woman to release more than she thinks is possible, for one is bound by the limited expectancies of the subconscious. She must enlarge her expectancies in order to receive in a larger way.

A woman so often limits herself in her demands.

For example, a student made the demand for $600 by a certain date. She did receive it, but heard afterward that she came very near receiving $1,000. She was given just $600 as the result of her spoken word.

"They limited the Holy One of Israel." Wealth is a matter of consciousness. The French have a legend giving an example of this. A poor man was walking along a road when he met a traveler, who stopped him and said, "My good friend, I see you are poor. Take this gold nugget, sell it, and you will be rich all your days."

The man was overjoyed at his good fortune and took the nugget home. He immediately found work and became so prosperous that he did not sell the nugget. Years passed, and he became a very rich man. One day he met a poor man on the road. He stopped him and said, "My good friend, I will give you this gold nugget, which, if you sell, will make you rich for life." The beggar took the nugget, had it valued, and found it was only brass. So we see, the first man became rich through feeling rich, thinking the nugget was gold.

Every woman has within herself a gold nugget; it is her consciousness of gold, of opulence, that brings riches into her life. In making her demands, a woman begins at her journey's end, simply by declaring she has already received. "Before ye call I shall answer."

Continually affirming establishes the belief in the subconscious.

It would not be necessary to make an affirmation more than once if one had perfect faith! One should not plead or supplicate, but give thanks repeatedly that she has received.

"The desert shall rejoice and blossom as the rose." This rejoicing that is yet in the desert (state of consciousness) opens the way for release. The Lord's Prayer is in the form of command and demand—"Give us this day our daily bread, and forgive us our debts as we forgive our debtors"—and ends in praise: "For thine is the Kingdom and the Power and the Glory, forever. Amen."

"Concerning the works of my hands, command ye me." So prayer is command and demand, praise and thanksgiving. The student's work is in making herself believe that "with God, all things are possible."

This is easy enough to state in the abstract, but a little more difficult when confronted with a problem.

For example, it was necessary for a woman to demonstrate a large sum of money within a stated time. She knew she must do something to get a realization (for realization is a manifestation), and she demanded a "lead."

She was walking through a department store when she saw a very beautiful pink enamel letter opener. She felt the "pull" toward it. The thought came, *I haven't a letter opener good enough to open letters containing large checks*. So she bought the letter opener, which the reasoning mind would have called an extravagance. When she held it in her hand, she had a flash of a picture of herself opening an envelope containing a large check, and in a few weeks, she received the money. The pink letter opener was her bridge of active faith.

Many stories are told of the power of the subconscious when directed in faith.

For example, a man was spending the night in a farmhouse. The windows of the room had been nailed down, and in the middle of the night he felt suffocated and made his way in the dark to the window. He could not open it, so he smashed the pane with his fist, drew in drafts of fine fresh air, and had a wonderful night's sleep.

The next morning, he found that he had smashed the glass of a bookcase and the window had remained closed during the whole night. He had supplied himself with oxygen simply by his thought of it.

When a student starts out to demonstrate, she should never turn back. "Let not that a woman who wavers think that she shall receive anything of the Lord."

A student once made this wonderful statement: "When I ask God for anything, I put my foot down, and I say, 'God, I'll take nothing less than I've asked for, but more!'" So a woman should never compromise. "Having done all—stand." This is sometimes the most difficult time of demonstrating. The temptation comes to give up, to turn back, to compromise.

"God also serves who only stands and waits."

Demonstrations often come at the eleventh hour because that's when people tend to let go, that is, stop reasoning, and Infinite Intelligence has a chance to work.

Dreary desires are answered drearily, and impatient desires, long delayed or violently fulfilled.

For example, a woman asked me why it was she was constantly losing or breaking her glasses. We found that she often said to herself and others with vexation, "I wish I could get rid of my glasses." So her impatient desire was violently fulfilled. What she should have demanded was perfect eyesight, but what she registered in the subconscious was simply the impatient desire to be rid of her glasses; so they were continually being broken or lost.

Two attitudes of mind cause loss: depreciation, as in the case of the woman who did not appreciate her spouse; or fear of loss, which makes a picture of loss in the subconscious.

When a student is able to let go of her problem (cast her burden), she will have an instantaneous manifestation.

For example, a woman was out during a very stormy day and her umbrella was blown inside-out. She was about to make a call on some people whom she had never met and she did not wish to make her first appearance with a dilapidated umbrella. She could not throw it away, since it did not belong to her. So in desperation, she exclaimed, "Oh God, you take charge of this umbrella. I don't know what to do."

A moment later, a voice behind her said, "Lady, do you want your umbrella mended?" There stood an umbrella mender. She replied, "Indeed, I do."

The gentleman mended the umbrella while she went into the house to pay her call, and when she returned, she had a good umbrella. So there is always an umbrella mender at hand on a woman's pathway when one puts the umbrella (or situation) in God's hands.

One should always follow a denial with an affirmation.

For example, I was called on the phone late one night to treat a woman whom I had never seen. She was apparently very ill. I made the statement: "I deny this appearance of disease. It is unreal, therefore cannot register in her consciousness; this woman is a perfect idea in Divine Mind, pure substance expressing perfection."

There is no time or space in Divine Mind; therefore, the word reaches its destination instantly and does not "return void." I have treated patients in Europe and have found that the result was instantaneous.

I am asked so often the difference between visualizing and visioning. Visualizing is a mental process governed by the reasoning or conscious mind; visioning is a spiritual process, governed by intuition, or the superconscious mind.

The student should train her mind to receive these flashes of inspiration, and work out the Divine Pictures, through definite leads. When a woman can say: "I desire only that which God desires for me," her new set of blueprints is given her by the Master Architect, the God within. God's plan for each person transcends the limitation of the reasoning mind, and is always the square of life, containing health, wealth, love, and perfect self-expression. Many a woman is building for herself in imagination a bungalow when she should be building a palace.

If a student tries to force a demonstration (through the reasoning mind), she brings it to a standstill. "I will hasten it," saith the Lord. She should act only through intuition or definite leads. "Rest in the Lord and wait patiently. Trust also in the Lord, and the Lord will bring it to pass."

I have seen the Law work in the most astonishing manner.

For example, a student stated that it was necessary for her to have a hundred dollars for the following day. It was a debt of vital importance that had to be met. I spoke the word, declaring that Spirit was never too late and that the supply was at hand.

That evening she phoned me about the miracle. She said that the thought came to her to go to her safe-deposit box at the bank to examine some papers. She looked over the papers, and at the bottom of the box was a new $100 bill. She was astounded and said she knew she had never put it there, for she had gone through the papers many times. It may have been a materialization, as Jesus Christ materialized the loaves and fishes. A woman will reach the stage where her "word is made flesh," or materialized, instantly. "The fields, ripe with the harvest," will manifest immediately, as in all of the miracles of Jesus Christ.

There is a tremendous power alone in the name Jesus Christ. It stands for Truth Made Manifest. He said, "Whatsoever ye ask the Lord, in my name, the Lord will give it to you."

The power of this name raises the student into the fourth dimension, where she is freed from all astral and psychic influences, and she becomes "unconditioned and absolute, in the same way that God is unconditioned and absolute."

I have seen many healings accomplished by using the words "in the name of Jesus Christ." He was both person and principle; and the Christ within each woman is her Redeemer and Salvation.

The Christ within is our own fourth-dimensional self, the inner-self made in God's image and likeness. This is the self that has never failed, never known sickness or sorrow, was never born and has never died. It is the "resurrection and the life" of everyone! This means that God, the Universal, becomes the Christ in us. So daily, we are manifesting Christ within.

A woman should make an art of thinking. The Master Thinker is an artist and is careful to paint only the Divine Designs upon the canvas of her mind; and she paints these pictures with masterly strokes of power and decision, having perfect faith that there is no power to mar their perfection and that they shall manifest in her life the ideal made real.

All power is given to each woman (through right thinking) to bring her heaven upon her Earth, and this is the goal of the "Game of Life."

The simple rules are fearless faith, nonresistance, and love!

May each reader be now freed from that thing that has held her in bondage through the ages, standing between her and her own, and "know the Truth that makes her free"—free to fulfill her destiny, to bring into a manifestation the Divine Design of her life, health, wealth, love and perfect self-expression.

"Be ye transformed by the renewing of your mind."

Affirmations

For Prosperity

*God is my unfailing supply, and large
sums of money come to me quickly,
under grace, in perfect ways.*

For Right Conditions

*Every plan God in heaven has not planned
shall be dissolved and dissipated, and
the Divine Idea now comes to pass.*

For Right Conditions

*Only that which is true of God is
true of me, for I and God are ONE.*

For Faith

*As I am one with God, I am one with
my good, for God is both the Giver and the
Gift. I cannot separate the Giver from the Gift.*

For Right Conditions

Divine Love now dissolves and dissipates every wrong condition in my mind, body, and affairs. Divine Love is the most powerful chemical in the Universe and dissolves everything that is not of itself!

For Health

Divine Love floods my consciousness with health, and every cell in my body is filled with Light.

For the Eyesight

My eyes are God's eyes. I see with the eyes of Spirit. I see clearly the open way; there are no obstacles on my pathway. I see clearly the perfect plan.

For Guidance

I am divinely sensitive to my intuitive leads and give instant obedience to Thy will.

For the Hearing

*My ears are God's ears. I hear with the ears
of Spirit. I am nonresistant and am willing
to be led. I hear glad tidings of great joy.*

For Right Work

*I have perfect work in a perfect way.
I give perfect service for perfect pay.*

For Freedom from all Bondage

I cast this burden on the Christ within, and I go free!

ABOUT THE AUTHORS

Florence Scovel Shinn (1871–1940) carried out her work in the first half of the 20th century. Through her teachings and numerous books, she was a profound influence on Louise Hay and other pioneers of personal transformation.

Louise Hay, author of the international bestseller *You Can Heal Your Life*, is a metaphysical lecturer and teacher with more than 50 million books sold worldwide. For more than 30 years she has helped people to discover and implement their full potential for personal growth and self-healing.
www.louisehay.com

BONUS CONTENT

Thank you for purchasing *The Game of Life and How to Play It* by Florence Scovel Shinn. This product includes a free download! To access this bonus content, please visit www.hayhouse.com/download and enter the Product ID and Download Code as they appear below:

Product ID: 5307

Download Code: ebook

For further assistance, please contact Hay House Customer Care by phone: US (800) 654-5126 or INTL CC+(760) 431-7695 or visit www.hayhouse.com/contact.

Thank you again for your Hay House purchase. Enjoy!

Hay House, Inc. • P.O. Box 5100 • Carlsbad, CA 92018 • (800) 654-5126

Caution: This audio program features meditation/visualization exercises that render it inappropriate for use while driving or operating heavy machinery.

Publisher's note: Hay House products are intended to be powerful, inspirational, and life-changing tools for personal growth and healing. They are not intended as a substitute for medical care. Please use this audio program under the supervision of your care provider. Neither the author nor Hay House, Inc., assumes any responsibility for your improper use of this product.

We hope you enjoyed this Hay House book. If you'd like to receive our online catalog featuring additional information on Hay House books and products, or if you'd like to find out more about the Hay Foundation, please contact:

Hay House, Inc., P.O. Box 5100, Carlsbad, CA 92018-5100
(760) 431-7695 or (800) 654-5126
(760) 431-6948 (fax) or (800) 650-5115 (fax)
www.hayhouse.com® • www.hayfoundation.org

***

Published and distributed in Australia by: Hay House Australia Pty. Ltd.,
18/36 Ralph St., Alexandria NSW 2015
Phone: 612-9669-4299 • Fax: 612-9669-4144 • www.hayhouse.com.au

Published and distributed in the United Kingdom by: Hay House UK,
Ltd., Astley House, 33 Notting Hill Gate, London W11 3JQ
Phone: 44-20-3675-2450 • Fax: 44-20-3675-2451 • www.hayhouse.co.uk

Published and distributed in the Republic of South Africa by:
Hay House SA (Pty), Ltd., P.O. Box 990, Witkoppen 2068
info@hayhouse.co.za • www.hayhouse.co.za

Published in India by: Hay House Publishers India,
Muskaan Complex, Plot No. 3, B-2, Vasant Kunj, New Delhi 110 070
Phone: 91-11-4176-1620 • Fax: 91-11-4176-1630 • www.hayhouse.co.in

Distributed in Canada by: Raincoast Books,
2440 Viking Way, Richmond, B.C. V6V 1N2
Phone: 1-800-663-5714 • Fax: 1-800-565-3770 • www.raincoast.com

***

Take Your Soul on a Vacation

Visit www.HealYourLife.com® to regroup,
recharge, and reconnect with your own magnificence.
Featuring blogs, mind-body-spirit news, and life-changing
wisdom from Louise Hay and friends.

Visit www.HealYourLife.com today!